Indigenous Peoples' food systems & well-being

interventions & policies for healthy communities

Edited by ❧ HARRIET V. KUHNLEIN ❧ BILL ERASMUS ❧ DINA SPIGELSKI ❧ BARBARA BURLINGAME

Food and Agriculture Organization of the United Nations

Centre for Indigenous Peoples' Nutrition and Environment

Rome, 201

BN 978-92-5-107433-6

 FAO 2013

r further information, please contact:

utrition Division,
ood and Agriculture Organization
 the United Nations,
ale delle Terme di Caracalla, 00153 Rome, Italy
mail: nutrition@fao.org / Web site: www.fao.org

arriet V. Kuhnlein
entre for Indigenous Peoples' Nutrition and
vironment (CINE), McGill University
,111 Lakeshore Rd.,
e. Anne de Bellevue, Quebec, Canada
x (1) 514 398 1020
mail: harriet.kuhnlein@mcgill.ca
eb site: www.mcgill.ca/cine

Table of content

Photographic section *(out of text)*

Indigenous Peoples have historically faced the denial of their rights, been victims of ancestral knowledge theft and experienced the destruction of their livelihoods. Nevertheless, they still possess rich and diverse cultural knowledge, language, values, traditions, customs, symbolism, spirituality, forms of organization, standards of living, world views and conceptions of development. These components form the basis of their cultural heritage, and allow them to interact with and have a positive influence on the economic, social and political dynamics of any region or country.

Traditional knowledge has been stored in the collective memory of Indigenous Peoples for centuries, and is seen through the day-to-day activities of men and women. This knowledge is expressed through stories, songs, folklore, proverbs, dances, myths, cultural values, beliefs, rituals, community laws, local language, agricultural practices, tools, materials, plant species and animal breeds. In essence, the natural environment is what makes the knowledge of each people unique and different from that of any other.

Indigenous Peoples' greatest and most laborious achievement so far is the United Nations Declaration on the Rights of Indigenous Peoples. This was adopted by the General Assembly of the United Nations in the belief that Indigenous Peoples' control over the events that affect their lands, territories and resources will maintain and strengthen their institutions, cultures and traditions and promote the development of individuals according to their aspirations and demands.

However, there is still need for targeted strategies and policies that facilitate and foster Indigenous Peoples' use, processing and management of their natural resources for food security and health through self-determination and autonomy. These policies should be effective at the local, state, national, international and regional levels if they are to be successful; they should stress the importance of using cultural knowledge to develop health promotion activities and improve overall health (mental, emotional, spiritual, physical) and well-being.

I find it important to mention some of the cultural indicators for food security, food sovereignty and sustainable development according to Indigenous Peoples, in order to understand the importance of linkages between traditional knowledge and traditional foods:

1. access to, security for, and integrity of lands, territories, natural resources, sacred sites and ceremonial areas used for traditional food production;

2. abundance, scarcity and/or threats regarding traditional seeds, plant foods and medicines, food animals, and the cultural practices associated with their protection and survival;

3. use and transmission of methods, knowledge, language, ceremonies, dances, prayers, oral histories, stories and songs related to traditional foods and subsistence practices, and the continued use of traditional foods in daily diets;

4. Indigenous Peoples' capacity for adaptability, resilience and/or restoration regarding traditional food use and production in response to changing conditions;

5. Indigenous Peoples' ability to exercise and implement their rights to promote their food sovereignty.

Indigenous Peoples' right to food is inseparable from their rights to land, territories, resources, culture and self-determination. An integral human rights-based approach will open constructive dialogue on what policies, regulations and activities are needed to ensure food security for all, regardless of adaptation. Encouraging meaningful participation by all parties may be the key to building trust and resolving ongoing resource conflicts. Needless to say, the United Nations Permanent Forum on Indigenous Issues will address this issue in upcoming sessions.

In brief, the importance of this book arises from its detailed documentation, using a participatory methodology for ten years in different agro-ecological contexts; it provides evidence on the nutrient composition of traditional diets, which supports the Indigenous Peoples' approach.

We hope the book motivates different actors to continue supporting traditional knowledge, to deal jointly with the crisis we face. Ultimately, we hope this book provides the first step in a detailed analysis of Indigenous Peoples' activities and influence in today's world.

Myrna Cunningham
Centre for Indigenous Peoples' Autonomy
and Development
Chairperson, United Nations
Permanent Forum on Indigenous Issues
Nicaragua, 2013

> Comments to: myrna.cunningham.kain@gmail.com

I live in the community of Ndilo and am a member of the Yellowknives Dene First Nation. Our territory is adjacent to Yellowknife, the capital of the Northwest Territories, Canada. As National Chief of the Dene Nation and Regional Chief of the Northwest Territories for the Assembly of First Nations my association with the Indigenous Peoples' Food Systems for Health Program has been a privilege and a responsibility, to ensure that the indigenous voice brings the credibility and relevance of this work to Indigenous Peoples everywhere. Having served for many years as Chair of the Governing Board of the Centre for Indigenous Peoples' Nutrition and Environment (CINE) at McGill University I know well the quality of the participatory processes and scientific work of CINE.

The Indigenous Peoples' Food Systems for Health Program is built on networks of partnerships at several levels. Each of the case studies we work with involves a network of partners, and the overall programme is built on these networks, which have met together at least once a year over the last ten years to share experiences and strategies on how to promote and use traditional food systems to enhance health. Each of the chapters in this book reflects understanding and commitment from collaboration among community and academic partners. Together we have worked to define processes and activities and how they are evaluated to improve food systems and health in the communities.

Working with Indigenous Peoples from 12 global regions has given important insights. Similarities in the circumstances faced by Indigenous Peoples are striking and have no boundaries. Indigenous Peoples have been colonized and are dealing with the forces of assimilation. Their land and resources have been assaulted and access to their own food has been threatened. Vastly different ecosystems and cultures of Indigenous Peoples have followed similar paths through time and development, witnessing similar changes in access to land and food. These include changes in and losses of animals and plants in the ecosystems where Indigenous Peoples live, and changes in health, which are manifested by both undernutrition and overnutrition (obesity) and the accompanying chronic diseases. There has been loss of cultural knowledge with the passing of elders in the circle of life and the imposed imperative to move from expressing the benefits of sharing and assisting others to embracing competition and the wage economy. Throughout this "modernization", Indigenous Peoples have been marginalized, vulnerable, living in extreme poverty and losing their sense of identity and self-determination.

The good news is that the programme described in this book has brought people together to work to improve these patterns. Indigenous Peoples are now expressing their human rights to self-determination and health, and are less isolated and demoralized. Instead, they are becoming increasingly aware that their original diets and local foods are very healthy. Indigenous Peoples have a great deal to share among themselves, and with all humankind. The wisdom inherent to their cultures and ways of knowing and doing demonstrate the all-encompassing connectedness of the land and food to their physical and mental health and spirituality.

Recognizing, understanding and dealing with the benefits and threats to Indigenous Peoples' ecosystems, and the biodiversity provided in their food helps them to know "who they are" and to "feel good about themselves", a confidence that is important to everyone. The way forward is to take things in hand, and work together at the community level with our partners in academia, government and non-governmental organizations, with valuable support from the United Nations.

In this book you will find an inspiring array of methods, processes and activities that can mobilize leaders and their communities. The book will assist them in understanding their ecosystems and, whenever and however possible, using their local food to experience the benefits it gives to cultural expression, food security and health. The book is a natural sequel and partner to the preceding book in this series, *Indigenous Peoples' food systems: the many dimensions of culture, diversity and environment for nutrition and health* (FAO & CINE, 2009).

Bill Erasmus
*Dene National Chief
and Regional Chief,
Assembly of First Nations
Yellowknife, NWT, Canada
2013*

> Comments to: berasmus@afn.ca

Writing this section is very difficult for us. There are so many people and agencies to acknowledge that the task is indeed daunting. Each chapter in this book has its own section recognizing contributors in their various roles. So here we note important contributions to the overall effort and to this volume.

Our work is grounded in understanding Indigenous Peoples' food systems for their scientific and social qualities, beginning with the very basics of scientific identification, nutrient composition and cultural value of the many foods contained in the rural ecosystems where Indigenous Peoples live. We have been fortunate to have the engagement of many interdisciplinary scholars within our team who have worked with us to focus on specific unique communities of indigenous people. This breadth in disciplines makes the research and its conclusions meaningful for recognizing and promoting Indigenous Peoples' food systems from many perspectives. Our results lead us to challenge the international community at all levels to give these food systems and the people who depend on them the respect and credibility they deserve for providing nutrition and good health. Therefore, we gratefully thank all of our partners for bringing this effort to fruition, and we also sincerely appreciate those who have responded to our work.

We thank McGill University and staff of the Centre for Indigenous Peoples' Nutrition and Environment (CINE) who have given immeasurably of their resources and time to undertake the work of the programme, including by providing the necessary management and office facilities. Several chapters have been developed with those who are or have been part of McGill/CINE as professors, affiliated members, graduate students and staff, or who are in leadership roles through the Governing Board of CINE. Kp-studios.com of Anacortes, Washington, United States of America provided most of the excellent photographs from the case study areas.

We thank the members of the International Union of Nutritional Sciences (IUNS) Task Force on Indigenous Peoples' Food Systems and Nutrition, recently renamed the Task Force on Traditional, Indigenous, and Cultural Food and Nutrition, who helped with the conceptual planning of the programme that has been implemented over the last ten years. Several lead authors of chapters in this volume were members of the task force and contributors to the meetings and satellite meetings of the International Congresses of Nutrition, sponsored by IUNS in Vienna, Durban and Bangkok in 2001, 2005 and 2009, respectively. IUNS and its executives have given exceptional support to the concepts of the task force and this programme, which evolved within it.

The Food and Agriculture Organization of the United Nations (FAO) has been a pillar of support for our work. In providing leadership for the publication process of our series of three books dedicated to the programme, FAO has given editorial and typesetting services as well as outstanding advice on aesthetic qualities in crafting the publication. FAO has provided Internet versions of the books and distributed hard copies free of charge through the United Nations publications distribution system, including FAO's regional offices, and university libraries. In addition,

FAO provided initial funding to kick-start several of the case studies described here. Case study leaders have presented their data and results in several venues sponsored by FAO. Without doubt, the world's attention to issues surrounding Indigenous Peoples' food systems, and our work in this realm have been fostered and promoted by FAO.

The Rockefeller Foundation's Bellagio Center on the shores of beautiful Lake Como in Italy provided grants to bring our team members together on three occasions, for meetings to stimulate our thinking about ways to make progress during the programme. This was essential for our within-programme communications and provided stimulation that was most welcome. We all extend a warm "thank you" to The Rockefeller Foundation for these opportunities in 2004, 2007 and 2008, and to the staff at the Bellagio Center for their unforgettably beautiful setting, hospitality, facilities and efficient service.

Primary funding for the research grants based from CINE was given by the Canadian Institutes of Health Research through the Institute of Aboriginal Peoples' Health, the Institute of Population and Public Health and the Institute of Nutrition, Metabolism and Diabetes. Funding for research and conference grants was given by the Canadian International Development Agency and the International Development Research Centre. Each of the case study intervention and evaluation efforts in this book gained additional funding from other agencies in the country concerned; these agencies are acknowledged in each chapter. We thank the Ernst Göhner Stiftung for assistance in the final months of preparing the publication, and for funds to prepare the videographic documentation of case studies at www.indigenousnutrition.org.

It is the people in the communities participating in the research described here to whom we are greatly indebted. Their wisdom, encouragement and enthusiasm emboldened the academic partners and led all of us to make the chapters reflect their thinking, their perspectives and their work. Thousands of Indigenous Peoples of all ages in seven countries are represented in this book. They built the knowledge base represented in all of the publications from the programme. Hundreds of community associates, many of whom were volunteers, contributed and continue to contribute to the ongoing activities in each of the case studies.

The collective intention of chapter authors is to dedicate this book to children in communities of Indigenous Peoples, who will continue to face challenges to protecting their traditional knowledge and use of their local foods for physical, social and environmental health. We know that our books will give them power and strength.

✾ **Harriet V. Kuhnlein**
Ph.D., F.A.S.N., F.I.U.N.S., LL.D. (hon.),
Founding Director, CINE, and
Emerita Professor, Human Nutrition,
McGill University,
Montreal, Canada

✾ **Bill Erasmus**
National Chief, Dene Nation, Yellowknife, and
Regional Chief, Assembly of First Nations,
Ottawa, Canada

✾ **Dina Spigelski**
R.D., M.Sc., Coordinator,
Montreal, Canada

✾ **Barbara Burlingame**
Ph.D., Principal Officer,
Nutrition Division, FAO,
Rome

2013

Overviews

Chapter 1

Why do Indigenous Peoples' food and nutrition interventions for health promotion and policy need special consideration?

HARRIET V. KUHNLEIN[1]   BARBARA BURLINGAME

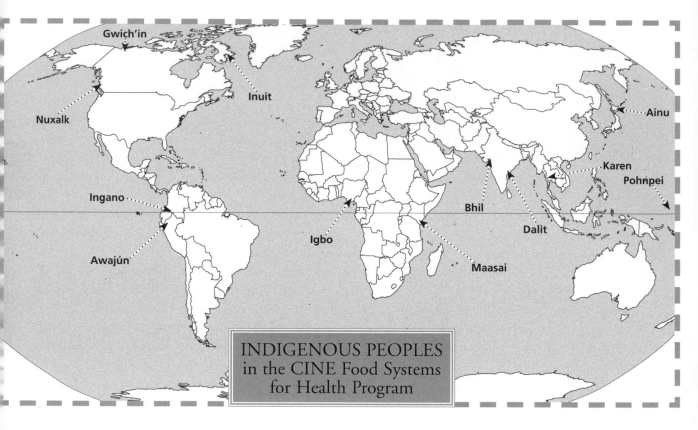

INDIGENOUS PEOPLES
in the CINE Food Systems
for Health Program

1
Centre for
Indigenous Peoples'
Nutrition and Environment
(CINE) and
School of Dietetics and   2
Human Nutrition,   Food and Agriculture
McGill University,   Organization of
Montreal,   the United Nations (FAO),
Quebec, Canada   Rome, Italy

Key words > Indigenous Peoples, food systems,
food-based interventions, nutrition interventions,
food policy, health promotion

> # "Traditional food system work is a reality check; it helps to not forget where you come from and who you are. It makes sense to us."
>
> Earl Nowgesic, indigenous health leader, Canada

Abstract

This book is about Indigenous Peoples' food systems and how important local knowledge about foods and the ecosystems that provide them can be used to improve health and well-being. The authors describe processes and activities in nine cultures of Indigenous Peoples, where interventions were developed and implemented with local knowledge, and explain how this information can benefit Indigenous Peoples everywhere, and all of humankind.

Food and nutrition insecurity and the burden of high incidence of non-communicable diseases reach all corners of the globe. This "nutrition transition" is driven by changing lifestyles, loss of livelihoods for all those engaged in food production, increasing poverty and urbanization, and sedentary lifestyles with changing dietary patterns. The result is increasing global obesity and non-communicable diseases, including malnutrition in all its forms. The situation is especially critical for Indigenous Peoples, who often experience the most severe financial poverty and health disparities in both developing and developed countries, particularly where they depend on ecosystems under stress to support their needs for food and well-being.

The programme of work that concludes with preparation of this book was developed from the view that Indigenous Peoples with cultural homelands in the most rural areas of developing regions experience common challenges in relation to their traditional food systems, food security and health. The authors' view is that Indigenous Peoples' existing resources and knowledge about their cultures and ecosystems can be used to develop and implement effective health promotion activities.

The programme process and progression

The chapters in this book describe the adventures and findings of more than 40 interdisciplinary collaborators who have researched Indigenous Peoples' food systems and health promotion interventions. It is the third publication from the Indigenous Peoples' Food Systems for Health Program originating from the Centre for Indigenous Peoples' Nutrition and Environment (CINE) and the Task Force on Indigenous Peoples' Food Systems and Nutrition of the International Union of Nutritional Sciences (IUNS). This excellent team of scholars and community leaders has conducted research with communities of Indigenous Peoples for about ten years, creating methodologies, food systems documentation, and unique interventions to improve health by using aspects of local food systems.

In this book, the findings are considered from several interdisciplinary roots by experts in nutrition, epidemiology, anthropology, human rights, nursing and ethnobotany, in partnership with leaders of indigenous communities. Nine unique food system interventions conducted in widely diverse cultures and ecosystems are presented, with overview chapters on the circumstances and challenges rural-dwelling Indigenous Peoples face in health, environment, child nutrition and human rights, and how intervention practices and policies can be developed. The book is written for readers from a variety of backgrounds, to share this diversity of perspectives from the unique case studies. As much as possible, the writing style and community perspective on intervention activities have been maintained, to enrich the chapters.

Work on traditional food systems is important for Indigenous Peoples. It provides a reality check and assurance that outsiders with diverse expertise value local foods and practices in the social contexts where Indigenous Peoples experience them. It helps people to realize the importance of maintaining their connections with nature and their own cultures, and between heart and mind, to reaffirm identity. Where people identify themselves with their culture and natural environment, knowledge and use of traditional food systems to improve health builds community support and engagement for holistic health and well-being. This affects the many aspects of physical, emotional, mental and spiritual health – for adults, children and elders, individually and in community and cultural collectives – in recognizing continuity from the past, into the present and towards the future.

The stories of the nine interventions presented here show how much the indigenous world has to offer through insights into the mysteries of successful food and nutrition promotion programmes everywhere. They demonstrate how special consideration to building cultural pride, cross-sectoral planning, enthusiastic and energetic advocates, and community goodwill can challenge the obstacles and barriers to knowledge transmission for action, through healthy behaviours for individuals and the community. Special focus has been on children and youth, and on finding ways to make local traditions relevant and useful to them. In fact, all the indigenous communities and leaders represented here recognized the programme's importance for the health of their children today and into the future – hence, the selection of a photo of an indigenous child and his traditional meal for the book's cover.

The journey began in Salaya, Thailand, to create the methodology for documenting the food systems of indigenous cultures. This methodology is now electronically available.[1] After identifying community and academic partners for the case studies, the authors proceeded to the documentation of 12 international case studies (Map on p. 4). This resource is available through the United Nations bookstore, and also online.[2] An extensive network of collaborators developed and exchanged information in annual meetings over several years to build the knowledge base resulting in this book.

All the participants remember the beautiful rainbow they saw over Lake Como, Italy during their first meeting in the conference room of The Rockefeller Foundation Bellagio Center. At that meeting, they recognized the commitment and strength of their collaboration, and how their aspirations, if realized, would benefit Indigenous Peoples everywhere. Participants remember Dr Suttilak Smitasiri's words that their "pot of gold" is the grounding of their nutrition work in the unique cultures in which they work. Several years and much hard work have passed, and the results of that labour are reported in this book, Book 3 in the series. Readers are invited to refer to the acknowledgements section, where the authors thank their teams of collaborators and many supporters.

Context of the interventions

Ways of identifying Indigenous Peoples have been described at length by United Nations agencies[3] and in earlier publications.[4] Throughout the world, there are more than 370 million Indigenous Peoples, speaking more than 4 000 languages and located in more than 90 countries. Those living in their rural homelands depend on traditional food systems rooted in historical continuity in their regions, where food is harvested with traditional knowledge from the natural environment, and prepared and served in local cultural settings. Foods purchased from markets, often through globalized industrial outlets, are also part of Indigenous Peoples' food systems today, and are among the considerations required for interventions to promote healthy diets in communities.

1   www.mcgill.ca/cine/research/global/
2   www.fao.org/docrep/012/i0370e/i0370e00.htm
3   www.un.org/esa/socdev/unpfii/documents/sowip_web.pdf
4   www.fao.org/docrep/012/i0370e/i0370e00.htm

Recognition of the vast food biodiversity in indigenous knowledge, and research to identify the composition of nutrients and other properties of this food are at the centre of the knowledge needed to build good nutrition promotion programmes for community members of all ages. This combination of traditional indigenous knowledge with "Western" scientific documentation was welcomed in the case study interventions. Information exchange is at the heart of this work. The reader will find many examples of successful knowledge sharing within communities, which stimulated good dietary patterns and provided the impetus for important evaluations based on food use and dietary quality.

All the programme's research throughout the last ten years has been greatly enhanced by state-of-the-art participatory processes, with indigenous community and academic colleagues collaborating equally in the decision-making for project activities. Each of the intervention chapters demonstrates how health promotion has been conducted successfully. The chapters present nine very different stories regarding types of intervention, local resources used and evaluation methods. Attention was always given to the right social settings, and to using social capital combined with capacity building. Logistical constraints and the limited availability of funds necessitated work with small population groups, where meaningful control groups do not exist. Intervention effectiveness was usually evaluated through before-and-after measurements and qualitative techniques, and involved active community participation and support. The interventions directed at improving food provisioning were most successfully evaluated through food and dietary measurements, often with qualitative techniques. The reader is directed to the interventions described in Chapters 5 to 13, and to the overview of interventions in Chapter 14.

The aim of this programme has been to build the scientific credibility of local food systems, to use this information to improve the health of the people directly involved, and to share success stories to influence policies at the local, national and international levels. Chapter 16 discusses how policies can be influenced to benefit Indigenous Peoples' use of their food systems,

particularly by increasing access to the range of biodiversity available. Several of the interventions increased their scope and dimension by scaling up activities to additional communities within the region and – in some cases – more broadly. For example, Chapter 12 notes the requests for and activities of the Pohnpei Go Local! project throughout the Pacific region. Such scaling up is surely the gold standard of a successful health promotion programme.

Common themes in interventions

The nine interventions capture themes that address the challenges Indigenous Peoples face in nutrition and health, access to their ecosystem food resources, and the social contexts in which food is prepared and consumed. The activities and local-level policy implications described are impressively case-specific and diverse. These interventions stress the necessity of working from the bottom up, using the indigenous community's perspective of what works and how to proceed, with evaluations that are meaningful to the people directly involved. Successful and less successful engagements with government are also described, calling attention to the benefits that government interaction can provide.

Interventions to improve the health circumstances of Indigenous Peoples in rural, often remote, settings can be very different from those for a country's general population. Activities must be in harmony with the local cultural and social settings, local personnel and local sources of food. There is financial poverty in the rural settings where these interventions were conducted, but communities prefer not to define their success in terms of money. Instead, they measure success through the benefits brought to local social, cultural and ecosystem contexts. For the Karen people in Thailand, for example, "food is a part of happiness" that cannot be measured with money. It is difficult to create behavioural changes to improve people's obesity or stunting status, but being "short/small" or "big" is not as important as changing the conditions that cause these conditions, with full self-determination for the communities directly involved.

Throughout the book, the text shifts between past and present tenses. This is because most of the interventions are still ongoing, and evaluations were completed at a designated point after their initiation. Although this style is awkward, it also reflects how the state of the case studies changes with time. In the Nuxalk project (Chapter 11), for example, the intervention and initial evaluation were conducted several years ago, while the current chapter refers to a recent revisiting of the project.

Implications for policy

Indigenous Peoples attach profound importance and commitment to protecting their land and access to food resources and promoting the benefits of local food to enhance food and nutrition security. Policies protecting Indigenous Peoples' right to food are centred on access to traditional food system resources and giving Indigenous Peoples priority in their use. Chapters 15 and 16 elaborate on these principles.

Much can be learned from studying the food systems of Indigenous Peoples. To start with, nations must be encouraged to disaggregate population data by culture and geographic location, to explore the circumstances faced by Indigenous Peoples in their home areas. The Pan-American Health Organization (PAHO) has identified the disparities faced by Indigenous Peoples in the Americas,[5] but other regions are not as diligent in uncovering such disparities for segments of their populations. Only when the extent of the problem is known can reasonable and meaningful action be taken to promote equality in food and nutrition security and well-being.

Nutrition improvements and health promotion interventions with Indigenous Peoples can be successful when they give full attention to the social context, social support, social capital and local food resources and provisioning. The book offers the perspective that understanding how to use local foods to improve Indigenous Peoples' health benefits them directly, and also gives new insights for nutritional health promotion initiatives in general.

The authors hope that the evidence supplied from this programme and its publications will stimulate others to promote traditional food systems for Indigenous Peoples in their regions, and to contribute to mainstreaming food-based approaches with local resources. In addition to the three major publications from the programme, there are hundreds of peer-reviewed articles from team members, documentary videos[6] and presentations given at local, national and international meetings. This issues a clear call for nutritionists and their colleagues in leadership roles in indigenous communities to experience the wide variety of unique foods and the social settings in which they are used, and to promote these important elements of local culture and ecosystems for their health benefits and their promise to provide sustainable solutions to food and nutrition security ◉

> **Comments to:** harriet.kuhnlein@mcgill.ca

---

5   www.unscn.org/layout/modules/resources/files/scnnews37.pdf

6   www.indigenousnutrition.org

Health disparities: promoting Indigenous Peoples' health through traditional food systems and self-determination

GRACE M. EGELAND[1]   GAIL G. HARRISON

1
Centre for Indigenous
Peoples' Nutrition and
Environment (CINE) and
School of Dietetics and
Human Nutrition,
McGill University,
Montreal, Quebec,
Canada

2
School of Public Health,
University of California,
Los Angeles, California,
United States of America

Key words > Indigenous Peoples,
determinants of health, thrifty gene,
undernutrition, overnutrition, type 2 diabetes,
food security

Abstract

Although there is considerable global diversity in indigenous cultures and ecosystems, one shared commonality is that Indigenous Peoples experience disparities across all dimensions of health. This chapter discusses the disparities in unhealthy body weights from under- to overnutrition, the emergence of type 2 diabetes mellitus, food security and micronutrient deficiencies, and longevity.

Indigenous Peoples living in arid or semi-arid areas that experience drought or seasonal fluctuations in food availability continue to suffer from underweight and malnourishment, while those living in remote, biodiverse areas and engaged in traditional activities with little reliance on market economies tend to be of normal weight. In sharp contrast, Indigenous Peoples living in developed countries have a risk of obesity that is generally 1.5 times greater than that observed for non-Indigenous Peoples residing in the same country or affiliated state. The thrifty gene, the thrifty phenotype and, more recently, the environmental programming hypotheses need to be researched within a context that does not ignore the profound contributions of other underlying causes of health disparities. Assaults on "indigeneity" and self-determination contribute to the health effects of disparities in poverty, education, nutrition, food security, household crowding, and poor access to and utilization of health care.

Multiple strategies are needed to help narrow the gap in nutrition-related chronic diseases. Successful programmes are likely to be those that improve health through the promotion of cultural strengths and self-determination, including traditional food systems. Although this is a challenge, improving the health and longevity of Indigenous Peoples is not an impossible task.

Introduction

For Indigenous Peoples, determinants of health take on the additional dimensions of collective assaults on "indigeneity", where the end results are profound and far-reaching and contribute to the wide gaps in indigenous health and well-being.

There are an estimated 370 million indigenous people worldwide, with considerable cultural diversity. However, one commonality is that Indigenous Peoples experience disparities across all dimensions of health indicators (Anderson *et al.*, 2006; Cunningham, 2009; Montenegro and Stephens, 2006; Ohenjo *et al.*, 2006). The key to understanding the underlying causes of these disparities lies in the current relationship of Indigenous Peoples to the larger society. While the social determinants of population health are now widely appreciated (Glouberman and Millar, 2003), for Indigenous Peoples determinants of health take on the additional dimensions of assaults on "indigeneity", including colonization and disassociation from their land, cultural and linguistic heritage and even families – when there has been forced residential schooling. In these situations, self-esteem and individual and group identity and self-determination have been eroded. The end result of these collective assaults on "indigeneity" are profound and far-reaching, and contribute to the wide gaps in indigenous health and well-being (Cunningham, 2009; King, Smith and Gracey, 2009; Ohenjo *et al.*, 2006; Stephens *et al.*, 2006).

The dimensions of health and well-being cover a broad range of health outcomes, but this chapter is limited to disparities in unhealthy body weights, emerging type 2 diabetes mellitus (DM), micronutrient deficiencies, longevity and food security. The chapter ends with indigenous perspectives on health determinants, as a guide for identifying health promotions and interventions that will make a difference by improving health through the promotion of cultural strengths and self-determination, including traditional food systems.

Micronutrient deficiencies

Micronutrient deficiencies among Indigenous Peoples tend to mirror those that are prevalent in the larger society, but Indigenous Peoples face increased vulnerability, particularly to the extent that they suffer disproportionately from poverty. Micronutrient deficiencies rank among the top 20 risk factors for morbidity and impaired quality of life, with particular burdens falling on populations in poorer countries, women of reproductive age and young children. However, they are sufficiently prevalent (among more than 2 billion people globally) to affect almost all population segments to some degree (Lopez et al., 2006). Amelioration of these deficiencies constitutes one of the most cost-effective public health interventions in terms of improving overall health, the outcomes of common infectious diseases and the quality of life (Jamison et al., 1993; Tulchinsky, 2010; Harrison, 2010). A list of specific micronutrient deficiency conditions and qualitative estimates of their global prevalence is provided in Annex 2.1.

Although deficiencies of particular vitamins and minerals can be conceptualized singularly, it is important to remember that they usually occur in combination. Diets limited in variety and in foods of animal origin are most frequently deficient in micronutrients. On the public health agenda, deficiencies of vitamin A, iodine, iron, folic acid and to some extent zinc have had the most attention because of the evidence base for the efficacy of correcting these deficiencies and the feasibility of effective interventions. However, there is abundant evidence that the problems

of micronutrient deficiencies are far from being solved globally.

Existing case studies provide evidence that Indigenous Peoples suffer from micronutrient deficiencies at least as severely as the larger populations from which they are drawn, and probably even more so, given their disparities in poverty (Carino, 2009). Micronutrient deficiencies are not inconsistent with high levels of overweight and obesity, as shown by vitamin A deficiency documented in almost one-third of children aged two to ten years in Mand, Federated States of Micronesia, where adult obesity is very high (Englberger et al., 2009). There is also evidence that as problems of food insecurity increase – through constraints in access to traditional food sources, lands and waterways – traditional practices that protect against micronutrient deficiencies may increase (see section on food insecurity on pp. 17 to 20). For example, the prevalence of total anaemia among Inuit preschoolers illustrates the potential for traditional foods to ameliorate food insecurity. Children who had consumed no traditional food the previous day and were food-insecure, based on the United States Department of Agriculture (USDA) assessment tool for food insecurity, had the highest prevalence of anaemia, while iron-deficiency anaemia prevalence was low among children who had consumed traditional food the previous day, regardless of food insecurity status (Egeland et al., 2013) (Figures 2.1 and 2.2).

From under- to overnutrition

Currently, the risks of underweight and obesity vary considerably among Indigenous Peoples living in diverse settings. In general, Indigenous Peoples who live in remote areas with considerable biodiversity and who are engaged in traditional activities with little reliance on market economies tend to be of normal weight: among Awajún women of Peru, 92 percent had a body mass index (BMI) in what is considered the normal range of 18.5 to 24.9 kg/m^2 (Creed-Kanashiro et al., 2009); among the Ingano of Colombia, 89 percent of women and 96.6 percent of men had normal healthy BMIs (Correal et al., 2009); and for

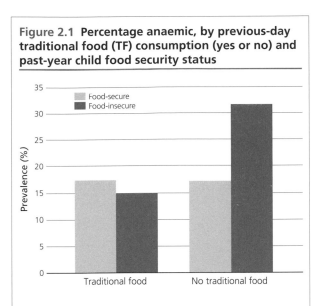

Figure 2.1 Percentage anaemic, by previous-day traditional food (TF) consumption (yes or no) and past-year child food security status

Prevalence (%)

- Food-secure
- Food-insecure

Traditional food   No traditional food

Anaemia defined as venous or capillary haemoglobin < 110 g/litre for three-to-four-year-olds and < 115 g/litre for five-year-olds. $p < 0.10$ for the interaction term TF by food security in logistic regression model adjusting for age, sex and sampling method (capillary versus venous).

Sources: Data from Egeland *et al.*, 2011; Nunavut Inuit Child Health Survey 2007–2008.

Figure 2.2 Percentage iron-deficient, by previous-day traditional food (TF) consumption (yes or no) and past-year food security status

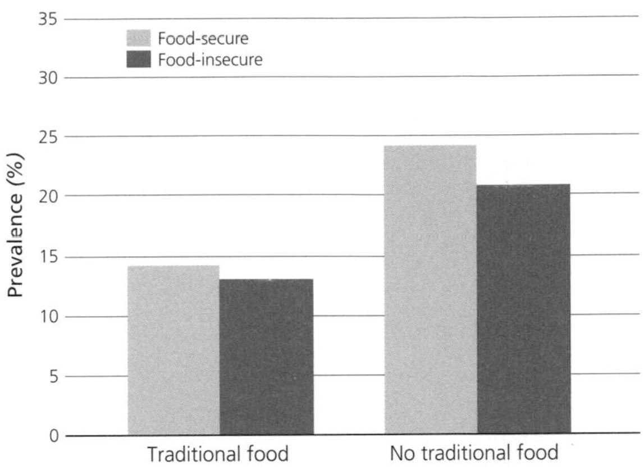

Prevalence (%)

- Food-secure
- Food-insecure

Traditional food   No traditional food

Iron deficiency defined as plasma ferritin < 12 µg/litre; those with hsCRP > 8 mg/litre were excluded. $p < 0.06$, indicating a borderline significantly lower percentage of iron deficiency among Inuit preschoolers who consumed TF the previous day compared with non-TF consumers in logistic regression analyses, adjusting for age and sex.

Sources: Data from Egeland *et al.*, 2011; Nunavut Inuit Child Health Survey 2007–2008.

the Karen of Thailand, overweight was rare among children aged 0 to 12 years (Chotiboriboon *et al.*, 2009). In arid or semi-arid areas with less biodiversity and drought or seasonal fluctuations in food availability, underweight and malnutrition are prevalent: among the Maasai experiencing drought, between 25.1 and 35.7 percent of children were underweight and 46.1 to 60.3 percent were stunted (World Vision Kenya, 2004; Oiye *et al.*, 2009); in India, among landless Dalit working as farm labourers, chronic energy deficiency affected 42 percent of women (Salomeyesudas and Satheesh, 2009); malnutrition is also the primary concern for Bhil and other tribal populations in India (Bhattacharjee *et al.*, 2009; National Institute of Nutrition, 2000; Hamill *et al.*, 1977); and among the Igbo of Nigeria, 42 percent of children were stunted, 25 percent were underweight and 9 percent were wasted (Okeke *et al.*, 2009).

In sharp contrast, Indigenous Peoples living in developed countries have a risk of obesity that is generally at least 1.5 times greater than that observed for non-indigenous peoples residing in the same country or affiliated state. Obesity-related chronic diseases have

increased from being rare to what is now considered an epidemic, particularly in type 2 DM.

In Pohnpei, Federated States of Micronesia, the prevalence of overweight and obesity has increased from almost zero immediately after the Second World War to its current levels, with a third of women being overweight (Pohnpei STEPS, 2002) and the population suffering unprecedented rates of obesity, hypertension and diabetes (Durand, 2007).

In Australia, where undernutrition was of paramount concern for Indigenous Peoples 30 years ago (Gracey, 1976), an epidemic in overweight and obesity has led to disabling and often fatal chronic diseases (Gracey, 2007). A survey found that aboriginal and Torres Strait Islanders were 1.3 times more likely to be obese than non-indigenous Australians (Australian Bureau of Statistics and Australian Institute of Health and Welfare, 2005). In New Zealand, 41.7 percent of Maori were obese in the 2006/2007 health survey: a rate 1.5 times the 26.5 percent observed among New Zealanders of European descent (New Zealand Ministry of Health, 2008). In 1996, the lifetime risks of developing type 2 DM for Maori were 26 percent for men and 32 percent for women, compared with 10 and 8 percent for men and women of European descent (New Zealand Ministry of Health, 2002).

In the United States of America, overweight and obesity are a notable health problem among American Indians and Alaska Natives. In a study of five-year-old children, 47 percent of boys and 41 percent of girls were overweight; 24 percent of the children attending 55 schools on 12 reservations were obese; and the risk of overweight and obesity increased with successive age groups evaluated (Zephier *et al.*, 2006). Indigenous Peoples in the United States were also twice as likely to suffer type 2 DM than non-Hispanic whites (Steele *et al.*, 2008), apart from in Alaska, where rates for type 2 DM were similar for indigenous and non-Hispanic whites (6 percent). There is evidence that type 2 DM is increasing among Alaska Natives and Greenlanders (Inuit), who have historically been spared from the epidemic observed among other Indigenous Peoples (Ebbesson *et al.*, 1998; Jørgensen *et al.*, 2002). The Pima Indians of Arizona suffer from an excessively high rate of type 2 DM, at five times that observed among the Pima Indians of remote mountainous northwestern Mexico, where a traditional lifestyle and diet and greater physical activity have been reported as accounting for reduced obesity and type 2 DM risk (Schulz *et al.*, 2006; Ravussin *et al.*, 1994).

In Canada, overweight and obesity rates are highly prevalent among Indigenous Peoples regardless of geographic location or ethnicity (Young and Sevenhuysen, 1989; Tjepkema, 2002; McIntyre and Shah, 1986; Kuhnlein *et al.*, 2004; Galloway, Young and Egeland, 2010). The prevalences of diabetes for First Nations Canadian men and women were respectively 3.6 and 5.3 times higher than those of the general Canadian population (Young *et al.*, 2000).

In addition to the high risk of overweight and obesity among Indigenous Peoples living in developed countries, there is evidence that overweight and obesity are emerging among Indigenous Peoples in low-income countries who are undergoing acculturation in the context of poverty. In these situations, indicators of overnutrition in adults coexist with indicators of undernutrition, particularly in children, indicating

rapid nutrition transitions. Among the Suruí of the Amazon, 60.5 percent of 20 to 49.9-year-olds were either overweight or obese (Lourenço *et al.*, 2008), and among the Ribeirinhos of Brazil (Piperata, 2007) and Andean populations of Argentina (Romaguera *et al.*, 2008) overweight and obesity are prevalent and coexist with indicators of poor growth such as stunting.

The thrifty gene hypothesis and environmental programming in context

The "thrifty gene" hypothesis might have seemed like a good idea many years ago in the absence of experimental-based knowledge in the pre-genomic era. But, current research suggests that in most cases a single mutation in a single gene is unlikely to predispose an entire group of people to a complex outcome like type 2 diabetes.

Dr Robert Hegele

A commonly cited underlying cause for the high rates of obesity and type 2 DM in Indigenous Peoples is the "thrifty gene" hypothesis, which postulates that there is inherited susceptibility for a biological incapacity to adapt to a modern sedentary lifestyle with a consistent supply of energy. However the thrifty gene hypothesis has been criticized as too simplistic (Paradies, Montoya and Fullerton, 2007; Fee, 2006), given that human beings are remarkably genetically similar, sharing 99.9 percent of their genomes, and have a relatively common and recent evolutionary history of hunting and gathering in periods of feast and famine, which essentially places all humans in a similar thrifty gene risk paradigm (Paradies, Montoya and Fullerton, 2007).

Although there has been progress in genetic research, the aetiology of type 2 DM is complex and multifactoral, and current research suggests that " in most cases a single mutation in a single gene is unlikely to predispose an entire group of people to a complex outcome like type 2 diabetes" (R. Hegele, personal communication, 2009). One exception occurs among the Oji-Cree of northern Ontario, in whom hepatic nuclear factor-1alpha (HNF1A) G319S has been found to be associated with a distinct form of type 2 DM with an earlier age of onset, a lower BMI and a higher

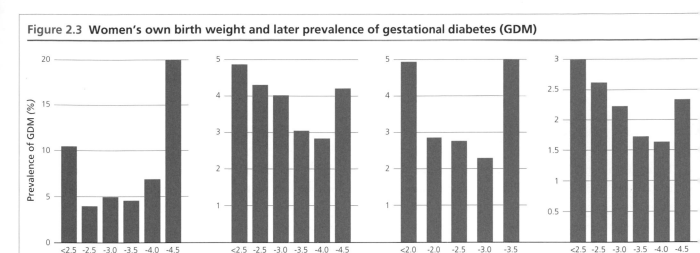

Figure 2.3 Women's own birth weight and later prevalence of gestational diabetes (GDM)

Prevalence of GDM (%)

Pima women | Norwegian women | African-American women | Women from New York State

Mother's birth weight *(kg)*

Source: Adapted from Pettitt and Jovanovic, 2007.

post-challenge plasma glucose level than usually observed (Hegele, 2001). However, the genetic variant HNF1A G319S is highly specific to the Oji-Cree and has not been found among other Indigenous Peoples or even among other Canadian Cree with high rates of diabetes.

In recent years, the "thrifty genotype" hypothesis has been supplemented by recognition that what may be happening proximally as populations undergo rapid changes in environment and lifestyle is intergenerational transmission of a "thrifty phenotype". This concept was introduced by Hales *et al.* (1991) and proposes that environmental factors acting in early (intra-uterine) life, particularly undernutrition, may influence later risk of type 2 DM and other chronic diseases of adulthood. The hypothesis is based largely on consistent observations of the inverse relationship between birth weight and risk of future ischaemic heart disease in adulthood (Huxley *et al.*, 2007). However, a growing body of evidence suggests there is a U-shaped data curve between birth weight and chronic diseases (with more chronic disease at the lower and higher ends of birth weight), especially for diabetes. Among Pima Indians, for example, a U-shaped curve was identified between women's own birth weight and later risk of gestational diabetes (Pettitt and Knowler, 1998). These findings have also been observed in other study populations (Egeland, Skjaerven and Irgens, 2000; Williams *et al.*, 1999; Innes *et al.*, 2002; Pettitt and Jovanovic, 2007) (Figure 2.3). The data suggest that maternal obesity and diabetes may create susceptibility to later chronic disease in the offspring, through hyperglycaemic and epigenetic mechanisms (Smith *et al.*, 2009; Egeland and Meltzer, 2010; Dabelea, 2007; Pettitt *et al.*, 1998; Silverman *et al.*, 1998).

Emerging data suggest possible epigenetic effects (changes in gene expression transmissible intergenerationally but not involving alteration of DNA base sequences); the concept has been termed "environmental programming" (Lucas, 1991; Lindsay and Bennett, 2001).

However, as health disparities between Indigenous Peoples and their non-indigenous counterparts are observed across a broad range of health outcomes – including intentional and unintentional injuries, psychological distress and mental illness, birth defects, cancers, perinatal and post-neonatal mortality (Stephens *et al.*, 2006) – scientific research into the thrifty gene, the thrifty phenotype and the environmental programming hypotheses must take into account the profound contributions of other underlying causes of health disparities. The literature on Indigenous Peoples' health identifies poverty, low education, marginalization and racism, disassociation from land, culture and family, inadequate health care access and utilization, and disparities in many other determinants of disease, such as food insecurity, as underlying causes for health inequities (Carino, 2009; Cunningham, 2009; Stephens *et al.*, 2006; King, Smith and Gracey, 2009; Ring and Brown, 2003).

Mortality disparities

Indigenous Peoples experience greater rates of health disparities and decreased longevity compared with non-Indigenous Peoples, regardless of the geographic area in which they live (Zinn, 1995; Stephens *et al.*, 2005; Ring and Brown, 2003; Ohenjo *et al.*, 2006; Montenegro and Stephens, 2006). However, incomplete data on indigenous status and health indicators make it difficult to estimate consistently the extent of disparities between indigenous and non-indigenous people by country (Stephens *et al.*, 2006). Indigenous Peoples in remote areas are usually underrepresented in the literature, and many countries do not identify ethnicity in their health statistics. In Latin America, studies indicate excessive infant mortality rates among Indigenous Peoples, which are three to four times as high as national averages of up to 30 per 1 000 live births with, for example, 99 to 100 per 1 000 live births for the Campa-Ashaninka and Machiguenga of the Peruvian Amazon, and 67 and 83 per 1 000 live births respectively for the Cumbas and Colimbuela of Ecuador (Montenegro and Stephens, 2006; Cardoso, Santos and Coimbra, 2005; Garnelo, Brandão and Levino, 2005). In Africa, excessive infant and under-five mortality have been noted among forest-dwelling Aka in the Central African Republic, the Twa in Uganda, and the Mbendjele in northern Congo, with a high death rate associated with infectious diseases (Ohenjo *et al.*, 2006).

Excess mortality and disease-specific mortality for Indigenous Peoples have been noted in developed countries with generally good health surveillance systems (Bramley *et al.*, 2004). Although indigenous status is underreported in death records in Australia, the existing data indicate a fourfold excess age-adjusted death rate among indigenous compared with non-indigenous Australians from 2000 to 2004 (Australian

Bureau of Statistics, 2002; 2003; 2004; 2005). In addition, while infant mortality has been decreasing over time, disparities in infant mortality between indigenous and non-indigenous Australians have increased, owing to greater reductions in mortality among non-indigenous infants (from 8.4 to 3.7 per 1 000 live births between 1980 to 1984 and 1998 to 2001) than among indigenous Australians (from 25 to 16.1 per 1 000 live births over the same period) (Freemantle *et al.*, 2006). The most recent data suggest that indigenous children are twice as likely to die than non-indigenous children (McCredie, 2008) and that infant mortality rates were three times higher for indigenous boys and 2.5 times higher for indigenous girls than for non-indigenous children between 2004 and 2006 (Australian Bureau of Statistics, 2007). At birth, Indigenous Peoples in Australia also face a 15- to 20-year reduction in life expectancy, with the leading cause of mortality now being cardiovascular diseases (Trewin and Madden, 2005). In New Zealand, the gap in life expectancy was eight years between Maori and non-Maori people in 1999, and although decreases in mortality have been noted over time, disparities persist (New Zealand Ministry of Social Development, 2004).

In the United States of America, American Indians and Alaska Natives born in 1999 to 2001 had a life expectancy that was 2.4 years less (at 74.5 versus 76.9 years) than that of the overall United States population (Indian Health Service, 2006). Infant mortality is approximately 20 percent higher among American Indians and Alaska Natives (at 8.3 per 1 000 live births in 2002 to 2004) than for the total United States population (at 6.9 per 1 000 live births in 2003) (Indian Health Service, 2009). In the Federated States of Micronesia, which is now independent but freely associated with the United States of America, improvements in infant mortality have been noted, but the rate remains excessively high at 33 deaths per 1 000 live births in 2007 (ESCAP, 2009).

In Canada most health indicators identify large disparities in morbidity and longevity between Indigenous Peoples and the general population, with gaps in life expectancy of 5.5 years for females and 8.1 years for males in 2000 (Health Canada, 2005). For

Inuit, the estimated average life expectancy is 15 years less than that of the general Canadian population. Infant mortality rates among Inuit, although decreasing over time, remain four times higher than those among the general population (Wilkins *et al.*, 2008).

Avoidable deaths and morbidity

A greater proportion of the mortality among Indigenous Peoples is avoidable than is among non-indigenous people (Ring and Brown, 2003). In New Zealand, the Ministry of Health estimates that the avoidable death rates among Maori are almost double those among New Zealanders of European or other descent (New Zealand Ministry of Health, 1999). In Australia, the risk of preventable infant deaths was 8.5 times higher among indigenous than non-indigenous Australians, with higher infant mortality rates attributable to infection and birth defects, particularly in remote areas (Freemantle *et al.*, 2006).

To reduce the gap in avoidable mortality, more emphasis should be placed on primary health care services for prevention and early diagnosis and treatment (Ring and Brown, 2003). In a global review of type 2 DM complication rates, Indigenous Peoples, regardless of their geographic location, experienced a disproportionate rate of complications relative to their non-indigenous counterparts (Naqshbandi *et al.*, 2008).

In addition to improving the availability of health care services and diagnostic screenings, there is also need for education to improve indigenous communities' awareness that services are available and accessible and that they can make a difference in promoting health (Reading, 2009).

Food insecurity

Food security is fundamental to population health and is a common theme of the case studies described in this volume. Populations are considered food-insecure when there is limited availability of, or ability to acquire, culturally acceptable, nutritionally adequate and safe foods on a sustained basis (FAO, 1996). Gradients of food security range from fully secure, to mild anxiety

about not having enough food, to outright hunger. Even in developed countries, food-insecure people suffer from various degrees of poverty and have poorer perceived health and lower nutrient intakes or nutrition status than the food-secure (McIntyre *et al.*, 2003; Ledrou and Gervais, 2005; Kirkpatrick and Tarasuk, 2008; Vozoris and Tarasuk, 2003; Rose and Oliveira, 1997; Skalicky *et al.*, 2006; Zalilah and Tham, 2002). Among the Orang Asli (Timian) of Malaysia, a study of 64 children identified that 82 percent of homes with children reported food insecurity, which coexisted with high prevalence of underweight, stunting and wasting (at 45.3, 51.6 and 7.8 percent, respectively) and dietary intakes noted at less than two-thirds the recommended dietary allowance (RDA) levels for energy, calcium and iron, but remarkably good intakes of many other nutrients (Zalilah and Tham, 2002).

For Indigenous Peoples, the current definitions of food security are inadequate as they rely entirely on the assessment of monetary access to market food (Nord, Andrews and Carson, 2006), whereas Indigenous Peoples also consume traditional foods. Given the role of traditional food systems and food sharing networks in contributing to food security, nutrient intakes and cultural identity, the definition of food security for Indigenous Peoples should include assessment of traditional food intake and the stability of access to traditional foods (Egeland *et al.*, 2010; Power, 2008; Lambden, Receveur and Kuhnlein, 2007). The contribution of traditional food to nutrition status can be substantial, and assessments of the impact of food insecurity can be hindered when assessment tools consider only monetary access to market foods.

Food insecurity, with resulting outright hunger and undernourishment, remains a global public health challenge. In the case studies included in the CINE series, the Maasai of Kenya, the Dalit and Bhil of India and, to a lesser extent, the Igbo of Nigeria showed evidence of severe food insecurity resulting in high prevalence of undernutrition.

Paradoxically, however, food insecurity has not been consistently related to undernourishment (Renzaho, 2004; Zalilah and Tham, 2002), and studies now report either a greater risk of obesity or no differences in adiposity by food security status (Casey *et al.*, 2006; Jiménez-Cruz, Bacardí-Gascón and Spindler, 2003; Gundersen *et al.*, 2008; Townsend *et al.*, 2001; Dinour, Bergen and Yeh, 2007; CDC, 2003; Rose and Bodor, 2006; Whitaker and Orzol, 2006). Variable gradients in the severity of food insecurity and differences in the local cultural and economic context of the amounts and types of food obtained and the extent of physical activity are likely to contribute to the conflicting findings between food insecurity and adiposity in the literature.

Food security for Indigenous Peoples is affected by changing environments, including environmental contamination and degradation, climate change, urban growth, modern farming and ranching and other infringements on traditional lands. Global market forces and colonization also play an omnipresent role in influencing dietary and lifestyle habits, as they generally increase dependence on highly processed food of poor nutrient quality, usually in the form of refined grains, and food with added sugar and fat, which increase the energy density of food consumed. All of these factors can work in tandem to reduce the viability of traditional food systems, and are serious threats to sustaining food security, especially for marginalized Indigenous Peoples (Thrupp, 2000). The dietary changes associated with globalization and colonization were cited as contributing to the recent epidemic of obesity and obesity-related chronic diseases in the Pacific Island countries and East Africa (Hughes and Lawrence, 2005; Raschke and Cheema, 2008).

An inverse relationship between the energy density of food and energy costs has been reported, and energy-dense and nutrient-poor food provides kilocalories at affordable cost; the high palatability of added sugar, sodium and fat in highly processed food can also lead to overconsumption of energy, and has been suggested as a mechanism by which poverty and food insecurity can lead to obesity and type 2 DM, both of which follow a socio-economic gradient in risk (Drewnowski and Specter, 2004; Drewnowski, 2009). In Guam, 49 percent of Chamorros were obese and Chamorros had a high dietary energy density, with 1.9 kcal per gram of food consumed compared with 1.6 kcal/g among

Filipinos, of whom only 20 percent were obese (Guerrero *et al.*, 2008). The high energy density of the diet in the Federated States of Micronesia is coupled with low physical activity, with 64.0 to 77.2 percent of youth not meeting recommended levels of physical activity in a recent survey (Lippe *et al.*, 2008). A shift in energy balance occurs when energy-dense processed market food is adopted, while physical activity is reduced as traditional activities are abandoned, both leading to weight gain.

From the biodiverse tropical areas to the far reaches of the Arctic, Indigenous Peoples' food security situation is highly variable, reflecting not only changes or degradation in ecosystems, but also geopolitical factors such as civil unrest, global economic forces, and urbanization and development. Environmental degradation can lead to loss of biodiversity, thereby threatening food security. In the highlands of Papua New Guinea, for example, the degradation of soil and vegetation has led to an overdependence on sweet potato on the high-altitude plateau and the dry grasslands, with women and children being more vulnerable to reduced dietary diversity (Bayliss-Smith, 2009). For the Maasai of Kenya, access to land for grazing their cattle (a cornerstone of their traditional diet) is increasingly limited as the growth of Nairobi and other large cities consumes more land, imposing barriers to free movement for Maasai and their cattle. In the biodiverse western Amazon area, Indigenous Peoples such as the Sacha Runa of Ecuador, the Ingano of Colombia and the Awajún of Peru actively cultivate biodiversity, and utilize both wild forests and cultivated fields for sustaining resilience in their ecosystem, to support their food security, medicinal care and cultural heritage (Garí, 2001; Creed-Kanashiro *et al.*, 2009; Correal *et al.*, 2009).

As proposed development projects involving oil, mineral and timber extraction and agrodevelopment encroach on to the lands of Indigenous Peoples in the Amazon, food security is threatened unless an alternative paradigm of development is promoted in which indigenous agro-ecology and biodiversity can continue to thrive. The Ingano made historic progress by being the principal actors in the development and management of a protected area fully recognized by the State, the Indiwasi National Park. Conserved land areas represent a means by which Indigenous Peoples can conserve biodiversity and cultural integrity, promote food security and reach the Millennium Development Goals (Pathak, Kothari and Roe, 2005). In addition, climate change is threatening small island states (Barnett, Dessai and Jones, 2007) and the Inuit traditional food system (Chapter 9 – Egeland et al., 2013), and is projected to have sweeping effects globally on food security, especially for Indigenous Peoples, who often live in vulnerable and extreme environments (FAO, 2008a).

Many indigenous communities, including those in Australia, Canada, the United States of America, Japan, New Zealand, India (the Bhil) and the Federated States of Micronesia, depend largely on market economies for access to basic foodstuffs, and supplement their market food diets to varying degrees with hunted, caught and cultivated traditional foods. In the context of cash-poor economies and a volatile global market of fluctuating food prices, the maintenance of cultural knowledge and traditions regarding food is a matter not only of cultural identity and transmission, but also of maintaining food security and nutritional health. Global market forces, including the utilization of grain for supplying biofuels, resulted in exorbitant price increases for basic food grain staples in 2007/2008, with direct effects on increasing hunger among the disadvantaged (FAO, 2008b). While grain supply costs declined somewhat in 2009, following a peak in mid-2008, the global economic crisis continues to undermine food security (Harrison, Tirado and Galal, 2010). Given the high rates of poverty among Indigenous Peoples, the current global economic crisis will undoubtedly put a disproportionate burden on the food security of Indigenous Peoples. The prices of food purchased are also heavily influenced by the costs of transportation, with effects on food security for remote communities.

Food insecurity is reported among Indigenous Peoples in developed countries. In New Zealand, women, Pacific Islanders and Maori were at greater risk of food insecurity than the overall population (Parnell et al., 2001). In Canada, poverty rates are twice as high among aboriginal peoples than non-aboriginal Canadians (Canadian Council on Social Development, 2003), and 33 percent of aboriginal people were food-insecure compared with 9 percent among non-aboriginal Canadian households (Willows et al., 2009). Inuit, who reside in remote communities, face even greater challenges to food security than other Indigenous Peoples in Canada. For Inuit, climate change threatens traditional food species reproduction and survival and limits the human navigation that is essential for hunting (Chapter 9 – Egeland et al., 2013). This in turn would limit access to traditional food and threaten food security and nutrition status (Egeland et al., 2009; Johnson-Down and Egeland, 2010; Kuhnlein and Receveur, 2007). In focus groups in six Inuit communities, participants stated that "a lot of people are living near poverty" and having difficulty obtaining enough food to eat (Chan et al., 2006). Despite large Canadian subsidies for a food mail programme (of CAD 46 million in 2006/2007) for northern communities, food costs in most remote communities are at least double those in southern Canadian cities or towns closer to food distribution routes (Indian and Northern Affairs Canada, 2007). The combination of inability to afford market food, lack of access to hunting and fishing and/or lack of a hunter in the family contribute to food insecurity in Canadian Arctic communities, particularly among households headed by women (Duhaime, Chabot and Gaudreault, 2002; Lambden et al., 2006).

Among the case studies in this volume, the one in which traditional food systems were closest to being entirely lost is that of the Ainu of Japan. Strong assimilation policies by the Japanese Government over a fairly long period have stopped food insecurity (in the usual definition of inadequate access to enough food) from being a problem, but have also resulted in loss of identification of traditional foods and dishes. A recent effort aims to reidentify traditional Ainu foods and culture before the relevant knowledge is lost. The Government of Japan has now officially recognized the Ainu as an Indigenous People and has publicly recognized the hardships and poverty they have endured (Ito, 2008), which represents progress for the Ainu.

Food security is complex, relying on local food culture and ecosystems, fluctuations in precipitation and climate, soil quality, and socio-economic aspects of trade and food purchasing behaviour. Where soils are poor and weather extreme, food diversity and adequacy are compromised, and macro- and micronutrient deficiencies may be evident, depending on the predominant dietary practices. For example, in areas with low consumption of animal food, low vitamin D, calcium, n-3 fatty acids and B_{12} would be expected, whereas in areas with low fruit and vegetable consumption, carotenoids and phytonutrients may be compromised (Wahlqvist and Lee, 2007). An additional consideration is the growing body of evidence indicating that micronutrient deficiencies are more common among obese than normal-weight people in a wide variety of populations and age groups (Garcia, Long and Rosado, 2009). This evidence is fairly robust and includes antioxidant nutrients, vitamins A, D, C, B_{12} and folate, iron and zinc. There are several plausible biological mechanisms to explain this, and it is not yet clear in which direction any causal pathways may be operating. However, it is clear that the presence of overweight and obesity is consistent with, and may even be associated with, higher risk for impaired micronutrient status.

Where do we go from here?

Indigenous Peoples' knowledge and food systems are fast disappearing but are of utmost importance, not only for sustaining Indigenous Peoples but also for providing alternative paradigms for coping with diverse ecosystems in a changing global environment.

The challenge is to retain traditional food knowledge and food systems, including market food, through sound governance for food security. Indigenous Peoples have been remarkably resilient, and collectively provide a vast tapestry of culturally diverse examples of human ingenuity in food systems that are adaptive to different and often harsh ecosystems. However, Indigenous Peoples' knowledge and food systems are poorly documented and fast disappearing, even though they are of utmost importance not only for sustaining

Indigenous Peoples but also for providing alternative paradigms for coping with diverse ecosystems in a changing global environment.

While the social determinants of health are now widely accepted, Indigenous Peoples suffer from additional assaults on "indigeneity" and self-determination, which contribute to disparities in poverty, education, nutrition and food security, household crowding, poor access to and utilization of health care, and preventable diseases. Thus, Indigenous Peoples' health problems "cannot be resolved solely through health interventions"; policies with a stronger emphasis on indigenous rights are needed (Stephens *et al.*, 2006; Gracey and King, 2009; UNPFII, 2009). In addition, governments need to collect health information on needs and conditions, and should allocate adequate resources to addressing the socio-economic inequities between indigenous and non-indigenous people, to narrow the disparities in health and disease (Gracey and King, 2009). Specific programmes that target mothers and children, nutritional deficiencies, improvements in sanitation and household crowding as a means of reducing infectious diseases, improving living conditions and opportunities for urban residents, and addressing diseases of acculturation are among the top priorities (Gracey and King, 2009). Opportunities for reducing micronutrient deficiency disease through food fortification (Harrison, 2010) should be examined carefully, with a view to identifying the potential effects on the diets of indigenous populations.

To identify the next steps in reducing the enormous health disparities that exist, Indigenous Peoples' conceptualization of health and the determinants of their health framework also needs to be understood (Mowbray and WHO Commission on Social Determinants of Health, 2007). Indigenous Peoples' view of health is not limited to individual health and the absence of disease, but also encompasses the health of the entire community and of the ecosystem on which it relies; this includes the concept of well-being, which is more than the absence of disease (King, Smith and Gracey, 2009). The traditional holistic view of health covers spiritual, mental, physical and emotional well-being. Where colonization is part of the history of indigenous groups, it features as a prominent determinant of poor health because it relates to the disruption of ties to the land and traditional food systems that had an omnipresent role in defining traditional social arrangements, self-identity with defined roles for community members, and systems of knowledge. The weakening or destruction of cultural practices and language, disconnectedness from cultural identity and ongoing marginalization in which Indigenous Peoples are not recognized or understood by society's institutions all contribute to health disparities. The disruptions associated with colonization, which lead to lack of autonomy and self-esteem, are linked closely to poor health status (Durie, Milroy and Hunter, 2009).

Reversing the effects of colonization therefore depends on efforts to encourage self-determination in all facets of life. This will promote collective and individual identity, self-esteem and a greater locus of control, which can improve a broad range of health outcomes for Indigenous Peoples. In addition, international collaboration among Indigenous Peoples can foster innovative health research and help identify solutions to commonly shared problems (Reading, 2009).

Traditional food systems provide a strong foundation for cultural identity, a basis for social support networks and medicinal remedies, and nutritional health. The promotion of traditional food and food systems assists Indigenous Peoples in gaining greater autonomy and self-determination and promotes health. Policies need to encourage sound environmental husbandry by all sectors of society, and provide opportunities for Indigenous Peoples to continue or enhance their utilization of traditional food systems. Nutrition and health education and making healthy market foods affordable provide a highly worthwhile complementary approach, which is needed to reduce the disparities in nutrition-related chronic diseases.

Improved primary health care and models of health care delivery in cross-cultural settings are also needed. Community-led or -partnered programmes will enhance acceptability among community members and are likely to improve the performance indicators of programmes' successes (Ring and Brown, 2003; Gracey and King, 2009). Indigenous Peoples are taking leadership in the development of community-based programmes that emphasize nutrition and physical activity, such as the Unity of First People of Australia innovative health promotion programme in Kimberley Region, where positive changes in knowledge about food, nutrition and exercise are having an impact on diabetes risk factors (Gracey *et al.*, 2006). Another success has been the community-led Kahawá:ke Diabetes Prevention Project, in which a decline in the incidence of type 2 DM between 1986/1988 and 1992/1994 coincided with the Mohawk community's mobilization of prevention efforts (Horn *et al.*, 2007).

The many initiatives described in the CINE case study series indicate how the collaborative ties that Indigenous Peoples have can make a difference in the health of their communities. There is no single successful strategy; multiple strategies are needed to help narrow the gap in nutrition-related chronic diseases. However, successful programmes will likely be those that include Indigenous Peoples' initiatives and perspectives and local food resources; partnerships among and within communities; community partnerships with nutri-tionists, health care providers and health care specialists; the involvement of government and non-governmental agencies; and locally operated points for health care screening and feedback tied to health education. Improving Indigenous Peoples' health is a "critical but complex challenge" (Stephens *et al.*, 2005), but not an impossible task ◉

> **Comments to:** g.egeland@isf.uib.no

Annex 2.1 Micronutrient deficiency conditions and their worldwide prevalence

Micronutrient	Deficiency prevalence	Major deficiency disorders
Iodine	2 billion at risk	Goitre, hypothyroidism, iodine deficiency disorders, increased risk of stillbirth, birth defects, infant mortality, cognitive impairment
Iron	2 billion	Iron deficiency, anaemia, reduced learning and work capacity, increased maternal and infant mortality, low birth weight
Zinc	Estimated as high in developing countries	Poor pregnancy outcomes, impaired growth (stunting), genetic disorders, decreased resistance to infectious diseases
Vitamin A	254 million preschool children	Night blindness, xerophthalmia, increased risk of mortality in children and pregnant women
Folate (vitamin B_6)	Insufficient data	Megaloblastic anaemia, neural tube and other birth defects, heart disease, stroke, impaired cognitive function, depression
Cobalamine (vitamin B_{12})	Insufficient data	Megaloblastic anaemia (associated with *Helicobacter pylori*-induced gastric atrophy)
Thiamine (vitamin B_1)	Insufficient data, estimated as common in developing countries and in famines, displaced persons	Beriberi (cardiac and neurologic), Wernicke and Korsakov syndromes (alcoholic confusion and paralysis)
Riboflavin (vitamin B_2)	Insufficient data, estimated as common in developing countries	Non-specific – fatigue, eye changes, dermatitis, brain dysfunction, impaired iron absorption
Niacin (vitamin B_3)	Insufficient data, estimated as common in developing countries and in famines, displaced persons	Pellagra (dermatitis, diarrhoea, dementia, death)
Vitamin B_6	Insufficient data, estimated as common in developing countries and in famines, displaced persons	Dermatitis, neurological disorders, convulsions, anaemia, elevated plasma homocysteine
Vitamin C	Common in famines, displaced persons	Scurvy (fatigue, haemorrhages, low resistance to infection, anaemia)
Vitamin D	Widespread in all age groups, low exposure to ultraviolet rays of sun	Rickets, osteomalacia, osteoporosis, colorectal cancer
Calcium	Insufficient data, estimated as widespread	Decreased bone mineralization, rickets, osteoporosis
Selenium	Insufficient data, common in Asia, Scandinavia, Siberia	Cardiomyopathy, increased cancer and cardiovascular risk
Fluoride	Widespread	Increased dental decay, affects bone health

Source: Tulchinsky, 2010, adapted from Allen *et al.*, 2006: Table 1.2, pp. 6–10.

Global environmental challenges to the integrity of Indigenous Peoples' food systems

NANCY J. TURNER[1]   MARK PLOTKIN[2]   HARRIET V. KUHNLEIN

1
School of Environmental
Studies, University of
Victoria, Victoria,
British Columbia, Canada
2
Amazon Conservation
Team, Arlington, Virginia,
United States of America

3
Centre for Indigenous
Peoples' Nutrition and
Environment (CINE)
and School of Dietetics
and Human Nutrition,
McGill University,
Montreal, Quebec,
Canada

Key words > Indigenous Peoples,
Indigenous Peoples' food systems,
environmental issues, biodiversity,
climate change, food security, food sovereignty

> "In less than 100 years since the colonization of Hokkaido, our land was changed to farmland and resort land, the mountains are ruined, rivers are covered with concrete and their flows were changed by dams."
>
> Koichi Kaizawa, Ainu community leader

Abstract

The integrity of Indigenous Peoples' food systems is intimately connected to the overall health of the environment. Recent declines in many aspects of environmental quality, from loss of biodiversity to environmental contamination, have combined with social, economic, political and cultural factors to threaten the health and well-being of Indigenous Peoples, and ultimately of people everywhere. This has affected the quality of indigenous food, restricted its availability or curtailed access to it.

All of the global case studies of Indigenous Peoples in the Indigenous Peoples' Food Systems for Health Program indicate concerns over environmental degradation as a major aspect of Indigenous Peoples' declining use of their indigenous food. Interconnected concerns include biodiversity loss of wild species and of cultivated species and varieties; hydroelectric dams and their impacts on fish and other foods; contamination of water and food from a host of chemical, radioactive and biological pollutants; and climate change, with its accompanying uncertainties and instabilities regarding food systems.

Reconnecting Indigenous Peoples with their traditional territories, and reversing some of the restrictive regulations against Indigenous Peoples' historical hunting and plant harvesting practices may help to restore and maintain traditional resources. More cooperative arrangements for co-management of habitats and resources should be instated. Collaborative research is recommended, such as that reflected in this volume in which environmental and other relationships among Indigenous Peoples' cultures, lands and resource stewardship are complemented with supporting work by academic partners. Ultimately, this will help to maintain and strengthen the resilience of ecosystems and cultural systems, including diverse and healthy food systems.

Introduction

Humans are completely dependent on healthy environments for their health and well-being. Global human food systems have been created and supported by a combination of the earth's multitudes of life forms and ecosystems and by human ingenuity, developed and shared over many thousands of years. Today, however, both the cultural diversity and the global biodiversity that gave rise to human food systems are threatened in many places, and Indigenous Peoples' food systems are particularly vulnerable (Davis, 2001; Carlson and Maffi, 2004; Wilson, 1992). To maintain the integrity of human food systems around the world, the environmental problems affecting biodiversity and biological productivity must be addressed, as the survival of the life forms that provide food, directly and indirectly, is fundamental to the well-being of human cultures and populations.

Almost daily, reports of environmental problems with impacts on human nutrition dominate the media. All of these influence human nutrition through:

- overexploitation of major fish stocks (Jackson *et al.*, 2001; Myers and Worm, 2003; Pauly *et al.*, 2000; Roach, 2006; Schindler *et al.*, 2002), forests (FAO and IPGRI, 2002) and terrestrial wildlife (Bennett and Robinson, 2000);

Table 3.1 Environmental impacts identified as affecting indigenous food systems of case study communities

Source/type of environmental impact	Examples/food system impact	References
Erosion of biodiversity (wild species)	Threats to caribou calving grounds from natural gas pipeline and oil drilling in Arctic regions; widespread loss of tropical forests; decreased yield and availability of certain foods (e.g., ooligan for Nuxalk; wild fish and shellfish species, and wild game in many places)	Egeland *et al.*, 2009 (Inuit, Nunavut); Kuhnlein *et al.*, 2009 (Gwich'in, northern Canada); Chapter 8 in this volume (Ingano, Colombia); Turner *et al.*, 2009; Chapter 11 in this volume (Nuxalk, western Canada)
Erosion of biodiversity (cultivated species)	Decreased use and loss of cultivated varieties (cultivars or landraces) (e.g., traditional cereals, banana varieties, taro, breadfruit); threats from large-scale monocultures and genetically modified food crops	Brookfield and Padoch, 1994; Chotiboriboon *et al.*, 2009; Chapter 10 in this volume (Karen, Thailand); Creed-Kanashiro *et al.*, 2009; Chapter 5 in this volume (Awajún, Peru); Englberger *et al.*, 2009; Chapter 12 in this volume (Pohnpei, Federated States of Micronesia); Salomeyesudas and Satheesh, 2009; Chapter 6 in this volume (Dalit, India); Turner *et al.*, 2009; Chapter 11 in this volume (Nuxalk, western Canada)
Deforestation and overexploitation of forest resources	Destruction of forests through logging and illicit crop cultivation; overharvesting of rubber; deforestation through charcoal making and fuelwood harvesting	Chotiboriboon *et al.*, 2009; Chapter 10 in this volume (Karen, Thailand); Correal *et al.*, 2009 (Ingano, Colombia); Creed-Kanashiro *et al.*, 2009; Chapter 5 in this volume (Awajún, Peru); Oiye *et al.*, 2009 (Maasai, Kenya)
Water shortages	Drought, desertification; acute shortages of water for livestock and household use	Correal *et al.*, 2009 (Ingano, Colombia); Oiye *et al.*, 2009 (Maasai, Kenya); Salomeyesudas and Satheesh, 2009 (Dalit, India)
Hydroelectric dam construction	Loss of salmon and other indigenous food; changes in environment; loss of access to indigenous food; loss of water quality	Iwasaki-Goodman, Ishii and Kaizawa, 2009 (Ainu, Japan)
Water pollution from domestic and livestock waste	Solid waste disposal problems; inadequate sanitation; faecal contamination of water and bacterial disease from poor waste disposal	Correal *et al.*, 2009 (Ingano, Colombia); Creed-Kanashiro *et al.*, 2009; Chapter 5 in this volume (Awajún, Peru); Englberger *et al.*, 2009; Chapter 12 in this volume (Pohnpei, Federated States of Micronesia); Oiye *et al.*, 2009 (Maasai, Kenya)
Contamination of food web, and threat of contamination, from industrial development, mining, herbicide spraying, nuclear power facilities	Pollution and chemical contamination from mining, oil drilling and petrochemical development; toxic residues in food	Correal *et al.*, 2009 (Ingano, Colombia); Creed-Kanashiro *et al.*, 2009; Chapter 5 in this volume (Awajún, Peru); Egeland *et al.*, 2009 (Inuit, Nunavut); Kuhnlein *et al.*, 2009 (Gwich'in, northern Canada)
Soil erosion and deterioration	Decline in soil fertility; soil loss; overgrazing and reduced carrying capacity for livestock; deterioration of pastures	Correal *et al.*, 2009 (Ingano, Colombia); Oiye *et al.*, 2009 (Maasai, Kenya); Okeke *et al.*, 2009 (Igbo, Nigeria)
Global climate change	Melting glacial ice and sea ice (in the north); changes in rainfall patterns; weather extremes, floods; raised sea levels	Correal *et al.*, 2009 (Ingano, Colombia); Creed-Kanashiro *et al.*, 2009; Chapter 5 in this volume (Awajún, Peru); Egeland *et al.*, 2009 (Inuit, Nunavut); Englberger *et al.*, 2009; Chapter 12 in this volume (Pohnpei, Federated States of Micronesia); Kuhnlein *et al.*, 2009 (Gwich'in, northern Canada); Oiye *et al.*, 2009 (Maasai, Kenya)

Chapters in this volume:
5 – Creed-Kanashiro *et al.*, 2013;
6 – Salomeyesudas *et al.*, 2013;
8 – Caidedo and Chaparro, 2013;
10 – Sirisai *et al.*, 2013;
11 – Turner *et al.*, 2013;
12 – Englberger *et al.*, 2013.

- habitat loss from urbanization and the industrialization of landscapes (Millennium Ecosystem Assessment, 2005; CBD, 1992);
- invasive species (Crosby, 1986; Wilson, 1992);
- pollution and degradation of lands, waterways and the foods they produce (WWF, 2004; Ross and Birnbaum, 2003; Kuhnlein and Chan, 2000);
- global climate change (Ashford and Castleden, 2001; IPCC, 2007; Salick and Ross, 2009; Thomas *et al.*, 2004).

Invariably, environmental impacts on food systems are cumulative and interconnected, and they interact at multiple scales of time and space. To understand and mitigate these impacts more effectively, it is necessary to recognize their pervasiveness, examine the origins of the problems and the processes involved, and address these at multiple levels. Looking at individual case studies of indigenous communities and their direct connections to local environments and food sources provides a solid and tangible starting point.

Widespread environmental deterioration leading to the erosion of biodiversity is not a recent phenomenon. However, because the world's population is increasingly urban and distant from the natural rural environment, the signs and signals that sources of food and clean water are imperilled have received little attention until recently (Ommer and Coasts Under Stress Research Project Team, 2007; Pollen, 2006). For example, most of the medicinal plants traditionally employed in East Africa come from forests that have been nearly eliminated throughout most of their original range (Cunningham, 1997). People living close to their food sources – who include many if not most of the world's Indigenous Peoples living relatively traditional lifestyles – have been firsthand witnesses to much of this environmental loss. For example, the Kogi Indians of the Sierra Nevada de Santa Marta in Colombia have been noting accelerated glacier melting and other associated climatic changes for decades (J. Mayr, personal communication to M. Plotkin, 2006). Far to the north, Canadian Indigenous Peoples of the polar regions, including Inuit, Gwich'in and Dene, have also been observing environmental deterioration: melting of sea ice, thawing of permafrost and siltation

of rivers, with a host of effects and impacts on wildlife and Indigenous Peoples' food systems (Berkes *et al.*, 2005; Krupnik and Jolly, 2002; Salick and Ross, 2009). In many cases, it is the observations, experiences, practices and cultural institutions of local Indigenous Peoples that help to determine the rates and causes of environmental loss, and Indigenous Peoples can often have some of the best ideas of possible ways to protect habitats, repair some of the damage and adapt to changing conditions (Turner and Clifton, 2009). This chapter focuses on the environmental aspects of Indigenous Peoples' food security, and discusses how the damage that threatens local and global food resources can be mitigated or possibly reversed.

Indigenous Peoples' food systems and environments

Investigations of the food systems of indigenous communities participating in the CINE Indigenous Peoples' Food Systems for Health Program (Kuhnlein *et al.*, 2006) sought to improve understanding of the environmental context of Indigenous Peoples' foodways. The state of each region's ecosystems and their capacity to support Indigenous Peoples' food systems is of fundamental importance. Indigenous communities participating in the programme identified several major environmental problems that negatively affect their overall food security and food systems (Table 3.1). These include specific concerns, such as declining populations of resource species: caribou in northern Canada, ooligans and salmon on the west coast of Canada, and crop diversity for bananas and other species in Pohnpei, Dalit and Karen communities in the Federated States of Micronesia, India and Thailand, respectively. They also incorporate some impacts that are more indirect but just as significant, such as deforestation, water deterioration, soil erosion and climate change. Each of these conditions and situations affects not only the case study indigenous communities, but also many other Indigenous Peoples and, eventually, all humanity and other species on the globe. As many Indigenous Peoples hold a "kincentric" worldview, in which all species are respected as close relatives, the notion of harm to species such as polar bears, salmon or orca whales is as alarming and upsetting as direct impacts on human communities themselves (Salmón, 2000; Senos *et al.*, 2006).

In the following sections, four of the overriding environmental problems that affect Indigenous Peoples' food systems are described in more detail to demonstrate the complex web of issues that are involved with each: biodiversity loss, especially of food species; hydroelectric dams and their effects; contamination of water and food; and global climate change.

Biodiversity loss

On every continent, Indigenous Peoples, other local peoples and biologists have noted alarming declines in the populations of many of the world's species (Wilson, 1992; Millennium Ecosystem Assessment, 2005). In recent times, many species have become extinct, for diverse reasons, most of which are directly or indirectly attributable to human activity. There are compelling examples of past human-caused extinctions or severe depletions of important food species, including the passenger pigeon in the Americas and the American bison (Davis, 1998). Today, with burgeoning human populations, globalization and increasing commodification of wild resources that were, and still are, major components of Indigenous Peoples' food systems, erosion of biodiversity is an ever-growing concern, and needs increased attention. For both wild species and crop varieties important to Indigenous Peoples, the largely negative role of large-scale commercialization and globalization of the marketplace cannot be ignored.

Many Indigenous Peoples have traditionally had strong protocols and culturally mediated prohibitions against overharvesting and towards the sustainable use and enhancement of food resources (Anderson *et al.*, 2005; Berkes, 2008; Deur and Turner, 2005; Johannes, 2002; Turner and Berkes, 2006). Today, however, species that were once carefully stewarded by local people – such as sea urchins, herring eggs and abalone for British Columbia coastal peoples in Canada – have become commodified, with global demands for immense

quantities. Without proper and careful constraints on the use of these species, this situation characteristically leads to overexploitation, to the ultimate detriment of the local peoples who rely on them (Berkes *et al.*, 2006). Similarly, the health and livelihoods of local and Indigenous Peoples in many countries are threatened by escalating unsustainable use of wild meat or "bushmeat" (Bennett and Robinson, 2000; Anderson *et al.*, 2005), and by industrial and government-sanctioned deforestation to meet a great world demand for timber and dominant agricultural crops (Mackenzie, 1993; Balée, 1994; Turner and Turner, 2006; 2008; Graham, 2008). Habitat loss, the impacts of introduced species and the loss of pollinators are a few of the many threats to Indigenous Peoples' food systems, beyond direct overharvesting (Porcupine Caribou Management Board, 2007; Kuhnlein, 1992; Nabhan, 1986). The story is repeated again and again, from flying foxes and tropical forests in Samoa to Pacific salmon and coastal temperate rain forests on the northwest coast of North America (Cox, 1997; Nabhan, 2006).

Salmon farming or marine net-pen aquaculture can cause many direct and indirect negative impacts on marine environments. Depletion of fish stocks used as fish feed, destruction of coastal ecosystems such as eelgrass beds that are important nursery grounds for marine species, potential invasion of introduced Atlantic salmon, eutrophication caused by nutrients from fish and excess food and faeces, use of antibiotics, and sea lice infestations are some of the challenges facing Indigenous Peoples on the northwest coast of North America, who rely on the annual runs of wild Pacific salmon for their nutrition and cultural integrity (Volpe, 2007). Globally, all marine systems are now showing deleterious effects of human-caused change (Pauly *et al.*, 2000).

Alongside the decline and extinction of native or wild species around the globe, crop varieties and special landraces (adaptations of domesticated species) of plants and animals have also been declining dramatically (Fowler and Mooney, 1990; Nabhan and Rood, 2004). Again, the reasons are complex, but political and industrial agendas are clearly implicated (Shiva, 2000), along with valid efforts to provide sufficient food for a burgeoning world population through a movement known as the green revolution. Increasing use of fertilizers, pesticides and herbicides, high fossil fuel inputs for ploughing, seeding and harvesting, and monoculture crop production are outcomes of the green revolution. The escalating production of genetically engineered crops has caused growing concern for Indigenous Peoples wishing to retain control over their own landraces and food systems (La Duke and Carlson, 2003; Pasternak, Mazgul and Turner, 2009; Kurunganti, 2006). Plantations of sugar cane, coffee, maize and other megacrops for export markets often give employment to Indigenous Peoples, but have widely replaced their diverse subsistence crops. Large-scale production of cattle and other livestock, with the accompanying pollution and degradation of pasturelands and deforestation, has also had severe negative consequences for Indigenous Peoples. Drought and desertification – often resulting from poor management practices, overcrowding and overgrazing – are also widely recognized as threats to Indigenous Peoples' food security.

One of the growing threats to subsistence food production is the biofuel industry. Biofuels are becoming a popular alternative and supplementary fuel for motor vehicles and heating. Although they tend to burn cleaner than fossil fuels and are theoretically a renewable resource, the market forces at play often result in the sequestering of lands formerly used for food production, to generate biofuels – often at the expense of Indigenous Peoples' well-being. Food security may decrease with cash cropping (Dewey, 1979; 1981): in Brazil, sugar cane, soybeans, castor beans and maize are being grown in increasing quantities to produce ethanol, reducing the nutrition opportunities for smallholder farmers (Conservation International, 2007; FIAN International, 2008; Graham, 2008).

Hydroelectric dams

Industrial-scale hydro projects provide power, but have proven destructive to Indigenous Peoples' ways of life and food systems; however, more dams are being planned and constructed. For example, the James Bay project of Hydro-Quebec in Canada put thousands of square kilometres of traditional Cree territory under water in 1983, not only cutting off access to Cree food resources, but also placing the Cree's health at risk from mercury contamination of the fish they consumed. The decomposing trees and other plants covered by the dam floodwaters produced methane, which converted natural mercury in the soil into a toxic form that entered the food chain, poisoning both the fish and those who eat them (Richardson, 1991; Kuhnlein and Chan, 2000). Another example is dam construction in Mato Grosso State of Brazil, which will severely restrict aquatic protein for 14 tribes whose main source of protein is fish (M. Plotkin, personal observation, 2009).

Iwasaki-Goodman, Ishii and Kaizawa (2009) have documented a wide range of impacts resulting from the construction of an immense hydroelectric dam on the Saru River, site of the Ainu homeland for at least 1 000 years. The Ainu resided along the riverbanks and obtained much of their sustenance from the river by fishing, while farming and hunting on the adjacent lands. Traditionally, the river also provided high-quality drinking-water. More than 100 years ago, non-Ainu Japanese began colonizing the area, and the Hokkaido Government started establishing regulations aimed at assimilating the Ainu and restricting their cultural traditions, including hunting and fishing. In 1997, the Nibutani Dam was completed in the heart of Ainu territory, against the wishes of the Ainu. A court challenge of the legality of this dam by two Ainu landowners eventually resulted in a judgment that the government had failed to assess the effect that the dam's construction would have on the local Ainu culture, thereby ignoring values that required serious consideration. This led to increased recognition of the importance of cultural impact assessments in any future developments, establishing an important precedent. In the same year that the dam was completed, the Law Concerning Promotion of Ainu Culture and Dissemination and Enlightenment of Knowledge about Ainu Traditions was enacted. This law has reinforced an ongoing movement to revitalize Ainu culture, and interviews about the impacts of the Nibutani dam have been part of the impact assessment programme required before any subsequent dams are constructed.

Respondents in these interviews identified many changes resulting from the dam construction:

- cooler weather and more fog and mist;
- increased siltation and significant shallowing of the river;
- undesirable flooding of the rice fields following a typhoon in 2003;
- loss of access to the other side of the river for food gathering;
- restrictions on children's play areas, fishing places and picnicking areas;
- loss of shallow ponds and riverbanks that were sources of fish and other cultural resources;
- loss of spawning areas for smelts (fish);
- disappearance of kelp, shellfish, flounders and octopus;
- muddying of the river and loss of clear drinking-water.

In short, the Nibutani Dam "killed the natural environment" for the Ainu (Iwasaki-Goodman, Ishii and Kaizawa, 2009).

Contamination of water and food

Consumption of and exposure to contaminated water is an ongoing and growing concern, especially for people living in rural areas. Worldwide, many indigenous communities have been adversely affected by contaminated water. For example, the Wayanas of southern Suriname have up to 17 times the recommended level of mercury in their hair samples, resulting from mercury pollution (C. Healy, personal communication to M. Plotkin, 2006; Nuttall, 2006).

Sewage pollution is another ongoing and related issue. Many small communities – and some large cities – discharge large amounts of raw or minimally treated sewage into rivers, lakes and coastal waters, which affects the foods in these systems. The city of Victoria, Canada, which used to have some of the best clam digging beaches on the coast, now has chronically contaminated beaches; for many decades, the local Straits Salish First Nations have not been able to harvest their seafood near the populated areas of the Saanich Peninsula and Victoria coastline.

Among the case study communities, the Awajún of Peru and their neighbouring communities face major problems relating to water quality and pollution from human waste: all homes in the region have precarious access to basic water and sewage services, most have no running water, and rubbish is thrown into the river. Human faeces are commonly seen in public areas, and a system of latrines installed by a government organization in the early 1990s – when there were concerns about cholera in the region – is generally considered a failure because of poor design, bad location and lack of training in maintenance. Many people, both children and adults, suffer from diarrhoea, parasites and other illnesses related to contamination, and there is concern about typhoid fever (I. Tuesta, M. Carrasco and H. Creed-Kanashiro, personal communication, 2008).

Environmental contaminants that biomagnify and concentrate in food webs are also a threat, and have been well studied in some places (Kuhnlein *et al.*, 1982; 2005; Kuhnlein and Chan, 2000; Chan *et al.*, 1996; Chan, Kuhnlein and Receveur, 2001; Thompson, 2005; Ross, 2000; 2006; Ross and Birnbaum, 2003; Ross *et al.*, 2004). As already mentioned, mercury contamination has been particularly insidious, causing health concerns such as nervous system disorders from eating local fish from affected rivers (Lebel *et al.*, 1997; Shkilnyk, 1985; Khaniki *et al.*, 2005), in addition to the more widely publicized phenomenon of mercury contamination of coastal ecosystems and large oceanic species such as tuna.

As well as mercury and other metals, a range of organic industrial compounds, classed generally as persistent organic pollutants (POPs), are also of concern. These are semi-volatile fat-soluble toxic compounds, including polychlorinated biphenyls (PCBs), polychlorinated dibenzo-p-dioins (PCDDs, also known as dioxins), polychlorinated dibenzofurans (PCDFs, also known as furans), polybrominated diphenyl ethers (PBDEs) polybrominated biphenyls (PBBs) and polychlorinated naphthalenes (PCNs) (Iwasaki-Goodman, Ishii and Kaizawa, 2009; Rayne *et al.*, 2004; Ross, 2006; Ross and Birnbaum, 2003; Ross *et al.*, 2004). The origins of these compounds are mainly industrial, and range from local sites such as pulp mills and discarded machinery, to diffuse, distant sources from which the contaminants are transported through the atmosphere, ocean currents, soil and waterways, including by migratory species such as whales and salmon that have been contaminated (Johannessen and Ross, 2002; Krümmel *et al.*, 2003; Lichota, McAdie and Ross, 2004). Arctic regions are particularly vulnerable to contamination from POPs whose sources are known to be very distant; many of the contaminants in northern Canadian, for example, come from industrial centres in northern Asia and Europe (Knotsch and Lamouche, 2010).

Because predator species such as tuna, salmon and seals are at the upper trophic levels of food webs, these "sentinel" species are particularly vulnerable to contaminants, which accumulate in their fatty tissues (Ross, 2000). Humans who use these species as food in any quantity are placed at risk: ingesting contaminated food is the principal means by which humans are exposed to these highly toxic environmental pollutants (Parrish *et al.*, 2007). This situation is of particular concern in the food systems of Indigenous Peoples who consume large amounts of seal, salmon or other predator species (Johannessen and Ross, 2002; Mos *et al.*, 2004; Ross and Birnbaum, 2003). For example, POPs can interfere with the immune function of animals and – potentially – humans, making them more vulnerable to infectious diseases (Ross, 2002; Ross, Vos and Osterhaus, 2003). They can also disrupt endocrine function, reproduction and vitamin A production in the human body (Ross, 2000; Simms *et al.*, 2000). Recently, researchers have been observing possible associations between diabetes and levels of POPs (Jones, Maguire and Griffin, 2008; Rignell-Hydbom, Rylander and Hagmar, 2007).

Many indigenous communities have expressed concerns about contamination of their food (and their medicines and basketry materials) from agricultural chemicals and pesticides and from the herbicides used in industrial forestry, factory farming and powerline rights-of-way (Wong, 2003; Pollen, 2006). Mining and its associated smelters and refineries also present contamination concerns. Centres of industrial activity, such as at Kitimat in British Columbia, Canada have affected the habitats and food systems of indigenous and other local people. In Kitimat, pollutants from an aluminium smelter and other industrial plants have contaminated many Haisla foods, such as oulachens (ooligans, a favourite fish of the north coast) (Chan *et al.*, 1996; Turner *et al.*, 2009; Chapter 11 in this volume – Turner *et al.*, 2013), shrimp, clams and other species in the vicinity of the smelter. The Haisla elders used to refer to this area along the Kitimat River as their "grocery store", because it was such an important source of food, but they can no longer harvest their indigenous foods there (G. Amos, personal communication, 2007).

Gold, uranium, diamond and other mines are common in regions such as northern Canada, where many of the miners and local residents are Indigenous Peoples (e.g., Deline Mine in Canada's Northwest Territories). These people are directly affected by contaminants from the mines, while the caribou, fish and other animals on which they depend for food are affected by mining pollution and the impacts of the roads, settlements and infrastructure built to support prospecting and mining (B. Erasmus, personal communication, 2008). Mining in Amazonia is notoriously destructive to Indigenous Peoples and their food systems (Roulet *et al.*, 1999). The Awajún in Peru are concerned about possible mercury pollution of their rivers from gold mines upriver in the mountains and from mines in neighbouring Ecuador, but tests have not yet been carried out to determine the extent of the threat (I. Tuesta, M. Carrasco and H. Creed-Kanashiro, personal communication, 2008).

Oil and gas exploration and extraction, together with the construction of pipelines and their corridors, present a range of environmental problems and concerns regarding Indigenous Peoples' food systems and health (Wernham, 2007). In northern Alberta, Canada, the tar sands development, in which oil and gas are extracted from heavy crude oil that is mined from the surface and treated with large quantities of heated water, has resulted in environmental devastation and large deforested areas, described as "a moonscape" (Griffiths, Taylor and Woynillowicz, 2006). Impacts on wildlife are of great concern, with reports of entire flocks of ducks being destroyed in the expansive oil sands tailing ponds contaminated with bitumen residues (Torys LLP, 2010; B. Erasmus, personal communication, 2008). Not only is such destruction harmful to people's food resources, but it is also emotionally and culturally devastating to witness. In the Amazon region of Peru, the Awajún are concerned about the development of large-scale hydrocarbon extraction south of their lands; such development can cause deforestation and environmental devastation, as the Awajún are already observing in neighbouring Brazil.

Airborne radioactive contamination of food is a concern for Indigenous Peoples in the Arctic, where lichens absorb airborne contaminants before being eaten by caribou and reindeer, which are then eaten by humans. Concerns about poisoning from radioactive compounds have diminished since the cessation of aerial testing of nuclear bombs, but the threat of contamination from accidents in nuclear power plants continues. A catastrophic nuclear power plant accident at Chernobyl in the Ukraine region of the former Soviet Union in 1986 resulted in a massive atmospheric plume of radioactive contaminants that drifted across the Russian Federation, eastern, western and northern Europe and into North America, affecting the Sami of Scandinavia and the Inuit and other northern peoples of Canada (Berti et al., 1997; Strand et al., 1998; Kuhnlein and Chan, 2000).

Political decisions from governments and other agencies outside indigenous communities often have unrecognized or unacknowledged impacts on Indigenous Peoples' environments, cultures and food systems (Turner et al., 2008b). In Colombia, for example, government-sponsored large-scale aerial spraying of herbicides to destroy illegal coca crops has had impacts on the Ingano's crops. The herbicides fall on to grazing lands and farms, killing food crops such as manioc and banana. If the crops are mature when this happens, people consume them immediately, risking their own health to utilize crops that would otherwise soon die. The Ingano also suffer when the waste from cocaine production, referred to as cocasa, contaminates the rivers and streams they use for drinking-water and household purposes (Correal et al., 2009).

Global climate change

Global climate change is cited as a major concern in the Inuit and Gwich'in case studies (Chapter 7 – Kuhnlein et al., 2013; Chapter 9 – Egeland et al., 2013; Kuhnlein et al., 2004), and is perhaps the most pervasive, overarching threat to the security of Indigenous Peoples' food systems, both regionally and globally (Damman, 2010; Krupnik and Jolly, 2002; Myers et al., 2005; DFO, 2009; Environmental Change Institute, 2007; Dinar et al., 2008; Keskitalo, 2008; Turner and Clifton, 2009). Whether people rely on agriculture, pastoral systems, hunting, fishing, wild plant harvesting or a combination of food production and harvesting practices, climate change is causing, or has the potential to cause, major disruptions to their food systems. Among the host of interrelated problems attributed to climate change are:

- constrained water availability and water quality;
- unseasonably high temperatures, with threats of desertification;
- droughts and fires;
- unpredictable weather events (blizzards, hurricanes, floods, ice storms);
- shifts in seasonal weather patterns;
- changing sea levels, with impacts on coastal ecosystems;
- retreating glaciers and changing species distributions in high mountains;
- soil erosion;
- melting permafrost;
- spread of insect pests and diseases;
- changing wildlife migration routes;
- impacts on pollinators.

All of these affect Indigenous Peoples' food systems. The direct, indirect and cumulative effects of these factors on human food security are only starting to be noted, with locally based Indigenous Peoples sounding alarms (Environmental Change Institute, 2007). One example cited by indigenous people in Pohnpei (Chapter 12 – Englberger et al., 2013) is that rising sea levels are destroying coastal giant taro gardens. Another is the effect of permafrost melt on the safety of hunters and the turbidity of rivers in northern Canada (B.

Erasmus, personal communication, 2008). There is great concern that global climate change, exacerbated by indiscriminate tree cutting in the Amazonian forests will lead to progressive deforestation (WWF, 2008).

Discussion: maintaining food security and environmental sustainability

> Food security exists when "all people, at all times, have physical and economic access to sufficient safe and nutritious food to meet their dietary needs and food preferences for a healthy and active life".
> *FAO, 1996*
> Food sovereignty is recognized as the "right of Peoples to define their own policies and strategies for sustainable production, distribution, and consumption of food, with respect for their own cultures and their own systems of managing natural resources and rural areas", and is considered to be a precondition for food security.
> *International Indian Treaty Council, 2002*

The multitude of interrelated impacts of global climate change and other environmental threats described in the previous section illustrate the interactions and cumulative effects of many different factors facing Indigenous Peoples in their efforts to maintain their food security and food sovereignty.

As well as the environmental constraints on food security and food sovereignty, a range of social and economic factors also influence food choices: the impacts of residential schools in preventing intergenerational transference of knowledge and skills relating to food and health; urbanization; lifestyle changes; increased availability of convenience processed, marketed foods; television advertising; and many other pressures that move people away from their healthy original foods (Turner and Turner, 2008; Parrish, Turner and Solberg, 2007; Turner *et al.*, 2008b; Kuhnlein, 1989; 1992; Lambden *et al.*, 2006; Lambden, Receveur and Kuhnlein, 2007; Wernham, 2007).

Addressing such complex, cumulative stresses on Indigenous Peoples' food systems is no simple task. The CINE Indigenous Peoples' Food Systems for

Health Program has worked to renew and revitalize indigenous food systems as a way of increasing the health and well-being of Indigenous Peoples. Participants at the 2009 International Congress of Nutrition suggested a number of interventions that would help raise awareness and facilitate and promote local environmental stewardship, good nutrition and the use and relearning of Indigenous Peoples' foodways (Kuhnlein *et al.*, 2006). These included actions under five broad topics:

- *Harvesting wild plant/animal food resources:* Stimulate more community hunting/gathering/fishing activities, along with conservation training; work to increase access to land and water; teach these activities to youth; share harvests with elders and women; create community-based processing and storage facilities; and work to develop political leverage and agreements to ensure access to harvest areas.

- *Agricultural activities:* Stimulate home and community gardens and local food production; plant more trees and other produce; train farmers and others about nutrient-rich crops; develop medicinal plant gardens; form cooperative community groups to undertake agriculture activities; work to enhance access to land; and improve water quality.

- *Activities in community schools:* Ensure that school curricula focus on food and nutrition; involve children in teaching their communities about food; develop appropriate teaching materials; hold local food classes; promote healthy indigenous and locally produced snacks; and target unhealthy foods such as high-sugar beverages for elimination from schools.

- *General community projects:* Involve elders and cultural committees; encourage participation and cooperative work; train community health workers; prepare educational materials, posters, workshops, etc.; hold community health assessments; and stimulate physical and healthy lifestyle activities, etc.

- *Links with health care, agriculture, education, government, business and non-governmental organizations (NGOs):* Engage local steering committees in proactive work; develop prenatal programmes with healthy indigenous and local food; and network with businesses, NGOs, churches and schools to promote local food and health.

Broadly, these suggestions can be characterized as activities for cultural renewal and ethno-ecological restoration, in which Indigenous Peoples' food systems play a pivotal role (Senos *et al.*, 2006). Figure 3.1 illustrates the links and factors affecting Indigenous Peoples' food systems, including the positive effect that various interventions, combined with strong support from community leaders, government and academic institutions and others, can have on the overall health of cultures, environments and food systems. Without such support, the interconnected culture and environmental productivity are lost, resulting in loss of food resources, health and well-being.

Efforts to promote ecosystem enhancement and healthy cultural food systems are under way. Communities are participating in the current case studies and CINE programme, and in other projects with indigenous communities in many different places. Indigenous food harvesting and agricultural activities require government cooperation and collaboration, as

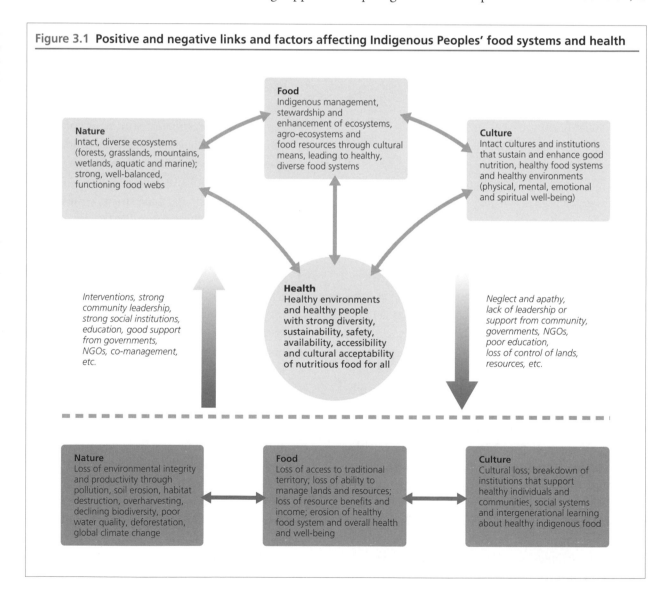

Figure 3.1 Positive and negative links and factors affecting Indigenous Peoples' food systems and health

Food
Indigenous management, stewardship and enhancement of ecosystems, agro-ecosystems and food resources through cultural means, leading to healthy, diverse food systems

Nature
Intact, diverse ecosystems (forests, grasslands, mountains, wetlands, aquatic and marine); strong, well-balanced, functioning food webs

Culture
Intact cultures and institutions that sustain and enhance good nutrition, healthy food systems and healthy environments (physical, mental, emotional and spiritual well-being)

Health
Healthy environments and healthy people with strong diversity, sustainability, safety, availability, accessibility and cultural acceptability of nutritious food for all

Interventions, strong community leadership, strong social institutions, education, good support from governments, NGOs, co-management, etc.

Neglect and apathy, lack of leadership or support from community, governments, NGOs, poor education, loss of control of lands, resources, etc.

Nature
Loss of environmental integrity and productivity through pollution, soil erosion, habitat destruction, overharvesting, declining biodiversity, poor water quality, deforestation, global climate change

Food
Loss of access to traditional territory; loss of ability to manage lands and resources; loss of resource benefits and income; erosion of healthy food system and overall health and well-being

Culture
Cultural loss; breakdown of institutions that support healthy individuals and communities, social systems and intergenerational learning about healthy indigenous food

there is often need to change regulations that prohibit food harvesting or prevent people from practising the management systems they used in the past to sustain their food resources (Posey, 1985; Anderson and Barbour, 2003).

Many environment-based regulations were established during the era of colonial, Euro-centric thinking, without clear understanding of Indigenous Peoples' conservation and management practices. For example, Straits Salish reefnet fishing was banned by the Canadian and United States governments because it was considered a form of "fish trap", and therefore assumed to be bad for conservation. However, this salmon harvesting technology is now recognized as an effective and sustainable management tool that reflects an entire way of life for the Saanich and other Straits Salish peoples (Claxton and Elliott, 1994; Turner and Berkes, 2006), and efforts are under way to reinstate this traditional fishery. Another example, affecting the Indigenous Peoples of western North America, is the banning of traditional landscape burning as being wasteful and destructive (Boyd, 1999; Anderson *et al.*, 2005; Anderson and Barbour, 2003). The positive ecological effects of mid-level human disturbance, including traditional burning practices, are now being revisited, and forestry officials are cooperating in experiments to explore the use of traditional fire regimes to renew huckleberry production and other resources for Indigenous Peoples in the region (Boyd, 1999). In all areas, regulations against Indigenous Peoples' historical hunting and gathering practices should be revisited and either revised or rescinded.

These restrictions could be replaced by more cooperative arrangements for co-management of habitats and resources. There are an increasing number of good co-management models, especially for parks and protected areas, and many have positive implications for Indigenous Peoples' food systems (Anderson and Barbour, 2003; Berkes, 2008; George, Innes and Ross, 2004; Hunn *et al.*, 2003; Nazarea, 1999; Turner, 2005). The United Nations Convention on Biological Diversity (CBD, 1992) and the Declaration on the Rights of Indigenous Peoples, adopted by the United Nations General Assembly in September 2007, contain explicit requirements for governments of Member Nations to respect the rights of Indigenous Peoples, and to consult and collaborate with them in all aspects of resource use affecting their lands and territories.

At the local level, many people are finding small but significant ways to alleviate the environmental problems they face. For example, the Awajún of Peru are starting to raise more chickens and to develop small-scale family fish-raising ponds. Developing these new protein sources has eased the impact on forest wildlife, allowing populations of wild animals to increase to the point where they can be hunted again, on a limited basis (Chapter 5 – Creed-Kanashiro *et al.*, 2013). Other peoples, such as those of the Pohnpei communities, are realizing that their traditional way of serving food on banana and other biodegradable leaves is more environmentally sound than using disposable plastic or other types of dishes (L. Englberger and M. Roche, personal communication, 2008). When practised by an entire community, the use of natural, biodegradable products to harvest, store, cook and serve food, and the recycling and reuse of more durable vessels and containers can have a positive impact on pollution and solid waste outputs (Wilson and Turner, 2004).

As indicated in the intervention ideas from the 2009 International Congress of Nutrition, education is another key factor in efforts to support Indigenous Peoples' healthy traditional food systems. A wide range of education processes should be supported: for indigenous youth and young adults, including the parents of young children, who may not be aware of the cultural or nutritional importance of their indigenous food (Beaton, 2004); for governments and decision-makers outside indigenous communities, and sometimes within them, who may not understand some of the issues regarding indigenous food loss; and for the general public, who could become allies and participants in efforts to restore ecologies and cultures and to renew healthy traditional foods for Indigenous and other local Peoples (Nabhan, 2006).

All Indigenous Peoples have their own educational needs and responses to different strategies for conveying the information required. Learning-by-doing is a well-tried method for developing knowledge and skills in food harvesting, processing and consumption. Providing children and youth with opportunities for hunting, fishing, berry picking and gardening, with their families or others and through science and cultural camps or school and college field trips, can be very effective in raising their awareness, enhancing their understanding and honing their skills. Participation in the development of demonstration food and medicine gardens and the creation of community and ethnobotanical gardens is also beneficial (Turner and Wilson, 2006). Finding ways for elders' voices to be heard and conveyed, directly in workshops and community meetings or indirectly through films and DVDs, is especially important, as they remember the most about historical food production and preparation. Many communities collaborating with NGOs or government agencies have been able to host cooking events, traditional feasts and other enjoyable, sociable and educational occasions that promote and educate people about the importance of indigenous food, while giving those who have not experienced it a chance to observe and taste such food; examples include Ainu food preparation classes (Chapter 13 – Iwasaki-Goodman, 2013) and community feasts with First Nations around Victoria, involving the Pauquachin, Tsawout, T'souke and Songhees nations (Devereaux and Kittredge, 2008; Pukonen, 2008; Turner *et al.*, 2008a). Programmes that support language and cultural renewal, including potlatches and feasts, dances, stories and ceremonies, are also important, as many indigenous food systems are closely linked to cultural practices and language.

For Indigenous Peoples whose food systems are based on agricultural crops, similar community activities aimed at renewing and reinstating traditional crop landraces and agro-ecology practices can be promoted. Many Indigenous Peoples' resource management systems are sound and sustaining; with cooperation from government and NGOs, these can often be reclaimed and applied to enhance soil fertility, water quality, crop diversity, biodiversity and the overall productivity of traditional food (Colfer, Peluso and Chung, 1997; Englberger *et al.*, 2006; Imhoff, 2003). Indigenous people – particularly women, whose role in conserving crop diversity is often overlooked – are often the best sources of knowledge about traditional landraces for crops such as maize, rice, manioc and many others (Hoyt, 1988; FAO and IPGRI, 2002).

Research undertaken in respectful, effective and collaborative ways is a key element in improving Indigenous Peoples' food systems. Research can help to document and characterize the local foods' contributions to the diet and to nutrient requirements that need special attention. Current dietary conditions can be used as a baseline for understanding dietary change and the environmental and social dimensions of this change. Research can identify the implications of dietary change in terms of threats from chronic disease; focus on issues relating to food safety and availability, including assessing the risk from food contaminants; and document strong cultural traditions and knowledge regarding natural resources, including unique food species held by Indigenous Peoples, as well as the risks of losing this knowledge. Participatory community-based research can build indigenous communities' capacity for improving their own health in the context of their own culture and language. It can identify patterns of land use and local food availability, and help clarify some of the controversies and issues that arise from government policies, such as the establishment of parks and protected areas. It can help to document the tremendous variation in species and varieties of food biota in indigenous areas, which is often unrecognized beyond a particular community, and provide scientific identifications and nutrient analyses of these foods. It can help to guide policy for environmental protection to ensure species habitats, and emphasize the value of cultural expression for retaining traditional knowledge and conserving species (Wyllie-Echeverria and Cox, 2000). It can also assist efforts to frame Indigenous Peoples' perspectives in ways that may be better understood by academics and policy-makers, such as use of the phrase "cultural keystone species" to emphasize the critically important nature of certain food and other species to particular cultural groups (Garibaldi and Turner, 2004).

Collaborative research on indigenous ecological knowledge systems can also help to create better responses to climate change and other forms of environmental change, through understanding socio-ecological adaptive processes and how these can apply to traditional land and resource management systems (Berkes, 2008; Turner and Berkes, 2006).

Conclusions: sustaining healthy food systems and environments in a changing world

Environmental threats to Indigenous Peoples' food security, food sovereignty and ability to maintain and utilize healthy foods from their own ancestral lands are very real. Although it is widely recognized that the food systems of Indigenous Peoples contain impressive levels of biodiversity (related to plant species, subspecies and varieties/cultivars and to animals and their subspecies), recent environmental impacts range from habitat loss to pollution and from erosion of biodiversity – including crop diversity – to increasingly evident climate change. These effects interact with each other in often unpredictable and insidious ways. The problems must be addressed at the local and global scales, and their complexity must be acknowledged and incorporated into solutions.

Indigenous foods benefit people's physical health, through both the consumption of good food and the physical activity of harvesting and preparing the food. In addition, these foods play a key role in maintaining diverse cultures, languages, heritages and identities – in short, in the mental, emotional, spiritual and physical well-being of Indigenous Peoples.

An important concept in maintaining environmental and cultural integrity – and therefore the integrity of Indigenous Peoples' food systems – is the inextricable linkage between the peoples and their territories. Indigenous Peoples' access to their lands, waters and resources, including genetic resources, is essential for their food security and for sustaining cultural

knowledge about traditional food and medicine systems (Laird, 2002; UNPFII, 2009; CBD, 2010). Wisdom and practical knowledge relating to the harvesting, processing, consumption and long-term management of food resources are tied to place and habitat. If these ties to territory are broken, the food system can no longer be maintained. Reconnecting the ties that have been frayed or severed is one of the major ways in which Indigenous Peoples' food security and environments can be enhanced and renewed.

Progress has been made, and Indigenous Peoples now have leading roles in the movement to protect their local food systems; they have participated in national and international controls on contaminant emissions, and initiatives promoting biodiversity in food systems. As is clear from all the case study projects in the CINE programme, the best way forward is to listen to and learn from indigenous elders, study original food systems as a baseline, and address current environmental and social challenges, thereby creating a symbiotic meld of ancient wisdom with modern knowledge and technologies. Such an approach – applied well and patiently – offers hope for not only Indigenous Peoples but also human societies everywhere. Indigenous knowledge can be applied to environmental protection, for example in protecting and conserving genetic resources of nutritious and pest-resistant crop varieties (cultivars or landraces), and in providing practical and effective strategies for sustaining crops, fish, wildlife, forest ecosystems, agroecosystems and other essential habitats. Indigenous worldviews can help other societies by creating a new ethic of respect for other life forms and other cultures.

The viability of Indigenous Peoples' cultures, food systems and ways of life is at stake. If the outside world had listened to the Kogi peoples of Colombia or the Inuit of North America several decades ago, when the impacts of climate change were first noticed, the challenges faced today may have been easier to resolve. Cultivars from indigenous agricultural systems have proved vital to global agriculture by increasing yields and decreasing pests and diseases. Regarding health too, it will be well worth the time and effort to look closely at the changing health circumstances of people living close to land where there are negative impacts of ecological change. Indigenous Peoples play an immense role, not only in reclaiming the food traditions of individual communities in culturally appropriate ways, but also in maintaining and strengthening the resilience of ecosystems and cultural systems, including the global diversity of healthy food systems ◎

Acknowledgements

The authors acknowledge and thank all of the individuals and indigenous communities participating in the Indigenous Peoples' Food Systems for Health Program: Ainu (Japan), Awajún (Peru), Baffin Inuit (Canada), Bhil (India), Dalit (India), Gwich'in (Canada), Igbo (Nigeria), Ingano (Colombia), Karen (Thailand), Maasai (Kenya), Nuxalk (Canada), and Pohnpei (Federated States of Micronesia). Their experiences and inputs were integral to the development of this chapter. We also thank the International Union of Nutritional Sciences and the Task Force on Indigenous Peoples' Food Systems and Nutrition it created in 2002, which was the impetus for the programme. We greatly appreciate The Rockefeller Foundation and all the staff at the Bellagio Center for supporting and facilitating our work. We are grateful to Chief Bill Erasmus, Chair of the CINE Governing Board, Regional Chief of the Assembly of First Nations and National Chief of the Dene Nation, for his leadership and encouragement, and to Dina Spigelski, Project Coordinator, Siri Damman, Gail Harrison, Peter Kuhnlein and James Thompson. We acknowledge assistance for preparation of the chapter from the Canadian Institutes of Health Research and The Rockefeller Foundation's Bellagio Center.
> **Comments to:** nturner@uvic.ca

Chapter 4

Infant and young child complementary feeding among Indigenous Peoples

GRACE S. MARQUIS[1] SUSANNAH JUTEAU[1]

HILARY M. CREED-KANASHIRO[2] MARION L. ROCHE[1]

1
Centre for Indigenous
Peoples' Nutrition
and Environment (CINE)
and School of Dietetics
and Human Nutrition,
McGill University,
Montreal, Quebec,
Canada

2
Instituto de Investigación
Nutricional (IIN, Institute
of Nutrition Research),
Lima, Peru

Key words > infant and young child nutrition,
complementary feeding, breastfeeding, indigenous
foods, traditional foods, Indigenous Peoples

> "In the Awajún culture a child who goes to the *chacra* [field] with the mother, often has a full belly by the time she comes home."

Awajún elde[r]

Abstract

This chapter examines infant feeding practices among nine of the 12 case studies that formed the Indigenous Peoples' Food Systems for Health Program of the Centre for Indigenous Peoples' Nutrition and Environment (CINE). Information was obtained from key informant interviews, case study reports and published literature. Traditional food practices in these indigenous communities are being altered as a result of environmental and land tenure changes and the influence of outside markets; these changes affect infant feeding practices. Local indigenous fruits, vegetables and animal-source foods can provide macro- and micronutrient-rich options for complementary feeding that may be less expensive and more nutritious than market foods. In all of these communities, there is need for consistent practical nutrition advice to promote exclusive breastfeeding for infants up to six months of age, and the timely introduction of complementary foods thereafter. Nutrition education interventions can help families provide optimal nutrition for their infants by integrating traditional food practices with the wise use of local market foods.

Introduction

The first two years of life involve rapid physical, cognitive and social development that require optimal nutrition. Adequate infant and young child (IYC) feeding practices are needed to support this development and provide protection from the risk of morbidity and mortality in low-resource environments. To achieve this, international IYC feeding recommendations include exclusive breastfeeding until six months of age, after which adequate complementary foods should be added; breastfeeding is recommended to continue for two years or beyond (PAHO and WHO, 2003). Complementary feeding is challenging because it requires selection of foods that are easy to prepare and commonly used in family meals; provide sufficient energy and nutrients for a growing child; are not contaminated; and are accessible, affordable, locally available and culturally acceptable (WHO, 1998). The international recommendations for complementary feeding of breastfed children give clear guidelines for timely, adequate, safe and appropriately fed complementary foods (PAHO and WHO, 2003). However, these recommendations have not been universally adopted by all countries, nor are they well adapted to many local situations. Traditional values and practices regarding food, food preparation and eating are key components of cultures and group identity and affect how infants and young children are fed. In many parts of the world, traditional indigenous food practices are being altered as a result of environmental changes and the influence of outside markets (Kuhnlein and Receveur, 1996). Factors that change access to and preparation and use of traditional foods also have an impact on IYC feeding practices.

It is generally accepted that breastmilk should be an infant's first food. At about six months of age, there is a gap between an infant's nutritional requirements and the energy and nutrients that can be obtained from breastmilk alone, and additional foods must be added to the diet (WHO, 1998). To develop nutrition messages for complementary feeding that are both nutritionally and culturally appropriate for a specific Indigenous People, there is need to understand traditional foods, their uses and the existing complementary feeding practices. While the nutritional values of many traditional foods (FAO, 2009) have been determined, there is little information on actual IYC feeding practices in indigenous communities. What are the first foods introduced to infants, and at what age are they given? With what frequency are infants fed complementary foods, and how are they fed? Does the feeding behaviour for complementary foods influence breastfeeding patterns? What combinations of foods are given, and are these combinations advantageous for nutrient absorption (Gibson, Ferguson and Lehrfeld, 1998)? Are traditional food processing methods such as fermentation that improve food safety and nutrient availability used at the household-level (Kimmons *et al.*, 1999)? Are there IYC feeding practices such as pre-mastication that expand access to nutrient-rich foods (Pelto, Zhang and Habicht, 2010)? Are there health concerns about these practices (Gaur *et al.*, 2009)? What changes have occurred in IYC feeding, and how can these be expected to affect IYC nutrition and health today?

In response to the pressure of the modern market and the loss of hunting and gathering opportunities, many IYC feeding practices among indigenous communities are expected to have changed. Indicators for complementary feeding are important for assessing current practices, to help determine which traditional and present-day practices may or may not be advantageous for children's well-being, to screen and target vulnerable populations, and to monitor and evaluate interventions (Ruel, Brown and Caulfield, 2003). Recently published indicators (Table 4.1) capture the adequacy of complementary feeding practices and include continued breastfeeding, age of introduction of complementary foods, minimal dietary diversity, minimal meal frequency, consumption of iron-rich foods, and a composite indicator for a minimum acceptable diet (WHO and Lippwe, n.d.).

This chapter examines examples of Indigenous People's IYC feeding practices today and, where possible, discusses changes in feeding behaviours that have occurred in the recent past. Information on indigenous communities was obtained from key informant interviews and direct observations previously collected for nine of the 12 case studies that formed CINE's Indigenous Peoples' Food Systems for Health Program: Awajún (Peru), Baffin Inuit (Canada), Dalit (India), Gwich'in (Canada), Igbo (Nigeria), Ingano (Colombia), Karen (Thailand), Nuxalk (Canada) and Pohnpei (Federated States of Micronesia) (FAO, 2009). The Igbo research team conducted additional interviews in 2008 with seven female Igbo elders aged between 65 and 80 years. Child feeding data from the 2004 baseline questionnaire for the Awajún are also included. Additional information about feeding practices in the

Table 4.1 Indicators for the feeding of infants and young children

Indicator	Definition
Early initiation of breastfeeding	Proportion of children born in the last 24 months breastfed within one hour of birth
Exclusive breastfeeding up to 6 months	Proportion of infants 0–6 months fed exclusively with breastmilk
Continued breastfeeding at 1 year	Proportion of children 12–16 months fed breastmilk
Introduction of solid, semi-solid or soft foods	Proportion of infants 6–9 months receiving solid, semi-solid or soft foods
Minimum dietary diversity	Proportion of children 6–24 months receiving foods from 4 or more food groups
Minimum meal frequency	Proportion of breastfed and non-breastfed children 6–24 months receiving solid, semi-solid or soft foods (including milk feeds for non-breastfed children) at least the minimum recommended number of times
Minimum acceptable diet	Proportion of children 6–24 months receiving a minimum acceptable diet (excluding contribution of breastmilk)
Consumption of iron-rich or iron-fortified foods	Proportion of children 6–24 months receiving iron-rich or iron-fortified food that is specially designed for infants and young children, or that is fortified in the home

Source: WHO, 2008.

indigenous communities was obtained either from published literature and interviews with case study partners, or from previous CINE case study reports for the region. It is hoped that this accumulated, albeit limited, knowledge can guide health promotion projects for encouraging the inclusion of appropriate traditional foods and cultural practices that will assist indigenous families in achieving optimal nutrition for their infants and young children.

Asia and the Pacific

Pohnpei – Pingelapese people of the Federated States of Micronesia

Timing of complementary feeding

Early introduction of complementary foods was reported by Englberger, Marks and Fitzgerald (2003). In the 1980s, 48 percent of Pohnpei infants were given solid foods by four months of age. In 2008, mothers based the timing of solid food introduction on advice from Pohnpei public health staff (Englberger *et al.*, 2009). Most mothers gave birth at the local hospital, where they were advised to breastfeed exclusively for the first six months before introducing complementary foods. The definition of exclusive breastfeeding includes the provision of oral rehydration salts and drops but not the introduction of *both* non-breast milk liquids and complementary foods. However, from the reports of public health staff, it appears that the exclusive breastfeeding message has been misinterpreted to mean only that no complementary foods should be offered in the first six months. Of the ten infants in the survey, four were younger than six months and had not been introduced to foods. Of the six who were older, four had been introduced to foods at six months, one before six months, and one at seven months. However, other liquids were given before the infants were six months old. One care giver reported that health staff had told her that it was all right to give water and coconut (*Cocos nucifera*) juice to her infant before six months. Coconut water was given most frequently, to seven out of ten infants, while water was given to five out of ten.

Complementary foods

The five most common foods given to infants in the 1960s were breadfruit (*Artocarpus altilis/mariannensis*), *Karat* (*Musa troglodytarum*, a vitamin A-rich banana cultivar), coconut embryo, other ripe bananas (*Musa* spp.) and ripe papaya (*Carica papaya*) (Englberger *et al.*, 2009). Englberger, Marks and Fitzgerald (2003) reported that in the 1980s, care givers did not believe that food had positive health qualities and so they did not encourage children to eat. Pohnpei mothers refrained from giving meat and fish until their infants were one year old, because they did not want the children to get used to meat and fish and then be sad when they were not available. Mothers also believed that these foods would cause diarrhoea. Late introduction of fish at about one year of age had already been reported by Marshall and Marshall (1980) in a study of 49 Peniyesene infants. In this 1974 to 1976 study, the first complementary foods included mashed banana, papaya and mango (*Mangifera indica*). Soft boiled rice and/or mashed cooked breadfruit were then introduced, along with taro, soft bread, flour soup, mashed sweet potato (*Ipomea batatas*) and coconut sauce.

Changes in feeding practices over the past 50 years were documented by Englberger *et al.* (2009). According to the Mand Community Working Group in Pohnpei, the most popular complementary foods given to infants in 2005 were ripe banana, giant swamp taro (*Cyrtosperma chamissonis*), coconut embryo, imported baby food and ripe papaya. Giant swamp taro and imported baby food were far more commonly used than they were in the 1960s. Additional imported foods that are now more frequently given to infants include bread, flour cooked in water, rice soup, doughnuts, ramen and biscuits. Among the six infants given solid foods, three were first introduced to grated boiled green banana, two to ripe *Karat* and one to imported cereal. Traditional rich sources of iron and zinc were not reported.

Dalit – scheduled caste of India

The Dalit farmers from Medak District of Andhra Pradesh in southern India were studied by Schmid *et al.* (2006; 2007). Most of the Dalit families were illiterate, had no land and worked as farm labour. The prevalence of mild to severe underweight (< -2 SD weight-for-age) was 63 percent for children aged six to 39 months among the *sanghams* (Dalit volunteer women's groups) (Salomeyesudas and Satheesh, 2009). A high prevalence of iron deficiency has also been found among children aged 12 to 23 months in the area (Schmid *et al.*, 2006).

Timing of complementary feeding

In 2003, a study on improving access to traditional Dalit foods reported feeding practices for infants (six to 11 months) and young children (12 to 39 months) (Schmid *et al.*, 2007). Season appeared to affect child feeding patterns: exclusive breastfeeding was more common in the summer season, and fewer children were fed complementary foods or weaned in the summer than the rainy season. Among the non-intervention children in the study, rainy season intakes were higher in energy, protein and iron than summer season intakes ($p < 0.01$).

Complementary foods

The main non-breastmilk food source for non-intervention infants in the Schmid *et al.* (2007) study was rice (*Oryza sativa*), contributing about one-third of energy and about 28 percent of protein from complementary foods. During the rainy season, sorghum (*Sorghum vulgare*) provided about one-third of iron (34 percent). Animal-source foods contributed only 4 percent of iron intake and 11 percent of vitamin A. In the summer season, fruits contributed about 24 percent of vitamin A. Some women working on farms gathered traditional leafy green plants and included these in the family diet. Overall, energy, protein, vitamin A and iron intakes were found to be below recommendations for young children.

Karen – Indigenous People of Thailand

The prevalence of stunting among the hill tribe Karen children was twice that of the general Thai population,

at 25 versus 12 percent (Panpanich, Vitsupakorn and Chareonporn, 2000). Among the 24 infants who were part of the case study in 2005, there was no incidence of underweight before seven months of age (Chotiboriboon *et al.*, 2009). Prevalence of inadequate weight gain increased with age, with 36 percent of young children being underweight by the second year of life. One-quarter of the infants younger than 12 months were stunted. Studies in older children suggest that nutritional problems intensify after one year of age. A recent survey in northern Thailand found very high rates of malnutrition among northern Karen children aged one to six years: 85.5 percent underweight, 73 percent stunting, and 48.4 percent wasting (Tienboon and Wangpakapattanawong, 2007).

Timing of complementary feeding

The case study data suggested that the Karen in Kanchanaburi Province usually breastfed exclusively until their infants were approximately three months old, when complementary foods were introduced (Chotiboriboon *et al.*, 2009). For some infants, however, complementary feeding was introduced later than the suggested six months of age.

Complementary foods

The case study also found that the Karen often fed a watery broth made by boiling vegetables mixed with rice (*Oryza sativa*) as a first food (Chotiboriboon *et al.*, 2009). Some mothers gave mashed rice with banana (*Musa sapientum*) or salt or clear soup as initial complementary foods. Ripe papaya (*Carica papaya*) or mangoes (*Mangifera indica*) might also be given to infants. Several leafy greens, vegetables and fish were also part of infant feeding. Most of the protein for infants came from breastmilk.

Poor-quality complementary foods led to an inadequate energy intake among infants aged six to 11 months, meeting only 58 percent of the Thai energy recommendations. Given the low micronutrient value of the foods, breastmilk continued to be an important source of nutrients for infants, particularly of vitamin A, for which breastmilk provided the total requirement.

Africa

Igbo – Indigenous People of Nigeria

The Igbo are located in the southeastern region of Nigeria. In the past, a new mother would learn about infant feeding from the experiences of her own mother, who would stay with her for one to three months after the infant's birth (Okeahialam, 1986). Among the Igbo living in urban areas, this tradition is disappearing. Today, urban mothers are less likely to receive this outside help because they often live and work away from their families and their own mothers are also more likely to be working. In the case study, seven Igbo elders described the infant feeding practices when they were children. One elder said that water was given immediately after birth, and breastmilk was first introduced a couple of days later, when the breast had been treated and washed. Today, although breastfeeding rates are still quite high, the Igbo introduce other liquids very early. In a study in a rural area of Igbo-Ora Nigeria, all of the 411 infants received water in the first week (Nwankwo and Brieger, 2002). Health care workers advised 97 percent of the women to give glucose-water shortly after birth, and 72 percent complied during the first week. Nearly half of the infants were given herbal teas in the first week, and 97 percent had received herbal supplements by four months of age. Mothers reported introducing liquids because of health care staff recommendations or beliefs about the inadequacy of breastmilk for their children's nutrition, or because exclusive breastfeeding would be too physically draining on their own health, owing to their inadequate diets.

Okeke *et al.* (2009) cited malnutrition prevalence data from the 2003 Nigeria Demographic and Health Survey for the Igbo Region (southeast Nigeria): 20 percent of children were stunted, 5 percent were wasted and 8.5 percent were underweight. They also cited data from the National Micronutrient Survey showing that 15 percent of children in this area were vitamin-A-deficient.

Timing of complementary feeding

In a 1970s study, Kazimi and Kazimi (1979) reported that 85 percent of mothers introduced complementary foods at between three and seven months of age. Igbo elders asked about complementary feeding when they had young children responded that it was initiated at between three and ten months (although one elder mentioned one week), reflecting variations in practice and the difficulty in recalling the past. Elders reported that infants are now given complementary foods by five to six months of age, but it was documented that infants younger than four months were sometimes introduced to food (Okeke *et al.*, 2009).

Complementary foods

Igbo elders reported that popular foods given in the past after the initiation of complementary feeding included roasted cocoyam (*Colocasia* spp.), cassava (*Manihot esculenta*), millet (*Pennisetum* spp.), cassava paste, cassava soup, maize (*Zea mays*) and beans, all of which are traditional Igbo foods. Elders reported that yam (*Dioscorea* spp.) and some edible insects were roasted or boiled to supplement infants' diets. Cocoyam was typically roasted and mixed with oil, while cassava flour was mixed with hot water and stirred to thicken into a soup. Elders considered these foods good for growth, easing hunger, health and satisfying the infant.

The primary animal-source foods given to infants under one year of age included chicken (*Gallus gallus*), chicken liver and edible insects such as crickets. Most elders said that animal-source foods were introduced at five months. Vegetables were most commonly introduced between three and six months, and included amaranthus (*Amaranthus* spp.) and eggplant (*Solanum macrocarpum*) leaves. Oranges (*Citrus* spp.) and apples were the most common fruits given. Other fruits were bananas (*Musa sapientum*), pears (*Canarium schweinfurthii*) and mango (*Magnifera indica*). These were said to be given at any time between two and ten months of age.

When the elders had their own children, additional complementary foods were introduced between three and eight months to ensure the child's health and growth. These included akamu (a semi-liquid porridge made from maize grains soaked for two to three days), rice (*Oryza sativa*), beans and Cerelac (a commercially processed cereal made from maize and milk). These new foods were introduced when there was exposure to them through the market, education about them, or an increase in household income making them affordable.

The elders from the case study unanimously reported a reduction in the traditional foods that are given to their grandchildren. Reasons for this reduction relate to: i) cultural beliefs about the foods themselves – "Giving traditional food will make the baby behave foolishly when grown"; ii) food preparation challenges – "Because of a time factor"; and iii) undesirable feeding practices associated with traditional foods – "Overfeeding will cause the child to be 'dull'". Negative perceptions or stigma relating to traditional foods and practices may have emerged recently as families move away from their communities.

According to previous work in the Igbo case study, the complementary foods used today are usually the same as those eaten by the rest of the family, with the addition of maize gruel. Foods that are considered to be healthy for infants are *akara* (bean balls), *ukwa* (*Artocarpus communis*, African breadfruit), *ukpo oka* (plantain pudding), African yam bean (*Sphenostylis stenocarpa*), plantain stew, boiled plantain, and *ujuju* (*Myrianthus arboreus*, a type of fruit) soup (Okeke *et al.*, 2009). In their questionnaire responses, the elders of the Igbo community reported new complementary foods as being milk (powdered, soy and liquid whole bovine) and commercial products, including Bournvita and Milo (chocolate energy drinks), Cerelac and Nan (an infant formula). The availability and use of traditional foods decreased as people had less access to land, new foods became available on the market, and families' income, exposure to external cultures and education increased.

North America

Baffin Inuit – Indigenous People of Canada

A study in Nunavut found that Inuit breastfed for a median duration of eight months (Schaefer and Spady, 1982). Key informant interviews with case study community partners found that the first foods introduced to infants were often meats, including seal meat, a rich source of iron.

Gwich'in Nation – Indigenous People of Canada

For the Gwich'in, the first complementary foods were animal-source foods such as pre-masticated meat, fish broth and fish table food (Kuhnlein *et al.*, 2009). A key food for the Gwich'in is caribou (*Rangifer tarandus granti*), an iron-rich wild meat. Leaders from the Gwich'in case study expressed concern about the migration of caribou due to unpredictable climate change, which will affect access to and availability of this meat in the future.

Nuxalk Nation – Indigenous People of Canada

The Nuxalk placed importance on infant feeding, with key foods including the highly valued ooligan (*Thaleichthys pacificus*), which has disappeared owing to offshore fish farming, among other reasons. Key informant elders mentioned the importance of wild berries and other fish, including salmon (*Oncorhynchus* spp.), for the diets of young children.

South America

Ingano – Indigenous People of Colombia

Although Colombia has made substantial progress in improving the nutrition status of its population, malnutrition remains a problem in some regions. Correal *et al.* (2009) reported a substantial prevalence of stunting among Ingano preschool-aged children (three to six years old), with 22 percent stunted, 19 percent underweight, and 2 percent wasted. Reflecting the nutritional transition in Colombia, 4 percent of preschoolers were obese.

Timing of complementary feeding
Among the Ingano, breastfeeding is the norm, but complementary feeding starts early, sometimes as early as one month after birth, and almost all the infants were given complementary foods by their fourth month

(Correal *et al.*, 2009). Field personnel working with the Ingano were interviewed as key informants. There was some variation in the age at which foods were first introduced, but early introduction appeared to be the norm. The key informants reported that fruits were introduced from two months, fish and banana soups as early as three months, and complementary foods in general by five months.

Complementary foods

Approximately 58 percent of children's diets were traditional Ingano foods (Correal *et al.*, 2009). Key informants also reported that much of the infant diet comes from traditional foods. The main fruits given were banana, *milpes* (*Oenocarpus bataua*, a fruit and palm tree), palm heart and papaya (*Carica papaya*). After five months of age, *cucha* (fish soup), chicken soup, *rallana* (banana and manioc soup), vegetables (such as squash, tomatoes, onions), cimarron or coriander leaves (*Eryngium foetidum*), *anduche* (banana drink), *arepas* (flour and maize mixture) and eggs were commonly given. Fish, chicken and meat from the forest were usually given next. *Mojojoy* (*Rynchophorus palmarum*, a nutritional and medicinal beetle larva) were added to the diet in late infancy. Field personnel reported that foods for Ingano infants were chosen for many different reasons (Table 4.2).

Table 4.2 Ingano complementary foods and value of food reported by staff key informants

Complementary food	Local importance of food
Cucha (fish soup)	Important for growth and to avoid malnutrition
Cimarron or coriander leaves (Eryngium foetidum)	To improve flavour and as a remedy for hepatitis B and anaemia
Banana (Musa sapientum)	For infants' growth
Milpes (Oenocarpus bataua)	As a remedy for coughs
Chontaduro (Guilielma gasipaes, a wild palm fruit), all fruits	Infants like them
Mojojoy (Rynchophorus palmarum)	For children with respiratory problems

The same local research team reported that the frequency of feeding for most Ingano traditional foods had not changed substantially from the past. The use of smaller quantities of some traditional foods, such as pheasant, can be attributed to communities no longer harvesting or hunting these foods. New complementary foods include powdered milk, rice, eggs and meat bought at local markets, lentils, beans, white sugar, pasta and snacks.

Awajún – Indigenous People of Peru

A recent study documented a high prevalence of stunting in the Awajún community, reaching 49 percent among children aged three to five years, 26 percent of whom were severely stunted (Roche *et al.*, 2007). Overall, 39 percent of the Awajún children up to five years of age were found to be stunted. Among children younger than two years, 44 percent were stunted, with a mean Z-score of -1.9 ± 1.0 height-for-age (Creed-Kanashiro *et al.*, 2009). Stunting is linked in part to inadequate early feeding.

Timing of complementary feeding

Information on the dietary intake of infants and young children aged 0 to 23 months was collected through interviews with 32 mothers in six Awajún communities in the northern Amazon rain forest (Roche *et al.*, 2010). More than half of the mothers exclusively breastfed their infants for the first six months. Among the remaining 48 percent of infants, complementary foods were introduced as early as two months of age.

Complementary foods

Liquids other than breastmilk were given soon after birth (Creed-Kanashiro *et al.*, 2009). A lightly fermented pre-masticated cassava beverage, *masato*, and *chapo* (roasted ripe banana drink) were the most common beverages given. Other milks (canned evaporated bovine milk) were given to 39 percent of infants, but usually after eight months and while still breastfeeding.

Key informants interviewed in the Awajún community in 2004 mentioned that *aves del monte* (wild birds), *majas* (small wild animal, *Cuniculus paca*), small fish from forest streams, *patarashka* (fish, including the organs and viscera, tomatoes, onion and sweet chilli baked in green leaves) were common complementary foods, but have become harder to obtain. Access to traditional and local foods such as these has been reduced as a result of increases in population and decreases in land availability. Encroachment on land and overhunting have reduced wild animal populations and increased the time and distance required for hunting. One Awajún key informant reported that the population had access to less biodiversity than before, so some foods were no longer available.

Among the first complementary foods now given to the Awajún are boiled foods (mainly banana), tubers and roots – cassava (*Manihot esculenta*), *sachapapa* (*Dioscorea* sp., a variety of potato) and turnip (Creed-Kanashiro *et al.*, 2009). Soups with added chicken or egg are also quite common. A key informant interview in the communities provided additional information about popular complementary foods: *chapo* (roasted ripe banana drink), ripe banana and *suri* (*Coleopterus* sp., beetle larva). Soon after these had been introduced, infants could be given almost anything else, including *chonta* (palm heart), eggs and *masato*. Tougher wild meats were not given to babies. Another key informant provided different information, reporting that fruits and vegetables were not given to infants younger than eight months because they are believed to cause diarrhoea and make children ill. Fruits were also not thought to have vitamins, so they were not considered important food for children, regardless of whether or not a child liked eating them. However, chonta was given to make children grow. This informant also mentioned that *perdiz del monte* (partridge) (*Tinamus tao* and *Steatornis caripensis*) was a common food given to infants. Other common foods for infants are fish, various animals and *paloma del monte* (*Columba subvinacea*, pigeon). When an infant is ill, *suri* is given.

Roche *et al.* (2010) used an infant feeding history questionnaire to document complementary foods that were commonly introduced. The most popular first foods mentioned were cassava and *chapo*. Following these, banana, palm heart, fish and egg were introduced.

For daily consumption, cassava, banana, *chapo*, *aguaje* (*Mauritia peruviana*, palm tree fruit) and seasonal fruits were the foods most often given. Fish, eggs and other animal-source foods were usually given not more than once a week. *Eep* (leafy greens) were usually given once a week. Infants aged six to 11 months were fed complementary foods an average of 0.9 to 1.6 times a day, which is less than the recommended minimum meal frequency, whereas children aged 12 to 23 months old were fed an average of seven times a day. Most Awajún infants were breastfed, making breastmilk a major source of dietary energy (not quantified).

Between six and 24 months of age, most (92 percent) of the energy intake from complementary foods was provided by local Awajún foods such as banana, cassava, *sachapapa* and *pituka* (*Colocasia esculenta*) (Creed-Kanashiro *et al.*, 2009). The remaining 8 percent was purchased or obtained from donations, including milk and rice. The median iron intake (4.9 mg/day) of infants and young children aged six to 24 months was about half the recommendation (7 to 11 mg/day).

Conclusions

The complementary feeding practices of Indigenous Peoples included in the nine case studies provided a spectrum of different geographical and cultural characteristics. However, the available data are limited, and provide only rough estimates of feeding practices. The overall picture to emerge from interviews and a review of statistics on IYC nutrition status at these locations suggests that there is wide variation in complementary infant feeding patterns among indigenous communities, and that these practices generally need to be improved to provide optimal nutrition.

Literature on other indigenous populations that were not included in the case studies concurs with these findings, with examples of both early and late introductions of foods. Early introduction of complementary foods was reported among Canadian First Nations in a study of 102 infants on Walpole Island (Kuperberg and Evers, 2006), where 19 percent of infants received solid foods before they were two

Table 4.3 Complementary feeding practices among Indigenous Peoples

Country	Age for CF introduction[1] in country (months)	Indigenous culture	Age for CF introduction in indigenous community (months)	Examples of nutrient-rich traditional foods excellent for CF	Practices that could be improved for better infant nutrition
Africa					
Nigeria	3.6	Igbo	5.5	Chicken liver, insects, cocoyam, yam bean	Exclusive breastfeeding for 6 months Use only market foods that are healthy choices
Asia and the Pacific					
Federated States of Micronesia	< 6	Pohnpei	6	*Karat*, mango, papaya	Exclusive breastfeeding for 6 months Use fish as CF
India	6–9	Bhil	10.5	NA	Introduce CF at 6 months
		Dalit	> 9	Wild dark leafy greens, sorghum, pulses	Feed enhancers of iron absorption (e.g., citrus fruit)
Thailand	2–4	Karen	3	Fish, papaya, mango, leafy greens	Exclusive breastfeeding for 6 months
North America					
Canada		First Nations		NA	Exclusive breastfeeding for 6 months
		Baffin Inuit		Seal meat	Continue breastfeeding after 6 months
		Gwich'in		Caribou meat, fish	
		Nuxalk		Salmon, wild berries	
South America					
Colombia	> 4	Ingano	5	Wild plants, insects, meats	Exclusive breastfeeding for 6 months Use only market foods that are healthy choices
Peru	3–4	Awajún	2–6	Local leafy greens (*eep*) and fruits (mango, *aguaje* and papaya), fish, *suri*	Exclusive breastfeeding for 6 months More frequent feeding of infants

NA = information not available.
CF = complementary food.
[1] The average age (months) for introducing complementary foods was obtained from the most recent individual country surveys available through the Demographic and Health Survey Statcompiler program, at www.statcompiler.com.

months of age and 57 percent before four months of age. The food most commonly introduced first was infant cereal, followed by pureed fruit. Infants were given cow milk at between six and 15 months of age, and more than half of the children had been given low-fat milk before two years of age (Kuperberg and Evers, 2006). In contrast, late introduction of complementary foods was common among the Bhils, the primary scheduled tribe in Jhabua District, India, where 1 percent of infants first received complementary foods at between four and six months of age, 18 percent at between seven and eight months, 37 percent at nine to 12 months, and 40 percent after 12 months (Taneja and Gupta, 1998). This was not a new infant feeding practice among these people; references from 1943 to 1954 described the introduction of solids at ten to 11 months among Bhil infants (Sellen, 2001). The consequences of this practice can be seen in the extremely poor nutrition status of the children: 25 percent wasted, 60 percent stunted, and 84 percent with anaemia (Sharma, 2007).

Table 4.3 shows the age of introduction of complementary foods at each of the indigenous sites surveyed, the comparative national rates in the respective countries, examples of nutritionally valuable indigenous foods that are available, and traditional practices that could be strengthened to improve infant nutrition status and health. Indigenous foods that are hunted, fished or gathered and provide a rich source of highly bioavailable iron and zinc or fat, for example, are important resources that can be prepared for infants and young children and should be preserved. Universally, there is a need for consistent practical nutrition advice (in communities as well as health care facilities) on IYC feeding. The promotion of exclusive breastfeeding during early infancy, with introduction of complementary foods starting at about six months of age is needed at all sites.

Historical information collected from diverse sources suggests that indigenous complementary feeding practices have changed over the years. To varying degrees, communities have incorporated market foods into their complementary feeding practices; such substitutes may be expensive and less nutritious than the local foods they are replacing. These changes have occurred because of necessity (e.g., to replace foods that have become unavailable with the loss of hunting grounds), but probably also because of individual choice (e.g., ease of preparation for market foods such as noodles or rice) and an absence of information about the nutritional and health values of local and market foods. Local indigenous fruits, vegetables and animal-source foods that provide macro- and micronutrient-rich options for complementary feeding exist and should be promoted. Nutrition education programmes can help families make informed decisions that will result in optimal nutrition for infants and young children, the preservation of appropriate traditional dietary practices, and wise use of local markets ◎

Acknowledgements

This chapter was written in collaboration with the Indigenous Peoples' Food Systems for Health Program case study teams for the Awajún (Peru), Irma Tuesta and Miluska Carrasco; the Igbo (Nigeria), Elizabeth Chinwe Okeke; and the Ingano (Colombia), Sonia Caicedo and Ana María Chaparro.

The authors would also like to acknowledge the contribution that other Indigenous Peoples' Food Systems for Health Program teams made to discussions and through sharing their knowledge about practices of child feeding within their communities.

> **Comments to:** grace.marquis@mcgill.ca

Case studies

Chapter 5

Promotion of traditional foods to improve the nutrition and health of the Awajún of the Cenepa River in Peru

HILARY M. CREED-KANASHIRO[1]   MILUSKA CARRASCO[1]

MELISSA ABAD[1]   IRMA TUESTA[2/3]

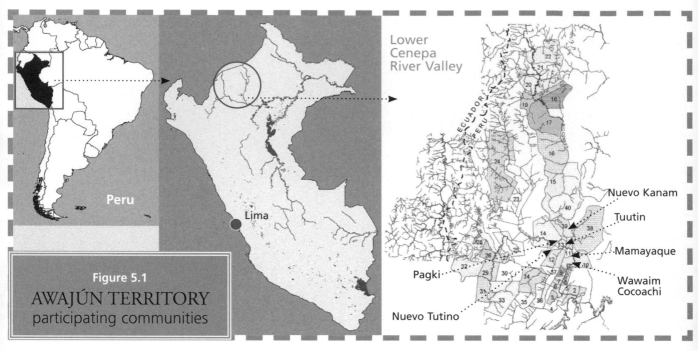

Figure 5.1

AWAJÚN TERRITORY
participating communities

Lower
Cenepa
River Valley

Nuevo Kanam
Tuutin
Mamayaque
Pagki
Wawaim
Cocoachi
Nuevo Tutino

Data from ESRI Global GIS, 2006.
Walter Hitschfield
Geographic Information Centre,
McGill University Library.
With addtion from
Instituto del Bien Común, *Perù.*

1
Instituto de Investigación
Nutricional (IIN, Nutrition
Research Institute),
Lima, Peru

2
Organización
de Desarrollo de las
Comunidades Fronterizas
de Cenepa (ODECOFROC,
Organization for
Development of the
Frontier Communities
of Cenepa),
Cenepa, Peru

3
Grupo de Trabajo
Racimos de Ungurahui,
Lima, Peru

Key words > Indigenous Peoples,
traditional food, Awajún, Peru,
Department of Amazonas, food security

Photographic section >> IV

"Before I didn't eat well; now I can get fish from my fish pond until the river is full of fish again and I can eat the food that I have in my farm."

Awajún mother

Abstract

This case study describes a project aimed at enhancing the nutrition and health status of Awajún communities of the Cenepa River through the promotion of key aspects of the traditional food system and culture. The project was built on previous participatory research and focused on increasing the production, accessibility, knowledge and use of nutritious traditional foods. Thirty-two elected nutrition and health promoters, representing 16 communities, participated in nutrition and production workshops. Monitoring of the project showed that these community promoters were effective in conveying the nutrition messages. The project created six community plant and seed nurseries and 400 fish farms.

Evaluation based on a transverse design, in which 64 families formed the baseline for comparison with 41 families post-intervention, found that knowledge about traditional foods' value and access to them had increased, with the traditional food diversity score increasing significantly, from 8.8 to 10.4 for women and children under five years of age. Ninety percent of energy was provided by locally produced or caught foods rather than market foods in both the baseline and the final evaluations. Although there was variation, mean energy and nutrient intakes – measured by 24-hour recalls for two non-consecutive days – were close to recommended daily intakes for children, but iron, zinc and calcium intakes were low.

Women's intakes of energy and some nutrients were generally lower than recommended. The percentage of energy from protein and fat increased significantly in children, and protein, iron and zinc intakes from animal sources increased for women and children. Increased amounts of meat (from hunting) and fish contributed to these changes. There were improvements in infant feeding practices, especially in the complementary foods given. In spite of these dietary improvements, however, the high prevalence of childhood stunting (53 percent) did not change over the two-year project period. Diarrhoea and parasite infections also remained high among both mothers and children.

The study demonstrates that the traditional food system can be maintained and accessibility increased by promoting local foods of high nutritional value, but more efforts are needed to address infant and young child feeding practices and the general health of the population.

The programme process and progression

The Awajún are one of the most important Indigenous Peoples of the tropical rain forest of Peru. They live mainly in the Department of Amazonas in northeast Peru, along the Upper Marañón River and its tributaries (Berlin and Markell, 1977). In 2007, the total population of the Awajún was reported to be 55 366 people, living in 281 communities (INEI, 2008) across an estimated area of 22 000 km^2.

The Awajún's main activities are subsistence farming, hunting and fishing (Ramos Calderón, 1999), and the lands and rivers supporting their traditional food system are their principal sources of livelihood (Creed-Kanashiro *et al.*, 2009). Currently, both the territory that they have owned from time immemorial and their traditional food system are threatened by mineral and petroleum extraction initiatives. The Awajún and other Indigenous Peoples of the Peruvian Amazon region are defending their territories to protect their livelihoods and environment.

Although a variety of traditional foods are available, the nutrition and health situation of the Awajún is not optimum. Studies have found high prevalences of infant and childhood malnutrition (stunting), and anaemia in women and children (Huamán-Espino and Valladares, 2006; Creed-Kanashiro et al., 2009; Roche et al., 2007). Reasons for this include changing ecological, cultural and food systems, and a high prevalence of infections and parasites (Huamán-Espino and Valladares, 2006).

The traditional food system of the Awajún of the Lower Cenepa River is described by Creed-Kanashiro et al. (2009) and Roche et al. (2007). Based on this information, which was gained through participatory research with the communities, this chapter describes a participatory project to promote the production and use of nutritious traditional foods that benefit the nutrition and health status of the population and preserve its food culture.

Context

Geographic, cultural and demographic characteristics

The district of Cenepa extends from the mouth of the Cenepa River, where it joins the Marañón, across the mountain ranges (cordilleras) to the frontier with Ecuador (Berlin and Markell, 1977). It is in the "high jungle" (ceja de selva), covered by dense rain forest vegetation, and has a tropical climate. Cenepa was legally recognized in 1941, and comprises three principal areas – Low, Middle and High Cenepa – with a total of 52 communities and an estimated population of 8 000 people (AECI, CIPCA and SAIPE, 2000). It is the only district in the Alto Marañón with no settlers from other parts of Peru.

The Awajún is one of four tribes of the Jivaroan linguistic family (Shell and Wise, 1971). Traditionally, they lived in widely dispersed hamlets, each consisting of several related households. Today the majority reside on or near the region's major rivers, in communities that range from 13 to 103 families each.[1] Awajún community organizations are headed by an Apu (Chief), with a Vice-Apu, a secretary, a police officer and one other voting member. The communities of Cenepa have formed the *Organización de Desarrollo de Comunidades Fronterizas del Cenepa* (ODECOFROC, Organization for Development of the Frontier Communities of Cenepa), which represents them within the local area and with the government and other institutions.

Awajún houses are built of local materials, especially canes from local *guayaquil* trees, with roofs of matted palm branches and earthen floors. Most houses do not have a separate area for cooking; the kitchen is part of the common living area. Wood is the principal cooking fuel, and houses are usually lit by candles or petrol burners. People obtain all their water for cooking and washing from the river. This water is not treated and there is no sewage system. However, since the early 1990s, latrines have been installed about 20 to 50 m from the houses.

Transportation and communication among Cenepa communities are mainly by river – in canoes or small motor boats – and on foot, along narrow, steep trails (AECI, CIPCA and SAIPE, 2000). Most communities do not have electricity; the few with a generator use it mainly for radio communication. Few communities have public telephones; some have government health posts, but access to health services is difficult for many; and some have a pre- or primary school, but there is only one secondary school in the area, which takes children from all accessible surrounding communities.

Overall health and nutrition status

The major nutritional and health problems of this population include high rates of childhood stunting, anaemia and heavy intestinal parasitic infection (MINSA/OGE, 2002). A study of schoolchildren aged six to 15 years in the Alta Marañón in 2002 showed that the most common parasites were *Entamoeba coli, Lodamoeba butschil, Anclystoma/Necartor* and *Ascaris lumbricoides* (Ibáñez et al., 2004). In the preliminary study for this chapter, 44 percent of children under two years of age (n = 32) and 50 percent of those aged two to 12 years (n = 39) were stunted (< -2SD height-for-age) (Creed-Kanashiro et al., 2009; Roche et al., 2007). Twenty-five percent of

---

[1] www.selvasperu.org

children under two years of age were underweight. National data on the nutrition status of children in the Department of Amazonas, including indigenous and *mestizo* populations (INEI, 2001), reported that 36 percent of those under five years were stunted and 42 percent suffered from anaemia. There are no data on the vitamin A status of this population, although researchers found no signs of the clinical deficiency that is typical in most of Peru. The national data for Amazonas reported high rates of childhood diarrhoea and respiratory infections (INEI, 2001).

The body mass index (BMI) of women in the preliminary study was generally within the normal range (Creed-Kanashiro *et al.*, 2009). Data for the Department of Amazonas (INEI, 2001) indicated that 32 percent of women of reproductive age were anaemic (below 12 g/dl). Psychological health is also a concern for this population, with suicides, particularly of young women, being a public health problem. The suicide rate has declined since implementation of ODECOFROC's Women's Programme, but remains a concern (I. Tuesta, personal communication, 2010).

The traditional food system is based on the major crops of cassava (*Manihot esculenta Euphorb*) and banana (*Musa balbisiana X Musa*), which are the main sources of dietary energy (Berlin and Markell, 1977; Creed-Kanashiro *et al.*, 2009; Roche *et al.*, 2007). Most of the foods consumed are cultivated, collected, hunted or fished locally, and there is a wide variety of local foods; the 2004 research identified a total of 215 local foods (Creed-Kanashiro *et al.*, 2009; IIN, CINE and ODECOFROC, 2005). Foods are prepared by boiling, roasting or smoking. Women make cassava beer and the popular drink *masato*, which is prepared from boiled manioc roots, masticated and fermented. Women also cultivate other roots and tubers and a variety of fruits and seeds in their fields (*chacras*).

Awajún men traditionally hunted for game animals and birds, but overhunting and community living have led to scarcities of these animals near river settlements. There is a wide variety of fish, which has traditionally been the major source of animal food, together with other river creatures such as frogs (*Colostethus*), snails (*Pomácea* sp.) and prawns/shrimp (*Macrobrachius*

brasiliensi). In the mid-twentieth century, missionaries introduced small domestic animals, including chickens (*Gallus gallus*) and pigs (*Sus scrofa*), which are raised mainly by women and provide a potential source of nutrients. Wild fruits, edible larvae (*suris*) and other insects are also collected for food. The consumption of vegetables, fruits and seeds varies by season.

The earlier research into traditional food systems (Creed-Kanashiro *et al.*, 2009) reported changes in the availability and use of traditional foods. Not all the 215 foods identified by participants were currently available, owing to changes in the environment, living and food activity patterns, overfishing, and more difficult access to wild animals. Intake of animal-source foods, particularly meat and fish, was generally low and infrequent, and depended on seasonal availability and hunting patterns. In recent years, the number of foods cultivated has also declined, to one or two varieties of roots and bananas and a few fruits, owing to people's reduced time for working in the fields and the government's donation of foods to families with young children. Nevertheless, relatively few foods were purchased from out of the area or were donated through government programmes.

By exploring the population's food availability, culture and perceptions, the 2004 study provided information about several nutrient-rich traditional foods that could be promoted through interventions to increase both production and consumption, especially among young children and women, thereby benefiting the nutrition and health status of the population.

Rationale

The project described in this case study was built on the participative research and results of the 2004 study, and was designed to enhance the nutrition, health and well-being of participating communities – especially women and children – through the promotion of key aspects of the traditional food system. It was delivered through close work with community organizations and community health and nutrition promoters, and focused on activities in three principal areas:

- *Food production:* Stimulate feasible production activities to increase the accessibility of traditional foods, emphasizing those with high nutritional value. Enhance women's roles in collecting and planting traditional fruit seeds and palms, raising small animals including *suri* (larvae) and participating in fish farms. Involve primary schoolchildren in seed planting projects. Promote full use of existing agricultural land, to avoid losing it to government appropriation.
- *Education:* Increase knowledge about the nutritional value, importance and worth of traditional foods within the communities and among schoolchildren, to maintain and recuperate the use of traditional foods and ensure that knowledge is not lost to future generations.
- *Participation and use:* Increase the use of a wide variety of traditional foods through activities in food preparation, recipes and diet, with special attention to young children and their nutrition.

Methodology and activities

Preparation for the participatory project

In 2005, the 2004 study results, the food list, composition information about 82 of the 215 traditional foods identified, and the project proposal were presented at the ODECOFROC assembly of representatives of the 52 communities. The project research agreement was signed, and the food composition information was distributed to ODECOFROC leaders, the Women's Programme and the six communities that participated in the earlier study. Contacts and agreements made previously with participating communities were renewed. Approval was obtained from the Ethics Committee of Peru's Institute of Nutrition Research (IIN – *Instituto de Investigación Nutricional*) and the Institutional Review Board of McGill University in Canada. All the people interviewed in the baseline and final evaluations or photographed for educational materials provided their prior consent.

The following subsections describe the methodology, results, monitoring and evaluation of project activities.

Intervention activities

Following the baseline evaluation in February 2006, a variety of activities were implemented over a two-year period to 2008, several of which were proposed by Kuhnlein *et al.* (2006). Project activities started with the training of 32 community nutrition and health promoters from 16 Cenepa River communities, who were elected by their communities and coordinated through the ODECOFROC Women's Programme to promote activities and messages in their own communities. Figure 5.1 on p. 54 shows the distribution of the communities along the Cenepa River.

Participation was voluntary. Most of the communities with reasonable access to the ODECOFROC centre (within two days of river travel) were invited to participate in the training, and all those that were interested were included. Activities were also conducted with pupils, parents and teachers of five local primary schools. The project focused on food production and empowerment, and included education and training activities on food, nutrition and cultural topics to support the promoters' role as nutrition and health leaders in their communities; these were the topics requested by the population. Activities were based on participatory workshops held twice a year in Cenepa and led by the IIN technical team, with follow-up and support through periodic community visits by the local Awajún health and nutrition team led by the ODECOFROC Women's Programme, which also provided translation from Spanish to Awajún when necessary.

Promotion and creation of plant nurseries and recuperation of traditional seeds

Production activities were an essential part of the intervention, and were conducted primarily by specialist plant cultivation and animal raising institutions. To support the conservation, recuperation and diversification of traditional plants, especially food plants for family use, the Women's Programme led an initiative involving the community nutrition and health promoters in collecting seeds and establishing plant and forestry nurseries in the communities. Training in appropriate management of these resources was provided by a forestry agriculturist at two workshops for the 32 promoters and at primary schools for 146 children and adolescents. A model/central nursery was created at the ODECOFROC centre, providing a source of seedlings for community and family nurseries. The Women's Programme maintained the central nursery and managed seedling distribution, while schoolchildren were responsible for their school nurseries.

In addition to the central nursery, five community nurseries were established. Table 5.1 lists the 16 seedling types planted at the central nursery. Fifty-seven mothers participated in community meetings organized by the promoters and the Women's Programme, receiving seedlings from the central nursery during the project, mostly of fruit and palm trees. Seeds of other vegetables, trees and medicinal plants were exchanged among communities.

Table 5.1 Inventory of plants at the central nursery, 2008

Local name	Scientific name	Tree type	No. of plants
Huasaí	Euterpe olerácea	Palm tree	230
Ungurahui	Oenocarpus bataua Mart.	Palm tree	113
Naranja	Citrus aurantium L.	Fruit tree	52
Macambo	Theobroma bicolor Humb	Fruit tree	123
Huevo de toro	–	Fruit tree	340
Pijuayo	Bactris gasipaes H.B.K.	Palm tree	130
Caimito	Pouteria caimito	Fruit tree	24
Namuk	Sicana odorifera	Fruit tree	9
Chonta	Bactris setulosa	Palm tree	27
Huacrapona	Iriartea deltoidea	Palm tree	11
Sampi	Inga sp.	Fruit tree	41
Naampi	Caryodendron orinocensis	Almond tree	53
Naam	Caryodendron orinocensis	Fruit tree	46
Cedro	Cedrela odorata L.	Forestry	4
Cedro rosado	Cedrela odorata L.	Forestry	25
Bolaina blanca	Guazuma crinita	Forestry	Seedlings

Table 5.2 Numbers of families with food production activities, by community

Community	No. of families in community	Fish ponds	Chickens	Guinea pigs	Pigs	Ducks	Turkeys	Cows
Bashuim	59	14	18	3	2	3	4	NA
Nuevo Tutino	27	16	all	3	NA	NA	NA	NA
Mamayaque	75	30	all	1	NA	NA	NA	7
Tuutin	50	47	all	2	NA	NA	NA	NA
Kusu Pagata	103	64	56	20	1	NA	NA	NA
Cocoachi	NA	10	all	NA	NA	NA	NA	NA
Nuevo Kanan	25	25	NA	NA	NA	NA	NA	NA
Canga	87	5	40	NA	2	4	NA	1

NA = No information available.
Source: Numbers of families from www.selvasperú.org

Workshops on raising fish and chickens

Community nutrition and health promoters and Awajún leaders received training in other parts of Peru and then held workshops on establishing and managing community or family fish ponds appropriate for local geographic conditions, and on raising chickens; both workshops were facilitated by the Women's Programme. The *Servicios Agropecuarios para Investigación y Promoción Económica* (SAIPE, Agrofishery Service for Economic Research and Promotion) Peru and the World Wide Fund for Nature (WWF) provided technical assistance for the fish ponds, and the project provided building materials. These initiatives complemented and reinforced production activities promoted by other agricultural institutions working in coordination with ODECOFROC and the communities.

By the end of 2008, there were an estimated 400 family fishponds in 32 Cenepa River communities. However, several of these were precarious and were not in continuous production, and there was demand for further technical and material assistance. Table 5.2 shows the animal production activities in the eight communities included in the monitoring of this intervention.

Food, nutrition and culture workshops

Four nutrition workshops were held over the two-year project period, focusing on the sharing of knowledge and experiences of local foods and their nutritional contributions, uses and preparation, and highlighting their positive characteristics, diversity and cultural identification. Special efforts were made to find ways of improving infant and young child feeding and nutrition using locally available foods, owing to concerns about the high prevalence of growth retardation observed in the 2004 study and baseline evaluation. The following topics were included in the workshops and reinforced during the project period:

- traditional local foods;
- food combinations for a balanced diet;
- infant and young child feeding and nutrition;
- feeding of preschool and schoolchildren;
- nutrition for pregnant and lactating women;

- care in food manipulation, hygiene and use of water;
- production and conservation of foods, from the field to the table: local and traditional foods versus market foods, traditional conservation methods;
- illnesses that result from not eating well (e.g. diabetes, hypertension).

Practical and participatory sessions included song, story and socio-drama creations, hand-washing classes, and elders' stories about their experiences and practices in the past. During the training, community nutrition and health promoters developed four or five key messages for each topic, and then promoted the most appropriate of these within their communities. This resulted in:

- 29 educational messages for dissemination in the communities by local radio and loudspeaker;
- three food/nutrition promotion posters;
- two songs in Awajún, disseminated in two primary schools;
- ten stories about good foods and hygiene, based on traditional stories;
- three socio-drama representations showing traditional good habits and customs reinforced with current knowledge;
- 58 food preparations (recipes) for infants and young children, which included animal products and other traditional foods such as cassava, *sachapapa* (*Dioscorea trifida*), *pituca* (*Colocasia esculenta*), *eep* (*Araceae philodendron* sp.), *ugkush* (*Piper* sp.) and palm hearts (various species) prepared in traditional ways, such as *patarashka* (steaming in banana leaves), smoking and boiling;
- preparation of lunch baskets for schoolchildren;
- a T-shirt designed by the promoters and distributed to them, the *Apus* and community leaders, with the message "Eating our own foods we shall be intelligent, strong and happy" written in Awajún.

To enhance dissemination of the nutrition messages, traditional foods and preparation methods were included in a cultural festival of traditional songs, dances and practices held by ODECOFROC, in which all the Cenepa River communities participated.

Workshop for the empowerment of promoters as community nutrition and health leaders

In 2007, the community promoters reported difficulties in transmitting the information and experiences they had obtained through the workshops, particularly regarding the organization and holding of community meetings to promote nutrition. In response, the Women's Programme coordinated a "Reflect-Act" workshop, led by an experienced educator, to facilitate the community promoters' role. At the workshop, the promoters reflected on their strengths, such as their own firm commitment; the recognition they received from their communities, including support from Apus and the community assembly; the population's enthusiasm about the production activities; and the valuable nutrition workshops and visits from the Women's Programme.

These discussions strengthened the promoters' self-esteem and emphasized the value of working in teams of two – often one woman and one man – in each community, to support each other. (There were 19 male promoters and 13 female.) The practical aspects of promoters' nutrition activities were reinforced, such as interpersonal communication through visits to individual families, the use of food preparations to illustrate messages, and the many opportunities for mentioning nutrition messages and practices, even briefly (e.g., when washing at the river, and at community assemblies).

Monitoring of project activities

Community activities were monitored throughout the two years of the project, usually by a nutrition promoter who had received special training in coordination with the Women's Programme and IIN. Three monitoring visits were made to each of the participating communities in Middle and Lower Cenepa, but it was not easy to visit communities higher up the Cenepa River owing to the difficulties and costs of transport. The project provided the Women's Programme with a canoe and outboard motor but the limited availability and high cost of fuel resulted in fewer visits than planned.

Six communities of Lower Cenepa were visited by the principal monitor, who obtained information about the activities implemented and mothers' perceptions about these, through observation and interviews with mothers, Apus and community promoters. These monitoring activities were complemented with information from the President of ODECOFROC and the community promoters attending workshops.

Tables 5.1 and 5.2 show the numbers of nurseries created, seeds sown, fish ponds operating and families raising small animals reported by the Women's Programme, ODECOFROC and promoters in 2008. The monitor visited eight cultivation areas in four communities, and noted that these families were cultivating a wider variety of food plants near to their homes. Fruit and palm trees were the most common crops, and all the families had used seeds (of eight different plants) that their promoters had brought from the central nursery, as well as those that they themselves had collected from the wild and from more distant fields. Mothers spoke positively about this activity, and commented that having these foods nearer to hand had benefited their families and made it easier to have food variety during rainy seasons, when reaching their main fields was difficult.

Of the 25 mothers interviewed in the six communities of Lower Cenepa, ten reported having a fish pond and consuming fish more frequently, thereby improving their families' diet; another ten expressed their intention to build a pond. Almost all the mothers (22) were raising chickens for family consumption. Seven reported improved feeding patterns for their families and children, through having a wider variety of foods in their diets, with more frequent use of animal products. Only four mentioned receiving information on infant feeding, advising them to feed infants breastmilk exclusively for the first six months, introduce the first complementary foods at six months, give foods of a thick consistency using local ingredients, and include animal products.

Evaluation

To evaluate the impact of the project on knowledge, practices and nutrition status a survey was conducted among mothers with small children before and after the project period, in February 2006 and February 2009, respectively.

The difficult socio-political situation of the Awajún in defending their territories from government appropriation and the arrival of resource extraction industries made it impossible to implement or evaluate activities in the second half of 2008 (AIDESEP, 2010).

Methodology

The baseline evaluation was conducted in six Lower Cenepa communities – Mamayaque, Tutino (Tuutin), Nuevo Kanan, Nuevo Tutino, Wawaim and Cocoachi – and the final evaluation in all except Wawaim and Nuevo Tutino, owing to their distrust of evaluation teams as a result of the socio-political situation.

The following information was collected from mothers in both evaluations:

- socio-demographic characteristics of the family and household;
- current and past health information about the mother and child/children;
- food security, using a version of the United States Department of Agriculture (USDA) Food Insecurity and Hunger Module (Vargas and Penny, 2010) adapted for Peru by IIN;
- consumption frequency of 30 traditional foods, selected for their nutritional value from the preliminary investigation of the traditional food system (Creed-Kanashiro et al., 2009);
- dietary intakes of the mother and child/children from 24-hour recalls on two non-consecutive days; portion sizes were determined by scales measuring up to 5 kg with accuracy of ±1 g;
- physical activity of the mother, using a short version of the International Physical Activity Questionnaire (IPAQ, 2002);
- anthropometry measurements of the mother and child/children, using a height/length board made locally, and bathroom scales measuring up to 150 kg with accuracy of ± 500 g; mothers were weighed with and without the child to assess the child's weight.

For the final evaluation, the following additional parameters were assessed:

- knowledge about nutrition and food production;
- activities of the mother, using an adaptation of the "Home questionnaire for child" (M. Penny, personal communication, 2008), in which mothers were asked to place small balls in containers representing each of their activities; the number of balls assigned to each container depending on the amount of time spent on that activity.

The evaluations were conducted in each community on three non-consecutive days, after consultations between the Apu and members of the Women's Programme. Each interview lasted about two hours and each IIN evaluator was accompanied by a trained translator.

Data analysis

Baseline and final survey data were entered into Visual FoxPro version 8 and analysed using SPSS version 17. Individuals' traditional food diversity scores (TFDS) (Roche et al., 2008) were calculated by assigning 1 point to each local food reported in the 24-hour diet recalls.

Nutrient and energy intakes were also calculated from the food intakes reported in the 24-hour recalls, using IIN's food composition tables (IIN, 2001) complemented with information from other food composition tables where necessary. If the composition of a particular food was not known, that of a similar food was assigned instead. This was done for 18 foods: three fish, three birds, four wild animals, three leafy vegetables and four fruits.

International recommended intakes were used to assess the adequacy of women's and children's intakes (FAO/WHO/UNU, 2004; 2007; FAO/WHO, 2002; Dewey and Brown, 2003). Energy requirements for moderate activity were considered for the women, to account for their agricultural activities.

Comparisons between the baseline and final evaluations were analysed using Mann-Witney non-parametric tests for each variable except for TFDS, for which ANOVA was used.

Results of baseline and final evaluations

Mothers with children younger than five years of age were invited to participate in the evaluations. The difficult socio-political situation meant that fewer families were able to participate in the final evaluation than in the baseline. Numbers of participants in each community are shown in Table 5.3.

Of the 41 families included in the final survey, 14 mothers had also participated in the baseline. Ten children were in both the baseline and final surveys, having been under two years at baseline and in the two-to-five-years age group in the final survey. Within each family, all the children under five years of age were included in the evaluation. The age distributions of the children differed significantly between the two surveys, as shown in Table 5.4 (p = 0.016). There were fewer children in younger age groups in the final survey and the mean age was 4.3 months older.

Family characteristics

Most characteristics of households and families were similar in the baseline and final surveys, indicating that although the sample was smaller in the final evaluation, the populations were comparable. Most of the families surveyed lived in their own houses (75 percent baseline, 85 percent final); others lived in relatives' houses. The majority of mothers lived with their married or common-law husbands (89 percent baseline, 80 percent final); others were single, separated or widowed. Most families lived as a nuclear family (70 percent baseline, 80 percent final); the remainder shared their homes with their extended families. On average, six family members lived together in a house of two rooms.

The mean age of the mothers was 25 years at baseline and 29 years in the final survey. The mean number of years of education was six, indicating completion of primary school. At baseline, 4.8 percent of mothers had no schooling and 63 percent had some primary school education. In the final survey, these figures were 7.7 and 54 percent, respectively. The majority of mothers in both baseline and final surveys reported working in their own fields as their major occupation (86 percent) as well as caring for their families, and 8 percent worked principally in the fields of others. Some mothers reported doing artisan work in the final survey (13 percent).

Knowledge about foods and nutrition

The surveys explored the changes in mothers' knowledge and attitudes resulting from the community health and nutrition promoters' education activities. At baseline, 47 percent of mothers reported having received advice related to food and nutrition, and 16 percent had attended an education session on the subject. In the final survey, 62.2 percent reported having received advice about nutrition, mainly from a promoter in the Women's Programme. The topics most remembered were infant and young child feeding, with mention of specific infant preparations, and how

---

Table 5.3 Numbers of families participating in baseline and final evaluations

Community	Baseline	Final
Mamayaque	10	16
Tutino	20	5
Wawaim	7	0
Nuevo Kanan	9	10
Nuevo Tutino	8	0
Cocoachi	10	10
Total	**64**	**41**

Table 5.4 Age distributions of children in baseline and final surveys

Age group	Baseline	Final
0–5.9 months	10	1
6–11.9 months	16	7
12–23.9 months	27	13
2–3 years	15	8
3–4 years	11	8
4–5 years	3	5
Total	**82**	**42**
Mean age of children	23.7 ± 12.8 months	28.0 ±15.3 months

to have a balanced and varied diet using local foods. Forty-nine percent of mothers reported putting the advice into practice.

Traditional and market foods

Mothers were asked about their knowledge and perceptions regarding traditional local food and market or donated foods from outside the area. At baseline, mothers reported that traditional foods were good (80 percent), natural (41 percent) and good for children and families (16 percent). In the final survey, more mothers mentioned that traditional foods were good for health (34 percent), natural and without chemicals (29 percent). Mothers mentioned their suspicion that market foods may contain chemical contamination – especially the canned foods distributed through government programmes – and thought that their traditional foods were better. However, they also mentioned certain foods from outside as good complements to the diet and useful in times of scarcity. Other market foods, such as sodas, biscuits and sweets, were recognized as not being healthy foods.

Regarding traditional foods, women stated that "they are natural, we know they do not contain chemicals". Regarding market foods they reported "we don't know what we are buying".

At baseline, 73.4 percent of mothers reported that the number of traditional food species to which they had access was diminishing over time, and mentioned specifically that animals were further away from communities: "before there were trees and animals near, now there aren't". In the final survey, 58 percent mentioned that the number of species had diminished, and 21 percent said it was increasing, as people were cultivating more foods: "before there were more foods, now there is more population and so there is scarcity, but we are now sowing in our fields"; "before we didn't sow much, we only collected from the wild, but now we are bringing plants from the wild and sowing them in our fields".

Mothers' feeding practices for children

The final survey included questions about knowledge of feeding practices. Eighty-nine percent of mothers said

that breastmilk was the first food to be given to a baby and the only food to be given for the first six months of life: "the baby's stomach is not yet developed and breastmilk is what the baby needs for development".

Forty-six percent of mothers said that the first complementary food should be introduced to a baby at six months of age, while 43 percent said this should occur at more than six months, as "[babies] will only then start to be hungry", "they do not know how to eat before this" or "[this is] when the teeth come". The mothers use these milestones as they do not necessarily keep precise track of their children's ages. Most mothers (76 percent) said that the first food should be thick food and fed after six months, according to the advice they had received on good infant feeding practices. Sixteen percent of mothers said that meat can be given at six months of age (the message given), but the majority (73 percent) considered that it should only be given at seven months or older, when the infant has teeth and is thus able to chew. Only 24 percent of mothers reported making food preparations specifically for their infants and young children.

The education sessions included nutrition for primary schoolchildren, especially regarding foods to be taken to school, as children generally only took cassava and banana to eat during the school day. In the final survey, 74 percent of mothers said that the lunch box should consist of a mixture of foods, including egg, fish and fruits, as well as the staples.

The education sessions also promoted healthy diets for pregnant and lactating women. In the final survey, 60 percent of mothers responded appropriately about the foods needed by pregnant women, with 68.5 percent stating that a lactating mother needs to eat more food of greater variety as "her body needs more".

Food production activities

Half of the mothers interviewed in the final survey had heard about producing food, cultivating seeds or raising small animals and fish ponds from the promoters of the Women's Programme and from other institutions supporting these activities in the area. Of these mothers, 86 percent reported that they had begun to raise chickens, 69 percent had established a household garden and planted seeds, and 44 percent had established a fish pond and were raising fish, mainly tilapia (*Tilapia melanopleura*). Thirty-eight percent of the mothers said they were now eating foods that they did not formerly eat. Of these mothers, 43 percent were eating new foods they produced or hunted themselves, 14 percent were eating new market foods, and 43 percent were eating both.

Eighty percent of mothers in the final survey said they preferred using their own produce to feed their families. They gave the following reasons:

- "They are the nicest tasting foods, as well as more healthy."
- "They are better quality because they are grown without fertilizer."
- "They are not contaminated, they are natural."
- "The plants are ours, they are not watered by others, we can eat them confidently, including without washing them."
- "I don't have money to buy, but I have my own land anyway so I can sow my food."
- "I don't need more food, I have sufficient."

- "Foods that we produce are free."
- "When you grow them and they are there you can eat them anytime."
- "We can vary our foods."

Food diversity

Table 5.5 shows the total numbers of foods consumed by families, as reported in their two 24-hour recalls for the baseline and final surveys. Although fewer families were included in the final than in the baseline survey, the total numbers of foods consumed were similar, at 101 and 97 respectively. The numbers of traditional/local and market foods were also similar, indicating that food variety and the use of local foods were maintained during the project period in spite of the introduction of more market foods into the area.

Traditional food diversity score

The TFDS indicating the number of traditional foods consumed by each individual in the 24-hour recalls are shown in Table 5.6. There were significant increases in the TFDS for both women and children after the project, showing that the use of traditional foods was maintained or increased.

Frequency of food consumption

The frequency of consumption of 30 traditional foods was explored in both surveys. In general, the consumption of seasonal local fruits was reported to have increased in the final evaluation, which recorded frequent consumption of *carotene aguaje* (*Mauritia fleuxosa Palmae*), *pijuajo* (*Bactris gasipaes*) and *sachamango* (*Grias Peruviana* Miers. [Lecythidaceae]). Traditional vegetables promoted during the project were eaten more frequently (weekly as opposed to monthly at baseline) as were some wild animals, such as frogs and wild pigeons, and organ meats. Children aged six to 12 months consumed fish more frequently, with 50 percent consuming it weekly after the project, compared with 23.5 percent at baseline; this was an effect of the family fish farms.

Food security

The food security questionnaire that is generally used to derive a score for food insecurity (Vargas and Penny,

---

Table 5.5 Numbers of different foods consumed in baseline and final surveys (24-hour recall)

	Baseline	Final
No. of families	64	41
Local/traditional foods	76	72
Market foods (from out of the area)	25	25
Total foods	**101**	**97**

Table 5.6 Traditional food diversity scores (24-hour recall)

	n	Mean (± SD)	Range
Baseline survey			
Women	64	9.4 ± 3.39	3–18
Children	67	8.1 ± 3.63	1–18
Final survey			
Women	40	10.7 ± 3.62[1]	4–17
Children	42	10.1 ± 3.72[2]	0–18

[1] Women, difference between baseline and final $p = 0.04$.
[2] Children, difference between baseline and final $p = 0.01$.

2010) was used. However, some of the questions were difficult to apply with the Awajún population, as they sounded similar or repetitive to both the translators and the mothers; the following results are therefore derived from only some of the questions on the questionnaire and are not total scores.

When asked about the food available in their homes during the past year, most mothers reported that they had sufficient, but not always what they would like (67.2 percent baseline, 60.5 percent final). Thirty percent at baseline and 34 percent in the final survey said they sometimes did not have sufficient food, but only 4 percent reported that this situation occurred frequently. The main reasons given were the scarcity of food in their fields at certain times and the difficulty of obtaining foods they would like, such as by hunting animals from the wild. In response to this situation, about 40 percent of mothers said that they sometimes gave less food to their children, but 86 percent at baseline and 95 percent in the final survey said their children never went without food. In both the baseline and final surveys, 90 percent of mothers said they were sure of their food supply for the whole year.

The respondents' principal concerns regarding food supply were that they may not have new fields to cultivate (19 percent baseline, 29 percent final) or access to quality foods (16 percent baseline, 13 percent final). Only 8 percent at baseline and 13 percent in the final survey mentioned uncertainty about whether or not they would have sufficient money for food. Mothers reported that when they lacked sufficient food at home, they collected it from their fields (67.2 percent baseline, 60.5 percent final), hunted animals in the wild (10.9 percent baseline, 12.1 percent final) or ate only cassava (12.5 percent baseline, 10.5 percent final). Obtaining food on credit from shops was mentioned by 5 percent in the final survey.

Mothers reported that they obtained their food mainly from their fields (93.8 percent baseline, 80 percent final); the use of money was mentioned by 3.2 percent at baseline and 20 percent in the final survey, indicating that purchasing food from others or from small local shops may have become more

common during the project period, as evaluated in this small and non-representative sample. Ninety-two percent of mothers at baseline and 80 percent in the final evaluation reported sharing food with their neighbours.

Families' participation in government social programmes

The government has several social programmes, mainly distributing food for populations classified as poor or extremely poor; Table 5.7 shows family participation in these as reflected in the evaluations. The Glass of Milk programme distributes canned milk or fortified milk powder from the municipality, for children aged six months to seven years and pregnant and lactating women. The community kitchens programme is run by the National Programme for Food Assistance of the Ministry of Women and Social Development and distributes foods such as beans, oil, rice and canned fish, which are prepared in community kitchens. JUNTOS is a conditional cash transfer programme implemented by the government for families classified as living in extreme poverty; it started in the Cenepa area in May 2008. Mothers of children under 14 years of age receive PEN (*nuevos soles*) 100.00 (approximately USD 35.00) a month on condition that their families take part in health, nutrition and education activities and that they present identity documentation for their children.

More than half of the families interviewed participated in a social programme, and participation generally increased over time (Table 5.7). Foods distribution to the Glass of Milk and community

Table 5.7 **Participation in government social programmes in baseline and final surveys**

Programme	% of population participating in social programme	
	Baseline	Final
All programmes	59.4	64.1
Glass of Milk programme	42.2	61.6
Community kitchens	42.2	18.0
JUNTOS	-	66.7

Several families participated in two or three programmes.

kitchens programme were not regular, owing to slow administration and the relative inaccessibility of the area, so participation in these programmes was reported to be variable. Mothers reported using the money received from JUNTOS to purchase school supplies for their children, market food and medicines.

Forty-seven percent of families at baseline and 29 percent in the final survey reported that their children had received foods from a community kitchen in the previous months. The principal reasons given were that it was a free government programme and that children "like to eat rice", a commonly served food. Those who did not participate said it was because they did not have this facility in their community. About half of the families reported that their children received milk from the Glass of Milk programme, but not all the time, only when it was available. The principal reason given was that milk is a good food for children and complements their diet (27 percent baseline and final); in the final survey 32 percent said it was assistance from the government. Those who did not receive milk said that the programme did not reach their communities – particularly at baseline, 27 percent compared with 5 percent in the final survey – while a few reported that their children did not like milk or the programme (12.5 percent baseline, 8 percent final).

Infant feeding patterns

All mothers reported breastfeeding their children, and breastmilk was the first food or liquid received by 95.3 percent of children at baseline and 97.4 percent in the

final survey, indicating that this was a prevalent practice and was reinforced during the project. Breastfeeding prevalence is shown in Table 5.8. Of the 11 children in the youngest age group at baseline, four (36 percent) were exclusively breastfed. However, as shown in Table 5.8, breastfeeding during the second year of life was not a common practice, and needs to be stressed in future interventions. Mothers said they breastfed because it was what they were taught to do by their elders, and must be good. Their reasons for stopping breastfeeding were that their children had teeth and bit their nipples, they felt they did not eat good enough food to be able to breastfeed, or they became pregnant with the next baby.

In the surveys, 85 percent of children had received water, infusions and *chapo* (a drink made from banana) as well as breastmilk. The mean age for introducing liquids was 6.5 months at baseline and nine months in the final survey, indicating that the timing for introduction was delayed after the project. Thirty-seven percent of children at baseline and 48.7 percent in the final survey had received milk other than breastmilk; the mean ages for introducing other milk were 8.9 months at baseline and 10.6 months in the final survey. Feeding bottles were used to give liquids to infants. *Masato* (a pre-masticated drink prepared from cassava) was given to 58 percent of infants at baseline and 67 percent in the final survey. The mean ages for introducing *masato* were 9.7 months (ranging from three to 24 months) at baseline and 13.2 months (six to 36 months) in the final survey. The *masato* given to young children is fresh (i.e., unfermented) and is considered good for young children as it can satisfy hunger, forms stools and – together with *chapo* – makes the child "chubby".

The average ages at which mothers reported first feeding their infants solid foods were five months at baseline (ranging from one to nine months) and 6.6 months in the final survey (four to eight months), showing reduced incidence of very early introduction of solids. The most common first foods given were watery broths or soups, and boiled banana and cassava, which were mashed or pre-masticated to make them soft and prevent the child from choking. Consumption of pre-

Table 5.8 Breastfeeding in baseline and final surveys, by age group

Age group	Baseline n	%	Final n	%
0–5.9 months	11	100	1	100
6–11.9 months	16	94	8	88
12–23.9 months	27	41	12	17
2–3 years	15	6	8	0
> 3 years	14	6	13	0

Table 5.9 **Median percentages of daily recommended energy and nutrient intakes among young children**

Nutrient	Infants and children 6–23 months			Children 2–5 years		
	Baseline	Final	p value	Baseline	Final	p value
	n = 38	n = 19		n = 29	n = 23	
Energy	116	109	ns	122	118	ns
Protein	179	256	0.064	182	178	ns
Vitamin A	93	167	0.056	232	200	ns
Ascorbic acid	297	405	ns	350	404	ns
Thiamin	79	103	ns	120	138	ns
Riboflavin	115	156	ns	239	190	ns
Folate	149	116	ns	99	93	ns
Iron	45	55	ns	62	83	ns
Zinc	49	80	ns	99	81	ns
Calcium	31	33	ns	56	43	ns
Dietary characteristics						
% energy from protein	7.4	10.3	0.009*	6.5	7.1	ns
% energy from fat	7.8	13.8	0.015*	6.3	7.5	ns
% energy from carbohydrate	89.0	75.3	0.023*	90.0	88.3	ns
% protein from animal sources	37	65	0.004*	28	41	ns
% iron from meat, fish, poultry	4.4	27.0	0.003*	5.2	11.3	ns
% zinc from animal sources	16.6	60.2	0.001*	11.3	31.6	ns

* Controlled for age owing to the difference in age distributions of children 6–23 months between the baseline and final evaluations.

masticated foods is discouraged by health personnel, and was reported less frequently in the final evaluation. The introduction of meat, commonly pre-masticated or mashed, was reported to occur at a mean age of 7.3 months at baseline compared with 6.5 months after the project. Twelve percent of children at baseline and 8 percent in the final survey had received an iron supplement at the health post to treat or prevent anaemia. In general, these results, and the interviews with mothers who reported using varied complementary food combinations, demonstrated improved complementary feeding practices after the project.

Energy and selected nutrient intakes

Tables 5.9 and 5.10 show the adequacy of energy and selected nutrient intakes for two age groups, compared with recommended intakes and calculated from the 24-hour recall data from the baseline and final evaluations. There was wide variability in intakes, so the tables present median values. Median intakes for children in both age groups met the recommendations for energy, protein, vitamins A and C, riboflavin and folate. However, median intakes for iron, zinc and calcium were well below recommendations, and although intakes of iron and zinc (in the younger group) appear to be higher after the project, they were still below recommended intakes. After the project, for all nutrients, the proportions of children consuming less than 80 percent of the recommended intake were lower than at baseline.

The energy contributions from protein and fat were low, but increased significantly in the youngest age group after the project, owing to the increased consumption of animal-source foods. The proportions of protein, iron and zinc from high-bioavailable animal sources also increased in this age group.

Table 5.10 Median percentages of daily recommended energy and nutrient intakes among lactating and non-lactating mothers

Nutrient	Lactating mothers		Non-lactating mothers	
	Baseline	Final	Baseline	Final
	n = 36	n = 11	n = 22	n = 25
Energy	79	73	88	98
Protein	60	65	76	80
Vitamin A	157	124	211	233
Ascorbic acid	338	298	442	510
Thiamin	61	55	60	87
Riboflavin	136	95	143	140
Folate	55	43	61	61
Iron	34	35	23	29
Zinc	57	87	79	103
Calcium	48	34	39	43
Dietary characteristics				
% energy from protein	5.9	5.3	7.6	6.1
% energy from fat	5.2	4.3	5.8	7.2
% energy from carbohydrate	92.3	93.6	91.0	87.6
% protein from animal sources	33.8	43.7	35.2	46.5
% iron from meat, fish, poultry	6.6	21.0[1]	10.3	20.0
% zinc from animal sources	14.2	24.0[2]	18.7	24.0

[1] $p < 0.05$.
[2] $p = 0.066$.

There was wide variation in daily energy intakes for women, but median intakes were lower than recommended intakes in both the baseline and the final evaluation, as shown in Table 5.10 (which does not present results for pregnant women as there were very few of them). Energy recommendations were calculated on the basis of moderate activity for these women, who frequently spend several hours a day doing agricultural work in their fields. Most women were within the normal range for BMI, as seen in Table 5.11, indicating that overall energy intakes were near to their requirements. Median protein intakes were lower than recommended. Women's median intakes of vitamins A and C and riboflavin were adequate, but – similar to children's – their intakes of iron, zinc, calcium and thiamine were low, although median intakes of zinc were significantly higher after the project, probably owing to eating more fish.

Table 5.11 Percentages of non-pregnant women carers with adequate BMI

BMI	Baseline %	Final %
	n = 56	n = 34
< 18.5	1.8 (1)	2.9 (1)
18.5–24.9	94.6 (53)	88.2 (30)
25–29.9	3.6 (2)	8.8 (3)

Food sources of energy and nutrients

The major sources of energy in the diet were cassava and banana. Together these provided more than half the energy in the diet, although for young children the proportion was less after the project (51 percent baseline, 44 percent final), indicating that more (non-breastmilk) energy came from other food sources. For young children, the proportion of energy from fruits (excluding banana) increased after the project; fruits are grown locally.

There was an increase in the proportion of protein from animals. Protein from wild animals increased from 8.4 percent of dietary protein at baseline to 14.1 percent after the project, for all women and children; the increase was even greater for young children (from 3.1 to 12 percent). These foods were consumed more frequently and in larger quantities after the project, including by young children. There was also an increase in the proportion of protein from fish, rising from 12.3 to 18.9 percent of dietary protein for all women and children, and from 13.8 to 22.7 percent for young children. Young children consumed more milk after the project.

Traditional and market sources of food energy
Most dietary energy for all groups was provided by traditional, locally produced or hunted foods, with only a small proportion coming from market or donated foods, as shown in Table 5.12. This proportion is higher for young children, mainly owing to the milk and food they receive from donation programmes, and to snack foods purchased from local shops.

Health situation – childbirth
The majority of children were born in the home, with the mother alone or with a family member present (80 percent baseline, 70 percent final); in both surveys, 17 percent were born at home with a health worker or midwife attending. In the final survey, 13 percent of children were born in a health facility, compared with 3 percent at baseline. The Ministry of Health is promoting institutional birthing to reduce maternal mortality, but accessibility to health facilities is very difficult in this remote area.

Child morbidity
There was high prevalence of illness among children in this population, especially younger ones, and particularly high incidences of diarrhoea, respiratory infections and parasites, which were recorded as higher in the final

survey than at baseline (Table 5.13). The project did not include testing or treatment for parasites; health promoters recommended herbal treatments and referred mothers to the health post. Mothers may have become more aware of ill health and parasites after the project, as they had had more contact with health promoters.

Women's health
Sixty-seven percent of mothers reported that they were in good health all or most of the time at baseline, and 63 percent in the final survey. Eighteen percent of mothers reported having diarrhoea during the past

Table 5.12 Median percentages of dietary energy derived from traditional/local and market foods, by age group

Age group	Traditional/local foods		Market foods	
	Baseline	Final	Baseline	Final
6–23 months	84.3	86.3	15.7	13.7
2–5 years	83.7	92.3	16.3	7.7
Lactating mothers	94.7	94.4	5.3	5.6
Non-lactating mothers	93.9	94.1	6.1	5.9

Table 5.13 Health status of infants and young children as reported by mothers

Health status	Age group in years	% children	
		Baseline	Final
		n = 64	n = 39
Child is healthy	< 2	64.1	61.9
	2–4	83.9	73.9
Child had diarrhoea yesterday	0–4	6.3	17.9
Child had fever yesterday	0–4	9.4	10.3
Child had cough yesterday	0–4	25.0	17.9
Parasites: present	< 2	26.4	47.6
Parasites: no/unknown	< 2	73.6	52.3
Parasites: present	2–4	54.8	47.7
Parasites: no/unknown	2–4	45.1	52.2
Diarrhoea in the last month	< 2	43.4	47.6
Diarrhoea in the last month	2–4	25.8	30.4

month at baseline, and 26 percent in the final survey. Similar to the children, the presence of parasites among women was high: 40 percent reported parasites at baseline and 50 percent in the final survey, with most of the others saying they did not know whether or not they had parasites. Other illnesses reported by mothers included malaria, dengue, tuberculosis and typhoid fever (with prevalences between 1.6 and 7.9 percent). In the final survey, two mothers reported diabetes and one hypertension. Forty percent of mothers reported that they had had a health problem during their pregnancies, and 31 percent reported mastitis.

Maternal mental health

Mothers were asked some simple questions regarding their mental health. About half of them did not answer these questions, perhaps because they did not understand what the translator was asking. Among those who did answer, 35 percent at baseline and 45 percent in the final survey said they felt animated during the day, whereas 44 percent at baseline and 50 percent in the final survey reported feeling tired, bored or sleepy. In both surveys about 90 percent of those who answered said they felt valued by their family; 70 percent at baseline and 91 percent in the final survey felt valued by their community, and 95 percent in both surveys said they considered that what they did was important.

Children's nutrition status

Table 5.14 indicates that there was no improvement in the nutrition status of young children during the project period, as measured by anthropometry. Although there appears to have been an increase in stunting in the younger age group and a decrease in the older group, no clear conclusions can be drawn, owing to the small sample size.

Women's nutrition status as measured by BMI

The mean BMIs of non-pregnant mothers in the survey were 21.8 at baseline and 22.3 at the final evaluation. The distribution of BMI shown in Table 5.11 shows that the majority of mothers were within the normal range, with no change after the project. Nevertheless, there was a slight (statistically insignificant) increase in overweight women in the final survey in this small sample.

Women's activities

Information about women's daily activities was derived from interviews with mothers and an interactive exercise of assigning balls to containers (see the previous section on Methodology). Typical daily activities involved rising very early, cooking (boiling) cassava and other food, eating and going to the fields (on only two or three days a week during the rainy season) to collect food for about six hours. On the days they did not go to the field, the mothers did housework, and in the afternoons they washed clothes (they estimated spending a total of 7.5 hours on housework and caring for family members) and engaged in recreational activities or sports (e.g., volleyball) for an estimated two hours. In the evenings they prepared the evening meal and went to bed early as they lacked light or electricity. Forty-two percent of

Table 5.14 **Proportions of children with ≤ -2SD for height-for-age, weight-for-age and weight-for-height Z scores**

Nutrition status	% children					
	0–23 months			2–4 years		
	Baseline	Final		Baseline	Final	
	n = 53	n = 20		n = 29	n = 21	
≤ -2SD height-for-age Z	43.4	55.0	ns	62.0	52.0	ns
≤ -2SD weight-for-age Z	20.8	20.0	ns	24.0	24.0	ns
≤ -2SD weight-for-height Z	0	5.0	ns	0	10.0	ns

the mothers interviewed did some paid work out of the house, mostly in other people's fields or caring for other people's children.

Discussion

The Awajún have always lived in close equilibrium with their natural surroundings. Their forests and rivers are their life-blood and livelihood: "…if the rainforest disappears, the Awajún disappear" (Chang and Sarasara, 1987). The food system was based on more than 200 traditional foods, but the nutritional and health situation of the Awajún is not optimum (Creed-Kanashiro *et al.*, 2009; Roche *et al.*, 2007). Increased nutritional knowledge – especially regarding the needs of infants and young children – greater access to disappearing high-quality foods, the integration of appropriate market products and improved health strategies are needed. This case study project, aimed at enhancing the nutrition and health of Awajún women and young children along the Cenepa River through the promotion of nutritionally appropriate traditional foods, led to increased knowledge about these foods. The project increased the production, accessibility and use of these foods for better nutrition, thus contributing to maintaining their essential role in the food system and in the population's diet, culture, ecosystem and environment.

Nutrition and health promotion requires access to appropriate foods and educational processes that lead to behaviour change. The project activities were based on the results of an earlier participatory study documenting the food system, and responded to community leaders' requests to address the nutritional situation. The project strategy focused on training elected community health and nutrition promoters following the ODECOFROC structure, and using innovative adult education methodologies. The community promoters then implemented food production and education activities in their communities, supported by intermittent visits from the leaders of ODECOFROC's Women's Programme. The promoters were enthusiastic and very interested participants, but had some difficulties in transmitting the information in their communities owing to their education level, logistics and gender issues, and the population's misperception that foods from "outside" are better than local ones. These issues were addressed in workshops, emphasizing practical aspects and key messages, but this resulted in the project being delivered differently in each community and delayed the communities' exposure to activities.

In spite of these difficulties, mothers remembered the food and nutrition messages; knowledge of the importance and benefits of traditional foods increased; the production of foods was greatly enhanced, through the local cultivation and recuperation of seeds and plants and the introduction of family fish farms; and more meat was obtained from hunted animals. The promoters found it easier to promote food production than impart nutrition information. The benefits of some of these activities, such as the cultivation of fruit, palm and other trees, will increase in the future, because of the time required for trees to grow and produce fruit, so the effects immediately after the two-year project were somewhat limited.

The total number of foods consumed by all the families evaluated, as measured by 24-hour dietary recall, was very similar in both the baseline and final surveys, in spite of the smaller sample size in the final survey. This was true of both traditional foods (76 baseline, 72 final) and market/government-donated foods (25 in both). At the individual level, the TFDS at baseline was similar to that described by Roche *et al.* (2008), but had increased significantly after the project, for both women and young children. These results indicate that food diversity and the traditional food system were maintained and enhanced despite the increased presence of market foods in the area, which has been shown to have a negative effect on the nutrition of other populations (Kuhnlein *et al.*, 2004; Kuhnlein and Receveur, 1996; Port Lourenço *et al.*, 2008; FAO, 2009).

The biodiversity available to populations with limited economic resources enhances their food security (Claverías and Quispe, 2001). The Awajún mothers interviewed generally expressed confidence in their current food supply, although it was not always of the quantity or quality that they would have liked, especially regarding animal products and fruits – use of both these foods increased after the project. However, the mothers expressed concerns about not having access to new fields to cultivate and about difficulties in reaching animals for hunting in the future.

Exclusive breastfeeding during the first six months of life and continued breastfeeding with complementary foods until two years of age provide optimum nutrition and protect against illness (PAHO/WHO, 2003). The final survey sample size was too small to evaluate the project's impact on exclusive breastfeeding, but the introduction of complementary drinks and foods was delayed until about six months of age, reflecting the adoption of improved practices. Good breastfeeding practices, especially during the second year, still need to be reinforced. There were high incidences of illness among infants and young children, and improved breastfeeding and complementary feeding practices are needed to help protect against illness. Foods were sometimes reported to be pre-masticated by the care giver, to facilitate acceptance and digestion by the child, and this may have a protective effect against illness (Pelto, Zhang and Habicht, 2010). This practice is decreasing, however, owing to health personnel's current recommendations to avoid pre-mastication.

With the exception of iron, zinc and calcium, children's median energy and nutrient intakes were estimated at close to recommended levels in both the baseline and final evaluations. For every nutrient, the proportion of children under 24 months not meeting 80 percent of the recommended intake decreased, indicating an improvement in dietary intakes. Some indicators of dietary quality improved significantly in children aged six to 23 months, including the proportions of energy from protein and of iron and zinc from high-bioavailable animal sources. Meat and fish intakes were higher after the project, a direct result of having more meat from hunted animals and fish from fish farming. These were foods distributed within the family to favour younger children. Fat intake also increased, which is important considering the low fat content of the children's diets.

The dietary intakes of energy and, especially, iron, zinc and calcium of adult women with moderate activity tended to be lower than recommended in both the baseline and final evaluations. As with the children, there were significant increases in the proportions of energy from protein and of iron and zinc from high-bioavailable animal sources among the women, indicating a similar increase in meat and fish intake after the project.

The case study recorded improved feeding practices and dietary intakes for children, but no impact on their nutrition status; rates of stunting were high, as in other studies of this and similar indigenous populations (Huamán-Espino and Valladares, 2006; Roche et al., 2011; Soares Leite et al., 2006), and remained high after the project. Besides diet, other factors also affect growth (WHO, 1998), particularly illness. The rates of illness (especially parasitic infections) reported in this study were extremely high, and are likely to have contributed to the lack of improvement in nutrition status. Mothers also had high rates of parasitic infections, which have also been reported in similar indigenous communities of the Peruvian and Brazilian rain forest (Ibáñez et al., 2004; Carvalho-Costa et al., 2007).

There were several limitations to this study:

- Health and nutrition promoters had variable success in transmitting activities and messages to the project communities.
- The limited access to communities in this remote area of difficult terrain made follow-up and monitoring activities very difficult and expensive.
- Although the project included home, food and personal hygiene information, it did not cover the evaluation or treatment of parasites and illnesses. Future interventions should include more on health problems, including water quality.
- Anaemia was not evaluated (owing to the population's concern about giving blood samples) or treated, although it is known to be a public health problem (INEI, 2001).
- The evaluation did not include a control group, which limits the validity of the results. (This is a frequent issue for interventions.)
- The small sample size in the final evaluation and the inclusion of some of the same families as in the baseline also limited the value of the results.

Despite these limitations, however, the project contributed to local awareness and knowledge about the value of traditional foods. The enhanced production of seeds and animal-source foods had a modest positive effect on feeding patterns, the distribution of food within families and dietary quality, particularly for young children. The Awajún people were aware of the changes that were occurring, and responded positively to the need to defend their culture and improve their food. They appeared eager to learn more about nutrition and better feeding for their families and young children. Further similar initiatives are needed to benefit the nutrition and health of the Awajún.

Lessons learned

Coordination with local community organizations was effective, and ODECOFROC considered the project leaders' interactions with ODECOFROC staff and the communities as setting a good example to follow, especially the research agreements and delivery of reference material on local foods prior to project commencement. However, the project and local health facilities would have benefited from greater coordination with government institutions such as the local Ministry of Health and food donation programmes. This could have increased the consistency of key messages and activities relating to traditional foods and nutrition in the area, and would also have facilitated the incorporation of health promotion into an integrated approach that included the treatment and prevention of illnesses in addition to the project's hygiene, nutrition and local food production components.

The project's community nutrition and health promoters were committed and involved in the project, and rapidly grasped information about local and traditional foods through the practical sessions on food selection and preparation. Working with promoters was an appropriate strategy for reaching communities along the riverbanks and in the hills, as designated by ODECOFROC and the Women's Programme. However, several promoters had difficulties in transmitting their knowledge and skills to their communities, although the project's increasing focus on practical aspects and key messages helped. Efforts are needed to remediate these aspects in future interventions of this nature. The IIN technical team was present for education activities only twice a year during the two-year project, which reduced the support it was able to give to the promoters. In future, more frequent technical support and accompaniment would benefit the promoters and their activities in the communities. Finding ways of compensating promoters for the time they dedicate to their community work may also allow them to spend more time on project activities.

The ODECOFROC Women's Programme was key to this project, especially its work to improve the quality of life in the communities, largely through promoting fish farms, reforestation and the raising of small animals; all of these were successful programmes that increased access to and use of local, traditional and nutritious foods. Without this, effective coordination with other institutions working in these areas could not have occurred. The Women's Programme was also key in coordinating and supporting the promoters, assuring continuity, assisting project implementation in the communities, and monitoring activities ✾

Acknowledgements

The project team members were: Irma Tuesta, National Coordinator, Women's Programme, ODECOFROC; Miluska Carrasco and Melissa Abad, IIN nutritionists; Ruben Giucam and Dola Chumpi, translators; Ruben Giucam, monitor; Rosita Chimpa, Dola Chumpi and Elena Sugka, Women's Programme leaders; Santiago, Guillermo and Julio, boat pilots; and Hilary Creed-Kanashiro of IIN.

The authors wish to express their sincere gratitude to all team members, to the Apus (chiefs) of the communities who gave their full support and, especially, to the community nutrition and health promoters of Cenepa and the women of the communities who participated in the intervention and the evaluations. They thank Margot Marin for data analysis, Carolina Pérez for data entry, Karla Escajadilla for secretarial support and Willy Lopez for logistical support.
> Comments to: hmcreed@iin.sld.pe

>> Photographic section p. IV

Chapter 6

The Dalit food system and maternal and child nutrition in Andhra Pradesh, South India

BUDURU SALOMEYESUDAS[1]   HARRIET V. KUHNLEIN[2]

MARTINA A. SCHMID[2]   PERIYAPATNA V. SATHEESH[1]   GRACE M. EGELAND[2]

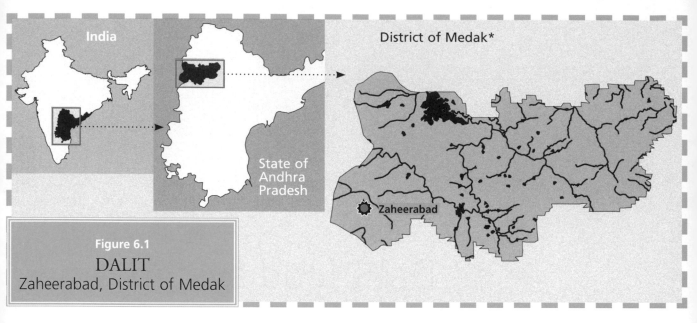

Figure 6.1

DALIT
Zaheerabad, District of Medak

India

State of Andhra Pradesh

District of Medak*

Zaheerabad

Data from ESRI Global GIS, 2006.
Walter Hitschfield
Geographic Information Centre,
McGill University Library.
*Digitized from
www.mapsofindia.com

1
Deccan Development
Society,
Hyderabad, India

2
Centre for Indigenous
Peoples' Nutrition and
Environment (CINE) and
School of Dietetics and
Human Nutrition,
McGill University,
St. Anne de Bellevue,
Quebec, Canada

Key words > traditional food, dietary intake,
micronutrients, anthropometry, biodiversity, Dalit,
India, maternal and child nutrition, food security

Photographic section >> VII

> " Today, if I look back, I can sense a sea-change in my life, and what is so exhilarating about it is the feeling of control that we are experiencing. Earlier, we were like drift-logs being swept here and there by external forces. We had to work for others on lands alien to us. We did not feel that anything belonged to us. We were just being used. But now, thanks to the *sangham*, we are shaping our life in a way that we have chosen on our own. "

Susheelamma, Raipally village, Jharasangam Mandal, Medak Distict, Andhra Pradesh State, India

Abstract

The food system of Dalit rural communities in the Zaheerabad region of south India includes 329 species/varieties of plants and animals, and unique patterns of food use. This chapter describes the effectiveness of using the local food system promoted through women farmers' organizations – called *sanghams* – in the area. The *sanghams* modified a national food distribution system for the poor, and named it the Alternative Public Distribution System (APDS). *Sanghams* worked closely with the Deccan Development Society to conduct many activities promoting local foods. A cross-sectional evaluation of APDS was conducted among pairs of a Dalit mother and her child (aged six to 39 months), from 57 villages in Medak District, 19 of which had active *sanghams*. Information was collected through dietary interviews in two seasons, socio-cultural interviews, anthropometry, and clinical examinations in the rainy season when health risks are highest. Results demonstrated that 58 percent of mothers suffered from chronic energy deficiency (CED), with higher rates among illiterate and active women; one-third of women were affected by night blindness during pregnancy; and 10 percent were identified as iron-deficient, based on pallor under the eyelid.

While children in all villages were similarly nourished, mothers in villages with the APDS programme had higher intakes of energy, protein, fibre, vitamin C and iron from greater consumption of sorghum, pulses, vegetables and animal-source foods. Traditional food fats, pulses and vegetables, roots and tubers showed protective associations against women's CED, after adjusting for number of children under five years of age and sanitation situation. Greater consumption of eggs and dairy products protected against night blindness, and uncultivated leafy greens were important for providing vitamin A during the rainy season. In conclusion, traditional Dalit foods were widely consumed, with associated positive health benefits in poor, rural communities in this district of India. These biodiverse, often unique foods should be promoted for their contributions to ecological farming and local culture and their benefits for food and nutrition security.

Context

Sorghum (*jowar*), pearl millet (*bajra*), finger millet (*ragi*) and foxtail millet (*korra*) are coarse cereals that have been mainstays of agriculture, diet and cultural systems in rural India. This is especially so in the vast dryland belts spreading across the Deccan plateau, north Karnataka, Marathwada, the deserts of Rajasthan, and many tribal areas in central India. These drylands are heavily populated with poor rural people. Agriculture is rainfed, and farming of these crops covers up to 65 percent of the geographical area, demanding few external inputs such as irrigation and fertilizer. For poor rural communities, such crops provide food security and sustainability at minimal cost. However, there is lack of political will in the country to achieve food security through such coarse cereals.

The Dalit are recognized as the "untouchables" (formally known as the "scheduled castes"), the fifth group below the four classes in the Hindu religion. There are more than 180 million Dalit in India; the majority are illiterate and landless, and work as farm labourers. Although the term "untouchability" has been abolished in law, Dalit are known to suffer severe caste-imposed discrimination that affects every aspect of life (IHEU, 2010; Minority Rights Group International, 2010), and Dalit women and children are likely to be the most poorly nourished sub-groups in India. Within the area studied, self-identified Dalit women are farm labourers, living and working close to the food growing areas of Zaheerabad.

Surprisingly, one of the major contributors to the problem of food insecurity for Dalit and others is the Public Distribution System (PDS). Possibly the largest affirmative action in the world, the PDS provides subsidized food at cheap prices to the poor, but concentrates on only two grains: refined rice and wheat. This massive programme covering all of India provides a regular and continued supply of these grains from the market and distributes them to the poorest people; the resulting steady prices make agriculture remunerative and attractive for rice and wheat farmers, who are already supported by subsidized irrigation, subsidized fertilizers and adequate crop insurance.

Deccan Development Society initiatives

This is the context in which the Deccan Development Society (DDS) operates. The 5 000-plus members of this grassroots organization are primarily agricultural labourers and marginal farmers owning 1 or 2 acres (about 0.4 to 0.8 ha) of farmland. In most cases, these lands are either *inam* (gift) lands given to the farmers by their landlord-employers, or lands assigned by the government as part of its land reform programme. Most

of these lands are degraded red soils, producing only about 30 to 50 kg of grains per acre (75 to 125 kg/ha). The grains produced by members of DDS *sanghams*[1] satisfy food needs for six to seven months a year. Addressing food insecurity for the other four to five months is a challenge, for both families and DDS.

The PDS has the potential to be a crucial policy instrument for food security and political stability in India. It plays an important role in averting famines by purchasing grain from surplus regions to be sold for a fair price in food-deficit/low-income regions. However, as part of the market-driven/irrigation-centred agricultural policies of the government and development agencies, the PDS has encouraged a new pattern of food consumption in the semi-arid tropical regions of India. The poor have shifted from a diet of locally grown rainfed cereals such as sorghum and millets to one of rice and other irrigated crops imported from distant areas, leading to a decline in the cultivation of dryland cereals and associated intercrops (pigeon pea, field bean, cow pea and other beans). Recognizing that the cheap rice-based PDS was destroying the agriculture of the poor in dryland areas, women members of DDS articulated the crux of the problem:

- "Cheap rice is attractive. But in the bargain we left our lands fallow."
- "We hanker after rice and neglect our own lands."

Extended discussions between women farmers and DDS since as far back as the mid-1980s have sought solutions to the problems of threatened agricultural practices and food insecurity for rural societies. When imports of subsidized rice through the PDS made it uneconomical for small farmers to cultivate their lands and grow the coarse cereals and pulses that are the backbone of their agriculture and traditional food, the plots owned by small and marginal farmers were left fallow, leading to the abandonment and degradation of productive land. The direct result of this increase in fallow land was a marked decline in the production and availability of traditional cereals, pulses and animal fodder. This affected the nutrition of rural people,

---

1 A *sangham* is a volunteer collective of poor women farmers, primarily Dalit, working within DDS in Andhra Pradesh, where each village has its own *sangham*.

especially women and children, and increased the shortage of draught animal power (Women's *Sanghams* of the Deccan Development Society, Satheesh and Pimbert, 1999).

One strategy suggested was to reclaim the fallow land, and to use deep ploughing and manure to grow food crops for family use. However, this approach required investments of approximately 2 600 rupees (INR) per acre (equivalent to about USD 83 per acre at the time). To obtain funding, women farmers of dryland crops campaigned to reverse unfavourable loan policies, and DDS approached the Indian Ministry of Rural Development. In 1994, this resulted in approval of the Community Grain Fund (CGF), which was distributed in agreement with the Government of India through the PDS. Under this arrangement, groups of largely illiterate and poor women set up and ran a community-managed Alternative PDS (APDS) based on coarse grains, which are produced, stored and distributed locally in 30 villages around Zaheerabad. The women deposit their excess grain in the CGF to repay for the loans they receive from the PDS.

Each *sangham* formed a committee of about five women to design and implement activities on about 100 acres (40 ha) of fallow land in each of the 30 villages. Overall, the committees found a total of 2 675 acres of suitable land, divided fairly equally among the villages. The women's committees then each selected about 20 acres (8 ha) on which to supervise the work of other women farmers, to ensure appropriate ploughing, manuring, sowing and weeding practices on the reclaimed fallow land. The committee members collect input support funds from the government and distribute them among the women's collectives managing the reclaimed fallow land. Under the current system, after each crop harvest, the committee members are responsible for collecting loan repayments in the form of grain from participants, and for storing the grain. Later in the year, during the season of food scarcity, the committee members sell this grain at greatly reduced prices to poorer households in the village, applying a quota system, with sale proceeds deposited in the village CGF account. Each of the 30 villages has its own account, controlled and managed by women committee members who are accountable to the villagers and to DDS.

By using the grain received from participating farmers to distribute to poorer households at a subsidized price, this system feeds the participating farmers' loan repayments back into the local village economy to support the very poorest, allowing them to obtain sufficient food and become more productive members of the community. Transparent procedures ensure that the money earned from sales of sorghum goes back into the CGF account held by the village committee. This money is used annually to reclaim more fallow land in the village, thus helping to increase productivity through diversifying farming systems and the use of locally available resources. More food is produced and sold locally, and job opportunities are created for people who would otherwise be excluded from the mainstream economy.

DDS documented the local food resources of the APDS communities in the Zaheerabad region of Andhra Pradesh. An amazing array of 329 species/varieties of local foods were recorded in their scientific, local Telugu and common English names, along with information about how they are prepared by village women (Salomeyesudas and Satheesh, 2009). Several previously unidentified plants were documented, as well as several plants recognized as "uncultivated greens". Nutrient composition was analysed in collaboration with scientists at the National Institute of Nutrition (NIN) in Hyderabad, and several food items were analysed for the first time. These food data were instrumental in creating the database for nutrient analysis of dietary data reported in this chapter.

Other major intervention activities

In addition to reclaiming land and redistributing it to women farmers, other community interventions that had a large impact on the food and nutrition security of families, mothers and children included:

- awareness campaigns focusing on traditional food systems;
- establishment of "Café Ethnic", serving millet-based foods;

- establishment of a community media trust, producing films on millets, recipes and uncultivated foods;
- cooking classes for family carers and hostel cooks;
- development of a millet processor;
- distribution of educational material to various agencies;
- promotion of food production systems based on agricultural biodiversity;
- food festivals at public places, fora, schools and colleges;
- formation of an organic farmers' association;
- formation of Zaheerabad consumer action group;
- provision of millet-based meals at day care centres;
- mobile biodiversity festivals;
- development of a mobile organic shop;
- networking with voluntary organizations from the local to the international level;
- establishment of an organic shop for sales of traditional foods;
- participation at national and state-level food festivals;
- publication of scientific information in the local language (Telugu);
- recipe competitions;
- screenings of recipe films;
- training in product packaging for traditional food crops;
- creation of a women's radio station by a community media trust.

Major impacts of the Alternative Public Distribution System

The main impacts of the APDS perceived by participants and others are:
- increased soil fertility;
- increased soil conservation;
- decreased crop disease, due to diverse cropping systems;
- increased availability of uncultivated greens from *sangham* fields;
- increased diversity of food for families;
- increased self-reliance;
- more work opportunities in the villages;
- reduced seasonal migration;
- increased animal fodder and livestock population;
- more dried plant materials for roofing, fencing, etc.;
- revival of rural livelihoods, such as blacksmithing and basket weaving;
- better and more food;
- improved health in families;
- increased knowledge about nutrition from local foods;
- more children attending school.

Rationale and research questions

The rationale for the study reported in this chapter was the need to evaluate the overall effects of APDS activities on the health of *sangham* households, especially among mothers and children, in the Zaheerabad Region of Medak District, Andhra Pradesh, south India. Mothers and young children agreed to take part to help increase knowledge about the seasonal use of traditional food crops; intakes of energy, protein, iron, vitamin A and other nutrients; and clinical signs of malnutrition. Comparative evaluation was conducted in villages with and those without the APDS. The following research questions were asked:

- Do Dalit mothers and their young children aged six to 39 months living in villages where the APDS is operating consume more traditional food during the summer season and the rainy season than Dalit mothers and their children from control villages (without the APDS)?
- Do Dalit mothers and their young children living in villages with the APDS have higher nutrient intakes (energy, protein, carbohydrates, fat, dietary fibre, iron, vitamin C and vitamin A) during the summer season and the rainy season, and better nutrition status during the rainy season than Dalit mothers and their children in control villages?
- Are consumption patterns and nutrient intakes predictors of nutrition status (chronic energy deficiency [CED], anaemia and vitamin A

deficiency) in Dalit mothers and their young children during the rainy season?

The research was epidemiological in nature. This chapter summarizes its results to evaluate the overall impact of DDS activities on food use and nutrition status among Dalit in the Zaheerabad Region of Andhra Pradesh.

Methods

Participatory research process

All the villages participating in the study were asked for their consent. *Sangham* leaders were consulted first, and the purpose of the study was explained to them at a meeting organized by DDS. Formal ethical approval was obtained from the Human Research Ethics Committee of McGill University (Canada). The village leaders discussed issues with the researchers and project leaders, and agreed to cooperate in the study. Researchers then visited individual villages to participate in *sangham* meetings and select individual study participants. Written consent was obtained from each of the identified mothers and from the *sangham* leaders. NIN trained six graduate students from Indian universities in interview techniques, which included a seasonal food frequency questionnaire, a 24-hour recall, a socio-cultural questionnaire, anthropometric measurement, and assessment of clinical signs. All interview schedules were translated into the local Telugu language and field tested to improve their validity. Data were obtained at the subjects' convenience, in their own homes and with their families' consent. The objectives and purpose of the study were clearly explained in *sangham* meetings to ensure maximum cooperation, and results were similarly presented. Two fruit trees were given to each participating household, in appreciation of its time and effort.

A cross-sectional sampling design was used in six rural townships (*mandals*) in Medak District, where a total of 263 Dalit mothers, each with a child aged six to 39 months, were found eligible, from 19 villages that had been implementing the APDS since its inception in 1995 and 18 villages without the APDS. All the participating households were members of their village DDS *sangham*, and only one mother per household was included. In households were there were two eligible mothers, the mother who was at home at the time of the survey was chosen. Mothers under 15 years of age and/or with twins were excluded. Of the 263 eligible mothers contacted, 223 participated in interviews in both the summer season and the rainy season of 2003. Of the 43 (16 percent) mothers who were unable to participate, 19 were working in the nearby city (7 percent), seven had recently given birth (3 percent), 16 were absent from the village at the time of the survey (6 percent), and one refused.

Summer season interviews administered an 83-item food frequency questionnaire (FFQ) and rainy season interviews one of 106 items. The FFQs were based on existing information from DDS on the seasonal availability of food species, personal preferences and market availability. Thirty-one items of foodgrains, nuts, oilseeds, pulses and animal foods were included in both seasons. Fruits, green leafy vegetables, other vegetables, roots and tubers were included according to seasonal availability. The summer season FFQ included 14 vegetables, six cultivated green leafy vegetables, ten wild green leafy vegetables, 11 cultivated fruits and 11 wild fruits. The rainy season FFQ included 15 vegetables, 14 cultivated green leafy vegetables, 36 wild green leafy vegetables, six cultivated fruits and three wild fruits. Frequencies were in number of days that the food item had been consumed during the season (each season was of two months or 60 days). Mothers' consumption frequencies of nuts and oilseeds, vegetables, roots and tubers, animal foods, green leafy vegetables, eggs, milk and milk products, meat and fruits were averaged across both seasons. The two-season average consumption frequencies of sorghum, rice and pulses were combined with the average amounts consumed, obtained from 24-hour recalls, to estimate average total amount consumed per day.

Mothers' nutrient intakes were calculated from a minimum of two 24-hour recalls per season. Recalls were obtained according to standard procedures adapted from the National Nutrition Monitoring Bureau surveys used by NIN. Each mother was asked to recall her own and her child's food intakes from the preceding day, and detailed descriptions of all the food and beverages consumed were recorded, including cooking methods and brands. Quantities of food consumed were weighed on digital kitchen scales (ATCO Model No D2RS-02-W) to the nearest gram, or were estimated with household measures or standardized vessels. For cooked dishes, such as dhal and curry, all the raw ingredients used for the family were weighed, and the volumes consumed by the mother and child were estimated from this. A standardized 12-vessel set was used to estimate volumes of cooked foods and liquids. The individual raw intake of each ingredient of cooked dishes was calculated, and standardized recipes and standard conversion factors for cooked rice were used for missing ingredients or missing volumes of dishes cooked for the family. Standard breastmilk consumption was assumed for breastfed children: 500 ml per day for children aged six to 12 months, and 350 ml per day for those aged one to three years (Belavady, 1969).

Nutrient values published by India's NIN (Gopalan et al., 1989) were used, with missing nutrient values or food items and the values of total dietary fibre taken from Association of Southeast Asian Nations (ASEAN) food composition tables (Puwastien et al., 2000) or European food composition and nutrition tables (Souci, Fachmann and Kraut, 1994). Food energy was calculated by assuming protein, carbohydrate and fat yields of 4, 4 and 9 kcal/g respectively. These were then converted into kilojoules (kJ) using the conversion rate of 4.2 J per calorie (4.2 kJ = 1 kcal). Pro-vitamin A carotenoids were converted into retinol equivalents (RE), assuming 6 µg β-carotene equals 1 µg RE. In the absence of β-carotene values, it was assumed that 6 µg total carotene equals 1 µg RE. The β-carotene values determined by high-performance liquid chromatography from recent publications were used whenever available (Bhaskarachary et al., 1995; Rajyalaksmi et al., 2001).

For some multivariate analyses, nutrient intakes from intervention (with APDS) and control (without APDS) villages were determined as population group totals for mothers and children, using a single 24-hour recall. Within each population group, nutrients were pooled into nine food groups, separately for the summer season and the rainy season: other food grains (wheat and maize); nuts and oilseeds (eight species); pulses (nine species); animal foods (nine items); green leafy vegetables (nine cultivated and seven wild species); vegetables (20 species); fruits (seven species); drinks (three items); and miscellaneous (11 items). The percentage contribution to each season's total nutrient intakes made by each of the food groups and single food items (sorghum, rice, cooking oil) was determined and ranked. Nutrient intakes were estimated using Candat (Canadian Nutrient Data Analysis Toronto, Version 5.1, 1988, Godin Incorporated, London, Ontario, Canada), based on nutrient values from the Indian food composition tables (Gopalan et al., 1989).

Anthropometric measurements

Anthropometric measurements, a socio-cultural questionnaire and an eye examination were administered during the rainy season. Portable height rods (Galaxy Informatics, Delhi, India) were used to measure the height of mothers, with an accuracy of 1 mm. Mothers' weights were measured on a digital balance (SECA BELLA 840, Hamburg, Germany), with an accuracy of 100 g. Women with a body mass index (BMI) of less than 18.5 kg/m² were classified as CED, using standard cut-off points (James, Ferro-Luzzi and Waterlow, 1988). Pregnant women (n = 14) were excluded from the CED analyses.

Young children were measured with portable infantometers (Galaxy Informatics, New Delhi, India), measuring lengths of 56 to 92 cm with an accuracy of 1 mm; the digital balance was a Tansi (Tamilnadu Small Scale Industry) hanging manual baby balance with a maximum measurable weight of 20 kg and an accuracy of 50 g; non-stretchable plastic tape (Dritz, Germany) with an accuracy of 1 mm was used to measure arm circumference.

Standard procedures based on international standards were used for all measurements (Lohman, Roche and Martorell, 1988). Weight and height measurements were taken without shoes and with minimal clothing. Mid-upper-arm circumference (MUAC) was measured on the left arm. The weight of the jewellery worn by mothers and children was recorded – mothers know this weight because of the economic value of silver. If a mother or child had more than 100 g of jewellery, which was rare, the weight was subtracted from the measured weight. Interviewers had one day of training, with practice measurements performed on children in a nearby village school.

Table 6.1 Variables of interest for determinants of CED, clinical vitamin A deficiency symptoms and iron deficiency in Dalit mothers

Variable	Index category
Variables for chronic energy deficiency (BMI < 18.5 kg/m²)	
Energy intake	kcal/day
Carbohydrate intake	% of energy
Fat intake	% of energy
Dietary fibre intake	g/day
Rice consumption	g/day
Sorghum consumption	g/day
Pulse consumption	g/day
Frequency of nuts and oilseeds	days
Frequency of vegetables, roots and tubers	days
Frequency of animal foods (meat, eggs, milk, milk products)	days
Variables for clinical vitamin A deficiency (Bitot's spot, conjunctival xerosis, night blindness)	
Nutritional supplement (enriched flours)	yes
Fat intake	% of energy
Vitamin A intake (RE)	µg RE/day
Sorghum consumption*	g/day
Pulse consumption*	g/day
Frequency of green leafy vegetables	days
Frequency of eggs, milk and milk products	days
Frequency of fruits	days
Variables for iron deficiency (under eyelid pallor)	
Nutritional supplement (enriched flours)	yes
Iron-folic acid tablets	yes
Energy intake	kcal/day
Dietary fibre intake	g/day
Iron intake	mg/day
Vitamin C intake	mg/day
Rice consumption*	g/day
Sorghum consumption *	g/day
Pulse consumption*	g/day
Frequency of green leafy vegetables	days
Frequency of meat	days

Dietary intakes and consumption frequencies for the summer season and the rainy season of 2003.
*   Obtained from averages of two to four 24-hour recalls and food frequency questionnaires during the summer season and the rainy season.
Source: Adapted from Schmid *et al.*, 2007.

Table 6.2 Variables of interest for determinants of stunting, wasting, underweight and iron deficiency in Dalit children aged 6 to 39 months

Variable	Index category	Reference category
Adjusting variables		
Age	6–12, 13–24 or 25–39 months	6–12 months
Sex	male or female	female
Housing	permanent house or traditional hut	permanent house*
Feeding status	weaned, complementary fed or breastfed	weaned
Duration of exclusive breastfeeding (including water)	≥ 6 months or < 6 months	≥ 6 months
Nutrition supplements (enriched flour)	yes	no
Vitamin A drops	yes	no
Protein energy malnutrition (stunting, wasting, underweight)		
Energy intake	≥ 1 220 kcal/day or < 1 220 kcal/day	≥ 1 200 kcal
Protein intake	≥ 21 g/day or < 21 g/day	< 21 g
Vitamin A intake (RE)	≥ 200 µg RE/day or < 200 µg RE/day	< 200 ±g
Frequency of sorghum consumption	days	-
Frequency of pulses consumption	days	-
Frequency of green leafy vegetables consumption	days	-
Frequency of animal food consumption	none, less than daily or daily	none
Iron deficiency (under eyelid pallor)		
Energy intake	≥ 1 220 kcal/day / < 1 220 kcal/day	≥ 1 200 kcal
Fibre intake	≥ 5 g/day / < 5 g /day	< 5 g
Iron intake	≥ 6 mg /day / < 6 mg /day	< 6 mg
Vitamin C intake	≥ 25 mg /day / < 25 mg /day	< 25 mg
Frequency of sorghum consumption	days	-
Frequency of pulse consumption	days	-
Frequency of green leafy vegetable consumption	days	-
Frequency of meat consumption	none/less than weekly/weekly	none

Nutrient intakes (from 24-hour recalls) and food consumption frequencies (from food frequency questionnaires) for the rainy season of 2003.
* House with or without a permanent roof.
Source: Adapted from Schmid *et al.*, 2007.

Eye examination

Mothers self-reported any night blindness (XN) they were suffering at the time of the survey and/or had suffered during their last pregnancy, noting the month(s) of pregnancy affected. Standardized terms for night blindness in Telugu were used, according to NIN procedures. The eyes of mothers and children were examined by trained interviewers who assessed the prevalence of clinical vitamin A deficiency, including Bitot's spot (X1B), conjunctival xerosis (X1A) and corneal xerosis (X1A), as classified by the World Health Organization (WHO) (McLaren and Frigg, 2001). Iron deficiency was classified according to whiteness or pallor in the inside lower eyelid (Gibson, 1990).

Statistical analysis

Data from mothers and children were analysed separately for the summer season and the rainy season. Chi-square analysis was used for dichotomous and categorical characteristic variables. When the expected count of the Chi-square was below 5, Fisher's Exact Test was used. The non-parametric (Wilcoxon) test was used for abnormally distributed data, including for all the nutrient intakes of children, and for the fat, dietary fibre, iron, vitamin C and vitamin A intakes of mothers. The paired Student's t-test was used to compare the means of normally distributed continuous variables between intervention and control villages. The paired Student's t-test and signed rank test were used to compare summer season and rainy season data in intervention and control villages. Differences in the intakes of each nutrient in each season between the intervention and control groups were tested for, and no significant differences were observed. A two-sided alternative hypothesis was tested with alpha at 0.05. Data from one child from a control village in the summer season were missing.

Descriptive statistics were used to provide means, standard deviations and percentages. Unadjusted relative risks and 95 percent confidence intervals (CIs) were calculated for categorical risk factors. Beta coefficients, standard errors and p-values were obtained from multivariate logistic regression analyses in which nutrient exposures were evaluated for their associations with outcomes, taking into consideration important determinants. Determinants for CED, clinical vitamin A deficiency and iron deficiency in mothers were examined separately. The following variables were considered in univariate and multivariate analyses: mother's age (years), number of children under five years of age (one or more), mother's physiological status (lactating, pregnant or neither), mother's activity level (moderate or low), household income above or below the poverty line (INR 1 000 per month), mother's literacy (ability to read and write), and household's lack of sanitation (i.e., an open field toilet). One woman was both pregnant and lactating, and was considered pregnant in all analyses.

Table 6.1 gives the nutritional variables explored for their association with CED, clinical vitamin A-deficiency symptoms and iron deficiency (pallor) in mothers. Nutritional factors that were correlated with each other in bivariate analyses were not entered together into multivariate models. Variables of interest for determinants of stunting, wasting, underweight and iron deficiency (pallor) in children aged six to 39

months are given in Table 6.2. Statistical analyses were conducted with SAS Version 8 (SAS Institute Inc., Cary, North Carolina, United States of America). Data sampling, rationale and analysis processes are described in greater detail by Schmid *et al.* (2006 and 2007).

Results and discussion

Mothers' dietary intake

The characteristics of mothers in intervention and control villages are given in Table 6.3. Mothers in both groups were similar with respect to most variables, but mothers from intervention villages (who were generally further from village centres) were daily labourers in the rainy season, so were assumed to have increased energy needs. More mothers were pregnant during the summer season than the rainy season, and about half of the mothers had taken iron-folate tablets while pregnant with the child included in the study.

Mothers from intervention villages had significantly higher intakes of energy (by about 1 000 kJ), protein (by about 8 g) and dietary fibre (by about 8 g) during both the summer season and the rainy season than mothers from control villages (Table 6.4). The percentage of total energy from fat was approximately 10 percent for all mothers in both seasons. The median iron intake of mothers from intervention villages was higher in both seasons, and significantly higher during the rainy season. Vitamin C intakes during the summer season were similar in both groups, but the median vitamin C intake in mothers from control villages was significantly higher during the rainy season. Mothers' fat and vitamin A intakes were similar in both intervention and control villages. In both groups, all nutrient intakes except for vitamin C were higher ($p \leq 0.05$) during the summer season than the rainy season.

Mothers' main sources of energy and protein were sorghum, rice and pulses, contributing 31, 48 and 9 percent, respectively, of total energy, and 35, 36, and 22 percent of total protein in intervention villages; similar values were found in control villages. Millet was consumed by fewer than 1 percent of mothers in both groups.

Table 6.3 Characteristics of Dalit mothers in intervention and control villages

Variable	Description	Mothers (n = 220)
Age	mean years (SD)	24.3 (3.9)
Weight*	mean kg (SD)	40.9 (5.7)
Height	mean cm (SD)	150.0 (5.0)
MUAC	mean cm (SD)	22.5 (2.0)
BMI*	mean kg/m² (SD)	18.2 (2.2)
BMI	grade 0 (BMI ≥ 18.5 kg/m²)	42%
	grade 1 (BMI < 18.5 kg/m²)	28%
	grade 2 (BMI < 17.0 kg/m²)	20%
	grade 3 (BMI < 16.0 kg/m²)	10%
Literate	read and write	79%
Open field toilet	lack of sanitation	94%
Woman's status	lactating	81%
	pregnant	6%
	neither	13%
Active	work in fields and other agricultural activities	71%
Clinical vitamin A deficiency		
None	no symptoms	84%
Night blindness (XN)	self-reported	7%
Bitot's spot (X1B)	examined	5%
Conjunctival xerosis (X1A)	examined	6%
Reported night blindness during pregnancy	yes	35%
Iron deficiency		
Inside of lower eye lid	white (pallor)	10%

Figures are means with SDs in brackets, or percentages from data collected during the rainy season of 2003.
* n = 207, pregnant mothers excluded.
Source: Adapted from Schmid *et al.*, 2007.

Primary sources of iron were sorghum, rice and pulses (Figure 6.2), contributing 56, 16 and 15 percent, respectively, to total iron intake in intervention villages, and similar percentages in control villages. Animal-source food contributed less than 2 percent in both groups. Cereals (sorghum, rice and wheat) contributed 79 percent in intervention villages compared with 68 percent in control villages. Cultivated and uncultivated green leafy vegetables contributed 2 percent of total iron intake and 11 percent of total vitamin C intake in intervention villages, and about three times as much to intakes in control villages.

Fruits and vegetables were major sources of vitamin A (Figure 6.3). During the summer season (mango season), fruits contributed 54 percent of vitamin A in intervention villages and 40 percent in control villages; in the rainy season, uncultivated green leafy vegetables contributed 43 percent of vitamin A in intervention villages and 36 percent in control villages. Vegetables, roots and tubers contributed 19 percent in intervention villages and 26 percent in control villages. Overall, sorghum and animal-source food items contributed 9 and 8 percent, respectively, of vitamin A in intervention villages, and similar percentages in control villages.

Mothers from intervention villages had higher energy and protein intakes in both seasons than mothers from control villages. Surprisingly, the difference in energy intakes was similar in both the summer season

Table 6.4 Nutrient intakes of Dalit mothers in intervention and control villages, by season

Nutrient	Summer season			Rainy season		
	Intervention villages (n = 125)	Control villages (n = 109)	p	Intervention villages (n = 124)	Control villages (n = 96)	p
Energy, kJ[a]	12 218 (3 511)	11 155 (3 347)	0.02[b, c]	11 189 (3 335)	10 193 (3 738)	0.04[b, c]
	11 437	11 117		10 769	10 038	
	(9 941 14 713)	(9 173 13 663)		(8 623 12 986)	(7 850 12 432)	
Protein, g[a]	77.5 (25.1)	71.1 (25.2)	0.05[b, c]	68.9 (22.6)	60.4 (23.8)	< 0.01[b, c]
	74.8	69.5		66.4	61.8	
	(60.2 98.6)	(50.5 87.8)		(53.5 82.2)	(42.6 75.2)	
Carbohydrates, g	578 (175)	519 (170)	0.01[b, c]	535(162)	490 (191)	0.06[b]
	549	525		513	479	
	(467 705)	(416 636)		(425 626)	(384 606)	
Fat, g	31.6 (14.4)	32.9 (15.5)	0.54	27.1 (14.7)	24.6 (11.7)	0.36
	27.6	29.2		24.2	23.9	
	(21.2 37.6)	(22.7 40.2)		(17.1 33.9)	(17.9 28.3)	
Dietary fibre, g	48.5 (23.2)	42.0 (23.1)	0.03[c]	40.8 (19.6)	32.5 (19.3)	< 0.01[c]
	46.8	39.4		41.8	33.6	
	(32.6 60.1)	(22.5 54.8)		(25.9 52.6)	(16.0 46.3)	
Iron, mg[a]	20.8 (12.0)	18.8 (12.1)	0.09	15.8 (6.6)	13.7 (9.1)	< 0.01[c]
	18.9	16.5		15.3	13.0	
	(13.0 24.5)	(12.1 22.4)		(10.8 20.5)	(7.6 18.2)	
Vitamin C, mg[a]	26.4 (31.0)	33.0 (66.4)	0.91	19.7 (35.5)	21.7 (26.1)	0.04[c]
	15.4	12.4		8.3	11.0	
	(2.1 38.9)	(2.3 33.6)		(2.1 22.1)	(4.2 31.4)	
Vitamin A, µg RE[a, d]	354 (629)	275 (503)	0.42	155 (271)	163 (250)	0.65
	110	103		73	75	
	(54 423)	(53 376)		(48 137)	(49 146)	

Figures are means with SDs in brackets, or medians with first and third quartiles in brackets. Non-parametric test (Wilcoxon).
[a] Recommended levels for pregnant and lactating women with moderate activity level, respectively (26): energy – 10 517 kJ, 10 937 kJ; protein – 60 g, 63 g; iron – 37 mg, 30 mg; vitamin C – 40 mg, 80 mg; vitamin A – 600 µg RE, 950 µg RE.
[b] p from two-sample pooled Student t-test.
[c] $p \leq 0.05$ statistically significant, all nutrient intakes (expect vitamin C) are higher ($p \leq 0.05$) in the summer season than the rainy season: intervention villages – summer season (17), rainy season (11); control villages – summer season (11), rainy season (12) repeated 24-hour recall.
[d] 1µg retinol = 1µg RE, 6 µg provitamin A carotenoids = 1µg RE.
Source: adapted from Schmid et al., 2007.

and the rainy season, perhaps because in both seasons more mothers in the intervention villages were pregnant. According to 1993/1994 data from the National Sample Survey, about 80 percent of India's rural population had energy intakes below the 10 080 kJ recommended for adults in rural areas. The poorest 30 percent of India's population consumed on average less than 7 140 kJ per day, and the poorest 10 percent less than 5 460 kJ (Measham and Chatterjee, 1999). In this study of the poorest segment of rural Dalit communities, mean energy intakes in both seasons and both groups were higher than 10 000 kJ, indicating the better provision of food sources in settings where poor rural women control their own agricultural production.

It was assumed that the activity levels of mothers in all villages were sedentary during the summer season, when labour demand was low. During the rainy season, mothers who were working as agricultural labourers

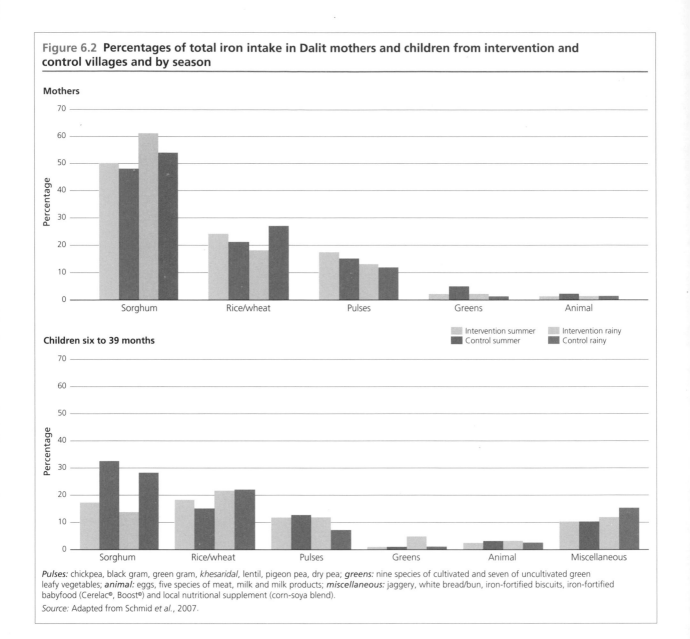

Figure 6.2 **Percentages of total iron intake in Dalit mothers and children from intervention and control villages and by season**

Mothers

Legend:
Intervention summer | Intervention rainy
Control summer | Control rainy

Children six to 39 months

Pulses: chickpea, black gram, green gram, *khesaridal*, lentil, pigeon pea, dry pea; *greens:* nine species of cultivated and seven of uncultivated green leafy vegetables; *animal:* eggs, five species of meat, milk and milk products; *miscellaneous:* jaggery, white bread/bun, iron-fortified biscuits, iron-fortified babyfood (Cerelac©, Boost©) and local nutritional supplement (corn-soya blend).
Source: Adapted from Schmid *et al.*, 2007.

were classified as moderately active. According to the Indian recommended dietary allowance (RDA), the energy requirement of an average Indian woman (weighing 45 kg and aged between 18 and 30 years) increases by approximately 1 500 kJ/day with moderate activity level, to reach 9 257 kJ, compared with the 7 792 kJ of sedentary women (ICMR, 1990). For both groups of women in the study, mean energy intake surpassed that calculated for the moderate activity level.

The Indian adult requirement for protein is 1.0 g/ day per kilogram of body weight, with an additional 15 g during pregnancy and 18 g during lactation, to give average standard requirements of 60 g of protein for pregnant and 63 g for lactating women (ICMR, 1990). In this study, mean protein intakes in the intervention and control groups were greater than 60 g in both seasons. According to Gopalan *et al.* (1989), the majority of Indians obtain 70 to 80 percent of daily energy needs and more than 50 percent of daily protein needs from cereals; intakes of cereals tend to be highest in low-income families. Marginal-farmer households in rural India are reported to obtain 72

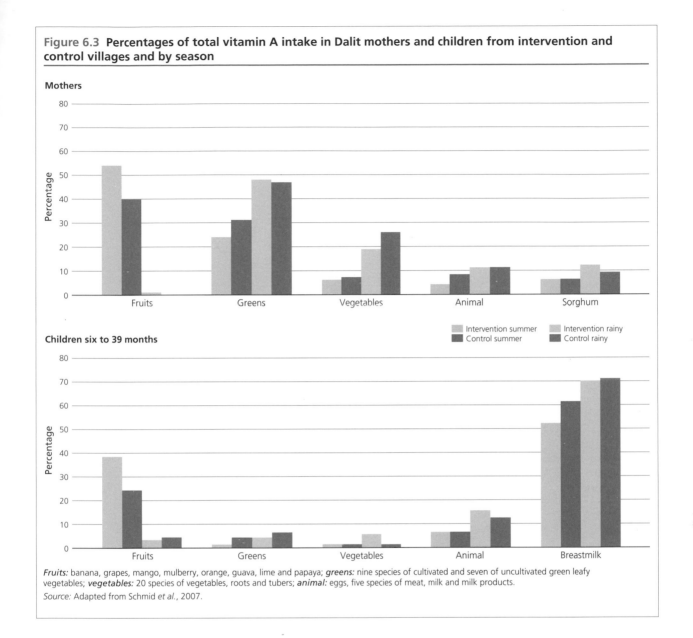

Figure 6.3 **Percentages of total vitamin A intake in Dalit mothers and children from intervention and control villages and by season**

Fruits: banana, grapes, mango, mulberry, orange, guava, lime and papaya; *greens:* nine species of cultivated and seven of uncultivated green leafy vegetables; *vegetables:* 20 species of vegetables, roots and tubers; *animal:* eggs, five species of meat, milk and milk products.

Source: Adapted from Schmid *et al.*, 2007.

percent of energy and 68 percent of protein from cereals, which are similar to the findings of this study, but only 4 percent of energy and 10 percent of protein from pulses, which are lower than this study's findings (FAO, 2002). This too demonstrates the relatively higher quality of the diet of impoverished Dalit Indian women.

India's iron RDAs are 30 mg for lactating and non-lactating women and 37 mg for pregnant women, based on an average absorption rate of 3 percent (ICMR, 1990). In North America, the estimated

average requirements for women aged 19 to 30 years are 8.1 mg for non-lactating women, 6.5 mg for lactating women and 22 mg for pregnant women, based on an absorption rate of 18 percent (Food and Nutrition Board Institute of Medicine, 2001). In this study, mothers from all villages had median intakes of less than 20 mg of iron in both seasons. Although cereals are not rich sources of iron, they are important contributors when they are consumed regularly. Mothers from intervention villages consumed more iron and dietary fibre during the rainy season, probably

Table 6.5 Characteristics of Dalit children aged six to 39 months in intervention and control villages

Variable	Description	Intervention villages (n = 124)	Control villages (n = 96)	p
Gender	Female	60 (48%)	50 (52%)	0.59
	Male	64 (52%)	46 (48%)	
Birth order	First	40 (32%)	36 (37%)	0.48
	Second	38 (31%)	34 (35%)	
	Third	25 (20%)	14 (15%)	
	≥ Fourth	21 (17%)	12 (13%)	
Nutrient supplement in 2003[a]	Yes	51 (39%)	41 (41%)	0.67
	No	80 (61%)	60 (59%)	
Vitamin A supplement since birth[b]	1–4 times	58 (44%)	53 (48%)	0.22
	No	73 (56%)	48 (52%)	
De-worming tablet since birth	Yes	12 (9%)	2 (2%)	0.03[c]
	No	119 (91%)	99 (98%)	
		Summer season (n = 125)	(n = 108)	
Age	6–12 months	47 (38%)	39 (36%)	0.57
	13–24 months	46 (37%)	46 (43%)	
	25–39 months	32 (27%)	23 (21%)	
Feeding status	Breastfed	27 (22%)	31 (28%)	0.25
	Complementary fed	70 (56%)	56 (52%)	
	Weaned	28 (22%)	21 (20%)	
		Rainy season (n = 124)	(n = 96)	
Age	6–12 months	30 (24%)	23 (24%)	0.20
	13–24 months	41 (33%)	42 (44%)	
	25–39 months	53 (43%)	31 (32%)	
Feeding status	Breastfed	12 (9%)	8 (8%)	0.62
	Complementary fed	74 (60%)	60 (63%)	
	Weaned	38 (31%)	28 (29%)	

Figures are counts with percentages of n in brackets. Chi-square test.
[a] Corn-soya blend and muruku (enriched chickpea flour).
[b] Vitamin A Prophylaxis Programme for children aged six to 36 months: six monthly doses of 100 000 IUs of vitamin A for infants and 200 000 IUs for young children.
[c] $p \leq 0.05$ statistically significant.
Source: Adapted from Schmid *et al.*, 2007.

because of frequent sorghum consumption. However, pulses, rice and grains are also rich in phytates and tannins, which interfere with iron availability. Vitamin C may enhance availability, but median vitamin C intakes in the study were less than 15 mg, which is lower than the Indian RDAs of 40 mg for non-lactating and pregnant women and 80 mg for lactating women (ICMR, 1990).

The Indian RDAs for vitamin A are 600 µg RE for non-lactating and pregnant women and 950 µg RE for lactating women (ICMR, 1990). Median vitamin A intakes for all mothers in the study were below these recommendations during both seasons, and fat intakes were fairly limited (at approximately 10 percent of total energy). During the rainy season, green leafy vegetables provided half of the vitamin A for mothers, mainly from uncultivated greens, especially in intervention villages. In a study of rural Andhra Pradesh (National Nutrition Monitoring Bureau, 2002), sedentary lactating women aged $\geq$ 18 years had lower mean energy (9 131 kJ), mean protein (50 g) and median iron (8.9 mg) intakes than those observed in this study. However, median fat (25.6 g) and vitamin A (106 µg) intakes were similar, and vitamin C intake (22 mg) was higher. In India (FAO, 2002), adults in marginal-farming households (with 0.5 to 1 acre/0.2 to 0.4 ha) were reported to have lower energy (9 500 kJ) and protein (59 g) intakes and a similar fat (33 g) intake compared with those reported in the current study.

Children's dietary intake

The characteristics of children in intervention and control villages (Table 6.5) were similar. Most were either the first- or second-born in their families, and were given complementary foods. About half received vitamin A drops at least once. More children in intervention than control villages were given de-worming tablets at least once.

Similar amounts of energy, protein, carbohydrate, fat, dietary fibre, iron, vitamin C and vitamin A were consumed by children in both intervention and control villages in both seasons (Table 6.6). These children had higher dietary fat, vitamin C and vitamin A intakes relative to requirements than their mothers. For all children, in both seasons, approximately 23 percent of energy came from fat. Among children in the intervention group, intakes of protein ($p \leq 0.04$), dietary fibre ($p \leq 0.05$) and iron ($p \leq 0.05$) were significantly higher during the rainy season, and vitamin A intake ($p \leq 0.02$) was higher during the summer season. In the control group, energy ($p \leq 0.01$), protein ($p \leq 0.01$), carbohydrate ($p \leq 0.01$), dietary fibre ($p \leq 0.01$) and iron ($p \leq 0.01$) intakes were higher during the rainy season.

Overall, in intervention villages, breastmilk, rice and pulses contributed 34, 33 and 5 percent, respectively, of total energy intake, and 29, 28 and 14 percent of total protein intake. Percentages were similar in control

villages (data not shown). Sorghum, rice and pulses provided most dietary iron (Figure 6.2). Miscellaneous food (including iron-fortified baby food) and animal-source food items contributed about 13 and 4 percent, respectively, of iron intake in both intervention and control villages. During the rainy season, sorghum contributed 41 percent of iron intake in intervention villages and 34 percent in control villages; cultivated and uncultivated green leafy vegetables contributed 6 percent of vitamin C in both groups. Breastmilk was the primary source of vitamin C, contributing about 70 percent in both groups.

Breastmilk, fruits and animal foods were main vitamin A sources for children aged six to 39 months (Figure 6.3). Overall, breastmilk and animal food contributed 61 and 11 percent, respectively, of vitamin A intake in intervention villages, and similar percentages in control villages. Sorghum contributed approximately 1 percent of vitamin A in both groups. During the summer season, fruits contributed 38 percent of vitamin A in intervention villages and 24 percent in control villages. During the rainy season, vegetables, roots and tubers contributed 5 percent of vitamin A in children from intervention villages.

Continued breastfeeding with delayed initiation of complementary feeding is common practice in India. In this study, 75 percent of all children aged six to 39 months were breastfed either exclusively or in combination with complementary food. In the summer season, more than half of the children aged six to 12 months had not yet begun to consume complementary food. Limited supplementation programmes distributing enriched flour, iron-folic tablets, iodized salt and vitamin A drops reached fewer than half of the rural Dalit families in this study.

Requirements for Indian children aged one to three years are estimated to be 5 208 kJ/day of energy and 21 g/day of protein (ICMR, 1990). In this study, 37 percent of children in the summer season and 24 percent in the rainy season were under one year of age, and their energy and protein intakes were below recommendations. Indian dietary standards recommend that young children aged one to three years have an iron intake of 11.5 mg/day (ICMR, 1990); the median

Table 6.6 Nutrient intakes of Dalit children aged six to 39 months from intervention and control villages, by season

| Nutrient | Summer season | | | Rainy season | | |
	Intervention villages (n = 125)	Control villages (n = 108)	p	Intervention villages (n = 124)	Control villages (n = 96)	p
Energy, kJ[a]	3 003 (1 625)	2 646 (1 449)	0.17	3 356 (1 856)	3 230 (1 491)	0.90
	2 780	2 352		2 881	2 940[b]	
	(1 365 4 166)	(1 365 3 734)		(1 835 4 670)	(1 953 4 200)	
Protein, g[a]	15.4 (10.8)	13.9 (10.0)	0.24	18.5 (12.4)	17.2 (10.0)	0.90
	12.0	10.1		14.6[b]	15.5[b]	
	(5.5 22.9)	(5.5 21.6)		(7.8 26.2)	(8.8 22.5)	
Carbohydrates, g	121 (84)	99 (72)	0.06	139 (95)	131 (76)	0.99
	101	78		117	117[b]	
	(37 191)	(35 154)		(58 204)	(67 189)	
Fat, g	17.5 (6.7)	17.7 (6.1)	0.87	17.9 (7.2)	17.9 (6.5)	0.62
	17.0	17.0		17.1	17.8	
	(14.6 20.1)	(14.6 20.0)		(13.7 21.2)	(15.0 21.0)	
Dietary fibre, g	4.7 (6.5)	4.0 (5.7)	0.20	6.7 (8.7)	5.6 (6.8)	0.81
	1.7	0.8		3.4[b]	2.7[b]	
	(0.1 7.4)	(0 6.5)		(0.4 10.6)	(0.5 10.2)	
Iron, mg[a]	2.8 (3.4)	2.5 (3.5)	0.19	3.4 (3.4)	3.3 (4.1)	0.86
	1.5	1.0		2.3[b]	2.2[b]	
	(0.3 4.2)	(0.2 3.9)		(0.7 5.0)	(0.7 4.7)	
Vitamin C, mg[a]	22.8 (14.2)	23.0 (14.1)	0.82	21.4 (16.8)	20.4 (11.8)	0.97
	25.0	25.0		20.0	20.8	
	(17.5 25.0)	(15.5 25.0)		(17.5 25.0)	(17.0 25.0)	
Vitamin A, µg RE[a, c]	248 (229)	215 (172)	0.17	163 (86)	165 (93)	0.64
	205[b]	205		179	182	
	(146 216)	(145 205)		(120 206)	(132 205)	

Figures are means with SD in brackets, or medians with first and third quartiles in brackets. Non-parametric test (Wilcoxon).
[a] Recommended levels for boys and girls aged one to three years (26): energy – 5 208 kJ; protein – 21 g; iron – 11.5 mg; vitamin C – 25 mg; vitamin A – 400 µg RE.
[b] $p \leq 0.05$ statistically significant higher intake during the summer season or the rainy season within group. Breastmilk consumption standardized at 500 ml for children aged six to 12 months, 350 ml for children over 12 months (20): intervention villages – summer season (8), rainy season (11); control villages – summer season (5), rainy season (11) repeated 24-hour recall.
[c] 1µg retinol = 1µg RE, 6 µg provitamin A carotenoids = 1µg RE.
Source: Adapted from Schmid *et al.*, 2007.

intakes of iron for all the children in this study were well below this in both seasons. Most iron came from sorghum, rice and pulses. Sorghum is not recommended as a major food source for children because of its poor digestibility (McLean *et al.*, 1981). As median vitamin C intakes were similar to the Indian RDA of 25 mg/day (ICMR, 1990), it may be assumed that iron is absorbed. However, the Indian vitamin A RDA for

children aged one to three years is 400 µg RE (ICMR, 1990), and median intakes in the study were less than 200 µg RE in the rainy season.

Limitations to the dietary evaluation in this study include likely overestimated intakes of vitamin A because, by necessity, they were derived from the values estimated by colorimetry representing the sum of total carotene, converted into β-carotene (no values were available for α-carotene and β-cryptoxanthin). RE was used instead of the retinol activity equivalent recently recommended for Indian dietary analysis. The finding that the bioconversion of carotene from dark-green leafy vegetables is less than previously thought has raised doubts about the efficacy of green leafy vegetables in improving vitamin A status (Castenmiller and West, 1998); nevertheless, epidemiological evidence from India implies there is good bioavailability of dietary carotenoids, because vitamin A deficiency is rarely seen in communities where many carotene-rich foods are consumed (Tontisirin, Nantel and Bhattacharjee, 2002). In addition, although it is well established that 24-hour recalls are most appropriate for assessing average intakes of food and nutrients for large groups, vitamin A is not easily estimated from this technique (Food and Nutrition Board Institute of Medicine, 2001), nor are the nutrient intakes of young children (Gibson, 1990). A further limitation arises from the assumption that breastmilk intakes were standard for age, which reduced the true variance of nutrient intakes and may have limited the validity of comparisons, particularly in children in the younger age groups.

Mothers' health

The average age of the mothers was 24.3 years (SD = 3.9) (Table 6.7). Some 16 percent of the mothers were less than 145 cm in height, and mean BMI was slightly below the cut-off point for a healthy body weight (BMI $\geq$ 18.5 kg/m²) (James, Ferro-Luzzi and Waterlow, 1988). Overall, 58 percent of mothers were classified as having CED, and 10 percent of these were classified as severely malnourished (BMI < 16 kg/m²); 1 percent were classified as overweight (BMI $\geq$ 25 kg/m²). The majority of women were illiterate (78.6 percent), used open field toilets (94.1 percent), and were lactating at the time of the study (80.9 percent). In addition, 41.4 percent of the women had a household income below the poverty line. The majority of women were characterized as moderately active owing to their work in agricultural fields (70.9 percent). Literate women were less likely to use an open field toilet and

be characterized as active than illiterate women were (Fisher's exact test, $p \leq 0.01$).

In evaluating CED (Table 6.8), mothers with only one child under five years of age had a 45 percent greater risk of CED than those with two or more children under five (rate ratio [RR] = 1.45, $p \leq 0.05$); and mothers without sanitation in their homes had a fourfold greater risk (RR = 4.1, $p \leq 0.05$) of CED. Illiterate and active women were more likely to have CED than literate and non-active women (RR = 1.6 and 1.4, respectively, $p \leq 0.05$), but literacy and activity level were not significant in multivariate analyses including sanitation and number of children under five years of age (not shown). Increasing levels of fat as a percentage of total energy were significantly associated with lower risk of CED (the RR of the lowest 25th percentile compared with that of the 75th percentile or above was 1.6, $p \leq 0.05$); these findings remained significant in multivariate analyses. Intake of pulses (g/day) was also inversely related to CED in univariate and multivariate analyses. Carbohydrate as a percentage of total energy was inversely related to percentage of energy from fat (RR = - 0.96, $p \leq 0.010$), and although positively related to CED in univariate analyses, carbohydrate consumption was not significant in multivariate analyses. Vegetable, root and tuber consumption was inversely related to CED in analyses adjusting for sanitation, having children under five years of age, and pulse intake (g/day) ($p \leq 0.05$). As consumption of vegetables, roots and tubers was significantly related to fat intake, these two variables were not considered in the same model. Intake of pulses (g/day) was highly related to energy intake (RR = 0.48, $p \leq 0.001$). Mothers' total energy intake, age, physiological status and income level were not related to CED in univariate or multivariate analyses. There was too little variability in the percentage of energy from protein to evaluate this variable as a determinant of CED (i.e., the 25th percentile was 9.8 and the 75th percentile 10.9).

At the time of the survey, 16 percent of mothers were suffering from one or more signs of clinical vitamin A deficiency, including night blindness, Bitot's spot and conjunctival xerosis (Table 6.3). Mothers'

Table 6.7 Characteristics of Dalit mothers in intervention and control villages

Variable	Description	Intervention villages (n = 124)	Control villages (n = 96)	p
Age	Years (SD)	24.5 (4.1)	24.0 (3.5)	0.30[a]
Total children per mother	1–2	74 (60%)	67 (70%)	0.30
	3–4	43 (34%)	25 (26%)	
	> 4	7 (6%)	4 (4%)	
Illiterate	Yes	101 (81%)	72 (75%)	0.25
	No	23 (19%)	24 (25%)	
Nutrient supplement in 2003[b]	Yes	51 (41%)	42 (44%)	0.70
	No	73 (59%)	54 (56%)	
Iron-folic tablet in 2003	Yes	28 (23%)	18 (19%)	0.49
	No	96 (77%)	78 (81%)	
Iodized salt usage	Yes	19 (15%)	17 (18%)	0.64
	No	105 (85%)	79 (82%)	
		Summer season (n = 125)	(n = 109)	
Physiological status	Non-lactating, non-pregnant	13 (10%)	9 (8%)	0.30
	Lactating	96 (77%)	92 (85%)	
	Pregnant	6 (5%)	1 (1%)	
	Pregnant and lactating	10 (8%)	7 (6%)	
		Rainy season (n = 124)	(n = 96)	
Physiological status	Non-lactating, non-pregnant	13 (10%)	14 (15%)	0.19
	Lactating	100 (81%)	78 (81%)	
	Pregnant	7 (6%)	1 (1%)	
	Pregnant and lactating	3 (2%)	3 (3%)	
Work as daily labourer	Yes	100 (81%)	55 (57%)	< 0.001[c]
	No	24 (19%)	41 (43%)	
BMI[d]	Body weight (kg)/height (m)²	18.2 (2.0)	18.3 (2.3)	0.71[c]
		Both Seasons (n = 140)	(n = 114)	
Participated in both seasons	Yes	110 (79%)	91 (80%)	0.81
	No	30 (21%)	23 (20%)	

Figures are counts with percentages of n in brackets, or means with SDs in brackets. Chi-square test.
[a] p from two-sample pooled Student t-test.
[b] Corn-soya blend and *muruku* (enriched chickpea flour).
[c] p ≤ 0.05 statistically significant.
[d] Pregnant women were excluded: experimental villages n = 114; control villages, n = 91.
Source: Adapted from Schmid *et al.*, 2007.

age in years and income were positively related to these signs of vitamin A deficiency, with 20 percent of women with incomes above the poverty line having symptoms, compared with 11.0 percent of women below the poverty line (X^2, $p = 0.07$), a difference that became significant in multivariate analyses including the mother's age (Table 6.8). Mothers' physiological status (lactating, pregnant or neither), activity level, number of children under five years of age, sanitation conditions and literacy were unrelated to vitamin A

Table 6.8 Prevalence, RRs and adjusted ORs of correlates of CED and vitamin A deficiency symptoms in Dalit mothers

Variable	Category	No.	%	RR	Adjusted OR (95% CI)
Chronic energy deficiency (BMI < 18.5 kg/m²)					
No. of children ≤ 5 years of age[a]	1	139	64	1.5[b]	2.54 (1.39–4.99)
	> 2	61	44	1.0	
Open field toilet[a]	Yes	194	61	4.1[b]	7.99 (1.66–38.81)
	No	13	15	1.0	
Fat intake[a]	% of energy				
	< 25th percentile	51	75	1.6[b]	2.97 (1.19–7.42)
	26th–74th percentile	104	54	1.1	0.99 (0.47–2.11)
	> 75th percentile	51	47	1.0	
Vegetables, roots and tubers[c]	Days/season	207	-	-	0.99 (0.98–0.99)
Pulses[a,d]	g/day	207	-	-	0.99 (0.98–0.99)
Clinical vitamin A deficiency[e]					
Age[f]	Years	220	-	-	1.12 (1.02–1.23)
Income > poverty line of INR 1 000/month[f]	Yes	91	20	1.8	2.41 (1.02–5.20)
	Below	129	11		
Sorghum consumption[f]	g/day	208	-	-	0.99 (0.99–0.99)
Dairy[g, h]	Low, intermediate, high	220	-	-	0.69 (0.42–1.12)
Night blindness during past pregnancy					
Dairy[g]	Low	56	42.9	2.0[b]	2.74 (1.26–5.94)
	Intermediate	94	39.4	1.8[b]	2.38 (1.18–4.82)
	High	70	21.4	1.0	1.0

[a]   Model includes number of children up to five years of age (one or more), sanitation situation (open field or toilet), fat intake and intake of pulses.
[b] $p \leq 05$.
[c]   Includes 20 species of vegetables, roots and tubers (green leafy vegetables excluded).
[d]   Includes seven pulses: chickpea, black gram, green gram, *khesaridal*, lentil, pigeon pea and dry pea.
[e]   Bitot's spot, conjunctival xerosis and/or night blindness.
[f]   Model includes age, income (above or below the poverty line) and sorghum intake.
[g]   Includes eggs, cow and buffalo milk, curd, butter milk and ghee.
[h]   Model includes age, income (above or below the poverty line) and dairy intake.
Source: Adapted from Schmid *et al.*, 2007.

deficiency in univariate and multivariate analyses. Analyses with dietary exposure showed that sorghum consumption was significantly and inversely related to vitamin A deficiency (β = -.004, SE =.002, p =.04), and the consumption of dairy products (coded as low, intermediate or high consumption of eggs, cow and buffalo milk, curd and ghee) was protective (but not statistically significant) against it (β = - 0.420, SE = 0.257, p = 0.10) in analyses adjusting for mother's age and income level. Sorghum consumption was also positively related to iron intake (RR = 0.49, $p \leq$ 0.001), and iron intake was positively related to intake of vitamin A (RR = 0.34, p < 0.001) and vitamin C (RR = 0.53, $p \leq$ 0.001) (not shown). No nutritional variables other than sorghum and dairy products were significant or of border-line significance in univariate or multivariate analyses.

Approximately one-third (35 percent) of the women reported having experienced symptoms of night blindness during their pregnancies. Of these, 75 percent said it occurred in the last trimester. This is of concern given the WHO recommendation that a prevalence of 5 percent of pregnant women with night blindness be considered of public health significance (Christian, 2002).

Mothers' age was positively related to night blindness during pregnancy, and consumption of eggs, milk and milk products was negatively associated. Women consuming low, intermediate or high amounts of dairy products had prevalence rates of night blindness during pregnancy of 42.9, 39.4 and 21.4 percent, respectively (providing RRs of 2.0 and 1.8, with the highest consumption group serving as the referent). These findings were significant in age-adjusted analyses (Table 6.8). No other nutritional variables were identified as significant determinants of night blindness during pregnancy.

Some 10 percent of women were classified as iron-deficient, based on pallor under the eyelid (Table 6.3). In univariate analyses, income, physiological status, literacy, activity level and number of children under five years of age were not significantly related to iron status. In multivariate analyses, mothers' age in years ($\beta = 0.18$, SE $= 0.06$, $p \leq 0.01$), physiological status ($\beta = 2.5$, SE $= 1.3$, $p \leq 0.05$) and activity level ($\beta = -1.1$, SE $= 0.58$, $p \leq 0.06$) were determinants of iron deficiency. No nutritional variables were identified as determinants of iron deficiency.

Correlates of diet with women's health

After controlling for important correlates of CED, including age, and adjusting for number of children under five years of age and sanitation situation, women's intake of traditional food (fat, pulses, and vegetables, roots and tubers) showed protective associations against CED. Paradoxically, women with only one child had higher prevalence of CED, perhaps reflecting higher infertility in the most malnourished women.

Women's consumption of sorghum showed protective association for clinical vitamin A deficiency symptoms after adjusting for age and income. This may be because sorghum (as *roti*) is consumed with vegetables and fat. Mothers with higher fat and pulse intakes and more frequent consumption of vegetables, roots and tubers had odds ratios (ORs) below 1 for CED.

For women during their last pregnancy and at the time of the survey, higher consumption of dairy products (buffalo and cow milk, curd and buttermilk) was protective against occurrence of night blindness, and no other variables were significant. The percentages of mothers suffering night blindness, Bitot's spot and/or conjunctival xerosis at the time of the survey, and especially the prevalence of night blindness during the last pregnancy, were higher than those reported in other Indian data (ICMR, 1990).

Other studies have shown that consumption of traditional foods improves vitamin A status. Among Bangladeshi men with low vitamin A diets, daily consumption of cooked and pureed Indian spinach (*Basella alba*) had a positive effect on vitamin A stores (Haskell *et al.*, 2004). In this study of Dalit women, the amount of sorghum consumed every day was negatively associated with clinical vitamin A deficiency symptoms; and low frequencies of egg, milk and milk product consumption were positively associated with night blindness during pregnancy. The surprising finding that mothers from households with incomes above the poverty line were at greater risk of clinical vitamin A deficiency than those with lower incomes may be explained by the higher-income households' lower intakes of traditional foods such as sorghum and the green vegetable dishes with which it is usually consumed.

WHO reported that 87 percent of pregnant women in India were suffering from iron deficiency in 1995 (WHO, 2000). Iron deficiency data distinguishing between moderate (7.0 to 9.9 g Hb/dl) and severe (< 7.0 g Hb/dl) deficiency reported prevalence rates in married or previously married rural women aged 15 to 45 years in Andhra Pradesh of 16 percent moderate deficiency and 2 percent severe, which were similar to findings from women in scheduled caste (IIPS and ORC Macro, 2000). The current study identified 10 percent of mothers as iron-deficient, using the subjective measure of eyelid pallor, but – as in other studies – this study was unable to support the hypothesis that nutritional variables are correlates of iron status.

Anthropometry and clinical data were collected during only the rainy season, whereas dietary data were collected in both the summer season and the rainy season and averaged between the two for data analysis. Ideally, data on seasonal variations in women's health and diet would have been collected for all the three major seasons of the Dalit annual cycle in Medak District, and averaged to give an annual estimation. Despite this constraint, however, the finding that consumption of traditional dietary items has a positive effect during the season of greatest health risk is important.

Dalit women are probably the most disadvantaged of Indian adults. Prevalence rates of CED and night blindness during pregnancy were higher among Dalit mothers in rural Medak District than the national data reported for rural women from scheduled castes. However, the consumption of traditional Dalit food items – including sorghum, pulses, vegetables, roots, tubers, eggs, milk and milk products – was negatively associated with the prevalence of CED and clinical vitamin A deficiency symptoms in the women in this study. Mothers from APDS villages had higher energy, protein, dietary fibre and iron intakes than mothers from control villages. Mothers in all study villages had mean energy and protein intakes above, and median iron, vitamin C and vitamin A intakes below the recommendations, in both study seasons. Despite the assumed higher energy needs during the rainy season, energy and nutrient intakes were higher during the summer season, confirming that food is scarce during the rainy season. For mothers, traditional food items including sorghum, pulses and green leafy vegetables were major sources of energy, protein, iron, vitamin C and vitamin A. Uncultivated green leafy vegetables were a particularly important source of vitamin A in the intervention villages during the rainy season.

Children's health

Anthropometry and clinical signs of vitamin A and iron deficiency are summarized in Table 6.9. Based on the 1977 National Center for Health Statistics growth curves, mild and severe stunting were reported in more than 30 percent of children; mild and severe wasting in more than 50 percent; and underweight in about two-thirds (National Center for Health Statistics,

1977). Clinical signs of vitamin A deficiency were seen in 4 percent of children, and pallor in the lower eyelid in 8 percent. These measures were addressed with multivariate models using the variables of interest for children shown in Table 6.2. ORs for the determinants of stunting, wasting, underweight and iron deficiency (pallor) are shown in Table 6.10. Children aged 25 to 39 months were at highest risk of stunting, followed by those aged 13 to 24 months, who were at greatest risk of underweight. Young boys were at higher risk than young girls. Children in houses classified as "permanent" were at higher risk than those in traditional huts (with earthen floors and plant materials for walls and roofs). Exclusive breastfeeding after six months also presented higher risk of wasting and symptoms of iron deficiency (pallor) (Schmid, 2005).

From 1998 to 1999, the second National Family Health Survey collected data on nutritional status in Indian children aged six to 35 months (IIPS and ORC Macro, 2000). It reported slightly greater percentages of severe (24 percent) and mild (24 percent) stunting, and much lower percentages of severe (3 percent) and mild (13 percent) wasting than found by the current study. Another survey of children aged one to five years from scheduled castes in rural Andhra Pradesh (National Nutrition Monitoring Bureau, 2002) reported more severe stunting (26 percent) among children, but less severe underweight (23 percent) and severe wasting (3 percent). Similar proportions of children were reported stunted in the current and the Andhra Pradesh studies, but this study reported higher percentages of severely wasted and underweight children.

It is well established that the prevalence of stunting increases in children up to 24 or 36 months, and then tends to level off (WHO Working Group, 1986). In the Andhra Pradesh study, the proportion of children who were underweight (< 2 SD) increased rapidly with age from 12 to 23 months (IIPS and ORC Macro, 2000). This is reflected in the current study's findings, where the highest risk of stunting was in children aged 25 to 39 months and the highest risk of underweight in those aged 13 to 24 months.

Protein-energy malnutrition may result from not only poor food supply and early growth faltering,

Table 6.9 Anthropometric measurements and the prevalence of clinical vitamin A deficiency symptoms and iron deficiency (pallor) in Dalit children aged six to 39 months

Variable	Description	Children (n = 220)
Anthropometric measurement		
MUAC	> 13.5 cm	110 (50%)
	12.5–13.5 cm	76 (36%)
	< 12.5 cm	34 (15%)
Stunted*	Normal (≤ -2 SD height-for-age Z-scores)	148 (67%)
	Mild (> -2 SD height-for-age Z-scores)	44 (20%)
	Severe (> -3 SD height-for-age Z-scores)	28 (13%)
Wasted*	Normal (≤ -2 SD weight-for-height Z-scores)	105 (48%)
	Mild (> -2 SD weight-for-height Z-scores)	55 (25%)
	Severe (> -3 SD weight-for-height Z-scores)	59 (27%)
Underweight*	Normal (≤ -2 SD weight-for-age Z-scores)	81 (37%)
	Mild (> -2 SD weight-for-age Z-scores)	49 (22%)
	Severe (> -3 SD weight-for-age Z-scores)	90 (41%)
Clinical vitamin A deficiency		
None		212 (96%)
Night blindness (XN)	Reported	6 (3%)
Bitot's spot (X1B)	Examined	0 (0%)
Conjunctival xerosis (X1A)	Examined	2 (1%)
Iron deficiency	White (pallor)	18 (8%)
Inside lower eyelid	Pink	202 (92%)

Figures are counts with percentages in brackets from data collected during the rainy season of 2003.
* Based on 1977 NCHS growth curves.
Source: Adapted from Schmid *et al.*, 2007.

but also low birth weight, inappropriate feeding practices and high morbidity rates (Gopaldas, Patel and Bakshi, 1988; Bhandari *et al.*, 2001). In India, delayed introduction of complementary foods, the use of foods with low energy and nutrient density, small servings at meals, and food restrictions due to cultural beliefs are all common (Bhandari *et al.*, 2004). The study results show that predominantly breastfed children aged six months and more had higher risks of wasting and iron deficiency. It is suggested that other factors, including the availability of health care facilities and a protected water supply, personal hygiene and environmental sanitation are also important determinants of nutrition status in preschool children (Laxmaiah *et al.*, 2002). The study found that children living in permanent houses were at greater risk of stunting than those living in traditional huts. This surprising finding might be explained by the more frequent consumption of traditional foods such as sorghum, pulses and wild fruits by children living in rural households with incomes below the poverty line.

The National Nutrition Monitoring Bureau (2002) reported absence of night blindness in rural preschool children in Andhra Pradesh and prevalence rates of 0.2 percent for conjuctival xerosis and 0.7 percent for Bitot's spot. A survey conducted in five states reported prevalence of 0.7 to 2.2 percent for Bitot's spot in children aged six to 71 months. Night blindness was reported in 1.6 to 4.0 percent of children aged 24 to 71 months (Chakravarty and Sinha, 2002). The current study reported higher percentages of children with night blindness and conjunctival xerosis than those reported

Table 6.10 Adjusted ORs and 95 percent CIs of significant determinants for stunting, wasting, underweight and iron deficiency (pallor) in Dalit children aged six to 39 months

Variable	Category	n = 220	%	OR (95% CI)
Stunting (< 2SD height-for-age)				
Age group	6–12 months	53	8%	1.00 (reference)
	13–24 months	83	33%	8.46[a] (1.93–37.14)
	25–39 months	84	48%	23.86[b] (4.50–126.41)
Gender	Female	110	25%	1.00 (reference)
	Male	110	41%	2.10[a] (1.08–4.09)
Housing	Traditional hut	64	19%	1.00 (reference)
	Permanent house[c]	156	39%	2.43[a] (1.07–5.54)
Wasting (< 2SD weight-for-height)				
Duration of exclusive breastfeeding (including water)	≥ 6 months	156	58%	1.98[a] (1.00–3.90)
	< 6 months	64	39%	1.00 (reference)
Underweight (< 2SD weight-for-age)				
Age group	6–12 months	53	40%	1.00 (reference)
	13–24 months	83	80%	3.67[a] (1.12–12.03)
	25–39 months	84	62%	1.27 (0.31–5.18)
Energy intake	≥ 1 220 kcal/day	38	42%	1.00 (reference)
	< 1 220 kcal/day	182	68%	3.33[a] (1.13–9.85)
Iron deficiency (under eyelid pallor)				
Duration of exclusive breastfeeding (including water)	≥ 6 months	156	10%	10.16[a] (1.05–98.2)
	< 6 months	65	2%	1.00 (reference)

Stunting, wasting and underweight were calculated from the NCHS growth curves (1977). Nutrient intake and food frequency consumption were obtained during the rainy season of 2003.
[a] $p \leq 0.05$.
[b] $p \leq 0.01$.
[c] House with or without a permanent roof.
Source: Adapted from Schmid *et al.*, 2007.

for Andhra Pradesh, but its results were similar to those from other smaller studies. More than 50 percent of the children did not receive vitamin A supplementation, which is normally provided by the Indian Government, so breastmilk and complementary foods were the primary sources of vitamin A. The second National Family Health Survey (IIPS and ORC Macro, 2000) reported that 46 percent of Indian children aged six to 35 months were moderately and 5 percent severely iron-deficient. This study found 8 percent of children to be iron-deficient.

Young children in both APDS and control villages had median energy, protein, iron and vitamin A intakes that were below recommendations, putting them at risk of malnutrition. No differences were seen between the two groups. Breastmilk was a major source of energy, protein and vitamin A, and traditional food items, including sorghum and pulses, were important sources of energy, protein and iron.

Conclusions

This study examined the highly diverse food system of the Zaheerabad Dalit to identify its significance for Dalit women and children according to whether they were or were not participating in a food intervention programme that promoted knowledge about and use of local food systems. The great diversity of Dalit food systems in rural India has still to be studied and compared with non-Dalit food systems in the same ecosystem, to develop understanding of and remedies for social injustice and disparities. In the meantime, existing studies already indicate the potential for improving health by promoting local traditional foods, as in the APDS applied by DDS.

The study found that traditional cultural food items were widely consumed and were the main sources of energy, protein, iron, vitamin C and vitamin A for both mothers and young children in all the study villages. However, mothers in APDS villages, which supported the traditional food system through the promotion of indigenous agricultural practices, had higher energy, protein and iron intakes than mothers in control villages. These findings provide evidence for evaluating, considering and promoting traditional food systems as a first step to increasing the intakes of critical nutrients in poor rural communities in India.

The prevalence rates of CED and night blindness during pregnancy were higher among Dalit mothers in rural Medak District than the national average reported for rural women from scheduled castes. The consumption of traditional food items including sorghum, pulses, vegetables, roots and tubers, eggs, milk and milk products was negatively associated with the prevalence of CED and clinical vitamin A deficiency symptoms in these women.

Severe stunting and wasting were important problems among preschool children and were found to be more prevalent than the usual for rural Andhra Pradesh. Above-recommended energy intakes were associated with lower prevalence of underweight among children. While there were no differences between children from intervention and control villages, the findings clearly show the importance of and potential for using the traditional food system and its dietary diversification to combat malnutrition.

When poor, rural Dalit mothers have access to agricultural land for home food production and to work as field labourers both they and their children have greater intakes of many foods that protect against chronic energy and protein deficiency and micronutrient deficiencies of vitamin A and iron. Sorghum, millet, wild fruits and uncultivated greens are of particular importance in this. These and many other foods –particularly those of animal source – in the traditional Dalit food system should be promoted to improve nutrition status. High frequencies of CED, night blindness and clinical vitamin A deficiency indicate a need for public health interventions for all Dalit women, which should include the promotion of locally available, traditional food for the women and their children ❋

Acknowledgements

The authors would like to thank the following individuals for their contributions to this study: the women *sangham* leaders of DDS, Pastapur, Medak District, Andhra Pradesh, India; the DDS *sangham* members who participated in the study; P.V. Satheesh, Director, DDS; the interviewers Kavitha, Uma, Raji, Sarita, Sesi, Jaya, Sheelpa, Fatima, Rupa and Anu; Nagraj, data operator, DDS, Krishi Vigyan Kendrea; Anwer, Pentappa and Tuljaram, DDS staff; and all staff members of Krishi Vigyan Kendra and DDS.

> **Comments to:** salomeyesudas@hotmail.com

>> Photographic section **p. VII**

Chapter 7

Gwich'in traditional food and health in Tetlit Zheh, Northwest Territories, Canada: phase II

🌸 HARRIET V. KUHNLEIN[1] 🌸 LAUREN GOODMAN[1] 🌸 OLIVIER RECEVEUR[2]

🌸 DINA SPIGELSKI[1] 🌸 NELIDA DURAN[3]

🌸 GAIL G. HARRISON[3] 🌸 BILL ERASMUS[1, 4] 🌸 TETLIT ZHEH COMMUNITY[5]

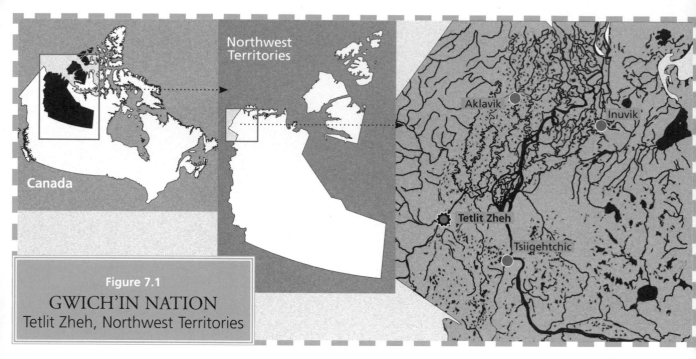

Figure 7.1

GWICH'IN NATION
Tetlit Zheh, Northwest Territories

Northwest Territories

Canada

Aklavik

Inuvik

Tetlit Zheh

Tsiigehtchic

Data from ESRI Global GIS, 2006
Walter Hitschfield
Geographic Information Centre
McGill University Library

1
Centre for Indigenous
Peoples' Nutrition and
Environment (CINE)
and School of Dietetics
and Human Nutrition,
McGill University,
Montreal, Quebec,
Canada

2
Department of Nutrition,
University of Montreal,
Montreal, Quebec,
Canada

3
School of Public Health,
University of California,
Los Angeles, California,
United States of America

4
Dene Nation and
Assembly of First Nations,
Yellowknife and Ottawa,
Canada

5
The Community of
Tetlit Zheh,
Northwest Territories,
Canada

Key words > Indigenous Peoples,
traditional food, Gwich'in, Dene Nation, First Nations,
Northwest Territories, food security

Photographic section >> X

Abstract

The First Nations Gwich'in community of Tetlit Zheh in the Northwest Territories of Canada has been undergoing a nutrition transition. Studies conducted in the mid-1990s indicated that the majority of the Gwich'in diet consisted of store-bought (market) food, a high proportion of which was calorie-rich but nutrient-poor. As part of the Indigenous Peoples' Food Systems for Health Program, Tetlit Zheh agreed to participate in activities to increase the consumption of traditional (local) food and healthier market food.

Pre-intervention assessment was carried out in winter (February to March) 2006 among youth aged ten to 15 years and young women aged 20 to 40 years. Compared with the overall Canadian population, data indicated a similar proportion of overweight/obese youth, but a greater proportion of overweight/obese women. Compared with overall Canadian youth, Tetlit Zheh youth spent similar amounts of their leisure time with television or computers. The majority of women were assessed as moderately active, and youth reported having participated in a wide range of physical activities throughout the year.

The most important traditional food species consumed by youth and women were caribou, moose and whitefish. The majority of both youth and women consumed at least one traditional food item regularly. Post-intervention activity assessments were not conducted because external forces precluded the documentation of behaviour and food consumption change.

Climate change and other factors that reduced access to traditional food species, and a sharp increase in market food and fuel prices (2008) were important challenges.

Introduction

Similar to many other indigenous groups worldwide, the Dene and Métis of northern Canada have witnessed many diverse changes in their recent history. Among the most dramatic have been transformations in their ecosystem, food use and lifestyle as they strive for a balance between traditional and modern practices. Before colonial contact, the Dene relied solely on traditional foods harvested from their local environment. Since the turn of the twentieth century, the overwhelming influence of the south has resulted in increased reliance on commercial market foods for Dene/Métis communities (Kuhnlein and Receveur, 1996; Receveur, Boulay and Kuhnlein, 1997; Kuhnlein *et al.*, 2004; Nakano *et al.*, 2005a; 2005b).

This nutrition transition has been marked by a well-documented loss of dietary quality among the Dene, as many market foods of relatively poor nutritional value are replacing nutrient-dense traditional foods (Receveur, Boulay and Kuhnlein, 1997; Kuhnlein *et al.*, 2004; Nakano *et al.*, 2005a; 2005b). Lifestyle patterns have also been interrupted, as many traditional methods of trapping, hunting and fishing are becoming less common. These rapid cultural changes have led to declining health status (Kuhnlein *et al.*, 2004). A sharp increase of many non-communicable diseases, including obesity, diabetes and cardiovascular disease, has emerged as a public health concern, with prevalence now disproportionately high among Canadian Indigenous Peoples compared with the national population (Health

Canada, 2001; 2003; Statistics Canada, 2003; First Nations Regional Longitudinal Health Survey, First Nations Centre, 2005).

At the same time, special concern has been expressed about climate change and its many effects on the quality of the local environment. In addition, elders have been concerned about a decrease in transmission to youth of traditional knowledge about the environment and how to use it effectively. Climate change has exacerbated the nutrition transition for Arctic peoples because it has a direct impact on the availability of local traditional food species. In 2004, the Canadian Government expressed the need for research on how climate change affects the sustainability, health, safety and food security of northern communities (Natural Resources Canada, 2004). Qualitative research among Inuit residing in Nunavik and Labrador documented potential direct and indirect climate-related health impacts. Inuit perceptions of climate-related health impacts were consistent with the Intergovernmental Panel on Climate Change Fourth Assessment Report and the Arctic Climate Impact Assessment, with additional insight into the negative impact of decreased access to traditional food on social and cultural values (Furgal and Seguin, 2006), and the need for coping and adaptation strategies. However, there has been little documentation on the impact of climate change and the adaptation to it by western Arctic Indigenous Peoples. Guyot *et al.* (2006) reported observations in two Yukon First Nations that demonstrated changes in water levels and species availability and the need to adopt new traditional food harvest strategies.

As the Dene continue to undergo cultural transformation, a balanced diet based on both traditional foods and healthy market foods will be important for supporting health. Lifestyles that incorporate physical activity also make significant contributions to well-being. Taking a proactive stance, the Dene have called for research to improve understanding of environmental change and to combat the negative effects of acculturation. This chapter addresses data, discussions and conclusions from the Gwich'in community of Tetlit Zheh of the Dene Nation.

Context

Totalling approximately 5 000 people, the Gwich'in First Nations live in communities across the northern interior of Alaska in the United States of America, and the Northwest Territories (NWT) and northern Yukon in Canada. The Gwich'in community of Tetlit Zheh (Fort McPherson) is located in the Gwich'in Settlement Area (Figure 7.1 on p. 102), which is on the east bank of the Peel River and in the Richardson Mountain range. Via the Dempster Highway, the community is accessible year round, apart from during the spring ice break-up and autumn freeze-up on the Peel and Mackenzie Rivers. About 800 people reside in Tetlit Zheh, and the majority are of indigenous descent. Community members continue to speak their traditional Gwich'in language dialect as well as English. The income of community residents has been documented as low, with an average family income of CAD 61 348 and 28.6 percent of households living on less than CAD 25 000 a year (NWT Bureau of Statistics, 2007). By contrast, during the same period, average family income in NWT – a region with high costs of living – was CAD 101 622, with 14.3 percent of families living on less than CAD 25 000. The Canadian average family income in 2007 was CAD 88 300.[1]

The Dene National Office and the Assembly of First Nations Regional Office are located in Yellowknife, NWT. The annual Dene National Assembly convenes Dene First Nations from the five regions of the Dene Nation: Akaitcho, Dehcho, Gwich'in, Sahtu and Tlicho. Concerns arising from land-use, resources, lifestyle and health issues have been addressed at annual assemblies.

As did all Dene, the Gwich'in traditionally led a nomadic subsistence lifestyle of hunting, fishing and gathering, which started to change in the mid-nineteenth century when a trading post was established. Small communities began to emerge, and the Gwich'in population settled year round in the 1960s. The Gwich'in retain extensive knowledge of their traditional food system, which consists of 75 to 100 species

---

[1] www.statcan.gc.ca/tables-tableaux/sum-som/l01/ind01/l3_3868_2812-eng.htm?hili_famil21.

of animals, fish and food plants. Approximately 60 percent of households in Tetlit Zheh were shown to consume most or all of their meat or fish as traditional food (NWT Bureau of Statistics, 2007). Of particular importance is the caribou subspecies, Porcupine caribou (*Rangifer tarandus granti*), which was recently documented as making significant contributions to the diet (Kuhnlein *et al.*, 2009).

Owing to southern influences, the majority of the Tetlit Gwich'in diet in recent years has been shown to consist of market food, with adults consuming an average of 33 percent of dietary energy as traditional food, and children aged ten to 12 years about 6 percent (Receveur, Boulay and Kuhnlein, 1997; Nakano *et al.*, 2005b). In 2006, two grocery stores in the community offered an array of food shipped from southern cities, including many convenient and processed options. Because of the high shipping costs, market food costs in Tetlit Zheh were more than 50 percent higher than in southern Canadian areas. In addition to the high costs, the lack of variety, availability and quality resulting from long-distance shipping can also be a barrier to purchasing food in northern communities (Ladouceur and Hill, 2001; Lawn and Harvey, 2001; 2003; 2004a; 2004b; Chan *et al.*, 2006; Skinner, Hanning and Tsuji, 2006).

Recently, the extreme escalation of food prices on global markets has exacerbated the high costs of purchased foods in northern Canada, and human rights aspects of the rising global food costs resulting from biofuel production and commodity and energy trading have been documented (FAO, 2008). The impact of surging food prices has forced millions of people into increased poverty and hunger, with the risk of health consequences, particularly for women and children (Shrimpton, Prudhon and Engesveen, 2009).

As traditional food use and lifestyle patterns decline, the Gwich'in are experiencing increases in many non-communicable health conditions influenced by suboptimal nutrition, as mirrored among other indigenous groups in northern Canada. Poor nutrition and high prevalence of obesity were also observed among northern indigenous children (Nakano *et al.*, 2005a).

Beginning in the mid-1990s, the Tetlit Gwich'in partnered the Centre for Indigenous Peoples' Nutrition and Environment (CINE) to research their traditional food system and the benefits and risks involved in consuming traditional foods (Kuhnlein *et al.*, 2006; 2004; Kuhnlein and Receveur, 2007; Lambden, Receveur and Kuhnlein, 2007; Lambden *et al.*, 2006; Receveur, Boulay and Kuhnlein, 1997; Receveur *et al.*, 1996). In 2000, this partnership continued with a study assessing dietary patterns among Dene children (Nakano *et al.*, 2005a; 2005b). In 2005, further cooperation led to Tetlit Zheh's participation in CINE's Indigenous Peoples' Food Systems for Health Program (FAO, 2009), which set in motion the project described in this chapter, with activities to promote health by improving access to traditional food and better-quality market food.

Objectives

The aim of the project reported in this chapter was to increase understanding of the food use and health status of the Tetlit Gwich'in and of how recent ecosystem changes have influenced access to traditional food. Activities were developed to increase awareness about local food resources, with particular focus on youth aged ten to 15 years and young women aged 20 to 40 years, as sentinels of the community's future.

In 2006, assessment of youth included measurement of anthropometry, physical activity and dietary parameters; assessment of women included measurement of anthropometry, physical activity, employment status, frequency of traditional food use, dietary self-efficacy, food security status and socio-cultural experiences, including food preferences and the accessibility/availability of food. Adult men and women were interviewed for their perceptions of the impacts of climate change on access to traditional food and of escalating food prices on access to healthy market food. All assessment activities were devised to enhance capacities and provide a foundation for community-driven activities to improve access to healthy food. A summary of results was returned to community leaders in June 2006, and suggestions for improving the use of traditional food, promoting purchases of fruits, vegetables and healthy beverages, and increasing physical activity were discussed.

The original aim was to complete before-and-after assessments following a multi-activity intervention programme guided by the community. However, two formidable external factors precluded this: the impacts of climate change and other factors, which seriously reduced access to traditional food species; and the highly publicized steep increases in food and fuel prices in 2006 to 2008, due to global forces. Although several excellent education activities were delivered within Tetlit Zheh, the community agreed that post-intervention activity assessments would not be carried out, because programme impacts would be imperceptible in the face of these external factors. This chapter describes the findings from the pre-intervention assessment in 2006 and perceptions about the impacts of climate change and high food prices. It closes with a description and discussion of the education activities undertaken to date, and possibilities for the future.

Methods and measurements

The project was encouraged by the Dene Nation and approved by the Tetlit Zheh Council as part of the activities of the Tl'oondih Healing Society. Approval for the research was granted by the Human Research Ethics Committee of the Faculty of Agricultural and Environmental Sciences at McGill University, Montreal, Canada, and a research licence was obtained from the Aurora Research Institute at Aurora College in Inuvik, NWT. All subjects gave their informed consent to participate.

CINE researchers and local community organizations collaborated to collect data on anthropometric, dietary and health indices of women and youth during the winter (January to February) of 2006. All community-resident Gwich'in women aged 20 to 40 years and youth aged ten to 15 years were invited to participate. Women were asked to attend the assessment at the research station, and youth were evaluated at their schools.

The diets of women and youth were assessed using the 24-hour dietary recall research tool. Participants were asked to recall all the food items they had eaten during the previous 24 hours. Measurement aids included cups, plates, bowls and food models; local food products were used as references to facilitate the quantification of all food items. A second 24-hour recall was requested of all participants, but administered to only 20 percent of women and 89 percent of youth on non-consecutive days. When available, two recalls from one individual were averaged for analyses, to maximize the dietary information provided.

To assess micronutrient intake, youth were divided according to dietary reference intake (DRI) categories based on gender and age. Adjusted median micronutrient intake values were determined using the Beaton adjustment technique (Beaton *et al.*, 1979). Unadjusted median values were reported when the group's intra-individual variation was larger than the inter-group variation. Where adjusted median intakes were possible, nutrient intakes were compared with the corresponding estimated average requirement (EAR) and adequate intake (AI), to determine the percentage of individuals falling below recommendations. The mean micronutrient, energy and macronutrient intakes of traditional food consumers were compared with those of non-consumers as recorded in the 24-hour recalls. A participant was defined as a traditional food consumer if she/he had consumed any traditional food (excluding bannock) on at least one recall day. Wilcoxon rank tests were performed to analyse the differences between consumers of traditional food and consumers who did not mention traditional food in dietary recalls. The Wilcoxon test was used because it is non-parametric and does not require data to be normally distributed, and because there were small sample sizes in some groups. Food group servings were derived from 24-hour recalls recorded from individuals.

Women also completed a traditional food frequency questionnaire to assess the consumption of traditional foods over the previous three winter months of November, December and January, when the community's traditional food consumption is known to be at its lowest level. A total of 58 traditional food species, including fish and sea mammals, land animals, birds and plant species, and various animal parts and organs, were included. The questionnaire asked participants to report the number of days each week that they had consumed the various traditional food items over the three months.

Different methods were used to assess the physical activity status of participating women and youth. Women's physical activity was measured with a modified version of the International Physical Activity Questionnaire (IPAQ, 2001; Craig *et al.*, 2003). Participants were asked to report the number of days in the last seven in which they had spent at least ten minutes being vigorously active, moderately active, walking and being sedentary, and how many minutes in one day they would spend on each of these activities. Based on these patterns, women were categorized into three levels of physical activity. Youth completed a physical activity questionnaire designed to capture their levels of activity and their television/video game/ Internet habits (Adams *et al.*, 2005). Types of activity and frequencies were recorded.

Anthropometric and clinical measurements included height, weight, waist circumference, blood pressure and body composition. Appropriate cut-offs were determined for body mass index (BMI), waist circumference, percentage of body fat and blood pressure of women and youth. All measurements were conducted by a trained staff member. Height was measured using a portable height rod with a horizontal headboard attachment; participants removed their shoes and stood as tall and straight as possible, keeping their heads level and their shoulders and upper arms relaxed at their sides. Height was measured at the maximum point of inhalation, repeated three times to a precision of 0.1 cm.

Waist circumference was taken with a flexible tape measure after participants had removed loose and bulky clothing. Clothes pins secured their shirts for access to the abdominal area. Participants stood straight with their arms relaxed at their sides and their feet together. The measuring tape was looped around the participant's waist at the midpoint between the hip and the bottom of the rib cage. Three consecutive measures were taken at the end of normal exhalations, and recorded with a precision of 0.1 cm.

Both weight and body composition were measured with a Tanita Bioelectrical Impedance Scale. Bulky clothing, shoes and socks were removed and a clothing reference weight of 0.5 kg was entered for each subject and automatically subtracted to provide the body weight. Gender, age, height and a standard build reference were entered for subsequent body composition calculations.

Blood pressure was measured as mmHg with a mercury sphygmomanometer and stethoscope, after ensuring participants were relaxed. Outer and tight-fitting clothing around the arm was removed by the participant, and an appropriately sized cuff was wrapped around the upper arm. Three measurements were taken one minute apart, and recorded to the nearest 1 mmHg. As part of the clinical measures, a short (ten-item) questionnaire was used to create a five-point scale of women's self-perceived health status.

CINE researchers developed a socio-cultural questionnaire for gathering women's views on the general use, accessibility, advantages, preferences and health benefits of traditional and market foods. The food security status of participating women and their households was evaluated with the Food Security Survey Module designed by the United States Department of Agriculture (USDA) for the Food Mail Project (Bickel *et al.*, 2000; Lawn and Harvey, 2003). This 18-item questionnaire evaluates households' food security situation for the previous 12 months. By combining a ten-item scale to measure the experiences of adults in the household and an eight-item scale for children under 18 years of age, the questionnaire provides a single measurement of overall food security. Correlations between women's food security status and frequency of traditional food use and the access score derived from the socio-cultural questionnaires were evaluated.

Interviews on perceptions of the impacts of climate change on access to traditional and market foods were conducted with men and women from Dene First Nations communities attending the 38th Dene National Assembly held in Fort McPherson in July 2008, and in Yellowknife, NWT, the following month. The interviews used 15 open- and closed-ended questions to explore whether individuals had enough traditional food, whether healthy foods could be purchased, and respondents' perceptions of how climate change and climate variation affected the use of traditional foods and people's health.

CANDAT (Godin London Inc., 2007) was used to assign nutrient values to all foods from the 24-hour recall data. Nutrient values for traditional foods were derived from a traditional Arctic food database created from laboratory analyses conducted over several years at CINE (Appavoo, Kubow and Kuhnlein, 1991; Kuhnlein *et al.*, 1994; 2002; 2006; Kuhnlein, 2001; Morrison and Kuhnlein, 1993).

There were no missing nutrient values in the foods included in the analyses. Anthropometric, clinical and interview data were analysed using SAS version 9.0 (SAS Institute Inc., 2003). A p-value of ≤ 0.05 was considered significant. Interview responses on traditional food access and climate change were summarized and tabulated.

Table 7.1 Characteristics of participating youth and women in a Gwich'in community

Age group	Youth (n = 65)		Women (n = 53)	
Characteristic	n[a]	Value	n[a]	Value
Age years (mean ± SD)	63	12.7 ± 1.7	53	30.2 ± 5.9
Height m (mean ± SD)	65	154.2 ± 10.1	48	1.6 ± 0.0
Weight kg (mean ± SD)	64	48.4 ± 13.5	46	74.9 ± 13.7
Body mass index				
kg/m² (mean ± SD)	62	20.0 ± 3.5		
BMI (> 85th percentile) kg/m²	15/62	24.2		
Normal (18.5–24.9 kg/m²) % of n			46	23.9
Overweight (25.0–29.9 kg/m²) % of n				30.4
Obese (≥ 30.0 kg/m²) % of n				45.7
Blood pressure				
mmHg (mean ± SD)	64	108 ± 9/69 ± 8		
Blood pressure (> 90th percentile) mmHg	14/62	22.6		
Normal (≤ 120/≤ 80 mmHg) % of n			47	85.1
Pre-hypertension (120–139/80–89 mmHg) % of n				14.9
Hypertension (≥ 140/≥ 90 mmHg) % of n				0.0
Waist circumference				
cm (mean ± SD)	56	73.6 ± 10.8		
Waist circumference (> 85th percentile) cm	11/54	20.4		
Not at risk (< 80.0 cm) % of n			41	4.9
Increased risk (≥ 80.0 cm) % of n				7.3
Substantially increased risk (≥ 88.0 cm) % of n				87.8
Body fat[b]				
% (mean ± SD)	57	19.7 ± 7.5		
Body fat (> 85th percentile) %	10/56	17.9		
Normal (21.0–32.0%) % of n			42	14.3
Overweight (33.0–38.0%) % of n				28.6
Obese (≥ 39.0%) % of n				57.1

[a] Not all participants completed all aspects of the study.
[b] From Gallagher *et al.*, 2000.

Findings

Description of participants

Table 7.1 describes the youth and women participating in the study. Households ranged in size from one to eight people, with an average of four. On average, only one person in each household was employed, either full- or part-time (data not shown). The mean ages were 13 years for youth and 30 years for women. Of the youth measured, 23 percent had systolic or diastolic pressure at or above the 90th percentile, according to the blood pressure reference percentiles for children based on height-for-gender and -age (National High Blood Pressure Education Program Working Group on High Blood Pressure in Children and Adolescents, 2004). According to the five-point self-perceived health status score, the majority of women (68 percent) described their general health as very good or good, and 26 percent as fair or poor. Eighty-five percent of women were found to have normal blood pressure, defined as systolic pressure ≤ 120 mmHg and diastolic pressure ≤ 80 mmHg (Heart and Stroke Foundation of Canada, 2008).

Anthropometry

According to BMI-for-age growth charts, 24 percent of youth were above the 85th percentile, the cut-off for overweight or obesity according to the Centers for Disease Control and Prevention (CDC, 2000; Ogden et al., 2002): 20 and 18 percent of youth were above the 85th percentile for waist circumference and body fat, respectively (Fernandez et al., 2004; McCarthy et al., 2006). Based on BMI and percentage body fat, 46 and 57 percent of women, respectively, were considered obese; according to measures of waist circumference, 88 percent were at a substantially increased risk of obesity-related health complications (Gallagher et al., 2000; WHO, 2000). In 2002/2003, BMI data for Canada showed that 32 percent of First Nations women were overweight and 40 percent obese (First Nations Regional Longitudinal Health Survey, First Nations Centre, 2005), compared with data for the overall population in 2005, which showed 29 percent of adult women overweight and 23 percent obese (Statistics Canada, 2006). In the Canadian Community Health Survey, 41 percent of aboriginal youth aged two to 17 years were overweight (21 percent) or obese (20 percent) (Shields, 2005), compared with 29 percent of overall Canadian youth aged 12 to 17 years in 2004 (Shields, 2006).

Physical activity

Youth's physical activity and use of television, Internet and video games were assessed. Thirty-nine percent reported watching two to three television programmes on schooldays. On Saturdays, watching television or movies for part of the morning (43 percent) or afternoon (52 percent) (two hours or less) was the most common activity. Playing video/computer games or surfing the Internet were less popular, with 68 and 49 percent of youth reporting that they would not do these on Saturday morning or afternoon, respectively. For comparison, young people aged 12 to 17 years in Canada in 2004 spent an average of ten hours each week watching TV. Adding the time spent on computers or playing video games increased the total to 20 hours a week in front of a screen (Statistics Canada, 2008).

A wide range of physical activities were reported by youth in Tetlit Zheh. The most popular in summer were bicycling (reported by 98 percent), swimming (90 percent) and soccer (80 percent). During the snow-covered school year, youth reported skidooing (97 percent), sledding (84 percent), hockey (70 percent), soccer (87 percent), basketball (81 percent) and hunting (70 percent). According to the seven-day physical activity recall, walking outside was the most frequent activity, with an average of six days per week. Not including activities in gym class, walking and jogging were the most frequent activities, with an average frequency of four days per week. In comparison, in 1994, 71 percent of children (aged eight to 12 years) in the Mohawk community of Kahnawake, Quebec, Canada, took part in physical activity for at least 30 minutes a day (Adams et al., 2005), and 36 percent watched less than two hours of television a day.

Table 7.2 Average daily intakes of traditional food (TF) in 24-hour recalls among Gwich'in youth (ten to 15 years) and women (20 to 40 years)

Traditional food item	Average daily intake for all individuals (g ± SD)	Average daily intake for individuals reporting food item (g ± SD)	Individuals reporting food item (number and % of TF consumers/% of total)
Youth TF consumers *(n = 53)* Total *(n = 64)*			
Caribou meat – cooked	113 ± 85	125 ± 80	48 (91/75)
Caribou meat – dried	12 ± 38	92 ± 60	7 (13/11)
Caribou stomach – cooked	1 ± 10	74	1 (2/2)
Moose meat – cooked	17 ± 47	113 ± 62	8 (15/13)
Moose kidney – cooked	1 ± 10	74	1 (2/2)
Whitefish – cooked	7 ± 35	180 ± 51	2 (4/3)
Women TF consumers *(n = 37)* Total *(n = 52)*			
Caribou meat – cooked	114 ± 124	185 ± 108	32 (86/62)
Caribou meat – dried	8 ± 27	68 ± 48	6 (16/12)
Caribou kidney – cooked	9 ± 56	237 ± 231	2 (5/4)
Caribou fat – cooked	0.1 ± 1	3 ± 0	2 (5/4)
Caribou stomach – cooked	3 ± 21	148	1 (3/2)
Caribou marrow – cooked	1 ± 6	45	1 (3/2)
Caribou heart – cooked	1 ± 10	74	1 (3/2)
Caribou tongue – cooked	1 ± 10	74	1 (3/2)
Moose meat – cooked	9 ± 41	164 ± 68	3 (8/6)
Moose liver – cooked	2 ± 14	99	1 (3/2)
Whitefish – cooked	9 ± 38	161 ± 15	3 (8/6)

Research has shown that the recalls of youth as young as grade 5 can be sufficiently reliable and valid for the assessment of physical activity (Sallis *et al.*, 1993). Physical activity of the youth in Tetlit Zheh was considered as falling within recommendations, as television viewing of two hours or less per day is within recommended levels for youth (American Academy of Pediatrics, 2001), and physical activity of at least 30 minutes a day is considered adequate (CDC, 2004).

Women's physical activity was assessed using the standardized IPAQ. Walking is an important physical activity for Gwich'in women, and therefore influences how comparison standards are used. If Gwich'in data include walking, the majority of women aged 20 to 40 years (60 percent) were considered moderately active, with 25 percent having low physical activity and 15 percent being very physically active. This is compared with data from the 2004/2005 Canadian Community Health Survey, which estimated that 52 percent of Canadian women (20 years and older) were inactive during their leisure time and 48 percent were moderately active (including walking) or more active (Canadian Fitness and Lifestyle Research Institute, 2005). When walking is not considered in the calculation, 33 percent of women in Tetlit Zheh met the United States recommendations of 30 to 60 minutes of moderate to vigorous activity on five days a week (Physical Activity Guidelines Advisory Committee, 2008).

Dietary intake

Dietary nutrients were examined using 24-hour recalls with youth and women, and the frequency of traditional food consumption was captured in a food frequency questionnaire with women. To identify

Table 7.3 Total energy and percentages of energy from macronutrients (mean ± SD) measured by 24-hour recalls in winter among Gwich'in boys and girls (ten to 15 years) and women (20 to 40 years): traditional food (TF) consumers versus non-consumers

| Energy/ macronutrient | Youth | | | | | | Women | |
| | Boys (n = 30) | | Girls (n = 34) | | Total (n = 64) | | (n = 52) | |
	TF consumers (n = 24)	TF non-consumers (n = 6)	TF consumers (n = 29)	TF non-consumers (n = 5)	TF consumers (n = 53)	TF non-consumers (n = 11)	TF consumers (n = 37)	TF non-consumers (n = 15)
Total energy (kJ)	6 721 ± 1 711	6 046 ± 2 014	6 577 ± 2 160	8 808 ± 6 482	7 003 ± 2 330	7 301 ± 4 573	7 472 ± 3 490	7 052 ± 4 385
% protein	19.9 ± 6.3*	13.3 ± 4.3	20.4 ± 6.4	13.2 ± 4.1	20.5 ± 6.2**	13.3 ± 4.1	24.4 ± 7.5***	11.9 ± 4.6
% carbohydrate	53.1 ± 10.4	63.0 ± 12.9	57.2 ± 9.5	61.0 ± 11.0	55.2 ± 10.0	62.1 ± 11.5	48.0 ± 10.9	58.7 ± 16.9*
% fat	28.1 ± 7.0	24.8 ± 9.2	23.2 ± 7.0	27.2 ± 8.3	25.2 ± 7.2	25.9 ± 8.4	27.3 ± 7.6	29.8 ± 12.9
% saturated	10.1 ± 2.8	8.1 ± 4.2	8.2 ± 2.8	9.7 ± 3.5	8.9 ± 2.9	8.9 ± 3.8	9.0 ± 2.8	8.5 ± 5.2
% MUFA	11.3 ± 2.6	9.9 ± 4.4	8.9 ± 3.0	9.9 ± 3.3	9.7 ± 3.0	9.9 ± 3.8	10.6 ± 3.4	11.2 ± 5.4
% PUFA	5.1 ± 2.6	4.2 ± 2.4	4.4 ± 2.2	4.9 ± 2.7	4.8 ± 2.3	4.5 ± 2.4	6.2 ± 2.8	5.8 ± 4.1

Wilcoxon rank test used to examine differences among groups.
* $p \leq 0.05$.
** $p \leq 0.001$.
*** $p \leq 0.0001$.
MUFA = monounsaturated fatty acid.
PUFA = polyunsaturated fatty acid.

potential underreporting, each participant's total energy, weight (kg) and height (m) were used in the Goldberg cut-off method, which compares participants' daily mean reported energy intakes with the intakes for a sedentary lifestyle recommended by FAO, the World Health Organization (WHO) and the United Nations Organization (UNO). The energy intake (EI)-to-basal metabolic rate (BMR) ratio x 1.5 was used as a cut-off for determining inadequate estimation of energy intake based on existing methodology (Goldberg *et al.*, 1991; Gibson, 2005). Energy intakes as recorded by 24-hour recalls for women were below the levels representative of habitual intake (means of 7 472 ± 3 490 kJ and 7 052 ± 4 385 kJ). As portion sizes of traditional food were substantial, often in excess of 100 g/day, it was estimated that underreporting was most likely due to beverage additives and purchased sweets and snacks. Research suggests that Goldberg cut-off values based on adults may overestimate the extent of dietary misreporting when applied to youth (Kersting *et al.*, 1998). Therefore, estimation of intakes was assessed for youth (using median intakes), and mean intakes for both youth and women were used for comparison of consumers and non-consumers of traditional food.

Traditional food consumption

Table 7.2 shows the average daily intakes (g) of traditional foods consumed by youth and women as reported in 24-hour recalls. Of the 64 youth with recalls, 53 reported at least one traditional food item, and of the 52 women, 37 did. The most important traditional species consumed by youth during the winter season assessed were caribou, moose and whitefish. Traditional food frequencies reported by women also emphasized the importance of traditional meats (caribou and moose) and fish. There were no significant differences in traditional food intakes between boys and girls (not shown). Table 7.2 gives the percentages of individuals reporting a specific item among traditional food consumers and among all respondents with recalls.

Macronutrients

Total energy and percentages of energy from macronutrients reported by youth and women are shown in Table 7.3. Among both youth and women, more individuals consumed traditional food than did not, and energy intakes were not significantly different

Table 7.4 Median intakes of micronutrients in winter diets compared with EAR and AI of boys and girls (ten to 15 years) in a Gwich'in community

Micronutrient	Boys (n = 30)		Girls (n = 34)	
	10–13 years (n = 21)	14–15 years (n = 9)	10–13 years (n = 24)	14–15 years (n = 10)
Vitamin A (µg)[a]	301.0*/445.0	425.0/630.0	346.5/420.0	294.5/485.0
Thiamin (mg)	1.2/0.7	1.7/1.0	1.4/0.7	1.1*/0.0
Riboflavin (mg)	1.6*/0.8	1.8/1.1	1.8/0.8	1.8/0.9
Niacin (mg)	16.4*/9.0	18.0*/12.0	16.2/9.0	18.7*/12.0
Vitamin B_6 (mg)	1.2/0.8	1.3/1.0	1.2/0.8	1.4/1.0
Vitamin B_{12} (µg)	6.1*/1.5	6.6/2.0	6.7*/1.5	7.1/2.0
Folate (µg)[b]	241.0/250.0	456.0*/330.0	279.5/250.0	262.5*/330.0
Vitamin C (mg)	94.5*/39.0	117.1/63.0	161.6*/39.0	80.3/56.0
Vitamin E (mg)[c]	3.0/9.0	5.0*/12.0	4.0/9.0	4.0/12.0
Iron (mg)	16.4/5.9	19.2/7.7	14.7/5.7	15.0*/7.9
Phosphorous (mg)	1 099.0/1 055.0	1 551.0*/1 055.0	1 008.5/1 055.0	1 081.0/1 055.00
Selenium (µg)	58.8/35.0	75.8/45.0	50.4/35.0	48.7*/45.0
Zinc (mg)	11.8/7.0	15.5*/8.5	10.5*/7.0	11.9*/7.3
Calcium (mg)	511.0/1 300.0	688.0/1 300.0	543.0*/1 300.0	509.5/1 300.0
Vitamin D (µg)	2.6/5.0	3.0/5.0	2.3*/5.0	2.5*/5.0
Total fibre (g)	8.1*/31.0	13.4/38.0	8.5/26.0	10.2/26.0

[a] Measured as retinol active equivalent (RAE).
[b] Measured as dietary folate equivalent.
[c] Measured as α-tocopherol.
* Not adjusted to usual intakes.

across categories. More protein, as a percentage of total diet, was consumed by the traditional food consumers among both youth (boys and total youth) and women. Similar amounts of total fat and carbohydrate were consumed by youth who consumed traditional food and those who did not, but among women the percentage of energy from carbohydrate appears higher among non-consumers. These data corroborate similar findings reported earlier, although the ages of participants differed and the sample sizes were larger for adults (Kuhnlein and Receveur, 2007).

Micronutrients

Youth were divided into DRI categories based on gender and age categories, and the EAR and AI of selected micronutrients were used as references because it was not possible to determine population-level adequacy for nutrients (Institute of Medicine of the National Academies, 2006). Table 7.4 shows the median intakes of boys and girls compared with the EARs and AIs. Among boys aged ten to 13 years (n = 21), 52 percent fell below the DRI for folate and 48 percent below that for phosphorus, but only 14 percent and 19 percent were below recommendations for selenium and zinc, respectively. Vitamin A, riboflavin, niacin, vitamin B_{12}, vitamin C, vitamin E, calcium, vitamin D and fibre could not be calculated for this age group because variation among individuals was less than within-person variation. Among boys aged 14 to 15 years (n = 9), percentages falling below the recommendations were 89 percent for vitamin A, 11 percent for vitamin B_6, 22 percent for B_{12} and 11 percent for iron. Boys in this age category also fell below the AIs for calcium (100 percent), vitamin D (67 percent) and fibre (100 percent).

Among girls aged ten to 13 (n = 24), none fell below the recommended intake for iron, but 13 percent fell below that for thiamine, 13 percent for riboflavin, 13

Table 7.5 Intakes of micronutrients (mean ± SD) in winter 24-hour recalls from boys and girls (ten to 15 years) and women (20 to 40 years) in a Gwich'in community: traditional food (TF) consumers versus non-consumers

Energy/ macronutrient	Youth						Women	
	Boys (n = 30)		Girls (n = 34)		Total (n = 64)		(n = 52)	
	TF consumers (n = 24)	TF non-consumers (n = 6)	TF consumers (n = 29)	TF non-consumers (n = 5)	TF consumers (n = 53)	TF non-consumers (n = 11)	TF consumers (n = 37)	TF non-consumers (n = 15)
Vitamin A (µg)[a]	398.2 ± 202.0	227.7 ± 128.2	400.6 ± 258.1	377.8 ± 416.1	399.5 ± 232.2	295.9 ± 289.1	660.9 ± 1916.7	222.6 ± 127.1
Thiamin (mg)	1.6 ± 0.7	1.3 ± 0.6	1.4 ± 0.6	1.4 ± 0.9	1.5 ± 0.6	1.3 ± 0.7	1.3 ± 0.5	1.2 ± 0.8
Riboflavin (mg)	2.2 ± 0.9**	1.2 ± 0.4	2.0 ± 0.9	1.4 ± 1.1	2.1 ± 0.9***	1.3 ± 0.8	2.6 ± 1.2***	1.5 ± 0.8
Niacin (mg)	18.4 ± 6.5	13.5 ± 6.5	17.5 ± 7.9	18.6 ± 12.5	17.9 ± 7.2	15.8 ± 9.5	23.9 ± 20.2*	15.3 ± 9.8
Vitamin B_6 (mg)	1.3 ± 0.5	0.9 ± 0.4	1.3 ± 0.5	1.1 ± 0.8	1.3 ± 0.5	1.0 ± 0.6	1.5 ± 1.0*	0.9 ± 0.5
Vitamin B_{12} (µg)	9.6 ± 5.9***	2.3 ± 1.3	8.7 ± 4.8**	2.3 ± 1.4	9.1 ± 5.3****	2.3 ± 1.3	11.2 ± 7.1****	1.6 ± 1.3
Folate (µg)[b]	332.3 ± 197.3	313.5 ± 139.7	300.0 ± 161.4	336.4 ± 266.2	314.6 ± 177.5	323.9 ± 195.6	309.6 ± 191.5	367.1 ± 311.2
Vitamin C (mg)	112.5 ± 65.6	87.7 ± 87.8	126.4 ± 85.4	141.4 ± 76.3	120.1 ± 76.7	112.1 ± 83.5	98.9 ± 97.0	122.3 ± 140.1
Vitamin E (mg)[c]	4.5 ± 2.7*	2.2 ± 1.0	3.6 ± 2.1	3.6 ± 3.0	4.0 ± 2.4	2.8 ± 2.1	4.3 ± 2.9	2.9 ± 2.9
Iron (mg)	18.9 ± 7.4*	12.0 ± 5.5	15.3 ± 4.7	13.3 ± 8.6	16.9 ± 6.3*	12.6 ± 6.7	19.0 ± 8.3****	9.2 ± 6.1
Phosphorous (mg)	1 196.0 ± 403.1	960.7 ± 437.1	1 143.0 ± 404.7	1 020.8 ± 573.6	1 167.0 ± 401.0	988.0 ± 477.6	1 171.5 ± 539.0*	778.7 ± 527.2
Selenium (µg)	69.2 ± 32.6	65.8 ± 28.9	56.8 ± 26.8	56.0 ± 42.7	62.4 ± 30.0	61.4 ± 34.3	99.2 ± 80.6*	53.7 ± 35.8
Zinc (mg)	14.2 ± 6.2*	7.1 ± 4.3	11.9 ± 4.8	8.4 ± 5.0	13.0 ± 5.5**	7.7 ± 4.4	15.7 ± 7.8****	5.7 ± 3.2
Calcium (mg)	610.5 ± 268.0	476.5 ± 222.8	628.0 ± 301.7	591.6 ± 444.1	620.1 ± 284.4	528.8 ± 327.6	425.8 ± 240.1	499.5 ± 417.4
Vitamin D (µg)	3.6 ± 3.4	2.3 ± 1.7	3.4 ± 3.1	2.2 ± 1.9	3.5 ± 3.2	2.3 ± 1.7	3.7 ± 5.5	2.6 ± 3.1
Total Fibre (g)	9.3 ± 4.9	10.0 ± 3.0	8.4 ± 3.4	10.6 ± 8.9	8.8 ± 4.1	10.2 ± 6.0	7.8 ± 4.7	8.6 ± 4.7

a   Measured as RAE.
b   Measured as dietary folate equivalent.
c   Measured as alpha tocopherol.
Wilcoxon rank test used to examine differences among groups.
*p ≤ 0.05.
**p ≤ 0.01.
***p ≤ 0.001.
**** p ≤ 0.0001.

percent for niacin, 21 percent for vitamin B_6, 46 percent for folate, 96 percent for vitamin E, 58 percent for phosphorus, 17 percent for selenium and 100 percent for fibre. Among girls aged 14 to 15 years (n = 10), 80 percent fell below the EAR for vitamin A, 20 percent for riboflavin, 20 percent for vitamin B_{12}, 40 percent for vitamin C and 50 percent for phosphorus. Girls also fell below the AIs for calcium (90 percent) and fibre (100 percent).

Despite apparent underreporting by women, most women in the sample were obviously consuming adequate amounts of thiamin, vitamin B_{12}, iron, phosphorus, selenium and zinc, undoubtedly owing to the large portions of traditional food animal protein reported (not shown).

Table 7.5 shows the mean micronutrient intakes of youth and women who consumed traditional food and those who did not report any traditional food in the 24-hour recalls. Boys who consumed traditional food were found to have significantly higher intakes of riboflavin, vitamin B_{12}, vitamin E, iron and zinc. Vitamin B_{12} was also found to be significantly higher among girls who reported consuming traditional food. Overall, the youth who consumed traditional food had higher intakes of riboflavin, vitamin B_{12}, iron and zinc. Similar findings for women demonstrated that traditional food consumers had significantly higher riboflavin, niacin, vitamin B_6, vitamin B_{12}, iron, phosphorus, selenium and zinc intakes. These findings are important even though some nutrients are

Table 7.6 Food group servings (mean ± SD) in winter 24-hour recalls of youth (ten to 15 years) and women (20 to 40 years) in a Gwich'in community compared with recommendations

Food group	Recommended servings	TF consumers	TF non-consumers	Total
Youth		(n = 53)	(n = 11)	(n = 64)
Fruits and vegetables	5–8	2.6 ± 1.9	2.6 ± 1.6	2.6 ± 1.8
Grain products	4–7	4.7 ± 2.7	4.7 ± 3.0	4.7 ± 2.7
Milk and alternatives	2–4	0.9 ± 0.7	0.7 ± 0.5	0.9 ± 0.7
Meat and alternatives	1–2	3.6 ± 1.8*	1.8 ± 1.5*	3.3 ± 1.9
Women		(n = 37)	(n = 15)	(n = 53)
Fruits and vegetables	7–8	1.7 ± 1.8	2.2 ± 2.5	1.8 ± 2.0
Grain products	6–7	4.4 ± 3.1	4.7 ± 4.7	4.4 ± 3.6
Milk and alternatives	2	0.5 ± 0.5	0.5 ± 0.5	0.5 ± 0.5
Meat and alternatives	2	5.4 ± 4.1*	2.2 ± 2.0*	4.4 ± 3.9

Recommendations from Canada's Food Guide to Healthy Eating for First Nations, Inuit and Métis.
Wilcoxon rank test used to examine differences among groups.
* $p \leq 0.01$.
TF = traditional food.

not being consumed in sufficient amounts. The animal food portion of the diet, primarily as traditional food, is the major contributor of many of these nutrients. Earlier results with Dene/Métis and Yukon youth aged ten to 12 years (n = 98) showed similar patterns of nutrient intake among consumers and non-consumers of traditional food (Nakano *et al.*, 2005b). Dene/Métis adults (n = 1 007) from all regions also showed similar patterns, except that non-consumers had higher intakes of energy, carbohydrate, total fat, saturated fat, vitamins D and E and several minerals (Kuhnlein *et al.*, 2004).

Food group servings

Table 7.6 shows the mean numbers of food group servings derived from the 24-hour recalls reported by youth and women, and compares these with Canadian Food Guide recommendations. Youth had fewer than the recommended number of servings of both fruits and vegetables, and milk and alternatives. When youth and women who consume traditional food were compared with those who do not the only significant difference was that non-consumers were found to consume fewer servings of meat and alternatives.

As noted in Table 7.2, 82 percent of youth reported consuming a minimum of one item of traditional food

in the 24-hour recalls. Caribou was the most frequent food item consumed, with moose and whitefish also being popular. Traditional food consumers consumed an average of 154 g of traditional food daily, or 2.1 servings according to Canada's Food Guide for First Nations, Inuit and Métis (Health Canada, 2007). Results from the traditional food frequency questionnaire showed that women consumed a total of 30 different traditional food species and 81 different traditional food parts during the three-month reporting period. Caribou (*Rangifer tarandus*), whitefish (*Coregonus* sp., *Prosopium* sp.), Labrador tea (*Ledum groenlandicum*), loche (*Lota lota*) and cranberries (*Vaccinium* sp.) were particularly prominent in the diet, being consumed by the largest percentages of women.

Selected sweet, salty and fat-rich dessert market foods consumed by youth and women accounted for a large amount of energy in the recalls; Table 7.7 shows the average numbers of mentions per person and the average serving sizes. For example, women consumed up to 36 oz of sweetened drinks, 48 oz of pop and four servings of sweet desserts within the recall period. Similar findings for Dene youth were demonstrated earlier, with more than 50 percent of market food energy coming from these kinds of food (Nakano *et al.*, 2005b).

Table 7.7 Frequencies of daily consumption of selected market foods by youth (ten to 15 years) and women (20 to 40 years) in a Gwich'in community: data from one 24-hour recall per person

Food item		Youth (n = 64)	Women (n = 52)
Juice/powdered drinks/hot chocolate[a]	No. of mentions/person	1.75	0.58
	SD	1.17	0.75
	Range	0–4	0–3
Pop/sports drinks/ sweetened iced tea[b]	No. of mentions/person	0.44	0.75
	SD	0.59	0.88
	Range	0–2	0–3
Salty snacks[c]	No. of mentions/person	0.39	0.38
	SD	0.68	0.60
	Range	0–3	0–2
Sweetened desserts and flour products (cookies, cake, doughnut, pie, ice cream, gelatine)[d]	No. of mentions/person	0.52	0.50
	SD	1.13	0.83
	Range	0–6	0–4
Candy (hard, chocolate)[e]	No. of mentions/person	0.30	0.10
	SD	0.58	0.36
	Range	0–2	0–2
Free sugar, jam/jelly, syrup[f]	No. of mentions/person	0.45	1.73
	SD	0.71	1.07
	Range	0–3	0–4

[a] Average servings: 1 cup or 8 oz/235 ml (youth); 1.5 cups or 12 oz/355 ml (women).
[b] Average servings: 1.5 cups or 12 oz/355 ml (youth); 2 cups or 16 oz/473 ml (women).
[c] Average servings: 50 g (youth); 60 g (women).
[d] Average servings: cookies = 40 g (youth), 20 g (women); cake = 75 g (youth), 80 g (women); doughnut = 80 g (youth), 115 g (women); pie = 200 g (youth), 100 g (women); ice cream = 330 g (youth), 145 g (women); gelatine = 320 g (youth), 285 g (women).
[e] Average servings: hard candy = 4 g (youth and women); chocolate bar = 65 g (youth), 45 g (women).
[f] Average servings: added sugar = 1.5 teaspoons (youth), 2.5 teaspoons (women); jam/jelly = 1 tablespoon (youth), 2.5 tablespoons (women); syrup = 5 teaspoons (youth), 4 teaspoons (women).

Table 7.8 Food security status among participating women's households in a Gwich'in community

Variable	Number	%
All households[a]	53	
Food-secure	24	45.3
Moderately food-insecure	26	49.1
Severely food-insecure	3	5.7
Children only[b]	43	
Food-secure	34	79.1
Moderately food-insecure	9	20.9
Severely food-insecure	0	0.0
Adults only[a, c]	10	
Food-secure	5	50.0
Moderately food-insecure	4	40.0
Severely food-insecure	1	10.0

[a] Based on ten-item USDA Food Security Scale for adults/households.
[b] Includes only respondents with children in the household; based on eight-item USDA Food Security Scale for children.
[c] Includes only respondents without children in the household.
Source: Goodman, 2008.

Food security

In January 2006, household food security was assessed from young women's responses to a survey using a modified version of the USDA Food Security Survey Module questionnaire. Table 7.8 shows that 55 percent of households (29 households) were found to be food-insecure, with 6 percent (three households) reporting severe food insecurity. Among households with children, 21 percent (nine households) had experienced food insecurity to the extent that children within the household were affected. However, in all households, children were affected moderately, with no cases of severe food insecurity being found. Only ten households reported no children in the home, and half of these reported food insecurity.

To explore the issue of food access, women were asked questions relating to their access to both market and traditional foods. Twenty-eight percent of respondents reported difficult access to their favourite traditional foods, due to lack of transportation, equipment and/or storage, lack of a hunter in the household, reliance on others to provide traditional food, and time constraints. A greater number (43 percent) reported difficult access to their favourite market foods, with lack of affordability and availability being among the main barriers. When traditional food was not available, 45 percent of women indicated that their households would not be able to buy all the food they needed from the store. Only 40 to 45 percent of women's households had access to hunting and fishing equipment, with 11 and 29 percent of women reporting that hunting and fishing, respectively, were too expensive for their families. Food security was

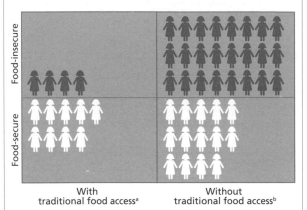

Figure 7.2 Relationship between food security status and traditional food access score among women in a Gwich'in community (n = 51)

Food-insecure

Food-secure

With traditional food access[a]

Without traditional food access[b]

$p (\chi2) < 0.05$.
Food security status based on ten-item adult/household scale from the USDA Food Security Survey Module.
[a] Positive responses to all traditional food access items included in the traditional food access score.
[b] Minimum of one negative response to a traditional food access item included in the traditional food access score.
Source: Goodman, 2008.

also found to be positively correlated with full-time employment (Goodman, 2008).

Traditional food access scores were developed from questions on the availability of hunting and fishing equipment, the affordability of hunting and fishing excursions, and access to traditional food in general. The food access scores were correlated positively with the frequency of traditional food use, and households with reduced traditional food access consumed less traditional land animal meats and total traditional food, quantified by serving sizes (Goodman, 2008). Figure 7.2 shows how the traditional food access score significantly predicted food security among those interviewed.

Climate change impacts on access to food and health

Because the impacts of climate change are thought to affect food availability negatively, key informants were interviewed to assess their perceptions of the impacts of climate change on access to traditional food species, market food and health. A total of 22 Dene First

Nations adults were interviewed, representing the five regions of the Dene Nation and all age categories; of these 13 men and eight women, eight were elders aged 60 years or more. Nine participants were from Gwich'in communities in NWT. Responses were categorized into three themes: perceived impact on the harvesting of traditional food; access to traditional foods and changes in health; and access to healthy market foods.

Theme 1 – impact on the harvesting of traditional foods

Sixty-eight percent (15/22) of the participants, including all Gwich'in participants, felt that climate change has affected their intake of traditional foods. Participants from northern (e.g., Gwich'in) and central (e.g., Akaitcho) regions observed changes in the species of fish and in the water. Changes in the temperature and cleanliness of the water were perceived as contributing to fish migration and health. Responses on changes in the fish included "discolouration", "smaller in size", "change in flesh texture", "unhealthy" and "less fish available to harvest". Comments included the following:

The water is not clean. I know there is something wrong with the water but I eat the fish anyway.
I see worms, parasites and bumps on the fish, but I still consume fish because it is traditional food.

Participants from all regions perceived a decline in numbers of caribou and moose, and attributed this to climate-related changes in migration patterns caused by warming temperatures, increased forest fires and reduced access to food sources. For example, caribou's access to food sources was reported to be limited by ice formation under the snow preventing them from breaking through the ice with their hooves to reach the lichen; as a result, caribou were reported as being thinner than previously. In addition, biting insects (mosquitoes, warble flies and bot flies) that harass caribou make it difficult for the animals to feed.

Changes in weather conditions were reported as affecting travel for hunting or fishing. An elder reported "it is difficult to read snow patterns". Participants observed changes in the snow and ice, with rivers and lakes freezing more slowly, which prevents hunters from moving across the land. Water under the snow also makes travelling dangerous for hunters. A hunter commented, "When on the land, I don't know if I will end up in open water where there wasn't any before."

Participants reported that there were fewer ducks to harvest; possible explanations included changes in flight patterns, with birds no longer stopping where they historically did so, and early migration from the north. Another perspective was "ducks do not lay as many eggs, and do not protect their young as well … decrease is seen in cranes and geese". New bird species entering the habitat were reported across all five regions.

Participants also noted that the rising cost of fuel is limiting their ability to go out on the land to harvest traditional food. The increased cost of fuel for heating homes was reported as reducing the funds available to go harvesting. The costs of supplies and equipment for hunting and fishing were reported to have increased, thus limiting harvesting. One respondent stated "In the past it would cost approximately CAD 700 to go out on the land and harvest (camp), but today it costs approximately CAD 1 200". Other factors affecting the harvesting of traditional foods included increased predators, outfitters and the tourist trade, and industries that interrupt migration patterns.

Theme 2 – access to traditional foods

Thirty-six percent (8/22) of key informants felt that they did not eat as much traditional food as they would like to, and that access to these foods was more difficult. Those who reported eating less traditional food than they would like stated that the costs of fuel and the time available for hunting were impediments, as well as reduced animal numbers in nearby hunting locations. Being able to purchase traditional foods from a store would improve their access to these foods.

Although participants reported that access to traditional foods had become more difficult recently, many also reported the cultural practice of sharing meat. Meat was noted as being given to single mothers and the elderly who cannot get to the land themselves. "Through harvesting traditional foods you practise your culture and live your heritage" was a comment that summarized the value participants placed on traditional foods. Participants viewed the changes

in traditional foods over the last two years as having had an impact on their families' health, by increasing diabetes and blood cholesterol in adults and dental caries in youth. Youth were reported as replacing traditional foods with high-sugar, high-fat foods. Trends in increasing intakes of high-sugar/-fat foods were also reported in adults, and were perceived to be a contributing factor in the rise in body weight, diabetes and cholesterol problems.

Changes in traditional foods were also perceived as affecting family members' behaviour and mood. One father stated "When there is no caribou, I feed my family hamburger, hotdog and pizza," and that he could see a difference in his family's behaviour. Another participant reported "A decrease in caribou affects their mood. Traditional foods strengthen their mood, self-esteem and attitudes, and increase their energy." Participants also consistently stated that traditional foods are healthy, and that they feel healthier when they consume them.

Theme 3 – access to healthy market foods

Participants observed that healthy foods cost too much in communities in northern regions. Seven of the nine Gwich'in participants stated that the healthy foods in stores are too expensive to buy, although the stores keep them in stock. Other factors reported as affecting access to healthy foods were lack of variety, the poor condition of fresh foods, and stores' inadequate supplies of healthy foods. Participants in remote communities stated that they travelled to the closest towns, often several hours' drive away, for their food shopping. Participants also noted that food costs are too high in the North: "for example, a litre of milk costs CAD 8 in the North, but only CAD 3 in the south". Participants stated that a reduction in the price of healthy foods, and timely deliveries of perishable foods to stores would make it easier to purchase healthy foods.

Project activities in Tetlit Zheh

In the context of activities being conducted by other agencies in the community, project assistants in Tetlit Zheh considered community activities for improving knowledge of and access to healthy traditional and market foods. After consultations, local personnel developed and delivered several activities over an 18-month period. One or two community health promoters based in local healing centres were supported throughout the project. The activities implemented included:

- food teaching events for community groups (youth, schools, young mothers, etc.), usually led by elders and Tl'oondih Healing Society staff;
- regular local radio announcements about project activities, recipes and traditional food availability and quality (in liaison with the Natural Resource Department);
- nutrition classes in schools;
- fitness events for women;
- classes and meals for pregnant women and young mothers;
- teaching schoolchildren about food labels (market food);
- classes on nutrition, the risk of diabetes, and traditional values from traditional food, at youth camps (grades 4 to 7) about 30 km up-river;
- regular updates to the Band Council;
- nutrition activities for the Moms and Tots programme;
- production of a DVD describing the current food situation and challenges, and emphasizing the importance of traditional food and healthy market food;[2]
- publication and distribution of a book about traditional Gwich'in food and health;
- Drop the Pop NWT.[3]

Future considerations

This chapter has provided a systematic evaluation of the food and health circumstances of youth and young women in the Gwich'in community of Tetlit Zheh, and has described awareness raising activities to improve health through the increased use of traditional

Gwich'in food and healthy market food. The Tetlit Gwich'in are similar to many indigenous communities in northern Canada, which have excellent traditional food resources but reduced access to them, while the resulting nutrition transition raises growing concerns about changing dietary patterns, reduced physical activity and increasing obesity and chronic diseases.

The focus of the research was on youth and young women during the deep winter, when traditional food use is at its lowest. This strategy was based on the belief that improving the dietary and activity behaviours of these two population segments in the lean season could lead to year-round change in the entire community, through activities planned and delivered to the community as a whole.

The researchers found that dietary patterns were short of several nutrients, despite important intakes of traditional food by both youth and young women. Those consuming traditional foods had better diets in several respects. Replacing market foods of low nutrient density with traditional food of higher nutrient density would improve nutrient intakes, even though food group analysis showed that servings of meat and alternatives exceeded Canadian Aboriginal Food Guide recommendations, while milk, and fruit and vegetable servings fell seriously short. In general, the traditional animal foods available to and consumed by youth and women were the best foods presented in the dietary recalls. If all parts of animals/fish and berries had been consumed more frequently, intakes of nutrients such as vitamins A, C, E and D would likely have met requirements among more youth and women (Kuhnlein et al., 2006).

Physical activity patterns were modest, and overweight and obesity affected both youth and young women, with BMI and body fat being higher among women. Fortunately, hypertension was non-existent, and pre-hypertension was present for only a small proportion of youth and young women. Youth expressed interest in and enthusiasm for several culturally relevant physical fitness activities, and did not spend excessive time with in-home entertainment media (television, video and video games), as has been reported in other aboriginal communities (Bernard et al., 1995). The

2 www.indigenousnutrition.org
3 www.dropthepopnwt.ca

extent of obesity among young women is especially concerning, as changing dietary patterns, reduced use of traditional food and increased use of food of lower nutrient density, coupled with a physical activity expenditure that is below the energy intake is the classic pattern of the nutrition transition and the root cause of increasing obesity and the onset of chronic disease (Kuhnlein *et al.*, 2004).

The food security of Tetlit Gwich'in women and households was associated with access to traditional food resources. Qualitative research showed that the impact of climate change and increased food and fuel prices had affected households' ability to hunt and fish for family subsistence. This led to perceptions of poorer health having resulted from a diet composed of less traditional and quality market food than people would prefer. It is telling that full-time employment was also correlated with food security, and that people with higher incomes would like to be able to buy traditional foods, which they cannot do locally owing to government controls on the sale of wildlife. From the extensive interviews reported here, it is obvious that Gwich'in adults are well aware of the nutritional and health qualities of their traditional foods, and that they must be proactive in improving their access to these foods. This was demonstrated in the response to creating a food and health book for this project, and in other recent ethnographic research on traditional Gwich'in food resources (Andre and Fehr, 2001; Parlee, Berkes and Teetl'it Gwich'in, 2005).

Project assistants presented diverse intervention strategies to the Tetlit Zheh community. These were primarily experiential activities focusing on traditional food preparation and the recording of recipes for a community food and health book, physical activities for youth and women, radio and Internet communications, and learning activities for schools and women's groups. The most successful activities tended to be those within the school structure, while the scheduling of activities targeting young women was more challenging, owing to the women's time commitments (child care, community responsibilities, etc.) and limited free time. The DVD and food book have been circulated in the community and have been very well received. In a small, informal survey, community members rated the food book as very useful. Individuals found the book useful for passing on traditional food knowledge to youth, and one school teacher was using the book in the classroom. Many of the people surveyed said that they had already learned a great deal regarding nutrition and the variety of ways of preparing traditional food.

A continuous flow of resources is needed to support sustained behaviour change strategies for all segments of the population. Of particular importance are mutually supportive activities that are coordinated from community offices and other settings – such as those sponsored from schools, the health centre, the Head Start centre for toddlers and their mothers, elders' committees, the Hunters' and Trappers' Association and the Band administration, and those promoted by researchers present in the community. If they are to be successful, interventions for preventing chronic disease by focusing on nutrition and physical activity in low-income settings must be interactive, culturally grounded and coordinated through primary care settings, with incentives for participation (Chaudhary and Kreiger, 2007).

Community food stores, of which two were operating during the project period, also offer potential for promoting better food environments. It has been shown that disparities in obesity prevalence depend on the quality of retail food environments in disadvantaged areas, especially when income is limited, and particularly among women (Ford and Dzewaltowski, 2008). Two examples of attempts to improve retail food outlets in the Canadian Arctic are the Government of Canada's Food Mail programme, which has worked to increase the availability of purchased foods in selected communities in the north (Lawn and Harvey, 2001), and Healthy Foods North. Neither of these programmes has been active in Tetlit Zheh, and Healthy Foods North had to close in 2010 owing to funding problems.

The objectives of the Gwich'in Traditional Food and Health Project in Tetlit Zheh were undermined by the impact of climate change and other forces that reduced the availability and presence of traditional animal food resources at a reasonable distance from the community, and by rapidly escalating food and fuel prices. The project could therefore not demonstrate that an intervention based on traditional food and healthy market food had improved the health of youth and young women. A post-intervention evaluation at the close of the project budget period was deemed unfeasible because of the constraints people faced with reduced access to both traditional and healthy market foods, despite popular and broad-based community education activities.

Indigenous Peoples throughout the world face circumstances that compromise their cultural food systems. In the case of the Tetlit Gwich'in, the need and wish of the community is to maintain traditional foods for their health and cultural benefits, and to be able to buy quality market foods that they prefer and can afford. Interventions that mitigate climate change and contribute to protecting the habitat of diverse food species will help meet this goal. It is also important to build on other social justice initiatives that protect the cultural food system and livelihoods, to assuage poverty in all its dimensions, and to continue to provide resources and tools for health promotion that develops and maintains a healthy community in the contemporary world ✺

Acknowledgements

The authors thank the women and youth participants, advisers, community members and families who kindly shared their knowledge and who warmly welcomed the research team into Tetlit Zheh. We especially acknowledge the elders whose expertise helped guide the project. Thank you also to Margaret Mcdonald and the Tl'oondih Healing Society, which provided key support through office space and project administration. We acknowledge the leadership and commitment of Elizabeth Vittrekwa in the initial stages of data collection, and the hard work of Mary Ross, Brenda Martin, Jayda Andre and Rhonda Francis in project completion. We thank Liz Cayen for her assistance in financial administration. Appreciation is extended to Jill Lambden, who worked with Dina Spigelski and Lauren Goodman conducting community assessments and completing data analysis, with support from Louise Johnson-Down and Rula Soueida. We are grateful to the Tetlit Gwich'in Band Council, Sharon Snowshoe, Hazel Nerysoo, Johnny Kay and the Hamlet for supporting this project. We thank Principal Bruce Spencer, Vice-Principal Shirley Snowshoe and staff of Chief Julius School for their gracious support in the collection of the youth data. We give special thanks to CBQM, the local radio station, and to Tena Blake and the Moms and Tots programme. Primary funding for this project was provided by a grant from the Canadian Institutes of Health Research (Institute of Aboriginal Peoples' Health and Institute of Nutrition, Metabolism and Diabetes).
> **Comments to:** harriet.kuhnlein@mcgill.ca

>> Photographic section p. X

Chapter 8

Inga food and medicine systems to promote community health

⚜ SONIA CAICEDO[1] ⚜ ANA MARÍA CHAPARRO[1]

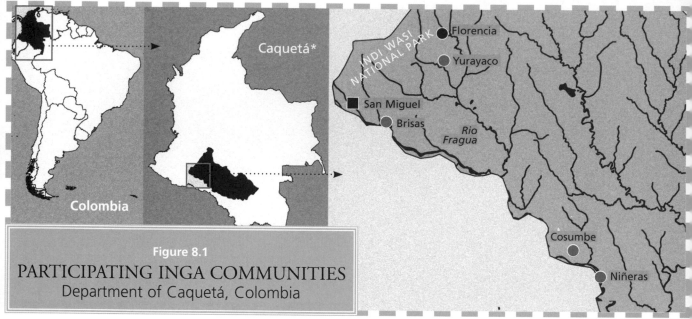

Figure 8.1
PARTICIPATING INGA COMMUNITIES
Department of Caquetá, Colombia

Data from ESRI Global GIS, 2006.
Walter Hitschfield
Geographic Information Centre,
McGill University Library
*www.maplandia.com/columbia/caqueta

Key words > Inga community,
Indigenous Peoples, traditional foods,
Andean piedmont, biocultural diversity,
nutrition, health, food security

Photographic section >> XIII

1
Amazon Conservation
Team Colombia,
Bogota, Colombia

Abstract

This chapter describes the activities and results of a project developed in five indigenous reserves of the Inga indigenous group from the Department of Caquetá in Colombia. The project's objective was to promote indigenous traditional foods and medicine as a strategy for ensuring community health.

Support to family and school vegetable gardens (*chagras*), and the establishment of farming projects and medicinal gardens – in addition to family visits, workshops and courses on nutrition and health, culinary festivals, seed exchanges, traditional recipe collection and health brigades – were essential elements of a process whose general purpose was to strengthen the health component of the Inga community's Life Plan (see text).

The participation of community elders and traditional healers strengthened the project's results by triggering and encouraging broad consensus on its benefits. Health recommendations were promoted through pamphlets and booklets on nutrition, health and traditional recipes and through radio programmes in the Inga language, as part of a communications strategy that aimed to build community awareness regarding the importance of nutrition and health.

The Inga ethnic group of the Caquetá region

Located in the eastern Andean piedmont (westernmost Amazon) in the Department of Caquetá (Figure 8.1), the project area, which is inhabited by numerous indigenous tribes, is known as one of the most species-rich sites in Amazonia and is considered a global conservation priority. Today, the piedmont is perceived as linking the mountains of the Andes to the plains of the great Amazon, "a staircase of earth that serves as a land bridge so that flora, fauna and people form a single landscape" (Ramírez, 2005). Characterized by flora and fauna that are globally diverse, the region displays unusual adaptations to dynamic environments, and has a high degree of local endemism. Located near the equator, it has no significant seasonal differences and only minor variations in average monthly temperatures.

The region's biological and cultural evolutions have taken place largely in parallel. Among its biodiversity, the botanical species *Banisteriopsis caapi* (Spruce ex Griseb.) C.V. Morton is important as a sacred plant for the region's original inhabitants, who call it *yagé* or *ayahuasca*. Some local tribes[1] use this "vision vine" as a sacrament in rituals, and are therefore sometimes referred to collectively as the "*yagé* culture".

The Inga ethnic group of Caquetá is part of this *yagé* culture, and its cultural practices include ceremonies in which *yagé* is consumed. Considered a gift from God, *yagé* is believed to provide not only the capacity to manage and understand the Inga culture, but also powers of healing. "These practices include a special relationship with nature, in which [the Inga] invoke the strength of their mythical animals such as jaguars, parrots, and snakes, and also achieve knowledge on the use of medicinal, food, psychotropic, stimulant, timber, and craft plants" (Ramírez, 2005).

The Inga's relative isolation from Western society has allowed them to conserve linguistic and cultural knowledge and to continue practising their indigenous traditions, including specific dietary habits. For the Inga, nature gives life by providing animals, plants and seeds that are used for family and community support. During the project's first phase, groups of these traditional foods were collected and characterized: vegetables, tubers and trees or palms such as *chontaduro* (*Bactris gasipaes* Kunth), *milpés* palm (*Oenocarpus bataua* Mart.) and yam (*Dioscorea* spp.); fruits such as *zapote* (*Matisia cordata* Bonpl.), papaya (*Carica papaya* L.), pineapple (*Ananas cosmosus* L.), banana (*Musa* spp.) and *arazá* (*Eugenia stipitata* McVaugh); small animals such as ants (*Atta* spp.), *churo* (a snail, *Pomacea maculata* Perry) and *mojojoy* (a grub, *Coleoptera* spp.); and larger animals such as *boruga* (an agouti, *Cuniculus paca* L.), *morrocoy* (a turtle, *Geochelone carbonaria* Spix), *churuco* (a monkey, *Lagothrix lagothricha* Lugens), deer (*Mazama* spp.),

cucha (a fish, *Hypostomus* sp.), shad (*Brycon* spp.) and *bocachico* (a fish, *Prochilodus nigricans* Spix & Agassiz). Correal *et al.* (2009) provide a fuller list of Inga traditional foods.

Inga leaders have stated their determination to defend their unique traditional lifestyle by asserting and implementing the five fundamental rights of Indigenous Peoples defined by the International Labour Organization's (ILO's) Convention 169: identity, participation, territory, autonomy, and autonomous development (ILO, 1989). These rights are exercised through the Inga Life Plan, an indigenous development plan designed in a participatory fashion by the Tandachiridu Inganokuna Association, supported by the Colombian Constitution – one of the most progressive in Latin America. The Inga Life Plan establishes the theoretical basis for future actions, the community's objectives, and the practical means of fulfilling these. Briefly, the plan seeks to achieve the holistic integration of all aspects of daily life, including education, health, agriculture, land tenure, nature and culture.

Organization of this Inga community started in 1988, when the Organization of Inganos of Southern Colombia (ORINSUC) was formed. In 2000, ORINSUC was transformed into the Tandachiridu Inganokuna Association of Senior Councils (Inga from the Caquetá), which has restored indigenous judicial and governmental structures, initiated a process for legalizing collective traditional lands in indigenous reserves, and established activities to improve communities' general health through primary health care programmes.

In February 2002, the Amazon Conservation Team and the Inga community, in partnership with the Colombian National Park Service, established the 77 380 ha Alto Fragua Indi Wasi National Park, located along the eastern Andean foothills of the Colombian Amazon at the headwaters of the Fragua River. The park was created to protect one of the world's greatest regions of biodiversity, as confirmed through inventories conducted by the *Instituto de Investigación de Recursos Biológicos Alexander Von Humboldt*. In addition to protecting several tropical Andean ecosystems, including highly endangered

---

[1] These include the Kametza in Sibundoy Valley; the Siona and Kofan along the Putumayo River; the Inga in Sibundoy Valley and the regions around Mocoa, Florencia and the Bota Caucana; and the Coreguaje in the vicinity of the Orteguaza River.

humid sub-Andean forests, the park also conserves endangered fauna, such as the spectacled bear, and sacred cultural sites designated by local Indigenous Peoples. Indi Wasi protects biological diversity in a vital area that links Andean and Amazonian biota and contains sites of high cultural significance for the Inga people.

The Inga community has also determined that younger generations must be educated under the guidance of traditional Inga authorities, many of whom have expert knowledge of the surrounding forests and their diversity. In response to the lack of suitable education for their youth, the Inga have developed a curriculum that embraces traditional knowledge areas such as botanical medicine and forest stewardship, as well as standard "Western" subjects at the innovative Yachaicurí Ethnoeducation School of Yurayaco, Caquetá.

At the school, approximately 90 Inga students aged five to 18 years are being trained to become conservation leaders. They participate in courses that emphasize sustainable agriculture, and record ancestral knowledge in their native language. Located on 55 ha, the school grounds include a natural science laboratory and an agro-ecological farm, where students learn first-hand the sustainable farming techniques that allow them to grow their own food, contribute to the food resources of surrounding communities, and provide an economic base for their school.

The Inga group that participated in the project is located in eastern Caquetá, 60 km south of Florencia, the capital of the region. Its territories cover approximately 19 778 962 ha, at 297 to 540 m above sea level, with an average temperature of 27 °C and relative humidity of 87 percent. In this region, the most readily available agricultural products, both cultivated and harvested from the wild, are banana, sugar cane, pineapple and maize. Other significant products are rubber, cocoa, *arazá* (*Eugenia stipitata* McVaugh), *cocona* (*Solanum sessiliflorum* Dunal), *copoazu* (*Theobroma grandiflorum* Schumann), *chontaduro* (*Bactris gasipaes* Kunth), *caimarona* (*Pourouma cecropiifolia* Mart), coffee and *borojo* (*Borojoa patinoi* Cuatrec) (Parra, 2004).

Cultural and environmental challenges

The eastern Andean piedmont of Colombia is characterized by uncontrolled land occupancy and immigration resulting from the advance of colonization. In recent decades, government development planning for these territories has focused on extensive cultivation through the use of credit and subsidies, placing a strain on sustainable use of the local environment. There is considerable logging activity in the region, and this is extremely difficult to monitor. Seismic exploration and exploratory oil drilling have also taken place. The available hydrocarbons appear to be of insufficient quantity and quality to merit further exploitation, but petroleum extraction may still represent a threat for conservation of the region's ecosystems and Indigenous Peoples. Mining is another potential threat.

These territories have been a refuge for paramilitary groups and armed militias such as the Revolutionary Armed Forces of Colombia–People's Army (FARC–EP) and the National Liberation Army (ELN). Local people face constant uncertainty in the face of external efforts to control the territory. Conflict has directly affected the Inga ethnic group and has increased economic, social and environmental instability. Violence against citizens, including abductions and assassinations, is increasing. Among adult men and women (between 15 and 64 years of age), violent conflict appears to be an important cause of death, although there are no differentiated data for the region's indigenous population, which represents 2 percent of the total (Departmental Health Institute of Caquetá, 2006). The cultivation of coca for cocaine production and the resulting activities of the narcotics industry have triggered fragmentation of the social infrastructure and devastation of local ecosystems. All of this has serious environmental and cultural implications for the region's indigenous communities.

Health and nutrition challenges

Through the partial loss of traditional lands and severe deforestation, access to the Inga's traditional cultivated and wild plants for food and medicine has decreased significantly. This has had significant negative consequences on local economies and indigenous food and medical systems. In addition, the shrinking of the area available for traditional rotation crop production has reduced the Inga's capacity for self-subsistence. Difficult access to health services and the scarcity of these services in ethnic territories, as well as poverty and social and geographical marginalization add to the challenges.

The United Nations Millennium Development Goals[2] provide the foundation for the Colombian Government's social policies for overcoming these challenges. With the goal of upgrading the coverage and quality of the general health and social security system, two legislative reform initiatives have been brought before the Colombian Congress, "seeking to promote the inclusion of currently uninsured low-income population sectors; improve efficiency in the provision of public services, including health; and increase capacity building and accountability at the regional (territorial) levels" (PAHO, 2007).

The 2005 Colombian National Survey on Health and Nutrition indicates slight improvements in the department's indicators, including those for nutrition (PAHO, 2007). For example, in 2005, chronic malnutrition in children under five years of age had diminished slightly (to 12 percent) since 2000 (13.5 percent); acute malnutrition was observed in only 1 percent of children under five years of age; and among those aged five to nine years, 13 percent showed stunting and 5 percent low weight-for-height.

Nevertheless, problems persist, and the health statistics for the country as a whole obscure large differences among regions, between urban and rural areas and across social levels. Minority groups are characterized by high poverty rates, markedly inadequate basic sanitation services and a higher degree of health problems than other population groups. In the Department of Caquetá where this research took place, the predominant health concerns are acute diarrhoeal diseases, acute respiratory infections, malnutrition and nutritional deficiencies, tuberculosis (TB), periodontal diseases and skin disorders (PAHO, 2007).

[2] www.un.org/millenniumgoals/

In 2006, the year of the most recent government study, the Department of Caquetá's infant mortality rate was 32 per 1 000 (compared with a national rate of 20). For children under five years of age, mortality was 41 per 1 000 (compared with the national 26). The department therefore has one of the highest infant and early childhood mortality rates in the country (Departmental Health Institute of Caquetá, 2006). Prevalent childhood diseases are the main cause of death among children under five, with acute diarrhoea and respiratory infections being the most prominent. Figures for maternal and perinatal mortality are also above average: a maternal mortality rate of 98 per 100 000 live births compares with a national rate of 79, and a neonatal mortality rate of 18 per 1 000 live births with one of 7.4.

Participatory research

Research for the project was developed in a participatory fashion using the Centre for Indigenous Peoples' Nutrition and Environment (CINE) methodology (Kuhnlein *et al.*, 2006). Working with the research team, the community determined the methods for collecting, recording and presenting the data. The data collection process also triggered and enabled community awareness building about the factors that influence nutritional and general health.

A multicultural team established at the beginning of the project was responsible for developing project objectives and recommendations. In addition to the project coordinator and support team, four indigenous local promoters collected and recorded key nutrition information and sensitized the communities to the importance of nutrition and health, using terminology and imagery common to the Inga. (Hereafter, the promoters will be referred to as "the team".)

The team participated in several meetings with leaders of the Tandachiridu Inganokuna Association to draft a cooperation agreement in which the role of each member was discussed, described and established. The signed agreement reflected the communities' expectations, guaranteeing a participatory decision-making process and protecting the rights of indigenous communities.

Project objectives and participants

The project's general objective was to contribute to protecting the health and food security of the Inga people by promoting the maintenance and recovery of their traditional agricultural production systems, as part of their Life Plan. Specific project objectives were to:

- improve and increase the availability of traditional foods that are important for the Inga's health and nutrition;
- promote the maintenance and application of the Inga's ethical beliefs, knowledge and cultural practices related to nutrition and health;
- offer students of the Inga Yachaicuri School a primary health care programme;
- carry out an anthropometric assessment of the Inga ethnic group;
- develop and assess a programme for improving the health and nutrition status of the community sustainably.

Intervention activities

The project lasted from 2005 to 2008. During the first year, meetings were held with the Senior Council of the Inga indigenous group of Caquetá, to establish a research framework and obtain permissions. Local health and nutrition promoters and traditional agriculture promoters were selected, and training workshops held for them. The project's team of four promoters conducted the fieldwork with participating families. The promoters were trained in and informed about health promotion issues. Among the subjects discussed were the definitions of health and illness; child care; health care for pregnant women, adults and elders; prevention of the most frequent illnesses within the community, including colds, diarrhoea, fever, malnutrition and anaemia; promotion of healthy foods; and the cultivation and use of medicinal plants.

In the project's first phase, research was conducted to define the sources and nutritional composition of key traditional foods. Nineteen traditional foods were selected for promotion, based on their cultural and nutritional values, ease of retrieval in the *chagras*[3] and

daily use for family nourishment (Table 8.1) (Correal *et al.* 2009).

During development of the project's information and educational strategy, the promoters visited families on each of the indigenous reserves. These visits were to collect information on the composition of Inga families and their health status and environment; the Inga's beliefs and knowledge regarding traditional plants and medicine; the availability of traditional foods; Inga food use frequencies; environmental, social and economic indicators; the distribution of traditional seeds and knowledge; breastfeeding practices; and recipes.

The promoters worked closely with family members, highlighting the positive aspects of their health status as well as those that needed improvement. Areas of emphasis included: i) promoting the development of *chagras* for cultivating traditional foods; ii) exchanging traditional seeds among families; iii) maintaining the cleanliness of housing and local environments; iv) using plants from medicinal gardens to help treat diseases; v) implementing recommendations for the improvement of health care; vi) implementing recommendations on nutrition and health care for senior groups; vii) promoting breastfeeding; and viii) discouraging the use of powdered milk for babies.

Each family was visited at least three times during each year of the project. The first visit was used to obtain information on the Inga's knowledge of health and their commitment to making specific improvements, especially by implementing recommendations for improving health care. The second and third visits assessed families' implementation of these recommendations. During the visits, promoters observed that the communities and families showed the most interest in improving their *chagras* and in seed exchanges, culinary festivals, and *yoco* and *yagé* ceremonies.

---

3   A *chagra* is a family and/or communally managed plot based on a diversified and sustainable production system that imitates the forest eco-system's dynamics by combining agricultural and forestry technologies. Its fauna and flora components are closely interrelated and selected to ensure the protection and sustainable use of the soil and other forest resources. The project's indigenous partners perceived the *chagra* as being like a market store that provides indigenous people with a supply of daily nutritional needs near their homes. The *chagra* is an example of the application of accumulated, inter-generationally transmitted indigenous knowledge about the harvesting and use of plants.

Table 8.1 Traditional foods promoted

Common name (scientific name)	Nutrients provided	Importance to health according to female and male indigenous healers
Chontaduro (Bactris gasipaes Kunth)	Protein, fat, fibre, vitamin A	Promotes proper growth of children; prevents malnutrition; protects against lung disease; helps maintain healthy skin and good vision
Milpés (Oenocarpus bataua Mart)	Fat, protein, fibre	Promotes proper growth of children and proper foetal development; provides energy for daily activities; prevents malnutrition; protects against heart disease; aids the digestive process
Mojojoy (Coleoptera spp.)	Protein, fat	Promotes proper growth of children and proper foetal development; provides energy for daily activities; facilitates weight gain; prevents malnutrition; protects against lung disease
Zapote (Matisia cordata Bonpl.)	Vitamin A, vitamin C	Protects against lung disease; helps maintain healthy skin and good vision; protects against colds; helps to heal wounds; protects against heart disease
Yoco (Paullinia yoco Schultes & Killip)	Not available	Mild stimulant and general health tonic
Cayamba (Auricularia auricular-judae (Bull.) Quel.)	Protein, fibre, minerals	Facilitates immune response; improves the digestive process; helps prevent the body from absorbing fats from foods, thereby protecting the heart and circulatory system
Ant (Atta spp.)	Fat, protein, niacin	Supports the functioning of the digestive system; protects the skin from infections; promotes a healthy nervous system; helps the body to produce energy
Snail (Pomacea maculate Perry)	Protein, phosphorus	Promotes proper growth of children and proper foetal development; prevents malnutrition; improves the body's defences against diseases
Cucha (Hypostomus sp.)	Protein, phosphorus	Promotes proper growth of children and proper foetal development; prevents malnutrition; improves the body's defences; facilitates the formation of bone and teeth
Cimarrón (Eryngium foetidum L.)	Iron	Prevents and treats anaemia; aids treatment of hepatitis
Ají (Capsicum L.)	Vitamin A, vitamin C, minerals, capsaicin, potassium	Protects against cancer; helps the digestive process; prevents bronchitis
Yam (Dioscorea spp.)	Carbohydrates	Prevents malnutrition; increases energy
Pineapple (Ananas cosmosus L.)	Vitamins, minerals, bromelain	Improves digestion and circulatory process; cleanses the intestines
Banana (Musa spp.)	Carbohydrates	Prevents low weight; increases energy
Sour cane (Begonia plebeja Liebm.)	Not available	Purgative and antipyretic
Nina Waska (not available)	Not available	Purgative; promotes internal cleansing
Papaya (Carica papaya L.)	Vitamin C, minerals, fibre, papain	Improves the digestive process; cleanses the intestines
Arazá (Eugenia stipitata McVaugh)	Vitamin C, fibre	Protects against colds; helps to heal wounds; protects against heart disease

The following were the recommendations for family and community health care:

- Drink something bitter once a week.
- Apply nettle.
- Do not eat sweets from the town.
- Eat abundant fruits.
- Women should take care during their menstrual period.
- Purge three times a year.
- People should take care when they have a cold.
- Consume aromatic plants in teas and juices, to avoid diseases.
- Eat only traditional foods and meals made from traditional recipes.

The following recommendations for different age groups are based on traditional Inga knowledge, which is shared by Inga shamans and elders during family visits.

Health care and nutritional support for children under two years of age

- Young children should be swathed in cloth diapers.
- To prevent skin ailments, bathe children with plants such as cane and maize leaves, *yarumo* (*Cecropia peltata* L.) and *balso* (*Ochroma pyramidale* Urban).
- Bathe children with herbs such as mint (*Mentha pulegium* L.) or wormwood (*Artemisia absinthium* L.), without rinsing.
- Leave children's clothes to dry in the air.
- Breastfeeding is recommended for at least the first three months. Bottle-feeding should be avoided.
- Remove parasites with mint (*Mentha pulegium* L.), purslane (*Portulaca oleracea* L.) and *paico* (*Chenopodium ambrosioides* L.).
- From one year of age, children should eat all categories of traditional nourishment. The only food that is not recommended for children under two years of age is chilli (*ají*).
- Animal heads should not be consumed by children under two years of age.
- Children can drink all kinds of *chicha*, as long as it is not too fermented.
- During teething, children should be given food with sufficient texture for biting and tearing.

Health care and nutrition for children aged two to 12 years

- Parasites should be removed by using purgatives such as *yoco* (*Paullinia yoco* Schultes & Killip).
- Lice should be prevented by applying *achapo* husk juice (*Prochilodus nigricans* Agassiz).
- All children should be taught respect for family and traditions, and to accept advice from their grandparents and elders.
- Children should be taught to honour the indigenous culture, customs, language, clothing, myths and legends, games and food.
- Children should be fed with all types of traditional food.
- Children should not be breastfed or bottle-fed.

- Give children traditional drinks such as *chicha*, *anduche* and *guarapo* that are sweet but not strongly fermented.

Health care and nutrition for adolescents, adults and the elderly

- Adolescents and adults can consume all types of animal.
- Traditional healers and apprentices should not consume pregnant animals or fish with teeth.
- Purgatives should be taken during a new moon.
- *Yagé* or *ambiwaska* (*Banisteriopsis caapi* [Spruce ex Griseb.] C.V. Morton) should be taken every two weeks.

Health care for women during their menstrual period

- Do not bathe in cold water, unless special plants such as lemongrass (*Cymbopogon citrates* Stapf.) or other herbs are added.
- Do not swim in the river: "coldness" may be passed to the uterus and the woman will suffer during labour.
- Do not lift heavy things.
- Do not drink milk or eat cassava or fruits, to avoid cystitis or burning urination.
- Women should avoid eating *danta* (*Tapirus terrestris* L.), *gurre* (*Cabassous unicintus* L.), *cerrillo* (*Tayassu tajacu* L.) and deer (which are irritating meats) and chilli (*ají*) (*Capsicum* L.).

Health care during pregnancy and breastfeeding

- Do not lift heavy things.
- Bathe in herbs such as basil (*Ocimum basilicum* L.), *altamiza* (*Artemisia vulgaris* L.), *ajenjo* (*Artemisia absinthium* L.) and orange leaf; these plants can be mixed.
- During pregnancy women should receive palpations from midwives or women healers to determine whether the baby is healthy.

The interviews conducted during visits were based on the individual physical health questionnaire (Annex

Table 8.2 Activity indicators with projected and final numbers achieved

Indicator	Projected number	Actual number
Families benefiting from the project	80	60
Hectares supported	N/A	81
Sustainable production projects for food security	N/A	19
Nutritional and health promoters trained and working with communities	4	4
People participating in *yoco* ingestion	100	140
Health brigades	5	5
People participating in health brigades	N/A	270
Information activities on health subjects (workshops on health and nutrition information)	5	7
People participating in health information activities	N/A	176
Traditional foods recovered	19	19
Schools using school *chagras*	4	4
Schools using traditional foods in their cafeterias	4	4
Radio programmes on health and food	N/A	50
Promotional literature products (flipchart and cookbook)	2	2

8.1), the dietary frequency questionnaire, a 24-hour recall, the infant food history survey (Annex 8.2) and the food security interview (Annex 8.3). The individual physical health questionnaire collected information on indicators including diminished visual perception at night, pallor, hair problems, oral lesions and bleeding of the gums. The dietary frequency questionnaire and 24-hour recalls were used to identify the main traditional and non-traditional foods and their frequencies of consumption in participants' families. The infant food history survey gathered data on breastfeeding practices and the health care status of indigenous children. The food security interview was used to assess families' perceptions regarding food availability. Among the questions asked and discussed were: Do you always have food? Do you buy food? Do you ever go hungry? Do you cook or provide food for others? Families' traditional food preferences were

identified through analysis of the information collected through this tool. Anthropometric data were collected for youth ≤ 18 years of age; information was classified into the indicator categories weight-for-age, height-for-age, and weight-for-height.

In addition, workshops and courses were conducted to build awareness and understanding both in the communities and among the students of Yachaicuri School. Workshops and group activities on nutrition, nourishment and health were developed in each indigenous reserve. Traditional food recipes were collected and prepared during culinary festivals. Promotional literature and visual materials regarding the recommendations for improving community health were prepared by the local promoters using information collected from community elders, particularly traditional healers.

Community and school *chagras*, farming projects and medicinal plant gardens were established to increase the availability of traditional foods and medicine. Students helped to create school *chagras* where cilantro, *cimarrón* (*Eryngium foetidium* L.), onion (*Allium* sp.), cucumber (*Cucumis sativus* L.) and other vegetables were grown for the children, and medicinal plants were cultivated.

Culinary festivals, seed exchanges and recipe collections were organized to promote the use of traditional foods and to identify the plants used and encourage the cultivation of traditional food crops in *chagras* and family gardens.

Promoters visited the communities to evaluate the menus of school cafeterias. For each school, the project prepared menus that included at least one traditional food preparation. Ways of preparing foods harvested in the school's *chagra* were recommended. The Colombian Institute of Family Welfare's menus were revised to include traditional drinks such as *anduche*, *chicha* and *chucula* (banana whipped with water) and foods such as *tacacho* (cooked and mashed banana).

The local promoters visited schools on the indigenous reserves to develop educational activities and introduce traditional foods, especially the 19 foods identified during the preliminary research. The nutritional and cultural importance of these foods

Table 8.3 **Numbers of people participating in project activities**

Age range (years)	Female	Male	Total
< 1	6	3	9
1–4	7	13	20
5–10	20	17	37
11–15	28	19	47
16–20	11	12	23
21–25	6	11	17
26–30	6	3	9
31–35	4	6	10
36–40	3	5	8
41–45	5	6	11
46–50	2	0	2
51–55	2	2	4
56–60	4	8	12
61–65	4	1	5
66–69	1	0	1
≥ 70	2	2	4
Total			**219**

Table 8.4 **Observed and self-reported health conditions (percentages)**

Condition	Baseline 2006 (n = 108)	Final assessment 2008 (n = 98)
Presence of oedemas	7.8	3.4
Pallor of the skin	48	44.8
Self-reported bleeding gums	13.9	11.5
Self-reported hair problems	13.1	13.8
Self-reported oral lesions	1.7	0
Self-reported night blindness	63	24.1

was highlighted, and recipes were prepared for the students to taste.

The Inga consider frequent invitations to drink *yoco* (*Paullinia yoco* Schultes & Killip) as being vital to their nutrition and health, along with the periodic drinking of cleansing plants. *Yoco* has traditionally been used as a stimulant, owing to the high caffeine content of its bark; it is also used as a laxative and in many other traditional indigenous treatments in the foothills. In each indigenous reservation, five health brigades provided services through a *taita* (traditional healer) and an apprentice.

Table 8.2 lists the project activities, with the anticipated and actual numbers reached.

Project results: improved health, nutrition and food availability

A total of 219 indigenous people from the five Inga indigenous reserves participated in project activities. Participants were from all age groups: 51 percent were women and 49 percent men. Children up to 15 years of age (51.6 percent) were the major population group participating in activities (Table 8.3).

Individual physical health questionnaire

Answers to questions on the self-perception of health status revealed that 60 percent of participants considered their health to be average; 36.7 percent considered themselves to enjoy good health; and 3.3 percent considered themselves to be in poor health. For the final evaluation, health status was deemed average in the presence of "pain in the bones", a common complaint resulting from work in the fields and, according to indigenous beliefs, snakebites. Noteworthy was that 100 percent of participants used traditional medicinal practices to prevent or treat health problems. An important improvement in night blindness was reported (Table 8.4).

Food frequency and 24-hour recall

All Inga families used plants and animals from their *chagras* to prepare their daily meals. Foods such as plantain (*Musa* spp.), yucca (*Manihot esculenta* Crantz), *chontaduro* (*Bactris gasipaes* Nunth), *píldoro* (*Musa* sp.) and *yota* (*Xanthosoma* sp.) were regularly prepared. Eighty-two percent of families consumed fruits weekly. The types of fruit consumed depended on the harvest cycle, and the most frequently used were guayaba (*Psidium guajava* L.), orange (*Citrus sinensis*

Table 8.5 Contributions of kilocalories, protein, iron and vitamin A to daily intake, traditional food (TF) versus non-traditional food, using 24-hour recalls (n = 58) (percentages)

	kcal		Protein		Iron		Vitamin A	
	TF	Non-TF	TF	Non-TF	TF	Non-TF	TF	Non-TF
Baseline 2006	47	53	60	40	14	86	80	20
Final assessment 2008	57	43	70	30	50	50	100	0

Table 8.6 Anthropometric nutritional evaluation of youth ≤ 18 years of age

		2006 (n = 227)		2008 (n = 127)	
Indicator	Age range (years)	No. participants	Participants < -2SD (%)	No. participants	Participants < -2SD (%)
Weight-for-height	< 5	41	0.8	29	0.6
Weight-for-age	< 5	41	3.6	29	3.5
	5–10	71	4.6	37	4.6
Height-for-age	< 5	41	14.5	29	14.4
	5–19	186	14.7	98	14.7

Osbeck), mandarin (*Citrus reticulata* Blanco), banana (*Mussa* sp.), *guama* (*Inga edulis* Mart), pineapple (*Ananas cosmosus* L.), *zapote* (*Matisia cordata* Bonpl.) and *arazá* (*Eugenia stipitata* McVaugh). Traditional drinks were prepared daily: maize *chicha*, *chontaduro chicha*, *anduche* (a banana drink) and cane *guarapo* were the most frequent. Fish was the most frequently consumed animal food (three or four times a week), while eggs were consumed daily among families with poultry farming facilities. Traditional foods that were not consumed frequently included beef, yam and hard-to-obtain foods such as snails and *milpes* (*Oenocarpus bataua* Mart).

The contributions to energy (kcal) and protein of traditional and non-traditional foods were calculated in the project's first phase and during the evaluation activities. Interviews based on 24-hour recalls were conducted with approximately 15 people from each community – the majority were students chosen randomly in schools – to observe variations between the percentage contributions of foods consumed and the nutrition recommendations provided by the Colombian Institute of Family Welfare (ICBF, 2005).

The contributions to kcal, protein, iron and vitamin A of indigenous traditional foods increased between the first and second phases of the project (Table 8.5). The kcal contribution of traditional foods was high because the traditional Inga diet is rich in carbohydrates and kcal, mainly from plantain, yucca and yam. Non-traditional foods in families' food baskets included rice, pasta and sugar. Traditional foods' contribution of protein increased by 10 percent among participating

Table 8.7 Duration of exclusive breastfeeding, mothers with children ≤ 2 years of age (n = 18)

Months from birth	Exclusive breastfeeding (%)
1	16.7
2	5.6
3	10.0
4	5.6
5	18.9
6	32.0
7	5.6
11	5.6

families, owing to increased consumption of the eggs and meat produced by families with poultry. This, combined with the use of *cilantro cimarrón* (*Eryngium foetidium* L.) in the preparation of foods also increased the amount of iron in participants' diets. Non-traditional foods such as beans and lentils were used three times a week at lunch and dinner, completing the families' nourishment. Traditional foods such as chilli (*Capsicum* L.) and fruits such as *zapote* (*Matisia cordata* Bonpl.) and papaya (*Carica papaya* L.) supplied vitamin A.

Anthropometric assessment

Despite improvements in health indicators and dietary assessments, there were minimal improvements in anthropometric indicators from 2006 to 2008. This was expected, because height indicators (stunting) are persistent. Few children had serious weight-for-age or weight-for-height deficiency (Table 8.6). This analysis emphasizes the need to generate specific parameters for comparisons of data from Colombia's indigenous communities, whose populations are on average shorter than those of the country as a whole.

Infant and child nutrition: infant food history survey

Inga people consult both traditional health agents, such as midwives and relatives, and Western doctors and nurses. During childbirth, 50 percent of the mothers surveyed used Western medicine, and the other 50 percent used midwives or relatives trained to help with childbirth. This is well known by the region's health entities: regional health action plans for improving mother-and-child health include the training of non-institutional midwives, indigenous health promoters, etc. to assist indigenous women during pregnancy and childbirth (Ministry of Social Protection, 2008).

As a rule, the indigenous women of the community breastfeed: 100 percent of survey participants stated that they started breastfeeding soon after the child's birth, with 78 percent starting one hour after delivery. This is significantly higher than the 47.2 percent of women who reported starting breastfeeding an hour after childbirth in the National Survey of the Nutritional Situation in Colombia in the Department of Caquetá (ICBF, 2005). For the current study, local promoters visited pregnant women and assisted with childbirth, teaching the proper position for breastfeeding and advocating against the use of traditional drinks and baby bottles during the baby's first month of life. Of the women interviewed, 57 percent stated that they breastfed exclusively for the child's first four to six months (Table 8.7). Twenty-seven percent stopped breastfeeding altogether at one year, and 18 percent stopped at two years; 64 percent of women breastfed for at least one year. The reasons for weaning are listed in Table 8.8.

In response to these findings, recommendations were developed to promote the "golden rule" of breastfeeding ("the greater the stimulus, the greater the production of milk"), the benefits for children's food security from breastfeeding, and the associated reduced food costs for the family. In indigenous families, complementary feeding begins between the third and sixth month of life, when mothers offer their children traditional foods such as fish, vegetables and plantain drinks. By ten months of age, children are ready to consume all the traditional and non-traditional foods eaten by the family. Table 8.9 summarizes the main foods used during the introduction of complementary foods.

During their visits, the local promoters worked closely with family members to collect information on

Table 8.8 **Reasons for stopping breastfeeding, mothers with children ≤ 2 years of age (n = 18)**	
Reason	*% mothers*
Child hungry	27.1
Work activities	20.0
Lack of milk production	19.5
Food support from government institutions	13.3
Introduced non-dairy beverages	13.4
Mother ill	6.7

Table 8.9 Complementary foods offered

Food or preparation	Traditional or homemade	Non-traditional or purchased	Age at introduction (months)
Fish soup	√		3
Chicken soup		√	3
Meat soup	√		4
Plantain soup	√		4
Vegetable soup		√	4
Plantain drink	√		4
Pumpkin	√		4
Meat	√		6
Guava	√		7
Orange	√		7
Chucula[a] and anduche[b]	√		7
Pineapple	√		8
Yucca	√		8
Yam	√		9
Pomo[c]	√		9
Cherimoya	√		9
Grape	√		8
Yota[d]	√		9
Cucumber archucha	√		9
Eggs	√		11
Sweet chichi[e]	√		14

[a] Cooked ripe plantain.
[b] Traditional drink made from fermented ripe plantain.
[c] Mountain apple (*Eugenia malaccensis* L.).
[d] Tuber (*Xanthosoma* spp.).
[e] Sweet drink made of fermented maize.

traditional food preparation practices. Of note is that the pre-mastication practices used by female ancestors to prepare traditional drinks for young children are no longer in use. Local promoters remembered drinking cassava *chicha* prepared through pre-mastication by their grandmothers and aunts. Following Inga traditional practices and beliefs, the *chicha* was prepared without sugar; the drink's sweet flavour resulted from the women's saliva mixing with the tuber. Younger mothers now believe that this is not hygienic, and have no memory of having participated in this type of food preparation. Occasionally, indigenous women chew lightly on foods of hard consistency (such as meat) to facilitate infants' consumption of complementary foods (from zero to four months). Today, no pre-mastication practices are used to prepare traditional drinks such as *anduche*, *chicha* or *chucula*.

Food availability: food security interview

Twenty-four households participated in the food security interview. One of the programme's most important achievements was to increase the availability of traditional foods, especially the 19 that had been selected, through activities that included *mingas*[4] at the *chagras* (Correal *et al.*, 2009).

All of the families interviewed (100 percent) stated that they considered traditional foods to be healthy and nutritious: 64 percent stated that they had access to the quality and types of food they preferred, while 36 percent stated that their families still had insufficient traditional foods for consumption. The factors that prevented indigenous families from achieving their preferred food consumption – shortage of traditional foods, decreased hunting or fishing stocks, and/or lack of financial resources for purchasing other goods – were resolved by families working to improve the quantity and quality of the food they grow and through conservation strategies for preserving fauna species in nearby forests and rivers.

It is noteworthy that 100 percent of the families stated that both youth and adults had regular access to certain quantities and qualities of traditional foods, and that during the project period no family members lost weight because of a significant reduction in their food consumption or a lack of food for an entire day. All families interviewed stated that traditional farm and garden foods – including meat, fish, yucca, plantain, onions, tomatoes, fruits and foods derived from livestock, such as cheese and milk – were always shared in their immediate community.

Family food security among the Inga is based on plant cultivation and harvesting and animal breeding. Following the establishment and strengthening of

---

[4] A *minga* is a work project that engages many of the people in the community.

Table 8.10 Production from *chagras* and farming projects, 2007 to 2008

Indigenous reserve	No. families	Area farmed (ha)		No. chagras		Area of chagras (ha)		Species in chagras*		Medicinal species*		Food species*	Species for other uses*
		2007	2008	2007	2008	2007	2008	2007	2008	2007	2008	2008	2008
San Miguel	15	16.25	29.88	18	25	11.5	13.75	42	61	0	24	34	3
Brisas	11	12	9.62	15	12	6.75	5	35	52	0	15	34	3
Yurayaco	9	0	11.44	9	12	7.25	7.5	28	47	0	7	38	2
Niñeras	14	33	29.5	21	30	11.75	12.51	70	106	0	27	63	16
Total	**49**	**61.25**	**80.44**	**63**	**79**	**37.25**	**38.76**						

* Some species were grown and used in more than one community.

family and school *chagras*, 71.4 percent of the families participating in project activities expressed decreased anxiety regarding food availability because food was always available in their *chagras*. Other goods (rice, pasta, butter, oil and salt) were purchased to complete the families' diet, but these products were not consumed daily. Seventy-two percent of the families had poultry farming facilities, and 37 percent produced milk and cheese for family consumption.

The families interviewed believed that the availability of traditional foods increased over the two-year project period. Families are now growing more species in larger *chagra* and garden areas. The diversity of species found in the average family chagra increased by 54 percent during the project. Each indigenous reserve now contains between 47 and 106 species, including the 19 traditional foods identified in the preliminary research (Table 8.10). There was also a marked increase in the harvesting of plants for medicinal use. Culinary festivals and seed exchanges were key to strengthening the use and cultivation of *chagras*.

The local promoters' activities to improve community nutrition and health care also supported the promotion, recovery, sowing and use of traditional seeds in accordance with traditional customs. Some traditional practices and knowledge resurfaced for the community's use. Families showed great motivation and willingness to continue recovering traditional foods and agricultural practices. Inga youth participated actively in educational activities such as tending school *chagras* and community plant nurseries. The strengthening

Table 8.11 Medicinal plants encouraged by local promoters

Plant	Use
Ambar (*Tetracera sessilliflora* Triana & Planch)	Infusion to calm nervous breakdowns and for fevers, headaches and kidney ailments
Chondur (*Cyperus* sp.)	For hair problems
Descancel (*Compositae* Bercht & J. Presl.)	For fevers and headaches, and to ease labour
Paico (*Chenopodium ambrosioides* L.)	For parasitic infection
Chiricaspi (*Bruntelsia grandiflora* Plum ex. L.)	For physical pain in general
Yawar chondur (*Cyperus* sp.)	For headaches
Ruda (*Ruta graveolens* L.)	For fevers
Limoncillo (*Cymbopogon cytratus* (DC Stapf.)	For menstrual pain and colic
Hojas de naranja (*Citrus aurantium* C. sinensis Osbeck)	For menstrual pain and colic
Bitter cane (*Costus spicatus* L.)	For fevers
Tabardillo/oreja negra (*Calliandra califorica* Benth.)	For fevers
Kalambombo (*Averrhoea carambola* L.)	For cuts and skin irritations
Hoja Santa (not available)	For headaches and acne
Nettle (*Urtica* L.)	To calm nervous breakdowns and for coughing
Sauco (*Sambucus mexicana* L.)	For eye irritations and as a purgative
Toronjil (*Melissa officinalis* L.)	To calm nervous breakdowns
Flor de muerto (*Cistus albidus* L.)	For stomach pains
Malva (*Malva sylvestris* L.)	For fevers
Achiote (*Bixa orellana* L.)	For cuts and skin irritations
Cat's claw (*Ucaria* spp.)	For kidney ailments and to clean the blood

of family and school *chagras* through the recovery of traditional seeds provided an additional benefit beyond fresh, healthy and nutritious foods, by providing venues where Inga elders, adults and youth can come together to share their knowledge of ancestral agriculture techniques.

Ongoing project activities by promoters include assisting the expansion of *chagras* and the production of organic fertilizers for soil restoration; providing technical assistance and onsite advice; assisting the continued recovery of traditional seeds; and encouraging the consumption of traditional rather that purchased foods.

Traditional medicine

Supplies of traditional seeds and oversight were provided during the project, to strengthen the communities' medicinal plant gardens. Awareness and understanding of the use of traditional medicine were increased through health brigades of *taitas* (shamans) and *mamas* (women healers) from the unions UMIYAC (*Unión de Médicos Indígenas Yageceros de la Amazonía Colombiana*) and ASOMI (*Asociación de Mujeres Indígenas: La Chagra de la Vida*). Local promoters encouraged the use of some medicinal plants (Table 8.11).

Conclusion

Barriers to implementation and data collection

Data were collected through informal interviews and discussions. Generally, the Inga feel more comfortable when activities are conducted informally. They believe that projects should support their Life Plan, rather than merely diagnosing or documenting their lifestyle. Notably, the Inga do not approve of the collection of blood samples and subsequent medical laboratory analysis.

The group of participants changed during the project. Some of the people who participated in the first phase left the indigenous reserve, while new indigenous people arrived in the research area during the course of the project. It was therefore difficult to develop precisely comparable information for before-and-after analysis of the intervention.

Armed conflict in the project territories occasionally hindered access to the Inga region during the project period. In addition, the Inga indigenous reserves in Caquetá are difficult to reach: travel is mainly by foot, so it can require hours to reach a community.

Project achievements

Traditionally, the main staples of the Inga group were locally obtained foods such as manioc and wild game. Increased contact with the outside world and the associated increased consumption of processed foods caused deterioration in the health status of Inga communities. This project's goal was to emphasize the importance of forest resources in supporting the Inga's nutritional health.

The project's promotion of a sustainable economy based on indigenous communities' traditional values resulted in improved health and nutrition in Inga families. The support provided to family and school *chagras* and the establishment of farming projects and medicinal gardens – as well as family visits, workshops and courses in nutrition and health, culinary festivals, seed exchanges, the collection of traditional recipes and health brigades – were fundamental to strengthening the health component of the Inga community's Life Plan.

Support to indigenous families and the local indigenous association in developing and implementing the health component of their Life Plan strengthened communities' governance and facilitated their engagement with traditional authorities, health promoters and healers. The Inga youth population became involved in promoting and preserving both indigenous culture and indigenous knowledge regarding environmental conservation. The Inga's holistic integration of all aspects of daily life – including education, health care, agriculture, land tenure, interaction with the environment and cultural expression – allowed the project implementers to integrate health and nutrition activities into an overall health improvement plan. The project strengthened the conservation and application of traditional indigenous knowledge and traditional food consumption.

In the Inga group of the Department of Caquetá, nutritional, health care and environmental challenges persist, despite the significant increase in local awareness. Specific health care models have still to be defined with the community, and more work is needed to bridge the communication gap between community leaders and youth regarding health issues.

The project built capacity through training workshops, enabling the Inga indigenous association to implement the health component of its Life Plan effectively, and ensuring project sustainability. In addition, the expansion of ancestral territories – particularly links between the Alto Fragua Indi Wasi National Park and the Yurayaco and San Miguel indigenous reserves – allowed the development of conservation strategies for the sustainable use of the natural resources that nourish the Inga community.

The project results define a path for further community outreach. Three other Colombian indigenous communities have improved their health status through food security projects initiated by the project's implementing agency, the Amazon Conservation Team. These encompass 638 families from 38 indigenous reserves. As well as the Inga of the Caquetá, the beneficiaries are the Coreguaje ethnic group in the vicinity of the Orteguaza River; the Inga of the Baja Bota Caucana; and the Siona community along the Putumayo River. The traditional agricultural activities implemented include seed exchanges and training workshops on health, sustainable production and conservation. A total of 797 traditional *chagras* have been established and supported, covering a total of 440 ha. Agroforestry plots and poultry and cattle farms have also been installed. Future activities will include food security projects for the Inga and Kofán communities of the Department of Putumayo and the Uitoto community of the Amazonas Region ✳

Acknowledgements

The authors extend their appreciation to the Inga ethnic group of the Caquetá region, particularly to all the Indigenous People of the following reserves: Cusumbe, Yurayaco, Brisas del Fragua, San Miguel and Niñeras. The leaders of the Tandachiridu Inganokuna Association provided invaluable assistance to the development of an appropriate approach and planning methodology for this project. Many individuals and organizations contributed their expertise. The authors would like to thank Eva Yela, Antonia Mutumbajoy, Mayorly Oliveros and Libia Diaz for their critical assistance in locating data and motivating the communities. Finally, much appreciation goes to the Amazon Conservation Team, with special mention of David Stone and Mark Plotkin, and to the Centre for Indigenous Peoples' Nutrition and Environment, whose financial and technical support made this work possible.
> **Comments to:** lmadrigal@amazonteam.org

>> Photographic section **p. XIII**

Annex 8.1 Individual physical health questionnaire

Date: **No.**

First name and last name:

Gender: **Age:**

Birth date:

Height: **Weight:**

Presence of oedemas:

Self-perception of health:

Do you have or have you had a disease?

What was the disease's cause?

Are you under treatment?

Are you taking *pharmacy medicines*, and which ones?

Are you taking *traditional medicines*, and which ones?

Can you see properly in the dark?

Do you have hair problems?

Pallor:

Do you have wounds or scars in your mouth?

Do your gums bleed easily?

Women

Are you pregnant?

How many pregnancies have you had?

How many childbirths have you had?

How many of your children are alive?

Are you breastfeeding?

Have you breastfed your children?

Children

Are they being breastfed?

How long were they breastfed?

Did they take different kinds of milk, which ones and when?

How old were they when you started giving them complementary food?

How was it given?

Do they take nutritional supplements or vitamins, which ones?

Family

Do you use iodized salt?

Do you drink alcohol?

Annex 8.2 Infant food history survey

Date: **No.**

Child's name and last name:

Age:

Mother's name and last name:

Indigenous reserve:

Local promoter's name:

Who helped you with childbirth?
(*midwife, local promoter, doctor, nurse, relative, other*)

When did breastfeeding start? Hours after childbirth?

For how long was the child breastfed?

Did you offer the child other types of milk, and why?

At what age and with what types of foods did you start complementary feeding? (*foods, age and preparation procedures*)

At what age did you stop breastfeeding your child? (*months*)

Why did you stop breastfeeding?

Respondent's name:

Date:

Indigenous reserve:

These questions make it possible to know your family's food security status over the past year.

Which of the following statements best describes what your family has eaten over the last 12 months?
1. In your family there is always the amount and type of food you want to eat.
2. Your family has the amount of food you want but not the kind of food you want to eat
3. Sometimes your family does not have enough food to eat.
4. Almost always your family does not have enough food to eat

If your family does not produce food, are you concerned that foods are gone before you have money to buy them?
a. Almost always
b. Rarely
c. Never

Are the foods that your family eats gone before you can obtain or produce more?
a. Almost always
b. Rarely
c. Never

If you do not produce food can you give the children in your family a balanced diet?
a. Almost always
b. Rarely
c. Never

Do you or another adult in your family eat less or skip a meal because there is not enough food for all the family?
a. Yes
b. No

If you answered "Yes" to the last question, how often does this happen in a year?
a. Almost every month
b. Some months but not all
c. Only one or two months

Have members of your family lost weight because they do not have enough food to eat?
a. Yes
b. No

Have you or another adult in your family not eaten all day because there is no food in the family?
a. Yes
b. No

If you answered "Yes" to the last question, how often does this happen in a year?
a. Almost every month
b. Some months but not all
c. Only one or two months

Has a child in your family had to skip a meal because there was no food in the house?
a. Yes
b. No

If you answered "Yes" to the last question, how often does this happen in a year?
a. Almost every month
b. Some months but not all
c. Only one or two months

Has a child in your family been hungry because there is not enough food in the house?
a. Yes
b. No

Has a child in your family not eaten all day because there is not enough food in the house?
a. Yes
b. No

Which of these factors or items prevents you from eating enough food or food that you would prefer to eat?
a. Age
b. Health problems
c. Lack of money
d. Lack of food in the area, or a place to buy it
e. Food markets are too far away
f. There are no traditional foods in the area
g. Do not know how to or cannot hunt or fish

Comments:

Do you and your neighbours share food?
a. Yes
b. No

If you answered "Yes" to the last question, which foods are shared?

List five foods that you buy for everyday use in feeding your family.

What foods do you produce for daily use in family meals?

What is good traditional food?

What are the problems with traditional foods?

In the last two years have you noticed changes in the amount of traditional food species? Please explain your answer.

In the last two years have you noticed changes in the quality of traditional food sources? Please explain your answer.

What are your favourite foods?

The value of Inuit elders' storytelling to health promotion during times of rapid climate change and uncertain food security

GRACE M. EGELAND[1] SENNAIT YOHANNES[1] LOOEE OKALIK[2] JONAH KILABUK[3]

CASSANDRA RACICOT[1] MARCUS WILCKE[4] JOHNNY KULUGUQTUQ[5] SELINA KISA[4]

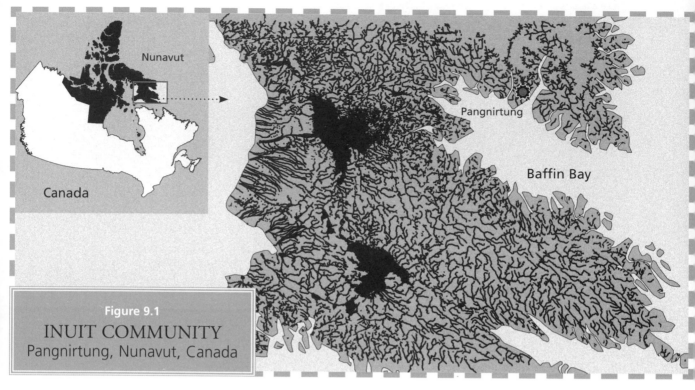

Figure 9.1
INUIT COMMUNITY
Pangnirtung, Nunavut, Canada

Nunavut

Canada

Pangnirtung

Baffin Bay

Data from ESRI Global GIS, 2006.
Walter Hitschfield
Geographic Information Centre,
McGill University Library.

1
Centre for Indigenous
Peoples' Nutrition
and Environment (CINE)
and School of Dietetics
and Human Nutrition,
McGill University,
Montreal, Quebec,
Canada

2
Health Programs, Inuit
Tapiriit Kanatami, Ottawa,
Canada

3
Community member,
Community Health
Promotion Steering
Committee, Pangnirtung,
Nunavut, Canada

4
Pangnirtung Health
Centre, Government of
Nunavut Health and Social
Services and Community
Health Promotion Steering
Committee, Pangnirtung,
Nunavut, Canada

5
Baffin Region Health
Promotion Office,
Government of Nunavut
Health and Social Services
and Community Health
Promotion Steering
Committee, Pangnirtung,
Nunavut, Canada

Key words > diet, health promotion,
storytelling, food security, climate change,
aboriginal health, Inuit, Indigenous Peoples

Photographic section >> XVI

Abstract

The ongoing nutrition transition in the Canadian Arctic is resulting in an epidemiologic transition towards the emergence of obesity and obesity-related chronic diseases. In response, the community of Pangnirtung in the Baffin Region of Nunavut, Canada, in partnership with the Centre for Indigenous Peoples' Nutrition and Environment, developed a community health promotion project in two phases. The first phase involved collecting health behaviour data from adults (2005) and youth (2006), and recording and transcribing elders' stories on the value of traditional food, including plants and remedies (2006 to 2007). In the second phase, the health behaviour survey data and storytelling were used to help develop an innovative pilot intervention in the community (2008 onwards). The intervention aimed to increase knowledge about traditional food and nutrition and improve nutritional health behaviours through the age-old Inuit tradition of storytelling. It targeted youth and young adults because of community members' concerns that youth were consuming more high-sugar drinks and "junk food" and less traditional food than older adults in the community.

The youth survey found that youth had consumed an average of 1.4 litres of sweet drinks a day, including two cans of pop, over the previous month. It also found that only five traditional food species had been consumed by more than 80 percent of the youth over the previous year, and that youth had a strong preference for caribou meat, with 98.7 percent of them consuming caribou in the past year, at an average of 87.2 g per day among consumers. No other traditional food was consumed to the same degree.

Elders' stories were incorporated into a DVD promoting knowledge and appreciation of a wide range of traditional foods. The stories were also incorporated with modern nutritional health advice for youth radio drama programmes aimed at reducing the high consumption of pop in the community. The DVD and radio programmes have already been pilot tested for effectiveness, cultural relevance and acceptability, and a broader community-wide evaluation of the community radio's nutritional health promotion is currently taking place.

In addition, elders' storytelling revealed elders' perceptions of climate change and its impacts on local flora and fauna, and their resulting concerns for the sustainability of subsistence food species. With climate change now outpacing projections, and potentially threatening favoured subsistence species, elders' storytelling can be a means of building youth's awareness and appreciation of the full range of traditional food available and increasing the diversity of traditional foods consumed. Elders' storytelling also provides opportunities for understanding changes in a historical context and, when combined with modern-day nutrition issues and modern media, may be a means of reaching youth, building social cohesion and promoting Inuit resiliency in a time of rapid climate change and uncertain food security.

Introduction

A nutrition transition has been documented in the Canadian Arctic, with increased consumption of processed market foods that are high in sodium, saturated and trans-fat and added sugars, and with reduced consumption of traditional food (TF) leading to consequences such as the emergence of obesity and chronic diseases (Johnson-Down and Egeland, 2010; Kuhnlein and Receveur, 1996; Kuhnlein *et al.*, 2004; Jørgensen *et al.*, 2002; Young and Bjerregaard, 2008). TF represents more than a superior source of nutrients and a contributor to dietary adequacy (Fediuk *et al.*, 2002; Egeland *et al.*, 2004; Kuhnlein *et al.*, 2002; 2006); it also symbolizes cultural identity, self-reliance, self-determination and connectedness to the land, and provides social cohesion through shared activities, all of which can contribute to health and well-being (King, Smith and Gracey, 2009; Egeland *et al.*, 2009). Thus, the promotion of Inuit TF is a central feature of the Inuit component of the Indigenous Peoples' Food Systems for Health Program of the Centre for Indigenous Peoples' Nutrition and Environment (CINE). The purpose of the case study described in this chapter was "to utilize traditional knowledge, Inuit storytelling, and country food to promote the health and well-being of community members" in Pangnirtung in the Baffin Region of Nunavut (Figure 9.1) (Egeland *et al.*, 2009).

Community steering committee members envisioned two phases of work. The first phase involved collecting health behaviour data on adults and youth (2005 to 2006) and recording and transcribing elders' stories on the value of TF, including plants and medicinal remedies (2006 to 2007). Community members wanted the elders' stories to be used to help promote TF as part of an effort to combat a dietary transition in their community, as adults and, to a greater extent, youth were thought to be eating more "junk" food and less TF. The health surveys were designed to quantify eating behaviours, to help guide interventions and provide a baseline for future evaluations of trends.

The second phase of the elders' storytelling project was developed to incorporate elders' stories into community health promotion activities targeting primarily youth and young adults. Storytelling is a strong Inuit tradition. The umbrella organization for Inuit in Canada, Inuit Tapiriit Kanatami ("Inuit we are united"), eloquently stated:

> Our past is preserved and explained through the telling of stories and the passing of information

from one generation to the next through what is called the oral tradition. Inuit recognize the importance of maintaining the oral tradition as a part of our culture and way of learning.[1]

Inuit *Qaujimajatuqangit* (traditional knowledge) is highly valued, and its incorporation into nutritional health promotion ensures cultural relevancy and acceptability.

This chapter provides details regarding the youth health survey and its findings, and the elders' storytelling interventions, which have been pilot tested and are currently undergoing a broader community-wide evaluation. During the elders' storytelling project, elders' observations of climate change and its impacts on local flora and fauna were captured, along with their concerns about the sustainability of TF species and food security. The chapter therefore draws on information on climate change, and the available literature related to climate change's potential impacts on subsistence food species and food security.

Youth health survey, Pangnirtung

Objectives and methods

As part of community and CINE collaboration, a youth health survey was conducted to document current dietary consumption patterns and body mass index (BMI) among youth aged 11 to 17 years. This information was to guide health promotion activities targeting youth in the community (Yohannes, 2009).

Ethics approval was granted by the McGill Faculty of Medicine Institutional Review Board; a research licence was issued by Nunavummi Qaujisaqtulirijikkut (the Nunavut Research Institute); and the Hamlet of Pangnirtung approved the research work. The age range of 11 to 17 years was selected for the convenience of conducting the survey at the local secondary school.

The youth survey took place over ten weekdays in May 2006, with all the youth attending school invited to participate. Community research assistants were hired and trained and conducted the interviews with

---

[1] www.itk.ca/publications/5000yearheritage.pdf

CINE staff members. Food model kits were used in the 24-hour dietary recalls to facilitate recollection and the reporting of portion sizes.

Parents were contacted through the school, and their informed consent and assent were obtained. The total of 75 students participating in the survey represented nearly all those attending school in May and approximately 50 percent of all the youth in their age range in the community, according to information from school administrators.

Each youth completed two 24-hour recalls on non-consecutive days, a semi-quantitative 38-item past-year traditional food frequency questionnaire (FFQ; Annex 9.1) and a five-item past-month market FFQ. The traditional FFQ was developed by CINE with guidance from hunters, trappers and other key informants. The market FFQ was designed to capture usual consumption of beverages and chips, based on community consultations indicating that chips would be a good indicator of the consumption of high-energy, low-nutrient-dense snack foods.

Harvest calendars collected through key informant interviews in the community were used to define the in- and out-of-season periods associated with each TF species, to estimate average TF consumption over the entire past year. Because of extremely high reports of TF consumption in the traditional FFQ, extreme intakes were truncated to the 90th percentile (i.e., all those reporting amounts higher than the 90th percentile were reassigned the 90th percentile value).

A database was developed in Microsoft Excel 2003. The dietary recalls were entered into CANDAT software (Godin, London, Ontario, Canada, version 2007), and nutrient intakes were estimated using the Canadian Nutrient File (Health Canada, 2007) and a file developed at McGill University with 2 000 additional foods, derived from food labels and standardized recipes. SAS version 9.1 (SAS Institute, Cary, North Carolina, United

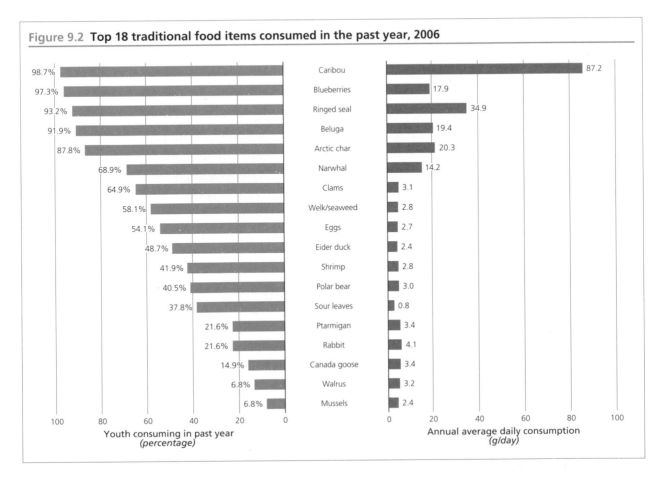

Figure 9.2 **Top 18 traditional food items consumed in the past year, 2006**

Food item	Youth consuming in past year (percentage)	Annual average daily consumption (g/day)
Caribou	98.7%	87.2
Blueberries	97.3%	17.9
Ringed seal	93.2%	34.9
Beluga	91.9%	19.4
Arctic char	87.8%	20.3
Narwhal	68.9%	14.2
Clams	64.9%	3.1
Welk/seaweed	58.1%	2.8
Eggs	54.1%	2.7
Eider duck	48.7%	2.4
Shrimp	41.9%	2.8
Polar bear	40.5%	3.0
Sour leaves	37.8%	0.8
Ptarmigan	21.6%	3.4
Rabbit	21.6%	4.1
Canada goose	14.9%	3.4
Walrus	6.8%	3.2
Mussels	6.8%	2.4

States of America) was used in all analyses. Macro- and micronutrient intakes were adjusted to reflect usual intakes, using procedures outlined by the Institute of Medicine (2000). When within-person variability was larger than between-person variability, calculation of usual intakes was not possible, and the unadjusted means of two days of dietary intake were reported.

Results

Traditional food frequency

Of the 75 youth participating in the survey, 39 were boys and 36 girls. Only five TF species had been consumed by 80 percent or more of the youth in the previous year: caribou (*Rangifer tarandus*), blueberries (*Vaccinium myrtillus*), ringed seal (*Phoca hispida*), beluga (*Delphinapterus leucas*) and Arctic char (*Salvelinus alpines*) (Figure 9.2). Among these five items, caribou meat was by far the most popular and heavily consumed TF, with 98.7 percent of the youth consuming it in the previous year, at an annual average daily consumption of 87.2 g per consumer, based on the truncated traditional FFQ data. No other TF was consumed to the same degree among youth. Although Arctic char was consumed by 87.8 percent, the annual average daily consumption was only 20.3 g among consumers. Narwhal (*Monodon monoceros*) was the sixth most popular TF species, consumed by 68.9 percent of youth in the previous year, with a daily average consumption of 14.2 g over the year. Much

lower percentages of youth reported consuming the other TF items, each with an annual average daily consumption of less than 3.5 g among consumers.

Market food frequency

Based on the market FFQ, youth reported consuming a median intake of 28 g of chips per day, equivalent to more than half a 44-g bag (Table 9.1). They also reported a high degree of daily sweet drink consumption, with median consumptions for fruit juice of 344 ml, powdered sugar drinks of 344 ml, and carbonated beverages of 710 ml (equivalent to two cans of pop per day). In contrast to the high soft drink consumption, at 149 ml/day, the youth reported a median milk consumption that was less than a third of the intake recommended by Canada's Food Guide for First Nations, Inuit and Métis (Health Canada, 2007).

24-hour dietary recalls and micronutrient intakes

Forty-four youth (58.7 percent) reported consuming TF in one or both of the 24-hour recalls, with TF contributing 11.8 percent (± 10.8) of total energy intake among consumers. Adjusted mean micronutrient intakes were evaluated for those consuming and those not consuming TF in one or both of the past 24-hour dietary recalls (Table 9.2). Iron, vitamin A, phosphorous, zinc, selenium and vitamin D were not significantly higher in boys who consumed TF than in those who did not. Girls who consumed TF had

Table 9.1 Median (interquartile range) past-month daily consumption of selected market foods by Inuit youth aged 11 to 17 years, by consumers and non-consumers of TF, Pangnirtung, Nunavut, 2006[a]

	Traditional food		
	TF consumed[a] (n = 43)[b] Median (25th–75th percentile)	TF not consumed (n = 31) Median (25th–75th percentile)	Total (n = 74) Median (25th–75th percentile)
Chips (g)	28.0 (12.3–80.0)	35.0 (8.00–56.0)	28.0 (12.0–56.0)
Fruit juice (ml)	344 (188–1 032)	258 (172–688)	344 (188–688)
Milk (ml)	158 (22.6–344)	125 (0–500)	149 (22.6–376)
Powdered drinks (ml)	344 (49.1–1 000)	376 (71.4–600)	344 (71.4–710)
Soft drinks (ml)	710 (355–1 376)	1065 (355–1 775)*	710 (355–1420)

* $p \leq 0.05$, Wilcoxon test.
[a] Consumed at least one TF item in one or both 24-hour recalls versus consumed no TF in either.
[b] One youth missing food frequency data; reported in one or both 24-hour recalls.
Source: Adapted from Yohannes, 2009.

significantly higher iron ($p \leq 0.01$) and vitamin A ($p \leq 0.01$) intakes. Both boys and girls who consumed TF had significantly higher protein intakes ($p \leq 0.05$) than those who did not. TF consumers reported greater protein intake as percentage of energy on the previous day than did non-consumers: among those consuming TF, 19.8 percent (± 7.3) and 17.3 percent (± 6.8) of energy was in the form of protein for boys and girls, respectively; whereas among those not consuming TF in the previous two days the equivalent figures were 14.0 percent (± 4.8) for boys and 11.7 percent (± 3.5) for girls (t-tests, $p \leq 0.05$). No differences in percentages of energy as carbohydrate or fat were observed between TF consumers and non-consumers (Table 9.3). However, boys who consumed TF derived significantly less energy from saturated fat ($p \leq 0.01$).

Based on the two 24-hour recalls, 92 percent of the youth reported consuming soft drinks on one or both days, with an average consumption of 2.5 cans (875 ml) per day among consumers. Those who consumed TF in either of the two 24-hour recalls reported significantly lower median intakes of carbonated beverages over the previous month (710.0 ml/day) than those who did not consume TF (1 065.0 ml/day) (Wilcoxon $p \leq 0.05$) (Table 9.1).

Discussion

Although the proportion of youth consuming any TF in either or both of the 24-hour recalls approached 60 percent, and the vast majority of youth had consumed some kind of TF in the previous year, there was a general lack of diversity in the species being consumed regularly, as indicated by the traditional FFQ. By far the strongest preference was for caribou meat, as indicated by the high average daily consumption over the previous

Table 9.2 **Usual mean micronutrient intakes among Inuit boys and girls aged 11 to 17 years, by consumers and non-consumers of TF, in two 24-hour dietary recalls on non-consecutive days, Pangnirtung, Nunavut, 2006[a]**

	Boys' mean intake		Girls' mean intake	
	TF consumed (n = 14) Mean (SD)	TF not consumed (n = 25) Mean (SD)	TF consumed (n = 17) Mean (SD)	TF not consumed (n = 19) Mean (SD)
Iron (mg)	18.7 (10.0)	14.3 (5.2)	19.8 (9.2)	13.1 (4.2)*
Vitamin C (mg)	107.7 (88.2)	128.6 (86.9)	160.6 (95.6)	156.5 (82.6)
Vitamin A (µg)	307.0 (101.0)	257.2 (59.9)	362.0 (144.2)	253.6 (79.9)*
Phosphorous (mg)	1 209.8 (419.2)	1 019.4 (317.3)	1 235.3 (222.7)	1 155.0 (279.8)
Selenium (µg)	65.9 (37.4)	63.0 (30.9)	75.0 (27.5)	66.6 (27.4)
Zn (mg)	10.5 (4.0)	8.8 (3.6)	9.3 (1.9)	8.1 (2.0)
B6 (mg)[b]	1.4 (0.6)	1.3 (0.6)	1.5 (0.7)	1.6 (0.7)
Niacin (mg)[b]	35.6 (19.0)	29.1 (11.0)	38.4 (14.8)	4.0 (16.3)
Thiamin (mg)	1.6 (0.4)	1.6 (0.5)	1.5 (0.3)	1.4 (0.4)
Riboflavin (mg)[b]	1.5 (0.3)	1.6 (0.3)	1.7 (0.6)	2.1 (0.9)
Calcium (mg)	572.5 (126.0)	570.6 (116.2)	553.2 (179.2)	587.2 (167.8)
Vitamin D (µg)[b]	2.2 (1.3)	2.7 (1.7)	2.8 (1.5)	3.4 (2.0)
Folate (µg)[c]	372.1 (187.4)	295.2 (135.7)	322.0 (105.2)	364.3 (82.3)
Total fibre (g)	8.7 (2.1)	9.5 (4.2)	10.3 (4.2)	11.0 (4.7)

* $p \leq 0.01$, Student's t-test.
[a] Adjusted usual mean (SD) intake unless otherwise noted.
[b] Designates when within-person variability among girls was larger than between-person variability; therefore, the unadjusted mean of two days' intake is presented rather than the adjusted mean.
[c] Designates when within-person variability among boys was larger than between-person variability; therefore, the unadjusted mean of two days' intake is presented rather than the adjusted mean.
Source: Adapted from Yohannes, 2009.

Table 9.3 **Total energy and percentages of energy from macronutrients among Inuit girls and boys, by consumers and non-consumers of TF, in two 24-hour dietary recalls on non-consecutive days, Pangnirtung, Nunavut, 2006[a]**

	Boys (n = 36)		Girls (n = 39)	
	TF consumed [c] (n = 25) Mean (SD)	TF not consumed [c] (n = 14) Mean (SD)	TF consumed [b] (n = 19) Mean (SD)	TF not consumed [b] (n = 17) Mean (SD)
Total energy				
(kcal)	2 128 (745)	1 922 (762)	2 213 (515)	2 416 (817)
(kJ)	8 910 (3 119)	8 047 (3 190)	9 265 (2 156)	10 115 (3 420)
% protein	19.8 (7.3)	14.0 (4.8)*	17.3 (6.8)	11.7 (3.5)§
% carbohydrate	57.8 (8.4)	60.4 (9.9)	58.5 (11.3)	63.8 (8.5)
% fat [c]	22.8 (3.9)	26.6 (6.0)	24.5 (3.4)	25.5 (5.9)
% saturated fat	7.0 (1.5)	8.5 (1.9)§	7.8 (1.9)	7.8 (2.2)
% MUFA [c]	8.7 (1.9)	10.1 (2.3)	9.9 (1.7)	9.5 (2.9)
% PUFA [b, c]	4.4 (1.3)	4.1 (2.2)	4.7 (1.6)	4.8 (1.7)

* $p \leq 0.05$.
§ $p \leq 0.01$, Student's t-test.
MUFA = monounsaturated fatty acid.
PUFA = polyunsaturated fatty acid.
[a] Adjusted usual mean (SD) intake unless otherwise noted.
[b] For girls, designates when within-person variability was larger than between-person variability; therefore, the unadjusted mean of two days' intake is presented rather than the adjusted mean.
[c] For boys, designates when within-person variability was larger than between-person variability; therefore, the unadjusted mean of two days' intake is presented rather than the adjusted mean.
Source: Adapted from Yohannes, 2009.

year. A preference for caribou was also identified among Inuit preschoolers in 16 Nunavut communities, where caribou consumption by 84.3 percent of the preschoolers far exceeded that of any other TF species in the previous month (Johnson-Down and Egeland, 2010). In CINE's previous dietary surveys across the Arctic too, caribou was a prominent component of the TF system (Kuhnlein and Receveur, 2007; Kuhnlein and Soueida, 1992). The results indicate that a broad variety of TFs could be promoted for consumption by youth.

The high consumption of sugar-sweetened beverages reported in the market FFQ was also observed in the 24-hour recalls. The consumption of soft drinks has been associated with obesity, and reduction in soft drink consumption has been related to weight loss (Chen *et al.*, 2009; Giammattei *et al.*, 2003; Sanigorski, Bell and Swinburn, 2007), although not all studies show consistent associations between soft drink consumption and weight gain or obesity (Gibson, 2008).

In the 24-hour recalls, the lower consumption of soft drinks among youth who consumed TF than

among those who did not was an unexpected finding. Although the mechanisms are still not clear, protein aids in satiety (Tome, 2004). Thus, greater protein intake among youth who habitually consume TF may, in part, explain the lower amount of pop consumed among TF consumers, as those consuming more protein may not have the same degree of cravings for sugar-sweetened beverages. Conversely, those who consume TF may be more traditional and may therefore avoid sugared beverage consumption. While additional research is needed on this topic, the current findings highlight the unexpected ways in which even small amounts of nutrient-dense TF may promote or be associated with a healthy diet in the contemporary context.

Food frequency data can both over- and underreport food items, and can vary by demographics such as age and sex (Marks, Hughes and van der Pols, 2006). Food items that are well liked by youth could be overreported because of the social desirability of reporting culturally valued foods. Therefore, the amounts recalled should be interpreted with caution. However, a strength of the research was that extreme reports of TF were truncated to the 90th percentile, which would limit the extent of overreporting in the data. Another strength of the study was that two approaches were used to assess dietary behaviours, the two 24-hour recalls and the FFQ, which provided opportunities to evaluate consistencies in the data.

In summary, the survey highlighted the low diversity of TF in youth's diet and the high intake of sugared beverages and high-fat snacks. In an effort to enhance youth's knowledge of TF, a storytelling project with elders was initiated.

Storytelling and health promotion research

Background

During storytelling interviews, community members asked elders about their experiences and knowledge of TF, including hunting and harvesting activities, what parts of animals were eaten by men and women, medicinal remedies, how TF differs from market food, and the elders' observations related to climate change. In 2006 and 2007, a total of 21 elders were interviewed in Inuktitut (by author Jonah Kilabuk), and their interviews were transcribed into English (by Looee Okalik). Interviews ranged in length from 20 to 45 minutes. The storytellers' informed consent was obtained for the use of their stories in publications and media.

Two pilot interventions were developed based on the stories: elders' storytelling in a DVD format; and youth radio drama incorporating elders' stories, to build appreciation of TF and encourage healthy food choices among youth.

Elders' stories in DVD format

Objectives and methods
The objective of this pilot intervention was to determine whether the use of elders' stories in a DVD would be an effective means of transferring traditional knowledge from Pangnirtung elders to youth (Yohannes, 2009). The elders were identified by community steering committee members and through the community Elder Centre. The DVD was developed by CINE (Sennait Yohannes) and steering committee members; Inuktitut with English subtitles was used throughout. Five of the 21 interviews with elders were used, as they provided in-depth information on three themes, each of which corresponded to a segment of the DVD: TFs and how they differ from market foods; what parts of the animal were eaten by men and women; and TFs that serve as medicinal remedies. The three themes were chosen by the steering committee and CINE, based on a review of the contents of the elders' stories.

Informed consent was obtained from the youth (students) participating in the pre- and post-DVD viewing assessments. CINE and the community health promotion steering committee developed a series of 28 true/false questions, based on the content of the five elders' interviews captured in the DVD (Yohannes, 2009). Four of the 28 questions had to be dropped from the assessments, as youth had problems understanding them. An increase in the number of correct responses to the true/false questions was considered an indication that viewing the draft DVD had improved knowledge of TF among the youth. The pilot intervention took place in May 2008.

During the first week, students completed the pre-DVD questionnaire and then viewed the first of two segments of the DVD; the second segment was played the following week, and the students then completed the post-viewing questionnaire. However, because of community events, not all students were present on both the pre- and post-DVD viewing occasions.

Results
The pilot intervention found that the mean knowledge score based on the true/false questionnaire among the 24 youth who took the post-viewing assessment was significantly higher than that among the 19 youth who took the pre-viewing assessment (post-viewing 15.8 ± 2.9, versus pre-viewing 13.8 ± 3.0; independent sample t-test $p \leq 0.05$). Similar results were obtained in the analyses of ten youth who took both the pre- and post-viewing assessments (post-viewing 15.5 ± 2.5, versus pre-viewing 13.3 ± 2.5; paired t-test, $p \leq$

Example of radio drama narrative

Maryann: When it comes to our health it is important to understand *piluajjaiqsimaniq* [Inuktitut for "moderation"]. Some foods and drinks are okay for us to have, but only in small amounts. We have to be very careful how much we eat or drink. Pop is an example of one of these foods. Too much pop can lead to serious health problems, including weight gain, cavities in our teeth, problems sleeping and concentrating at school or work. Let's listen to a story told by Josephee Keenainak who talks about country food and how it affects his life.

Taped elder: This excerpt from *Josephee's story* explains the traditions of country food, the importance of country food in our culture, and eating certain foods in *piluajjaiqsimaniq*.

"Our parents kept us informed and taught us. With *mattaaq* (whale skin) being very delicious, we were told not to overindulge eating *mattaaq* if we hadn't eaten any meat prior, we were also told not to eat whale meat if its oil had yellowed, for it had affected the meat. Those were always told and there weren't many to be cautious of. I had heard that one individual had overindulged on *mattaaq* upon having craved it. That is why we were reminded to eat a good amount.

As I was raised decades back and was informed of food scarcity that we could confront during life. We tend to eat meat wisely, give thanks upon gain [i.e., obtaining it] and give some meat to other people for their fulfilment. God appreciates it when we share our foods with others. My fellow Inuit and younger generations need to care for food well and on healthy eating. When there is country food at hand, take pleasure in enjoying it. Our bodies will be healthier. It was said that Inuit had healthier blood at the time of traditional food consumption. Cut down on the intake of junk foods. I was always told that my blood was healthy for I have minimal consumption of food with sweets. Let us be conscious, although we do our best, to live healthily for our bodies to be healthy."

Maryann: In his story, Josephee Keenainak talks about eating certain country foods only in *piluajjaiqsimaniq*, the example that he uses is *mattaaq*. We can use this knowledge that Josephee teaches us to be careful about how much pop we drink. A little bit once in a while is not bad, but if we drink too much pop too often, it can make us unhealthy, just like eating too much *mattaaq*. We hope that you have enjoyed our message today and thank you for listening.

0.01). Qualitative comments from the youth indicated a positive reaction to the elders' stories and a desire to learn more about TF and traditional ways. The assessment indicated that elders' storytelling in DVD format was a successful approach for transferring knowledge from elders to youth participating in the pilot assessment. The qualitative feedback was helpful in revising the DVD for future use and evaluation (Yohannes, 2009).

Youth radio drama

Objectives and methods

Radio drama with community youth was developed to help improve youth's nutrition and health knowledge and to enable them to make healthier food choices. Messages were designed to increase appreciation of TF and to target selected high-risk behaviours, such as the high consumption of carbonated beverages, which includes an average of 1 litre of pop per day. The messages were designed to link the themes of the elders' stories to modern nutritional advice involving healthy food choices, including market food choices.

The specific messages were developed by CINE (Cassandra Racicot), with advice from the community steering committee and youth. Before airing the dramas on local radio, youth were recruited for focus group tests of their messages, to ensure cultural relevance and acceptability. Box 9.1 gives an example of a radio drama that utilizes Inuit traditional knowledge through the elders' stories, youth dialogue and modern nutritional advice, with actors recruited from among community youth.

The radio drama pilot tests used key informant interviews and focus group discussions to ensure relevancy and acceptability. The results of a broader community-wide evaluation of the radio programmes, which were aired several times a week, were not yet available at time of writing. However, the approach holds promise for capturing youth's attention and engaging youth in building on the knowledge that already exists in the community, strengthening ties between youth and elders, and ensuring health promotion that is culturally relevant.

Food security, climate change and promotion of TF

> If ice doesn't form anymore, our traditional food system might disappear.
>
> *Elder Pauloosie Veevee, Pangnirtung,*
> *24 June 2006*

During the elders' interviews, elders' observations of the impacts of climate change on local flora and fauna and their resulting concerns for food security were noted. A sample narrative is provided in Box 9.2, by Elder Pauloosie Veevee from Pangnirtung. The temperature has risen three times faster in the Arctic than elsewhere in the northern hemisphere, outpacing climate change predictions, reversing a 2 000-year cooling trend (Kaufman *et al.*, 2009) and substantiating earlier claims by the former President of the Inuit Circumpolar Conference, Sheila Watt-Cloutier, who stated that "the Arctic is the world's barometer of climate change. We are the early warning system for the world. What is happening to us now will happen to others further south in years to come" (Alaska Native Science Commission, 2005). As predicted (Haines and McMichael, 1997), extreme weather events are being noted. Given the melting of the pack ice in the Arctic (Stoll, 2006) and climate changes that are outpacing predictions (Kaufman *et al.*, 2009), it is imperative to evaluate potential impacts on TF and food security.

TF contributes significantly to nutrient intakes, even when consumed in small amounts (Kuhnlein and Receveur, 2007; Egeland *et al.*, 2011; Johnson-Down and Egeland, 2010). Thus, the effects of climate change on the availability of and access to TF species and on food safety could have significant impacts on nutrition status and food security in the Arctic.

There are various definitions of food security; the World Food Summit defined it as being present when "all people, at all times, have physical and economic access to sufficient, safe and nutritious food to meet their dietary needs and food preferences for an active and healthy life" (FAO, 1996). The prevalence of food insecurity is high in Inuit communities, with estimates from individual community surveys using the 18-item United States Department of Agriculture

(USDA) module revealing prevalence of 40 percent in Kangiqsujuaq, Nunavik (INAC, 2004), 83 percent in Kugaaruk, Nunavut (INAC, 2003), and 88 percent in Igloolik, Nunavut (Ford and Berrang-Ford, 2009). Based on a three-item questionnaire, the prevalence of food insecurity in Nunavut, which has a predominantly Inuit population, was 56 percent, in contrast to 14.7 percent in southern provinces, 21 percent in the Yukon, and 28 percent in Northwest Territories (NWT) (Ledrou and Gervais, 2005). Some 69.6 percent of Inuit households with children aged three to five years were food-insecure (Egeland *et al.*, 2010).

Food security depends on the geography, economy and ecology of Inuit communities, which are in a dynamic state of change, with impacts on access to and utilization of both market food and TF resources. Geographical considerations relate to the remoteness of Arctic communities, which require market food to be flown, shipped or – in a few cases – transported by ice-roads in the winter, contributing to high food costs, especially for perishables, in Canadian Arctic communities (INAC, 2007). Geographical considerations also relate to the historical shift from small nomadic groups dispersed across the Arctic to larger settlements, which often results in the need to travel considerable distances to reach productive hunting areas.

Economic considerations relate to monetary access to market food, the costs of fuel and equipment, and the economic gains associated with harvesting and utilizing local species (Ford, 2009; Lambden *et al.*, 2006; Chan *et al.*, 2006). Food sharing networks are an important component of social support networks and should be considered when seeking to understand local economies. Local Inuit economies also rely largely on government or private sector service industry jobs, arts and crafts, tourism and, depending on the community, jobs related to resource extraction. Because the costs associated with hunting and harvesting are notable (Ford, 2009; Lambden *et al.*, 2006) and hunting requires time and flexibility, short-term, high-paying jobs are prized (Nuttall *et al.*, 2010). Hunting tourism can be a lucrative source of income for Inuit households, with foreign clients bringing in as much as CAD 15 000 per polar bear. In

Box 9.2
Elder Pauloosie Veevee reports on climate change in the Arctic

From my knowledge, climate change is most evident, with the ice conditions being affected most from how it used to be.

When digging for a seal hole on ice or to set nets, the ice is much softer today, whereas yesteryear, it used to make sounds when chiselled, when the ice was harder. Then, the ice used to be slightly softer than the lake ice. Today the sea ice is much softer. Even with the extreme cold temperatures, the ice conditions don't freeze up as hard; the ice is softer, like shortening. This has been occurring more recently.

If I can recall the year, I would be able to name it. As I am not too cognizant of time, the year is recent as of when the sea ice has softened.

Post-2000, that is when the sea ice conditions have changed. That is when that has occurred. The scattered water that used to be visible atop some ice doesn't occur as much now. In the spring, there used to be scattered water atop some ice. Today, water will overflow over the whole ice with minimal scattered water atop some ice. A lot of water overflows over the sea ice now, allowing for the sea ice to soften more quickly.

Before, the scattered water atop ice would have dust build-up along the edges, now that doesn't even occur. The sea ice doesn't pack as thick any longer. It will thicken but not in consistency. Before, the snow ice used to pack thick and build on to the land, and it would remain for longer periods of time. It occurs today but for a shorter time. Changes are very pertinent to the sea ice.

It is June, nearing the end of the month. We would be hunting for the infant seals with blizzards occurring at times. That is how it was back then by June. Although, it wasn't always like that. By June, the ice would break on the southerly winds.

The ice never broke off on its own. Now, the ice breaks up without any wind or with minimal wind. There was a word used to describe the southerly wind breaking off the ice, *nunningiarasuttuq*.

In *Paurngaturlik* (camp), that is how it used to be.

At times, the wind would break the ice in chunks and leave a channel of water. It was a perfect time to catch seals. The sea ice would ground into the sea bed at times in the Panniqtuuq Fiord. We would anchor where there were most seals at high tide. By high tide, we would take the dog-team while the sea ice seemed connected. Today, these don't occur anymore. The sea ice melts so rapidly. We used to wait for ships in the fall for they were our source of food rations. We had to paddle to return to our camps from trading. October was a hasty month to travel in, for the ice was forming and we had to return to our camps. At one time, we got caught when the ice formed by October, so we had to winter in Usualuk. Boating was impossible. October and November were the times the ice formed for the winter.

Today, the ice doesn't form until the Christmas month.

The sea ice totally solidifies by January. At times, Christmas arrives when there is still water visible.

I haven't observed any drastic changes with the snow. The snow doesn't compact itself as hard as it used to. On the southerly wind, snow would form atop the snow, called *naannguat*, not snow-drifts. This type of snow shudders the skidoo as it's going. When the wind comes from the west, snow forms into *naannguat* now too. That is evidence of the snow formation change.

It is different today. I didn't construct many iglus [igloos] but I had to build them when hunting by dog-team, as that was the practice then. It is different today to build an iglu, for we have to search harder to find the appropriate snow. Even with the wind blowing, the snow doesn't harden as much today.

We hear of climate change occurring today. I have a slightly different perspective on climate change effects on the ice. My view is that the weather is not totally warmer, I think the change is evident through the seawater.

The seawater doesn't get as cold anymore, although it isn't changing by warmth as much either. It gets extremely cold outside still at times. I can see the changes, for we lived a nomadic lifestyle with dog-teams. Living harmoniously, we used whips with our dog-teams. In the spring, we would dunk our whips into the water as deep as we could and when we'd pull it out, the tip of the whip would freeze up. This is not common any longer.

I know very well that the wind has a more cyclone pattern today. I have noticed the changes in the wind over the past two years. In the summer of 2005, it was most evident. It was windy with barely any calm days. On the calming of wind, it would pick up from the other direction.

In our youthful days, when we were paddling, the days were calm for long. Now, we don't see that kind of days. The wind is more frequent now, with fewer calm days, and I don't know why that is so. The westerly wind always calms down by evening, always has and still does. At sunset, the day is calm but as the day picks up the wind begins to hustle.

We have a word that describes this, as we have dialectal differences, the word *saqijaal-latuinnaqtuq* is the term we use for when the wind begins to pick up from the other direction. This is very new, the wind picking up from the opposite direction, it never was like that before.

Times have changed. Then, the weather seemed to be cold for one whole season. The cold today is as sharp as a blade with an intensity to it. It'll be cold but it won't freeze up or thicken the ice as it used to. When we had cold seasons, the weather would freeze up the ice fully. Today, it is a sharp cold. The changes in the cold weather are evident.

I don't go hunting as frequently anymore, but I keep informed. We all have perspectives on things. As the ice break-up is earlier now, seals' fur is browner now. They don't have as much time to bask in the sun. When you are used to seeing the full colour of seal fur, it is regrettable to see them brown. The ice they bask on is no longer accessible, so they have less ice and sun time. Polar bears will have no more feed left with seals being speedier and they get to places faster than the bears. I have thought of things I shouldn't even consider, that if ice doesn't form any more, our traditional food system might disappear. With the warming of the sea, we may have no more seals.

There have been some slight changes to the seal fur. It is evident if there is no more sea ice, as seals with the change of their fur colour go through a phase of dandruff release. Seals like to bask in the sun as they go through a colour change. But if there is no longer any ice left, seals will remain brown in the future.

I haven't noticed any effect on the meat. Seals rely on their food chain. I don't know when they have eaten something, their meat isn't as tender as it used to be. In the olden days, we used to leave the seal out for a day or two before butchering and then it would be so tender with the blood clotting a bit, they were so good. My thoughts are – as we have time to reflect on things when alone, although not all the time – in the year 2006, the seals were almost left with no time for birthing with the ice breakage being much earlier. Seals have birthing seasons in March and April. I feel if there is no natural habitat for the seals and their pups, the pups will die off. They will die off from cold. This almost occurred recently, barely stretching the luck. The seals always birth their pups within the snow.

There is so much noise pollution today, with boats roaming back and forth. Back in the day, there was barely any noise to disrupt the game. There is quite an abundance of seals still, they are just not as close to our homeland. As the game prefer solace, I feel they are keeping a distance.

There has been an abundance of change! Back then, the melted ice foliage remained. It would melt only when its time arrived. It would turn into water. Today, it melts even before its season arrives. I have no idea why that is. It is worrisome now for young hunters too, as ice breakage can appear to be melting ice when it is not.

I haven't noticed much change in the water currents. We have always sought for clams at low tide during full moon. I am not sure if this is a recent trend or whether it is common or not, the high tide will draw in high but the low tide doesn't extend as far as it used to. When full moon is drawing near, we always delightfully say, "umm, it'll be clam season soon". The water current is noticeable in that sense.

I haven't seen any changes in the waves. It's likely that others have observed them but I haven't seen any changes in the waves.

I noticed immense changes in the years 2005 and 2006, especially in 2006. It is as if the earth is in a rush. The plants have been turning green a lot earlier in the spring, even before their season has arrived for growth: our rare, earth's rare plants in the Arctic. I haven't noticed any foreign plants to date. But I have noticed that the plants green a lot sooner than they used to. The weather contributes to the growth of the plants. They wouldn't grow by themselves, the weather controls their growth.

There is an old saying among Inuit that if the snow melts earlier, the birds will nest their eggs earlier as well. On the other side of Cumberland Sound there is an abundance of birds that nest eggs. It is said the egg laying birds go with the cycle of summer. If the snow takes longer to melt, the birds will take longer to lay eggs. There hasn't been much change with this. It is more than likely that with snow melting sooner, the birds will lay eggs a lot sooner.

What is evident is the *sirmiit* [Inuktitut for "blue glaciers with melting ice"] that have running water throughout the winter are no longer here. There used to be *sirmiit*, and they would remain cold. They remained frozen. They melt now. Atop the mountains are glaciers. The ones that are not at the mountain top and have running water we call *sirmiit*.

I am not totally certain of how climate change has affected Inuit. But how I observe it is that Inuit easily get colds. It wasn't like that back then. People are more susceptible to colds now. I think with the varying physical demands, that may be a contributor. Or it's likely that the community is more populated and that could be the main contributor to the colds. Or the weather may be the factor to all the colds.

Safety is the top of the list today with being aware of the environment. We have always had to be cautious, but caution is required more today. Before, we always had to check the ice conditions with our hunting tools to avoid danger or accidents. Today, tools will be a necessity to check the ice conditions. We utilize the harpoons/tools to check the ice conditions. That was a requirement in our days. It'll be more of a prerequisite today, as climate change has big effects on the ice.

The last thing I'd like to say is, I am not praising myself, I am a seasoned elder having lived the times of dog-teaming. I am not getting any wiser. I have listened to many storytellings. I don't contribute to the storytellings. Although in interviews, I can provide some historical knowledge, I don't like to give second-hand knowledge. There isn't enough evidence in second-hand stories, so I prefer listening to first-hand stories. That is all I have to say. *Qujannamiik.*

Elder Pauloosie Veevee in an interview with Jonah Kilabuk, Pangnirtung, 24 June 2006

one year alone, the Baffin community of Clyde River earned CAD 212 000 from polar bear hunting tourism and most of this money went directly to Inuit households, which in turn purchased equipment to facilitate their own subsistence hunting as well as future hunting tourism (Nuttall *et al.*, 2010). Thus, although polar bear meat itself is not a major contributor to the diet (Kuhnlein and Soueida, 1992), threats to polar bear populations could have a profound effect on food security in the Arctic through reduced economic opportunities in tourism. Political lobbying regarding the banning of the sale of seal furs has already had devastating effects on the economy of Inuit communities, as seal furs represented a significant source of revenue (Carino, 2009). Thus, changes in the availability of subsistence species and in market-place policies can have profound economic effects on isolated communities with limited options for income generation, and serve as examples of the diverse ways in which decreased availability or utilization of TF species can threaten food security.

The food chain involves exposure to a variety of threats, including viruses, bacteria, biotoxins and parasitic pathogens found in subsistence species and other foods (Parkinson and Evengård, 2009; Hotez, 2010; Van Dolah, 2000); lead-shot micro-fragments and dissolved lead in game meats, particularly fowl (AMAP, 2009; Dewailly *et al.*, 2000); and trace metals and persistent organic pollutants, which are atmospherically transported to the Arctic and biomagnified in the food chain (AMAP, 2009). Surveillance and targeted interventions that support the consumption of TF while reducing exposure to harmful agents include community education to eliminate the use of modern air-tight containers and bags when fermenting and storing TF, to prevent botulism (McLaughlin *et al.*, 2004); the banning of lead-shot in Quebec, which has resulted in blood lead levels decreasing by a half (Dewailly *et al.*, 2007a; 2001); and *Trichinella larvae* testing of meat prior to community consumption (Proulx *et al.*, 2002). Biomonitoring indicates that exposure to methylmercury and persistent organic pollutants has declined, owing in part to reductions in environmental levels and in consumption of TF (AMAP, 2009). However, because the effects of contaminant levels observed in the Arctic are subtle and there are many competing benefits to eating TF (Egeland and Middaugh, 1997; Dewailly *et al.*, 2002; 2007b; Jacobson *et al.*, 2008; Kuhnlein *et al.*, 2002; Mozaffarian and Rimm, 2006), a review of the full scope of evidence led the Arctic Monitoring and Assessment Program (AMAP) to advocate for continuing the consumption of TF in the Arctic (AMAP, 2009).

However, there is evidence that climate change and related impacts can alter food safety in the Arctic. Water warming increases biotoxins, such as *Saxitoxin*, and the presence of pathogenic bacteria, such as *Vibrio parahemolyticus*, with respective consequences for paralytic shellfish poisoning (Van Dolah, 2000) and increased risk of bacterial-related food-borne illness (McLaughlin *et al.*, 2005). Floods, erosion and the thawing of permafrost can threaten community sanitation infrastructure, resulting in release of pathogens into the environment, and floods and erosion can increase distant agricultural pesticide runoff into streams and tributaries that ultimately reach the Arctic. Higher global temperatures would increase the volatization of contaminants, resulting in increased transport and deposition of contaminants in the Arctic (Kraemer, Berner and Furgal, 2005). Floods and erosion would increase inorganic mercury in water, and water warming would increase the methylation of inorganic mercury and, over time, the methylmercury burdens in subsistence species (Booth and Zeller, 2005). These are only a few of the numerous pathways by which climate change may alter food safety in the Arctic (Parkinson and Evengård, 2009; Kraemer, Berner and Furgal, 2005). Inuit also mention that the heavy use of tranquilizers for research in the Arctic has made polar bear meat inedible. Whereas Elder Jamesie Mike reported that frozen polar bear meat was edible 60 years ago, today polar bear meat must be cooked for hours to rid it of toxins (KP Studios, 2009).

In addition to potential impacts on food chain safety, climate change is having effects on Arctic ecosystems, with implications – which are not yet fully understood – for access to and availability of subsistence species that are important for food security. Climate change can alter access to TF species, as travel to hunting areas requires navigation, often of considerable distances over rough terrain, streams and inlets, and is safer when the landscape is frozen. Extreme weather conditions also represent threats to navigation and safety, with further implications for hunters' access to subsistence species (Krupnik and Jolly, 2002; Furgal, Martin and Gosselin, 2002; Ford, 2009; Ford and Berrang-Ford, 2009; Ford and Pearce, 2010; Guyot *et al.*, 2006).

Climate changes thus affect TF species; the Arctic has already witnessed the encroachment of non-Arctic flora and fauna species due to these changes (Simmonds and Isaac, 2007; Meier, Döscher and Halkka, 2004; Ferguson, Stirling and McLoughlin, 2005; Humphries,

Umbanhowar and McCann, 2004; Vors and Boyce, 2009). Although there have always been fluctuations in caribou populations over time, the decline now being noted (Vors and Boyce, 2009) is particularly important given the heavy reliance on caribou meat in the Canadian Arctic (Johnson-Down and Egeland, 2010; Kuhnlein and Receveur, 2007; Kuhnlein and Soueida, 1992). In addition to the scientific literature, communities too have reported that caribou have been scarce in the last couple of years, with migration routes considered to be off the usual ones taken. Given the historical fluctuations in herds and migration routes, elders state that the caribou will return to their usual migration path in time (L. Okalik, personal communication, 2010).

There is also evidence that early ice melt and reduced snow fall and snow thickness have an impact on populations of ringed seal (*Phoca hispida*) pups in Western Hudson Bay, and are projected to continue to diminish the species (Ferguson, Stirling and McLoughlin, 2005). The lack of ice floes in eastern Canada resulted in the deaths of thousands of harp seal (*Phoca groenlandica*) pups in 2007, and a similar occurrence was reported in 2002, when the Department of Fisheries and Oceans estimated that 75 percent of seal pups in the Gulf of Lawrence died coincident with a year of very little ice (MacKenzie, 2007). With the ongoing shrinkage of pack ice in the Arctic, seal populations will likely be greatly threatened. In addition, although current data are contradictory regarding whether polar bear (*Ursus maritimus*) populations are diminishing or in abundance (Aars, Lunn and Derocher, 2006; Dowsley and Wenzel, 2008), polar bears rely heavily on seals for their sustenance, raising concerns for the bears' propagation and survival if seal populations diminish. As caribou, seal and polar bear are a central component of the TF system and economy, the changes are potentially important for Inuit food security.

Inuit have historically been highly adaptive to changes in their environment, but current constraints in adaptive capacity have been noted (Nuttall *et al.*, 2010; Ford, Smit and Wandel, 2006; Ford and Pearce, 2010). Given that Canada's Action Plan for Food Security (Agriculture and Agri-Food Canada, 1998) listed TF acquisition as one of its ten priorities for dealing with food insecurity, understanding the impact that continued climate change will have on food insecurity should be a high-priority research area.

As traditional knowledge and strong social support networks have been listed as factors contributing to the adaptive capacity of Inuit communities (Ford *et al.*, 2006), elders' storytelling may be one of many strategies communities can utilize to help meet the challenges of climate change. Elders' storytelling regarding their knowledge of a full range of TF species and parts of species may be a means of enhancing youth's skills in and knowledge and acceptance of hunting and harvesting a wide range of subsistence species and utilizing diverse parts of species, and could be one of the much-needed strategies for building resiliency in a time of uncertainty and rapid climate change.

Summary

With caribou, seal and polar bear populations potentially in peril (three mainstays of the Inuit TF system and economy), the promotion of a wide range of TF is needed to build adaptive capacity in Inuit communities.

This case study reports on the development of innovative nutritional health promotion in Pangnirtung, where health promotion messages built on existing knowledge and cultural conceptualizations of health and well-being through Inuit elders' traditional knowledge combined with two culturally relevant modes of communication: community radio and storytelling. The intervention engaged youth by involving them in developing and testing messages prior to airing on community radio, and in conducting the radio programmes. The health promotion programme was developed in partnership with the community, and its main elements came directly from community steering committee members. Health promotion programmes developed locally are likely to be more acceptable, relevant and, ultimately, successful than programmes that are imported from non-Inuit communities. At time of writing, the results of a post-survey community-wide evaluation of youth and young adults were not yet known, but the community-CINE model developed in Pangnirtung holds promise for helping to prevent the negative consequences of acculturation and nutrition transition, and could be adapted to other indigenous communities in Canada and globally.

Storytelling also revealed elders' observations of climate change and its impacts on local flora and fauna, and the elders' resulting concerns for food security. As food security is a fundamental component of a population's health, this chapter has highlighted the economic and ecological context of food insecurity in Inuit communities. While the true impact of climate change is not yet known, enough information exists to suggest that research in this area should be a high priority. Given that climate changes are outpacing projections, health promotion programmes need to take into account the broader and likely future realities that will challenge Arctic communities.

Conclusion

The pace of change "has been breath-taking and has few parallels in the developed world" (Inuit Tapiriit Kanatami President, Mary Simon) (Simon, 2009).

Clearly, changes in the Arctic ecosystem are happening rapidly, and current changes follow the recent 60-year history of a rapid transition from nomadic life to the establishment of settlements throughout the Canadian Arctic, and the ensuing ratification of four land claim agreements (Egeland *et al.*, 2009). The ongoing changes will not stop with the loss of pack ice, as this will usher in an era of Arctic exploration and development projects to extract the vast wealth the Arctic holds (Yalowitz, Collins and Virginia, 2008). As the Northwest Passage becomes commonly used for international shipping and transportation, the effects on water and noise pollution will further disrupt game and their migratory paths.

It is worth pausing to consider the implications of the coming changes for Inuit communities, which are already strained by an unprecedented pace of change and unresolved social justice issues of poverty, household crowding, low educational attainment, lack of opportunities, and disparities in health and longevity (Egeland, Faraj and Osborne, 2010; Veugelers, Yip and Mq, 2001; Wilkins *et al.*, 2008; Standing Committee on Aboriginal Affairs and Northern Development, 2007). For Indigenous Peoples, cultural and environmental dispossession are among the top determinants of poor health (Richmond and Ross, 2009).

Efforts are needed on multiple fronts to promote Inuit health and resiliency and Arctic ecosystem sustainability. In the context of rapid changes, storytelling with elders is important in building knowledge of the past that may otherwise be lost to future generations. Firm roots in this knowledge and in elders' wisdom will help strengthen social cohesion and support, which is recognized as having beneficial associations with health and well-being (Richmond, Ross and Egeland, 2007). Strengthening the ties between elders and youth may also have other benefits in a time of rapid changes that are affecting all dimensions of life in the Arctic. Elders' storytelling that links to modern-day nutritional issues and uses modern media may be a means of reaching youth, building social cohesion, and promoting Inuit resiliency and adaptive capacity in a time of great uncertainty and rapid changes.

As Inuit Tapiriit Kanatami President Mary Simon stated, "with focused and responsible efforts we can harness the enormous potential of our youth and direct it towards a positive outcome" (Simon, 2009) ✳

Acknowledgements

The authors would like to acknowledge the financial support of Canadian Institutes of Health Research, the Social Sciences and Humanities Research Council of Canada, and the Max Bell Foundation. They also acknowledge the sage advice of the Inuit Tapiriit Kanatami, the elders of Panniqtuuq, the Hamlet of Pangnirtung, the Health Centre, the Government of Nunavut, Nunavut Tunngavik Inc., and the Nunavummi Qaujisaqtulirijikkut (Nunavut Research Institute). Finally, they appreciate all the help provided by community members, especially Mary Ann Mike, O. Nashalik, J. Alivaktuk, B. Akpalialuk, J. Akpalialuk, J. Alivaktuk, D. Inshulutak, M. Qumuatuq, M. Shoapik, P. Qiyuaqjuk, P. Kilabuk, L. Nauyuk, J. Qaqasiq, T. Kunilusie, R. Akulukjuk, G. Akulukjuk, T. Kunilusie, and T. Metuk.
> **Comments to:** g.egeland@isf.uib.no

>> Photographic section **p. XVI**

Annex 9.1 Traditional foods included in the traditional FFQ, Pangnirtung, Baffin Region, Nunavut, 2006

Marine mammals
- Beluga meat (fresh, cooked, frozen)
- Beluga meat (dried)
- Beluga *mattaaq* with blubber (raw or boiled)
- Beluga *mattaaq* without blubber
- Beluga blubber (raw or cooked)
- Beluga oil
- Narwhal blubber (raw or cooked)
- Narwhal *mattaaq* with blubber (raw or boiled)
- Narwhal *mattaaq* without blubber (raw or boiled)
- Ringed seal blubber (raw or boiled)
- Ringed seal liver (raw or cooked)
- Ringed seal meat (raw, cooked or frozen)
- Walrus blubber
- Walrus meat

Fish and seafood
- Arctic char
- Halibut
- Turbot
- Mussels
- Clams
- Shrimp

Land mammals
- Caribou meat (raw, frozen, baked, cooked and aged)
- Caribou meat (dried)
- Caribou liver
- Caribou heart (raw, boiled)
- Caribou kidney
- Caribou tongue (raw, cooked)
- Caribou stomach (walls and content)
- Polar bear meat (raw, boiled)
- Rabbit meat

Game birds
- Ptarmigan
- Canada goose
- Eider duck
- Eggs of goose or duck

Plants and berries
- Blueberries, crowberries, cranberries, other picked berries
- Sour leaves
- Welk (seaweed)
- Other flowers and plants (please specify)
- Other (specify)

Chapter 10

Culture-based nutrition and health promotion in a Karen community

SOLOT SIRISAI[1] SINEE CHOTIBORIBOON[2] PRAIWAN TANTIVATANASATHIEN[2]

SUAIJEEMONG SANGKHAWIMOL[3] SUTTILAK SMITASIRI[2]

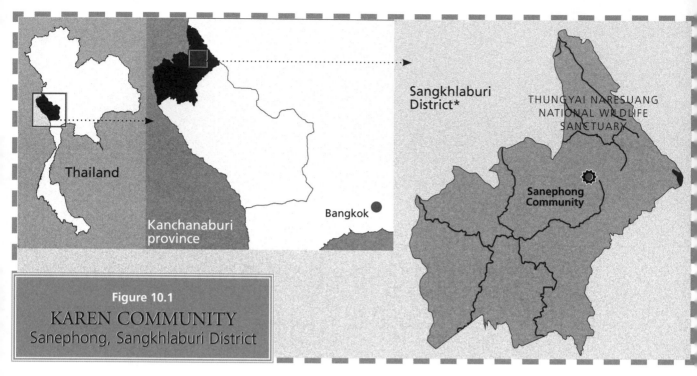

Figure 10.1
KAREN COMMUNITY
Sanephong, Sangkhlaburi District

Data from ESRI Global GIS, 2006
Walter Hitschfield
Geographic Information Centre
McGill University Library
*Digitized from
www.kanchanaburi-info.com

1
Research Institute of Language and Culture for Asia, Mahidol University, Salaya, Nakhon Pathom, Thailand

2
Institute of Nutrition, Mahidol University, Salaya, Nakhon Pathom, Thailand

3
Sanephong Village, Kanchanburi, Thailand

Key words > Indigenous Peoples, traditional foods, Karen, nutrition education, health promotion, culture-based approach, food and nutrition security

Photographic section >> XIX

> ## "We work together. We understand each other; much more than when we started the work."
>
> Suaijeemong Sangkhawimol (Sanephong traditional healer)

Abstract

Sanephong, a Karen community in western Thailand with a population of fewer than 700 people, benefits from the availability of traditional food, but the food system is deteriorating rapidly. The overall objective of this participatory intervention research project (2005 to 2009) was to use expert and community knowledge of the traditional food system and culture as a platform for working to improve community nutrition and health by: i) increasing awareness of the importance of traditional food sources; ii) promoting traditional food production and consumption, with a focus on children; and iii) increasing local people's capacities, knowledge and skills for enhancing children's food and nutrition security.

A culture-based approach, metaphors and social dialogue were used to develop four main intervention inputs: i) increased production of traditional foods at home; ii) motivation and nutrition education for schoolchildren; iii) women's empowerment; and iv) capacity strengthening for community leaders, local researchers and youth.

Results indicate that the project raised awareness and increased the availability and consumption of traditional foods, especially among children. Nutrition outcomes were also noted. Local change agents' capacity for continuing nutrition and food development was enhanced. The community became more aware of the importance of traditional foods and nutrition. High availability and use of local food sources confirmed that traditional food consumption remains common in the community. Researchers learned that successful nutrition and health promotion in an indigenous community relies heavily on processes that build trust and commitment among multiple stakeholders.

The design of intervention activities should be based on the community's own priorities, with academic researchers playing the roles of project catalysts and coordinators. Outside assistance is likely to be crucial in enabling indigenous communities such as Sanephong to achieve their development goals within a rapidly changing environment.

Introduction

Work in Sanephong, a Karen village in Thailand, began with a study to explore the indigenous food system, nutrition and health status. Results indicated that this community's food system has both strengths and weaknesses. Although plenty of nutritious traditional foods are available, the local food system is deteriorating rapidly, owing to economic and external market influences. The presence of 14 underweight and 20 stunted children indicated both acute and chronic undernutrition, and most children were consuming low levels of iron. Perhaps even more important, the traditional culture was being challenged. These results suggested that culturally appropriate nutrition and health improvement was essential for the community, especially for children (Chotiboriboon *et al.*, 2009).

On 21 January 2005, the Mahidol University Committee on Human Rights Related to Human Experiment granted ethical approval to the intervention research project. The signing of an official contract between external academic researchers and local community researchers and leaders is not common in Thai indigenous cultures. Through an anthropological approach, good communications and an open relationship helped to establish good rapport with the community (Langness and Frank, 1981). When good relations and trust had been built, community leaders and representatives were invited to join a seminar at Mahidol University, Salaya in August 2005. At this meeting, participants were encouraged to use the strengths and opportunities within their community to improve nutrition and health using traditional food sources. They affirmed their commitment to enhancing community food and nutrition security, especially for children, in collaboration with the academic research team and under the following principles:

- Indigenous culture, local food systems, traditional knowledge and the community's goals and expected outcomes are the foundation for the work.
- Intervention activities are jointly designed, are contingent on the results of the project's first phase (Chotiboriboon et al., 2009) and build on the community's strengths and opportunities for development.
- Researchers' knowledge and other relevant information are used to raise the community's awareness and support its decisions.
- Social dialogue is used to encourage communication and learning among the community, other stakeholders and research partners.
- Use of vernacular language and metaphor is encouraged to provide a communication tool for wider sharing of the project's vision and successes.
- Relationships among the community, other stakeholders and research partners are on equal terms and based on Karen tradition.

Context

Sanephong village is located in a mountainous tropical forest region of the Thungyai Naresuan National Wildlife Sanctuary, 336 km northwest of Bangkok, adjacent to the Myanmar border, at latitude 14° 55 and 15° 45 and longitude 98° 25 and 99° 05 east (Figure 10.1). Lying about 12 km east of Sangkhlaburi Municipality, the community can be reached only by four-wheel drive vehicle or on foot. A 2005 census reports the population of Sanephong at 661, with 345 males and 316 females (52 and 48 percent, respectively) living in 126 households. There is no electricity, but solar panels provide an energy source for charging batteries that are used in homes for television[1] and lighting. Social organization is based on kinship. Vernacular Karen is the everyday language, but Thai is also spoken, particularly among community leaders and the younger generation. Although community people identify themselves as Buddhists, indigenous animistic rituals are common. People worship Mother Earth and the Rice Mother to empower their indigenous spirits and ensure food security and well-being.

Most people live in extended families, with three or four generations in the same household. Men clear the forest for traditional farming, while women take charge of the household (e.g., gathering and cooking food and caring for children). Complementary activities are also carried out, and during periods of peak labour demand, whole families work on their farms, weeding and harvesting. At the household level, children are taught life skills by older siblings, and are brought to the Buddhist temple to observe community rituals from an early age. Food sharing, labour exchange and food dhana[2] for monks contribute to more equitable access to food. Socialization in the community occurs through day-to-day interaction at homes, local shops and the temple. Watching soap operas on television is the most popular leisure activity. From 2005 to 2008, the proportions of households with television sets increased from 18 to 39 percent of the total, with motorcycles from 15 to 21 percent, and with mobile telephones from 12 to 26 percent.

The food system

Sanephong people benefit greatly from the availability of local food. Domesticated local rice varieties, maize, taro roots and potatoes are the main sources of carbohydrates, while animals, particularly fish, are the main sources of protein, fat and oil. Vitamins and minerals come from traditional foods such as wild seasonal vegetables, fruits, cereals and animal

---

[1]   Televisions are more common in affluent households. They are bought with income from selling crops (e.g., chilli, coffee), while some are gifts from family members who work in town or city centres.
[2]   *Dhana*, meaning to give, provides merit in the Buddhist tradition, through gifts of living things such as food to monks.

sources. Rice is the staple food, supplemented by food items from traditional farming and seasonal natural resources. Most people still value traditional foods as medicines that make them strong and healthy. Local food species are grouped into four categories: i) cereals and roots (14 percent); ii) animal proteins – aquatic animals, insects and small mammals (17 percent); iii) vegetables and mushrooms (53 percent); and iv) fruits (16 percent). Although animal food sources are plentiful, people tend to prefer fishing, as hunting is strictly prohibited in and around the community. Vegetables and mushrooms from outside sources are now available in small local shops, as are chickens from industrial farming, pork, fish and canned fish, string beans, eggplant, cabbage and snacks of low nutritional quality.

Traditional farming techniques suit the local topography. Slash-and-burn for field rice cultivation is used on higher levels where irrigation is not possible. Paddy rice farming is practised on the alluvial floodplain. Harvested rice is stored for household consumption, sharing, offerings and occasional sales to neighbouring households. Food items such as cucumbers, gourds, sesame, taro roots and vegetables are grown on the same plots as the rice. Seeds are preserved for the next growing season. These traditional farming practices are also found in other Karen communities in northern Thailand (Ganjanaphan *et al.*, 2004). It should be noted that although most households grow rice for their own consumption, rice yields are low in Sanephong. Fewer than 30 percent of households produce sufficient rice for their yearly consumption.

Project objectives

The overall objectives of this intervention project (which is ongoing) are similar to those of other projects described in this book: to use expert and community knowledge of the traditional food system and culture as a platform for working with the community to improve its nutrition and health. Specific objectives are to: i) increase awareness about the importance of traditional food sources among community people; ii) promote traditional food production and consumption, especially among children; and iii) increase community capacities, knowledge and skills to take action on children's food and nutrition security.

Intervention concepts and methods

Figure 10.2 illustrates the project in a form that helps the community, researchers and other stakeholders develop a common understanding of the project vision. This communication method was found especially helpful for working with community people whose way of thinking is more concrete than abstract.

The project encourages villagers to grow more traditional foods in household and school backyards. Villagers also gather foods from natural sources (the forest, rice fields, ponds, canals, etc.) or buy them from small shops in the village. Both men and women are encouraged to increase the range and year-round availability of local foods, based on economic and nutrition considerations. Mothers and daughters are encouraged to use fresh food and a variety of foods, and to practise clean and safe food preparation and cooking that is both economical and nutritious. Families are encouraged to ensure that all family members, particularly children, have enough food and eat wisely (for food and nutrition security). All villagers are encouraged to take action to improve nutrition and health. In this way, the people of Sanephong can become healthy, strong and intelligent through improved nutrition and food security. For sustainability, traditional Karen culture and values are promoted among the local people, especially children and youth.

The project adopted Prawase Wasi's (2009) culture-based development approach. He compares this people-centred approach to building a pagoda, which normally starts from the ground, not from the top. He holds that it is wrong to initiate a community development programme from the "ivory tower" of expert knowledge while ignoring local knowledge. Wasi urges all those involved in community development to build a deep understanding of the local people, to learn from them, and to use their local knowledge as a driving force for development. Demonstrating respect for local people right from the start makes it easier for project workers

Figure 10.2 Conceptual framework of Karen culture-based traditional food intervention

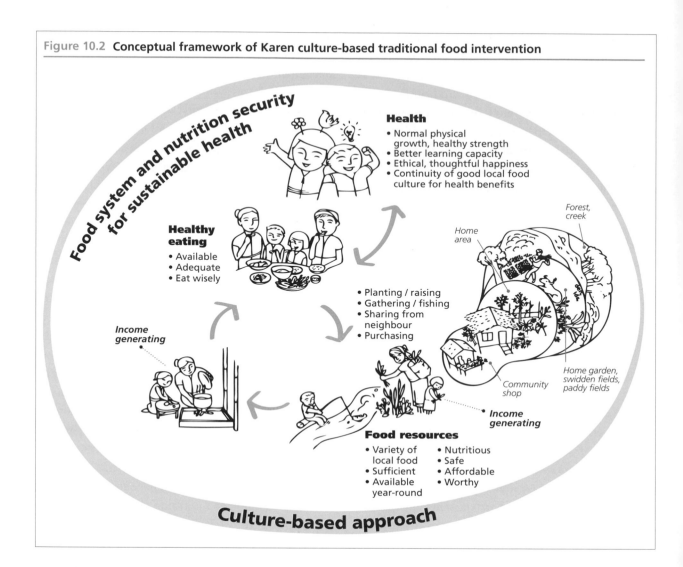

to win trust. Diverse stakeholders (community people and outsiders) should think together, and local resources and knowledge – or "cultural capital" – should be combined with adaptive knowledge, building on the community's strengths to improve the quality of life of its members (Wasi, 2009).

Using this culture-based approach, the participatory project integrates indigenous knowledge and researchers' knowledge, to promote food and nutrition security by using local foods and to carry out interdisciplinary research based on community collaboration. The project is implemented within the context of the community's culture, through dialogue among community people, local authorities and academic partners. All are key players throughout project processes (equitable

involvement): the project is guided by the community and its cultural preferences; and project implementation uses resources and strengths available in the community (Israel *et al.*, 1998).

For outside researchers, a culture-based approach is challenging because it requires them to build trust, find culturally appropriate ways of communicating with indigenous people, and ensure that intervention activities are in tune with cultural preferences (Grenier, 1998). The project uses social dialogue as a communication strategy because it allows participants to listen to one another and express themselves and their wishes freely. Indigenous partners are not seen as underprivileged or victimized, because all share equally in the dialogue and all views are recognized. Trust usually develops easily in such a context (Yankelovich, 1999; Isaacs, 1999; Wheatley, 2002).

Bohmian dialogue (BD) (Bohm, 1996) is used for team learning and decision-making. This method has been found to be particularly effective in situations where wide diversity among actors leads to a tendency for certain academic disciplines to dominate, thereby creating a strong social hierarchy. BD can briefly be described as an attempt to tackle fragmented and diverse thoughts and ideas. Innovative interventions and community empowerment are hard to achieve because individuals usually try to impose their own mental models of reality on others. If not properly handled, advocating and arguing tend to block the tacit development of thoughts and consensus. In groups or organizations – including rural communities – this generally results in forced acceptance, which leads to silent resistance. BD's effectiveness relies on having a natural and peaceful setting that allows equal opportunities for each individual to reflect on her/his experiences and generates collective thinking about the group's future. Mutual respect and trust are important elements. With appropriate facilitation, BD can lead to collective innovation and a sense of community. Project meetings are therefore conducted in peaceful, open and – when possible – culturally symbolic settings, such as Buddhist temples, mosques or mountain tops. To maximize the potential for a creative surge of insight and innovation, rituals and images such as prayers or sacrifices are also included when appropriate.

The project design is therefore based on a clear vision of collective action and shared expectations, a culture-based approach with multi-stakeholder participation, and good communication platforms. To facilitate collective action in an indigenous cultural context, metaphor has been found to be a simple and effective tool. For example, when working with an indigenous group in Latin America, Friulian lexicalized plant species by assigning them animal attributes that brought out their properties and colours (Pellegrini, 2006). A similar practice occurs in corporate organization culture, where people with different backgrounds often use metaphor as a tool in knowledge management and the generation of common goals and thinking, using imagination and intuition to develop a shared vision (Nonaka, 1998). Metaphor helps people to understand complex realities by using the common intuition they have developed from day-to-day experience and encapsulated in their vernacular language (Lakoff and Johnson, 1980). The community and research team developed "*SWA*" as a metaphor for communicating the project's shared vision among stakeholders (Figure 10.3).

"*SWA*" is a Karen vernacular term denoting a fish sanctuary in a running stream. The fish in a *SWA* are safe from predators and can resist the current of the flowing stream. In the metaphor, the *SWA* is the community, while the community members are the fish. Living in a *SWA* gives Sanephong people a safe space for their own culturally appropriate lifestyles, but it must also allow them to develop resistance so they can face rapid changes outside their community. Living in a *SWA* does not mean that people are trapped in their community for safety. Those who want to leave the *SWA* need to prepare themselves physically, mentally and spiritually to survive well outside. The *SWA* metaphor thus conceptualizes the "how" of the project. Working in Sanephong, for Sanephong people, intervention efforts should assist the community in

Figure 10.3 *SWA* – a visual model used to communicate shared vision among project stakeholders

building a strong *SWA* (i.e., in preserving an ecological niche and creating a living space that is suitable for them and future generations).

Intervention activities to date

Based on these concepts and methods, the community, other stakeholders and the academic research team planned and implemented four main intervention inputs: i) increased production of traditional foods at home; ii) motivation and nutrition education for schoolchildren; iii) women's empowerment; and iv) capacity strengthening for community leaders, local researchers and youth.

Increased production of traditional foods at home

A home garden is analogical to having a *SWA* at home, and this activity was designed to put the *SWA* concept into practice. First, 50 participating children were asked to list the traditional vegetables and fruits they found in a demonstration garden, and then to draw them, to create a visual model (Figure 10.4). Next, they explored the traditional vegetables and fruits in their own home gardens. Rare and endangered traditional plants were cultivated in a nursery and transplanted to home gardens. To stimulate the children's learning and action, they also visited other home gardens and recorded and drew pictures of what they found there. To endorse these activities, community adults advised the children on how to select good traditional food varieties and grow them. This was followed by a competition in which the children grew varieties of traditional vegetables and fruits in their home gardens. Three important criteria for selecting competition winners were easy access to traditional plant foods, self-reliance and diversity of plants. Agriculture experts were brought in to advise on the value of traditional agricultural techniques, such as growing several varieties of food in the same plot, storing seeds for the following season, and respecting nature, which sustains the environment and home food security. As water shortage can be an important constraint during the dry season, the research team also assisted the community with writing a proposal for the building of a small community dam, to present to the local administration.

Motivation and nutrition education for schoolchildren

To increase schoolchildren's confidence in their culture and traditional foods, four motivation and education camps were organized (Henderson, Bialeschki and James, 2007; Bialeschki, Henderson and James, 2007; Goldstein *et al.*, 2004) for a total of 350 children aged eight to 15 years. The children were conceptually seen as small fish in a *SWA*. Topics at the camps included: i) re-establishment of traditional food crops at the household level; ii) environment and natural resource conservation; iii) nutritious cooking of traditional foods; iv) basic health care; and v) nutrition. The process emphasized holistic learning development (language, music, arts and life skills), and included an introduction to traditional food sources from forests, waterways and households; food and nutrition songs; drawing; lectures; brainstorming; presentations for friends and community members; and "edutainment" (educational entertainment) activities, which included traditional indigenous ways of learning by doing. The children were encouraged to cook local dishes. Based on traditional knowledge, they were also encouraged to plant more local vegetables in upstream areas, to safeguard community water resources.

Information about the importance of traditional foods, and nutrition education were later integrated into primary school curricula. This involves parents, elders, community leaders and school teachers, to ensure continuation in schools. Education tools emphasizing the importance of food diversity include games and poems promoting traditional foods. Elders are also involved in educating children in school. Community women volunteered to share their cooking knowledge and skills, emphasizing the concept of "fresh, clean, nutritious and safe" food preparation with the schoolchildren. During these sessions, children were taught to be careful about money, especially regarding their habit of buying sweets

Figure 10.4 Household traditional food plant diversity in Sanephong

and snack foods from community shops. Nutritious local sweets were promoted as alternatives. After each session, schoolchildren were encouraged to share what they had learned with their families. Interpersonal communications and home visits were used to encourage the community to endorse the children's activities and to promote healthy traditional food sources. Scientific information about local food and nutrition helped stimulate discussions on local foods and enhanced the community's knowledge, attitudes and participation.

Women's empowerment

Fifty women volunteers were trained in techniques of "fresh, clean, nutritious and safe" cooking. This resulted in the creation of nutritious traditional food snacks, such as glutinous rice mashed with sesame, which the volunteers then prepared for the monthly community meetings where project activities are discussed. Two local shop owners were invited to join the women's group activities and to help organize a competition for local nutritious recipes, such as traditional vegetables in noodle soup and other common local dishes. Through the competition, the women's group gathered 20 nutritious local recipes and one modified healthy local snack, which they promoted in the community. The women volunteers were also trained in monitoring the nutrition status of their children and themselves, including education on achievable daily nutrition intakes from local foods, to ensure good nutrition and health for their families.

Capacity strengthening for community leaders, local researchers and youth

To strengthen the confidence and communication capacities of community leaders and local researchers, the academic research team provided opportunities for them to share their knowledge and experiences on issues such as subsistence practices, health and the community's future with non-governmental organization (NGO) officers working in the area, border patrol police, local district administrators and others. Study visits were also organized for the local leaders and researchers to learn more about organic farming, ecotourism and ecomarketing. Five local researchers were trained in financial management and traditional food recipe development. In addition, 100 community youth were trained in conserving traditional knowledge and culture, and good communication skills. Nine community leaders and the local researchers participated in a national seminar at Princess Sirindhorn Botanical Museum, Bangkok. Five community leaders and researchers joined international workshops in Italy, Japan and Canada, to exchange ideas and worldviews with international indigenous leaders, academics from universities and activists from NGOs.

Indirect project contributions

These four main intervention activities were not the project's only inputs for food and nutrition security in Sanephong. Following the capacity strengthening activities, community and local leaders, youth, women and children can now help bring about changes that will contribute to the continuing achievement of project objectives. For instance, community leaders – supported by the local administration – are now mobilizing the community and managing the budget for a community garden project, to provide vegetables for consumption and income generation. Project activities were designed not only to effect direct changes but also to generate additional changes through training and capacity building of community researchers and other agents.

Intervention results

This participatory research in a Thai Karen community was conducted through collaboration among the community, other stakeholders and the academic research team. These groups therefore represented

both the research network and the tools of the research. To collect the necessary data, both qualitative and quantitative research methodologies were applied in Sanephong community between 2005 and 2009. The process for studying indigenous food systems described by Kuhnlein *et al.* (2006a) provided guidelines for this. In-depth interviews, focus group discussions, participatory observations, community walks and surveys, and nutrition and health assessments were used. The following subsections summarize the evaluation results.

Trust and commitment

One important indicator of successful work with an indigenous community is the building of trust between collaborators in and outside the community. Throughout the project, the external research team generated community participation and involvement. Results of a community survey[3] indicated that people in the community recognized the project's contributions to providing knowledge (65 percent, n = 44), changing attitudes (15 percent, n = 10) and changing behaviours (20 percent, n = 14). Opinions from community people and leaders, recorded from interviews and group discussions, suggest that the community trusted the research team:

> Many community people participated in the project activities … more than half.
>
> *(Villager)*
>
> This group does not deceive us … they are good and diligent … they talked to us nicely … and they explained rather well.
>
> *(Deputy Village Headman)*
>
> I talked with them [community people] before but they were not interested … no action. It's heartbreaking for me again and again … When the research team came in and worked with us, things became clearer … I felt such relief.
>
> *(Sub-district Headman)*
>
> The research team and the community are like husband and wife now, we have good relationship.
>
> *(Villager)*

---

3 Survey respondents were the mothers of 68 primary school children in Sanephong.

Another important aspect for sustainable food and nutrition development in an indigenous community is that local commitments were strengthened. At least 20 community people volunteered to continue promoting traditional foods after the intervention. A group of these volunteers started to rehabilitate community waterways, to make them more environmentally favourable for aquatic animals such as fish, snails, crabs, frogs and shrimps.[4] They also established community rules prohibiting dive-harpooning to catch fish. Another group volunteered to cook for children at the school and the day care centre. These activities indicate that community people might carry the project achievements forward. Although community members have various levels of awareness of the importance of traditional foods, there is a core group of people who are willing to take action to improve the situation.

Traditional food awareness and availability

It was clear that the research team had created awareness about the importance of traditional foods in Sanephong. Results of a community survey[5] indicated that the research team contributed to increased awareness of traditional foods' role in making people strong and healthy (97.3 percent, n = 73). Only 14.5 percent reported that their families had cooked purchased vegetables over the previous year. Many (73.7 percent, n = 56) reported that over the previous four years, community people had collected vegetables from the surrounding forest to grow around their households, but when asked whether their own families had grown traditional vegetables over the previous four years, more than half (59.2 percent, n = 45) reported that they had not, and only 36.8 percent (n = 28) that they had (Figure 10.5).

Nevertheless, it was found that the number of vegetable and fruit varieties in community households had increased from 81 to 137 after the intervention,

Figure 10.5 **Opinions regarding traditional vegetables after the intervention**

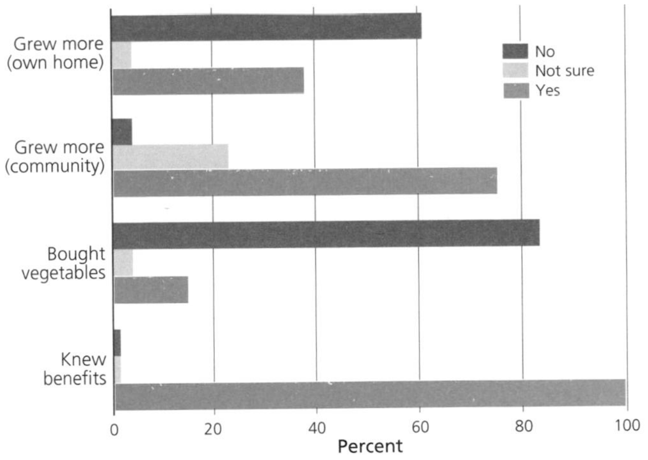

Table 10.1 **Numbers of vegetables and fruits grown by households, before and after intervention**

Type of vegetable and fruit	Before	After	Change
Traditional	75	119	+ 44
Others	6	18	+ 12
Total	81	137	+ 56

with 119 of these 137 items identified as traditional. However, the number of non-traditional vegetables and fruits had also increased, from six to 18 (Table 10.1). The vegetables *Pak Man Moo* (*Gnetum nemon* L. var. *tenerum* Markr.) and *Pak Kood* (*Diplazium esculentum* [Retz.] Sw.) have high nutritive values, and their use in households increased. However, after the intervention, only about 10 to 14 percent of the households were growing them; this slow uptake was due to traditional beliefs and lack of water. It seems likely that the project built awareness of nutrition, particularly from vegetables and fruits, and of the importance of traditional vegetables and fruits.[6] This view is supported by the community's continuing frequent use of traditional foods.

---

4   The community survey revealed that 68.4 percent of families (n = 52) cooked fish from community waterways and did not have to buy it.
5   Survey respondents were the mothers of 75 primary school children in Sanephong.

6   The Karen in this area do not usually raise traditional animal food sources, but they started raising aquatic food sources following a government initiation.

Figure 10.6 Children's activities in the traditional plant activity programme, as reported by parents

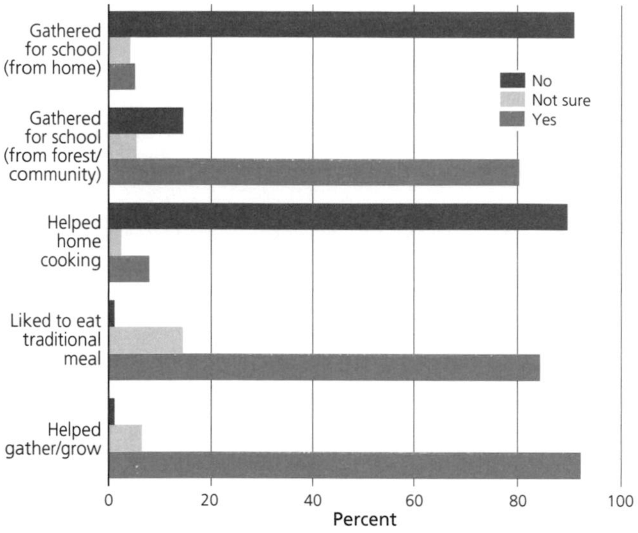

Gathered for school (from home)
Gathered for school (from forest/ community)
Helped home cooking
Liked to eat traditional meal
Helped gather/grow
No
Not sure
Yes
0
20
40
60
80
100
Percent

Table 10.2 Reported food preferences for families, children and community festivals

	% preferring item (n = 77)		
Dish	**For family**	**For children**	**For festivals**
Chilli paste and vegetables	61.5	1.3	17.9
Vegetable soup	12.8	14.1	–
Fish and vegetable soup	14.1	28.2	–
Chicken/pork and vegetable soup	1.3	15.4	2.6
Fried foods (stir- or deep-fried)	5.1	34.6	1.3
Curries (e.g., pumpkin with chicken, pork in coconut milk)	5.1	–	66.7
Dessert	–	–	3.8
Others (e.g., *Kanom Jeen*/Thai vermicelli)	–	6.4	6.4

Children as change agents[7]

Children are the future. To preserve Karen culture and traditions, community leaders gave priority to children's education. Figure 10.6 shows that community people were aware of the many activities for teaching children about traditional vegetables and fruits. According to the survey, children had become more involved in growing or gathering traditional vegetables over the previous year (92 percent, n = 69). They also gathered traditional vegetables from the forest and waterways to cook for school lunches, although traditional vegetables and fruits grown in household gardens did not contribute to school lunches.[8] Most respondents (89.5 percent, n = 68) mentioned that children helped with home cooking, and 84.2 percent (n = 64) agreed that "community children like to eat traditional meals very much".

Traditional food uses

Food preferences in the community indicate that people in Sanephong still have a rather simple lifestyle. Table 10.2 shows that there are only five categories of dish eaten by families and children and at festivals: chilli paste with vegetables, soups, fried foods, curries, and desserts. Those interviewed (n = 77) reported that chilli paste with vegetables[9] was the most popular food for the family (61.5 percent, n = 48) and was also used at festivals (17.9 percent, n = 14), but not for young children. Soups were the most popular dish among children, who also liked fried foods (34.6 percent, n = 27). Curries such as pumpkin with chicken or pork in coconut milk were the most popular festival foods. Based on these food preferences, it can be seen that family diets could rely on traditional food sources (i.e., traditional vegetables and fish), but children's preferences for chicken/pork soups and fried foods indicate a need for families to purchase some foods (chicken, pork and cooking oils). If this trend continues, family expenses on food can be expected to increase in the community.

Respondents often mentioned health and safety (32.1 percent, n = 25) as their reason for maintaining traditional diets: "eating our food makes us strong and healthy". Some (29.5 percent, n = 23) simply liked

7 A change agent is an individual with capacities for influencing changes in the community. By empowering the children through active participation in project interventions, the project team helped them become valuable project advocates and important players in future development.

8 This indicated that there were still plenty of traditional vegetables in the area around the community, and that households probably grow traditional vegetables in small amounts for their own consumption.

9 Chilli paste is prepared by pounding red chillies and garlic together with salt or shrimp paste. Karen chilli paste is generally very strong, so it is usually eaten together with raw or boiled vegetables.

traditional foods: "eating chilli paste and vegetables makes me happy". Self-sufficiency (23.1 percent, n = 18) was another reason: "eating food grown by ourselves is marvellous". Some (14.1 percent, n = 11) felt at home and proud: "I am proud to eat our local foods like my grandparents". Thus, increased exposure to mainstream communication and food marketing had not changed community dietary patterns, and positive attitudes towards traditional diets remained intact at the end of intervention.

Although the project promoted the use of traditional foods in general, it focused on the food items identified in the phase 1 study as being of high nutritional value and available in the community. For example, *Pak Man Moo* is a good source of vitamin A and folate, and an excellent source of vitamin C; *Pak Kood* is a good source of iron and an excellent source of vitamin C; and the shellfish *Khlu-mi* is a good source of calcium and an excellent source of iron (Chotiboriboon *et al.*, 2009). Mothers (82.4 percent, n = 61) mentioned that they cooked more of these foods (some up to two or three times per week) during the intervention.

Change agents as examples for others

Local researchers reported that persuading Sanephong people to take action to improve their food and nutrition security by using traditional foods takes both time and effort. After the nutritive values of traditional foods (the phase 1 results) were presented to the community, one of the local researchers decided that she would act as an example. Similar to many others in the community, her family included three adults (her husband, herself and a grandmother) and four children aged two to 12 years (three boys and one girl). Her husband played a major role in traditional farming while she took care of the household by gathering and cooking food and looking after the children. Her husband occasionally earned some extra cash from daily wages.

By January 2008, she was growing 58 different traditional vegetables and fruits in an area of about 0.2 acres (0.08 ha) around her house. From this home garden, she was able to pick 15 traditional vegetables in about 15 minutes, to cook a good meal for her family. This meal included stir-fried roselle with canned fish, a mung bean noodle soup with ivy gourd omelettes, pumpkin and canned fish curry and chilli paste. In terms of dietary reference intakes (DRIs) (Banjong *et al.*, 2003; Changbumrung, 2003) these four dishes with steamed rice were found to provide adequate energy and macronutrients (carbohydrate, protein, fat), at more than 40 percent of the DRIs. The dishes also provided 30 percent of the DRIs for iron and vitamin C, 28 percent for vitamin A, 78 percent for vitamin B_1, 26 percent for vitamin B_2, 18 percent for calcium, and 13 percent for niacin. The calculations of vitamin A content did not include the vegetables eaten with chilli paste, which is a rich source of carotene, so the family's vitamin A intake may have been more than 30 percent of the DRI. These percentages of DRIs represent good dietary contributions, especially of macronutrients, iron and vitamins C, A and B_1. However, intakes of other nutrients, especially calcium, vitamin B_2 and niacin, were inadequate. This meal required the purchase of canned fish, eggs and vegetable oil, and was estimated to cost about baht (THB) 6 or USD 0.16 per person.

This demonstrated that growing a variety of traditional vegetables and fruits is not only feasible but could also lead to increased dietary variety at minimum cost to the family. This local researcher was able to share her home-grown vegetables with neighbours, which is a highly regarded practice in the community. Even more important, her children's nutrition and health improved significantly. For instance, her eldest son's weight-for-age increased from 89 to 95 percent of the Thai standard, his height-for-age from 94 to 96 percent, and his weight-for-height from 105 to 107 percent. His haemoglobin level also improved, from 9.8 to 12.7 g/dl. As is the case for many women in developing countries, these positive changes were possible because of the local researcher's contributions to food production, food access and nutrition security (Quisumbling *et al.*, 1995). Based on this success, she has become an effective promoter of traditional foods in Sanephong.

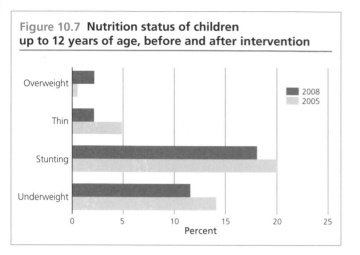

Figure 10.7 **Nutrition status of children up to 12 years of age, before and after intervention**

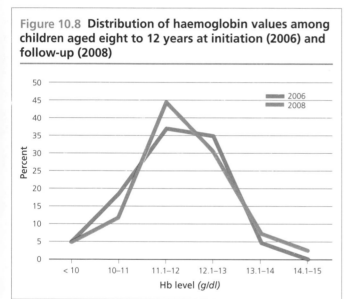

Figure 10.8 **Distribution of haemoglobin values among children aged eight to 12 years at initiation (2006) and follow-up (2008)**

Local capacity strengthening

The community and the research team made capacity strengthening their highest priority as a way of helping to maintain and sustain the changes long after the project ended. Throughout the project, change agents – including community leaders and local researchers – were observed to become more confident in their communications on food and nutrition security within their community, at the national and international levels. In January 2009, Her Royal Highness Crown Princess Sirindhorn visited Sanephong and the project's change agents gave a

good presentation. The Crown Princess encouraged them to continue their efforts to protect traditional food sources and to eat traditional foods with their families. The efforts of Sanephong change agents and the community at large are now widely recognized, and a Web site[10] reports on progress to the outside world. In addition, community leaders have allocated THB 200 000 (about USD 6 000) to developing homestead gardening activities, and a local youth group has organized its own learning camps, with support from elders and local officers.

Nutrition outcomes

The project focused on increasing awareness about the importance of traditional food sources, and strengthening local capacities, knowledge and skills to take action on children's food and nutrition security, but it did not have resources to influence directly the quality or quantity of diets. Changes in nutrition status thus rely on the community's abilities to adjust within their own means. It was observed that nutrition among children aged 0 to 12 years improved during the intervention. Figure 10.7 shows that the overall incidences of underweight among children decreased from 14.1 percent (n = 26) to 11.6 percent (n = 22), of stunting from 20 percent (n = 37) to 18 percent (n = 34), and of thinness from 4.9 percent (n = 9) to 2.1 percent (n = 4). The proportion of overweight children increased from 0.5 to 2.1 percent, reflecting the global nutrition trend (Khor, 2008).

Iron nutrition status among children aged eight to 12 years (n = 43), as measured by haemoglobin, indicated average levels of 11.7 mg/dl (SD = 0.9) before and 11.9 mg/dl (SD = 0.9) after the intervention, and a tendency for improved distribution of haemoglobin concentrations was also noted (Figure 10.8). According to World Health Organization (WHO) criteria (Gleason and Scrimshaw, 2007), the proportions of children with normal concentrations increased from 65.1 to 69.8 percent, those with mild shortages decreased from 30.2 to 25.6 percent, and those

---

[10] www.rdpb.go.th/rdpb/front/news/rdpbnewsdetail.aspx?rid= 123&catid=1

with moderate shortages remained at 4.1 percent. Tackling iron nutrition with traditional food sources in Sanephong remains a challenge because most local iron-rich foods (i.e., *Khlu-mi*, shellfish) are available for only a few months a year (March to May). Similar to experiences elsewhere, the shortage of iron-rich foods underlines the importance of ensuring that traditional foods are diverse and evenly distributed through the seasons (Arimond and Ruel, 2004). This issue needs further work in the community. Raising community awareness of and participation in traditional food rehabilitation can improve nutrition and health indirectly (Kuhnlein and Receveur, 1996; Damman, Eide and Kuhnlein, 2008; Khor, 2008), but more direct interventions are necessary if immediate results are expected, especially for iron nutrition improvement.

Conclusion

Living in a tropical forested area in harmony with nature, Sanephong people benefit greatly from the availability of local foods such as cereals and roots, animal foods, vegetables, mushrooms and fruits. However, the food system is deteriorating rapidly owing to environmental degradation and external socio-economic influences. This participatory intervention research project aimed to capitalize on indigenous culture and food sources as the basis for improving nutrition and health in the community, particularly among children under 12 years of age. The project was implemented with a culture-based approach, in close collaboration with stakeholders of diverse backgrounds (community people, primary school teachers, local health officers and academic researchers from different disciplines). Metaphor and social dialogue were used as communication tools for promoting equitable participation. Based on these concepts, four main intervention activities were initiated.

Increasing families' production of traditional foods at home was designed to promote easy access to traditional food sources, self-reliance and greater diversity of local foods. Children were key actors in this, with support from adults. Motivation and nutrition education for schoolchildren increased their confidence in their culture and traditional foods. The process emphasized holistic learning through "edutainment", consolidating language, music, arts and life skills into actions. Women's empowerment activities helped increase women's capacity to take care of their families through the use of fresh, clean, nutritious and safe food, as demonstrated by a woman volunteer change agent. Capacity strengthening for community leaders, local researchers and youth was designed to bring indirect improvements to nutrition and health through the use of local foods. This intervention focused on creating awareness and enhancing traditional knowledge, communication skills and management capacities, so that community people can handle nutrition and health issues themselves.

The project successfully raised awareness and availability of traditional food sources in Sanephong. Community people recognized the project as a partnership to provide the knowledge needed for attitude and behavioural changes. Working together, community-led changes occurred in all age groups. Children and youth cultivated and gathered more traditional vegetables, and helped with home cooking and the care of younger siblings. Community women developed and promoted local dishes using nutritious traditional food sources. Creative menus were used at the community child care centre, for school lunches, in home cooking and at community events such as learning activities, community meetings, hospitality and annual festivals. Community men set up new rules for protecting the *SWAs* that provide habitats for aquatic resources. All project activities were approved and supported by community leaders. Strong awareness and community efforts to increase traditional food sources indicated the important role that traditional food use continues to have in Sanephong. During the project, community people gathered more traditional vegetables from the surrounding forest and their home gardens. The project change agents explained the value of traditional foods and exchanged ideas and world-views on food and nutrition security at the local, national and international levels.

Lessons learned

While nutrition and health problems related to environmental change and the deterioration of local food systems are emerging in indigenous communities all over the world, enhancing traditional food systems can serve as a strategy for coping with malnutrition and sustaining development (Kuhnlein and Receveur, 1996). The participatory project in Sanephong demonstrates this (Chotiboriboon *et al.*, 2009). The project team set out to work with and for the people of Sanephong community, by combining team members' knowledge of community food systems with local knowledge and wisdom about Sanephong's traditional food system, to improve nutrition and health, especially of children. Although a culture-based approach was used, academic partners were crucial at the start of the project in ensuring that the process adopted allowed all community members to participate and identify their own priorities (Kuhnlein *et al.*, 2006b). As outsiders working with the consensus of community people from the outset, the academic researchers were catalysts and project coordinators. Multi-stakeholder partnership was key to project success. To work well in this context, members of the participatory research team needed trust, openness and deep respect for each another. Outsiders had to accept that changes would come only at the community's own pace.

Similar to experiences of the global network (Kuhnlein *et al.*, 2006b), the project team learned that using local foods to improve the nutrition and health of children requires more than technical knowledge. Project success relied heavily on human processes and factors, such as establishing long-term relationships, building mutual trust and developing effective communication among multiple partners. Individuals tend to see things from their own perspectives, which are derived from fragmented thoughts and personal experience. People's actions are driven by their own perspectives, and occasionally conflicts emerge. The project team learned that the different ways of thinking of the indigenous people and the urban-based researchers could be connected through socializing activities. Partnerships were created through exchanging ideas and sharing experiences from working together. Once mutual trust had emerged, appropriate implementation could follow. This process took both time and effort for all stakeholders. It is therefore recommended that future work to improve food and nutrition security in an indigenous community such as Sanephong should consider applying a culture-based, free-formed and organic approach rather than a mechanical and controlled approach with expected outcomes that are set by others without the community's participation.

Although the project turned out to be positive for both the Sanephong community and the academic research team, the road to better nutrition and health using traditional foods in this community is long. Similar to many other indigenous communities around the world, Sanephong is no longer isolated. It is connected to the global community through socio-economic and cultural aspects, together with the newly introduced market economy, media and transportation. Sanephong people should still be able to choose how they wish to live; the community has made great efforts and has expressed a strong interest in maintaining its own culture and traditions, which may include preserving traditional food diversity and traditional farming. Community people still firmly believe that retaining biocultural heritage nourishes the spirit of the Karen people. Nevertheless, the community needs outside assistance to help its members achieve their goals. Interested outsiders should act like fish from other *SWAs*, coming in to help people in the Sanephong *SWA* prepare their younger generation for a brighter future, so they can live proudly in their own *SWA* within the running stream of the rapidly changing social and global environment ❋

Participatory research team

A team of community and academic partners made this work possible. Major contributors from Sanephong community were Anon Setapan, Mailong-ong Sangkhachalatarn, Nutcharee Setapan, Sompop Sangkhachalatarn, Benchamas Chumvaratayee, Sanu, Jongkol Pongern, Plubplueng

Kamolpimankul and Suwatchai Saisangkhachawarit. The academic team included Charana Sapsuwan, Sopa Tamachotipong, Prapa Kongpunya, Pasamai Eg-kantrong, Saifon Phonsa-ard, Waragon Khotchakrai, Suwarin Yunaitum, Prangtong Doungnosaen, Kamontip Srihaset and Tharaporn Graigate of the Institute of Nutrition, Mahidol University; Winai Somprasong, Pramote Triboun and Bordintorn Sonsupab from the Division of Plant Varieties Protection, Ministry of Agriculture and Cooperatives; Rattanawat Chairat of the Faculty of Environment and Resource Studies; and Ariya Thanomsakyuth of the Faculty of Medicine, Ramathibodi Hospital, Mahidol University.

Acknowledgements

Heartfelt gratitude to the people of Sanephong community. Their trust and openness brought all project participants together for this action and learning. Thanks for collaboration and support from Sundaravej Border Patrol Police School, Laiwo Tumbol Organization Administration and Sangkhlaburi Hospital.

The research team is grateful to the United States Agency for International Development (USAID), FAO, the Bio-Thai Foundation and the Thai Health Promotion Foundation for their financial support. Without the support of Mahidol University for inter-centre and interdisciplinary research, this work would not have been possible.

Advisers and colleagues have been more than generous in their support. Thanks to Sakorn Dhanamitta, Chaweewon Boonshuyar, Orapin Banjong, Prapaisri Sirichakawal, Prapasri Puwastein, Tippan Sadakorn, Sujaritlak Deepadung, Kunchit Judprasong, Opart Panya, Orasuda Charoenrath, Kulvadee Kansuntisukmongkol, Kanlayanee Atamasirikul and Harriet Kuhnlein, whose advice was key to the research team's technical contribution to the community. Lastly, thanks to Richard Hiam, Harold C. Furr and Gene Charoonrak for editing support, and Sompong Ondej and Suvaraporn Maneesrikum for their assistance in documentary research.
> Comments to: nussm@mahidol.ac.th; lcssr@mahidol.ac.th

>> Photographic section p. XIX

Chapter 11

The Nuxalk Food and Nutrition Program for Health revisited

NANCY J. TURNER[1]   WILFRED R. TALLIO[2]

SANDY BURGESS[2, 3]   HARRIET V. KUHNLEIN[3]

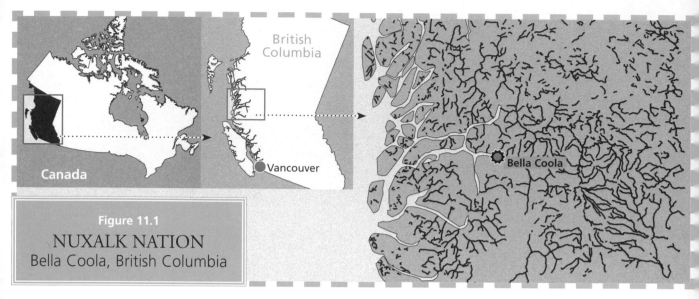

Data from ESRI Global GIS, 2006.
Walter Hitschfield
Geographic Information Centre,
McGill University Library.

1
School of Environmental
Studies, University of
Victoria, Victoria,
British Columbia, Canada

2
Nuxalk Nation,
Bella Coola,
British Columbia, Canada

3
(Retired) Salmon Arm,
British Columbia, Canada

4
Centre for Indigenous
Peoples' Nutrition
and Environment (CINE)
and School of Dietetics
and Human Nutrition,
McGill University,
Montreal, Quebec,
Canada

Key words > Indigenous Peoples, food systems,
traditional food, Nuxalk Nation, British Columbia,
intervention

Photographic section >> XXII

"They came out in droves!"

Rose Hans, in recollection of the feasts for youth that were part of the Nuxalk Food and Nutrition Program,

as remembered in 2006

Abstract

The original diet of the Nuxalk Nation incorporated a range of nutritious fish and seafood, game and various plant foods, including greens, berries and root vegetables. However, early research underlying the Nuxalk Food and Nutrition Program demonstrated a dramatic shift in diet during the twentieth century, with less use of traditional food and greater reliance on processed and less healthy food, combined with a more sedentary lifestyle. Documentation of the Nuxalk Nation's food system underlined the imperative of using community resources and local cultural foods as the platform for health education and promotion to improve food use and nutrition status.

The Nuxalk Food and Nutrition Program was conducted in the mid-1980s, with changes in food use and nutrition status determined through measurements taken before and after the interventions. This was the first programme of its kind in First Nations communities, and led to many similar initiatives in Canada. More than 350 activities were developed with input from community elders and leaders, and attracted thousands of individual participations from the population of about 500 on-reserve Nuxalk. Popular activities were feasts, food excursions and two widely distributed books on traditional food systems and recipes. Evaluation activities included interviews on food use and diet, and measurement of anthropometry and physiologic indicators of key micronutrients (vitamin A as carotene and retinol, folate and iron), dental health, and process indicators of programme success and participation.

Improved use of traditional food resources was shown, with increasing numbers of families using these foods, particularly fish, and increased amounts of food used per family. Dietary status improved with the increased use of fruit and vegetables and better intakes of vitamin A, folate and iron. Nutrition status regarding carotene, retinol and folate improved in all age and gender categories, and iron status improved among youth. Dental health, measured through examination of children's tooth decay, improved dramatically.

A follow-up consultation in 2006 examined long-term programme impacts, changes in traditional food availability due to environmental shifts, and concerns about increasing obesity and chronic disease within the Nuxalk Nation.

Introduction

The Nuxalk Food and Nutrition Program was conceived in the early 1980s and began officially in 1983. It was a collaborative research project involving the Nuxalk[1] Nation (including the community health centre, Band Council and regional leaders, elders and youth) and academic research partners. It represented one of the first comprehensive community-based projects to document Canada's Indigenous Peoples' traditional food systems and how these relate to health. Funded through grants from Health Canada, the project received ethics review and approval from the University of British Columbia, the main participating academic institution when the project began, and later from McGill University. The details of how the research was established, an inventory and nutrient analyses of traditional Nuxalk foods, and specific initiatives embraced within the overall programme are provided elsewhere (Kuhnlein, 1986; 1989; 1992; Lepofsky, Turner and Kuhnlein, 1985; Kuhnlein and Moody, 1989; Kuhnlein and Burgess, 1997; Turner et al., 2009).

This programme represented best practices in community-based nutrition research, and has served as a model for many other initiatives in Canada and other parts of the world. In all aspects of the programme, community researchers and health promotion staff were primary participants and collaborators, undertaking interviews, assessments and promotion activities. Throughout the course of the programme, the researchers followed agreed research protocols, seeking informed consent for interviews and photographs, maintaining the confidentiality and anonymity of interview respondents, and ensuring that results from the project were first shared with and reviewed by Nuxalk Nation participants.

---

[1] Pronounced "Noo-halk".

Context

The Nuxalk (formerly known as the Bella Coola) are a community of Indigenous People of the central coast of British Columbia, whose home territory is set within a network of deepwater inlets, channels, islands, river estuaries, floodplains and valleys, and rugged mountainous terrain. The Nuxalk language is classed in the Salishan language family. At the time of European contact in the late eighteenth century, the Nuxalk Nation included approximately 30 permanent villages extending along the Bella Coola River Valley and along the coasts of North and South Bentinck Arms, Dean Channel and Kwatna Inlet (Figure 11.1). By the early twentieth century, the numbers of Nuxalk people had dwindled significantly from their original (estimated) population of more than 2 000, largely owing to epidemic diseases brought by the European newcomers (Boyd, 1990). Those who remained came together at a village on the north side of the Bella Coola River, then moved to their present location at Bella Coola in 1936, after a major flood forced them to abandon their earlier village. In the 1980s, a new housing subdivision was established about 8 km east of Bella Coola, and many Nuxalk people now live there. At the time of the research reported here, reserve residents numbered approximately 800 people in 150 homes.

The Nuxalk territory is bounded by the territories of Indigenous Peoples from other language families: Wakashan peoples (Haisla and Hanaksiala, Heiltsuk and Kwakwaka'wakw) surrounding the Nuxalk lands and waters along the coast; and Athapaskan people (Ulkatcho Dakelh, or Carrier) on the inland side (Boyd, 1990). The Nuxalk are culturally similar to their Heiltsuk neighbours, having a complex social organization with hereditary leadership, strong ceremonial traditions and an oral history that reflects a deep relationship with and knowledge of the marine environment, rivers and associated habitats in which the Nuxalk people have dwelled since time immemorial (McIlwraith, 1948; Kennedy and Bouchard, 1990). Today, the Bella Coola Valley is also home to many non-native settlers.

The climate of the Nuxalk territory is typical of coastal British Columbia: high precipitation and relatively mild winters and summers, at least in the lowlands. The vegetation reflects this regime, as part of the coastal temperate rain forest. At lower elevations, the forests are dominated by western hemlock (*Tsuga heterophylla*), western red cedar (*Thuja plicata*), amabilis fir (*Abies amabilis*) and Sitka spruce (*Picea sitchensis*), all coniferous species. Some lodgepole pine (*Pinus contorta*) and Douglas fir (*Pseudotsuga menziesii*) occur in drier places. Black cottonwood (*Populus balsamifera* spp. *trichocarpa*) forms immense stands along the river valleys, and red alder (*Alnus rubra*) and Pacific crab apple (*Malus fusca*) are other common deciduous species. Higher-elevation forests, which receive more of their precipitation in the form of snow, include mountain hemlock (*Tsuga mertensiana*), yellow cedar (*Chamaecyparis nootkatensis*) and subalpine fir (*Abies lasiocarpa*). Industrial logging has removed much of the original old-growth forests, but there are still remnants of the giant trees that once covered the valleys and lower mountain slopes. The Bella Coola River and adjacent river valleys were formed through tremendous scouring of the original rock by glaciers; the geological history of the region is reflected in the steep-sided mountains, many of which are bare rock supporting little vegetation, with ice-capped peaks, waterfalls and streams tumbling down into the valleys. The diversity of the landscape leads to biological diversity; there is a wide range of different habitat types – coastal salt marshes and tidal flats, rocky shorelines, bogs, marshes, river estuaries, small prairies, gravel outwashes, rocky scree slopes, and deep-soiled forest habitats. Together with the rivers, lakes and ocean, these support the diversity of plants and game animals that have nurtured the people of the Nuxalk Nation, providing them with food, clothing, tools, shelter and medicines, generation after generation (Lepofsky, Turner and Kuhnlein, 1985). In particular, the Bella Coola and other rivers in Nuxalk territory have provided spawning and rearing habitats for all five species of anadromous Pacific salmon, a nutritious and staple food, while outer sea channels provide other fish and shellfish.

Similar to other Northwest coast peoples, the Nuxalk used to reside in permanent villages only over the winter, spending most of the year following a seasonal harvesting cycle, procuring springtime foods along the coast, estuaries and valley bottoms; travelling further up the valley and out towards the open ocean during the summer months, to harvest berries and hunt game; and harvesting root vegetables, game, fish and shellfish in the autumn. For each type of food, specialized processing – drying, smoking or other preservation – was required to prepare for winter storage. There was a general division of labour, with men hunting and fishing while women undertook the cutting and processing of meat and fish, and the harvesting and preparation of various plant foods – greens, berries and roots – which were an important part of the diet. All of these activities required special skills and knowledge that children and youth learned through participation, observation and instruction by working with adults, including elders.

In the first two years of this research with the Nuxalk in the early 1980s, 102 Nuxalk families (82 on-reserve and 20 off-reserve in urban centres such as Vancouver) were interviewed to identify traditional foods and the cultural patterns of food use (Kuhnlein, 1984). Another key undertaking was a series of elders' meetings organized by the research team to consult the elders about which traditional foods they considered important, which were still being used, and how these foods were traditionally harvested, processed and prepared for consumption. Changes in food use frequency were identified through another interview study (1982/1983), conducted with the grandmothers, mothers and daughters of families living on the Nuxalk reserve. In 1982, a food availability assessment was undertaken, and from 1980 to 1986 the nutritional values of traditional foods and diets were assessed, using standard practices for sample collection and nutrient analysis (Kuhnlein, 1986). Health status assessments were undertaken in 1983. This research set the stage for the food-based intervention activities conducted by the Nuxalk Food and Nutrition Program, which are summarized in this chapter, and the programme evaluation, which was conducted in 1986.

Food system and health change

Details of the food system and the health and nutrition status of the Nuxalk, as reported in baseline assessments for the Nuxalk Food and Nutrition Program, are summarized in Turner *et al.* (2009). There was ample evidence that families and individuals were using different foods and living different lifestyles from those of even a few generations previously (Kuhnlein, 1992). The study documented that 13 species of fish, eight species of shellfish, seal and seal lion were still being used. One of these fish was the ooligan (eulachon – *Thaleichthys pacificus*), a type of smelt used by the Nuxalk for countless generations, especially for the nutritious fat (grease) rendered from it through a sophisticated traditional process and used widely as a condiment (Kuhnlein *et al.*, 1982; Kuhnlein *et al.*, 1996).

The grandmother-mother-daughter interviews documented marked declines in Nuxalk traditional food use over the three generations (Kuhnlein, 1992; 1989). Not only was there a distinct drop in the diversity of traditional foods used, but a general decline in the frequency of use was also noted. Significantly, declining food use was linked to food availability and taste appreciation; when a food's frequency of use declined, taste appreciation also declined. Providing younger generations with opportunities to sample traditional foods and develop taste appreciation was therefore one of the strategies embodied in the programme.

In the 1980s, the Nuxalk were using only about 20 species of plant and animal foods from their traditional diet – a huge decline from the 70 or so traditional food species used earlier in the twentieth century. The continuing importance of salmon and ooligan was a main finding of earlier studies, which revealed that the Nuxalk were still eating significant amounts of salmon, ooligan grease and seafood – as much as they were able to get. They were also still hunting deer and some other animal foods at the time of the study. However, it has recently been noted (B. Tallio, personal communication, 2008) that the mule deer hunted in the 1950s and 1960s, along with the black-tailed deer, had declined to very low numbers by the 1980s; today, only black-tailed deer are available. In addition, mountain goats, which were formerly very important for not only their meat and fat, but also their wool, skins, horns and bones, are no longer seen in the Bella Coola area. In the 1980s, some people were continuing to harvest traditional wild berries (29 original species were documented), especially huckleberries, blueberries, soapberries and salmonberries. However, many traditional plant foods – root vegetables such as springbank clover, Pacific silverweed, northern riceroot and woodfern (seven species documented) (Kuhnlein, Turner and Kluckner, 1982); greens such as cow parsnip and fireweed (12 species); and the inner bark of black cottonwood and other trees (three species) – were largely unused by the latter part of the twentieth century. Significantly, these healthy vegetable foods were being displaced by less healthy purchased, processed food in contemporary diets.

Investigating the reasons for this dramatic shift away from many local traditional foods was among the aims of the Nuxalk programme. The causal factors are very complex and relate to, among many other factors, an array of regulations; time constraints due to wage jobs, school requirements and other obligations; loss of ability to pass on knowledge about foods from older to younger generations; prevalence and easier availability of marketed foods; concerns about pollution, declining populations and productivity; and loss of easy access to traditional food (Kuhnlein, 1984; 1989; 1992; Nuxalk Food and Nutrition Program Staff, 1984; Turner and Turner, 2008).

In British Columbia, aboriginal people have lower health status than the general population. Recent provincial data show life expectancy of 7.5 years less for aboriginal people, with higher rates and younger ages for all causes of death, although there has been steady improvement over the years. Between 1980 and 2002, life expectancy rose, for men from 58 to 68 years, and for women from 66 to 76 years. There

Table 11.1 Health promotion activities undertaken by the Nuxalk Food and Nutrition Program, 1983 to 1986

Activity	Numbers of events/participants	Notes
Food events involving elders, adults and youth	47/391	Included fishing, fish cutting and preserving, berry picking, greens and root gathering, bark gathering, ooligan grease preparation, children's food summer camps
Feasts and other meals featuring healthy traditional foods with advice and direction of elders	19/1 456	Included salmon barbecues, wild berries, greens and roots, ooligan grease. Often directed to specific populations groups (e.g., feasts for youth were very popular)
Public awareness and adult education	21/370 Not counted: many attending public events; weekly flyers to all homes	Included Mom's Time Out, adult nutrition classes, displays at fairs, weekly flyers. Included use and preparation of both traditional and available market food of good quality
School class presentations promoting traditional foods, good nutrition, dental health and hygiene, healthy lifestyle habits	94/2 716	Delivered in the Nuxalk Nation nursery school, 2 elementary schools and the high school
Fitness classes (regular and light aerobics) and fun runs	190/1 708 Not counted: fun run participants	Conducted by programme staff and David Bogoch of Bogie's Fitness, Vancouver. Included sessions at schools and sessions for overweight adults, adults with arthritis, and the general public
Designing and installing a Nuxalk food demonstration garden in the health centre yard	Visitors were not counted	Included a range of traditional food plants, from salmonberries and highbush cranberries and soapberries, to riceroot and *puuy'aas* (Labrador tea), so that people could see what these plants look like
Publication of a Nuxalk food system handbook and a recipe book prepared by programme staff	Distributed to all homes on reserve, and to many school classes	Included an overview of traditional Nuxalk and healthy market foods, safe preparation methods, and recipes contributed by community residents. These popular books were reprinted several times

Sources: Adapted from Nuxalk Food and Nutrition Program Staff, 1984; and Kuhnlein and Moody, 1989.

has also been reduction in infant mortality, which in 2001 was 4.0 per 1 000 live births, compared with 3.7 in the general population (Kendall, 2002). At the time of programme activities, unemployment was more than 30 percent, and formal education rarely exceeded ninth grade; fishing and logging industries provided seasonal employment for men (Census of Canada, 1981). In 2005, the situation was similar, with 30 percent of aboriginal peoples (versus 12 percent of the overall provincial population) having no formal education certificate, diploma or degree. Moreover, 30 percent of aboriginal peoples were unemployed compared with 20 percent of the general population (BC Stats, 2006).

A health-related quality of life survey conducted in 2001/2002 recognized lower scores for quality of life for Bella Coola Valley aboriginal people than for non-aboriginal people. The most prominent diseases were hypertension, depression, hyperlipidaemia, diabetes, chronic back/neck pain and osteoarthritis, with many co-morbidities. In particular, there was more diabetes among aboriginal valley residents, and they reported the worst scores for quality of life (Thommasen and Zhang, 2006).

Background objectives

The overall objectives of the Nuxalk Food and Nutrition Program were to understand and document the Nuxalk traditional food system, and to use this information as a platform for stimulating community activities to improve nutrition and health. From the project's initial stages, workshops and interviews to gather information about food and nutrition served as a culturally appropriate means of raising the profile of traditional food, and provided opportunities for elders to teach younger community members about the tastes and ways of preparing traditional foods. Explicit food, nutrition and health promotion activities were built into the programme. Assessments before and after the three-year intervention were conducted in 1983 and 1986, to assess shifts in household food use and health status. Follow-up after 20 years was conducted through consultations on the reserve in 2006.

Intervention methods and activities

The programme's health promotion activities are summarized in Table 11.1. Two project assistants were based at the on-reserve Nuxalk health clinic to facilitate the programme, with supervision from the community health nurse, the community health representative and other health staff. Activities were broad-based, emphasizing traditional food and lifestyles and contrasting these with current diet and physical activity practices. Process indicators were maintained to track participation and impressions of success for each activity (Kuhnlein and Moody, 1989).

More than 350 activities were conducted during the programme period, with thousands of individual participations. Many individuals and families participated in many events, and the programme was regarded as highly successful, with requests from other coastal British Columbia communities to provide advice and guidance for initiating similar programmes in their areas. Feasts, either as pot-luck or prepared by programme staff, were very popular events. The *Nuxalk Food and Nutrition Handbook* and the Nuxalk recipe book (*Kanusiam A Sncnik* "Real good food") described techniques for handling and preparing traditional and market foods. These were distributed to all families, and reprinted several times for use in school classes; they are still used today. Although approximately 25 percent of Nuxalk residents did not choose to participate – in particular, older men were rarely seen at events other than feasts – most people were aware of the programme and participated in several activities. Activities with the highest community participation were feasts, nutrition and dental education in schools, fitness activities, and activities based on the nutrition and health evaluation assessments (Kuhnlein and Moody, 1989).

During the course of the programme, the community's inventory of traditional food processing equipment increased, including through the building of ooligan grease bins and fish smokehouses, and collective purchases (at reduced prices) of food dryers, pressure and water-bath canners, and jars and tins for canning food. Teaching resources added to the health centre included a barbeque pavilion, an equipped fitness room and a

Table 11.2 Changes in Nuxalk family food use before and during the health promotion intervention programme

Food type	1981 % families (n = 73)	1981 Average/family/year[a]	1985 % families (n = 98)	1985 Average/family/year[a]	Difference 1981 to 1985[a]
Steelhead	49	56.9 lb	77	156.3 lb	+ 99.4 lb*
Spring salmon	64	85.4 lb	90	349.2 lb	+263.8 lb*
Sockeye salmon	79	61.1 lb	90	195.8 lb	+134.7 lb*
Pink salmon	23	4.1 lb	25	58.8 lb	+54.7 lb*
Chum salmon	22	76.4 lb	48	143.3 lb	+66.9 lb*
Coho salmon	37	138.1 lb	76	187.0 lb	+48.9 lb*
Ooligans	75	122.4 lb	78	38.9 lb	−83.5 lb*
Cod	4	11.7 lb	47	23.0 lb	+11.3 lb*
Other fish/shellfish	11	15.1 lb	64	26.5 lb	+11.4 lb*
All fish roe	7	27.2 lb	76	72.5 lb	+45.3 lb*
Ooligan grease	46	62.5 qt	61	8.2 qt	−54.3 qt*
Game	30	76.3 lb	73	196.2 lb	+119.9 lb*
Wild berries	56	41.5 qt	87	49.1 qt	+7.6 qt
Wild greens	14	1.0 lb	64	17.3 lb	+16.3 lb*
Garden vegetables	38	533.9 lb	61	288.6 lb	−245.3 lb**
Garden fruits	7	132.9 lb	82	167.0 lb	+34.1 lb

1 lb = about 0.45 kg.
1 qt = just under 1 litre.
a   Quantities only for families using the food.
* $p \leq 0.001$.
**$p \leq 0.01$.
Source: Adapted from Kuhnlein and Moody, 1989.

traditional plant food demonstration garden (Kuhnlein and Moody, 1989). Project staff also supported health clinic programmes for diabetes education, prenatal nutrition and fitness, and general healthy lifestyles.

Intervention measurements

In addition to the process indicators, which tracked activities and the participation they attracted, interviews were conducted in 1981 and 1985 to assess households' traditional food use and grocery store expenditures. These interviews were conducted by trained reserve-resident interviewers and completed by 65 to 70 percent of on-reserve families, with the woman in charge of the family's food usually being interviewed. At the time of these interviews, the prices for Agriculture Canada's "nutritious food basket", comprising 78 standard food items, were compiled for the one grocery market in the village. In an attempt to assess bias, participants were selected on the basis of their representativeness of food use in the entire Nuxalk community, as judged by Band Council members.

Quantitative measurements were collected from nutrition status assessments conducted in 1983 and 1986. These assessments included anthropometry measurements, dental health examinations, dietary evaluations by 24-hour recall, and blood tests for evaluating retinol, carotene, haemoglobin/ferritin and red cell folate levels. As a service to the community, vision and hearing evaluations were also completed during the assessment period, and referrals were made for eyeglasses and hearing aids when needed. The assessments were completed among males and females in three age groups: 13 to 19 years, 20 to 40 years, and more than 40 years. The entire community was invited to attend the assessments; 370 individuals participated in 1983, and

Table 11.3 Improvements in Nuxalk physiological status for retinol, carotene, folate and iron before and after the health promotion intervention programme, ages 13 to more than 60 years

Nutrient	no.	Test 1 (before)	Test 2 (after)	Test 2–Test 1	p*
Beta carotene (µg/dl)	102	38.1 ± 1.4	60.0 ± 1.7	21.9 ± 1.5	≤ 0.05
Retinol (µg/dl)	101	23.9 ± 0.6	41.2 ± 1.8	17.3 ± 1.8	≤ 0.05
Ferritin (ng/ml)	104	41.0 ± 3.3	46.4 ± 3.9	5.5 ± 3.5	NS
Red cell folate (ng/ml)	92	221.2 ± 11.5	267.8 ± 11.7	46.6 ± 13.2	≤ 0.05
Haemoglobin (g/dl)	104	13.7 ± 0.2	13.8 ± 0.2	0.1 ± 0.1	NS

* Paired t-test.
Source: Adapted from Kuhnlein and Burgess, 1997.

477 in 1986 (Kuhnlein and Moody, 1989). Throughout the programme, interviews were conducted by local Nuxalk staff; university research assistants contributed to the health assessments in 1981 and 1986.

Intervention results and discussion

A key finding was the change in Nuxalk family food use from 1981 to 1985, which was assessed through interviews to record the numbers of families using each food and the quantities used by each family (Table 11.2). While the grandmother-mother-daughter interview study clearly demonstrated that fish was a mainstay in Nuxalk diets during the twentieth century, use of plant foods had declined (Kuhnlein, 1989; 1992). One important finding was that the percentages of families using each food increased, often doubling, which demonstrated the programme's effectiveness in enhancing participation in traditional food harvesting and preparation. The significant increases in use of several species of fish and game were seen as improved use of these mainstay resources. The numbers of families using wild berries and greens, and garden fruit also increased. A notable exception was family use of the ooligan (*Thaleichthys pacificus*) and ooligan grease, which declined because of poor spawning conditions for the fish in 1985, although many families still had access to these resources during the period. More families produced garden vegetables in 1985 than in 1982, but the average weights produced declined significantly, primarily because potatoes were being grown by more people but on smaller plots (Kuhnlein and Moody, 1989).

The interviews also demonstrated that families were reducing their expenditures at the grocery store, from an average of CAD 104 a week to CAD 83 over the project period, while the cost of the nutritious food basket for a four-person family increased from CAD 105 to 125. Interview reports noted that the reduction in expenditures was directly due to increased use of home-harvested and -preserved food, as well as new knowledge of economical shopping practices (Kuhnlein and Moody, 1989).

The programme was evaluated before and after intervention activities, using nutrition assessments and venous blood tests for three micronutrients: vitamin A, folate and iron (Kuhnlein and Burgess, 1997). Table 11.3 summarizes the findings from these assessments. Adults of both genders had increased levels of carotene, retinol and folate in their blood. From paired comparisons (among those participating in both assessments), significant improvements were shown for carotene, retinol and folate in the community at large. Youth were at risk of iron (ferritin), carotene and folate shortages, but improved their status for these nutrients during the programme.

Dietary change was assessed from 24-hour recalls among women aged 20 to 40 years, conducted in 1981 and 1986. Table 11.4 shows changes in intakes of fruit, vegetables and selected nutrients. Significant increases are shown for the amounts of all fruits and vegetables and for the nutrient intakes of vitamin A, iron and folate. During this period, the percentages of women achieving less than 50 percent of the recommended intakes (for that time) of vitamin A, iron and folate

Table 11.4 Improvements in dietary intake for Nuxalk women aged 20 to 40 years before and after the health promotion intervention programme

Nutrient	1981 mean ± SD (n = 31)	1986 mean ± SD (n = 62)	p
Vitamin A (IU)	2 267 ± 1 810	5 599 ± 9 198	0.008
Iron (mg)	7.66 ± 4.01	10.36 ± 3.81	0.002
Vitamin E (µg)	2.35 ± 1.75	4.57 ± 3.09	0.000
Folate (µg)	78.03 ± 53.3	132.92 ± 101.64	0.001
Fruit (g)	123 ± 145	289 ±324	0.001
Vegetables (g)	93 ±110	143 ±116	0.050

Source: Adapted from Kuhnlein, 1987.

declined. Improvements in intakes of several other nutrients were also noted (Kuhnlein, 1987).

Dental health education was conducted by project assistants at prenatal classes and on-reserve schools; dental evaluations and referrals for dental treatment were made by the same Health Canada dental team in the pre- and post-test periods (1983 and 1986). Table 11.5 shows that all age groups of children, except two-year-olds, had fewer teeth recorded as "decayed, extracted, missing or filled". Improvements were also noted in the numbers of Nuxalk children and young adults aged 20 to 29 years who were free of periodontal disease (assessed as oral health category 1) during the course of the programme (not shown) (Kuhnlein, 1987).

Anthropometric assessments of both adults and children did not change significantly between 1983 and 1986. The same proportions of overweight and obesity existed, in both children and adults, and there was negligible underweight in all categories. The extents of chronic diseases were not assessed in either survey. Although there was modest concern about diabetes at the time, the programme did not emphasize weight loss or diabetes control, except by encouraging more physical activity.

In summary, impressive changes were made in the food use and health status of Nuxalk from 1981 to 1986. These were reflected in increased numbers of families using Nuxalk traditional foods, and larger amounts of most of these foods being used by each family. Dietary intakes of young Nuxalk women showed increased fruit and vegetable use, and better nutritional intakes of vitamin A, folate and iron. As expected from this better diet, all members of the community registered improved health status for vitamin A, folate and iron, which have numerous health benefits. Dental examinations revealed substantial improvements in dental health and hygiene.

Although the Nuxalk Food and Nutrition Program was prominent in the community during this period, it is not clear which activities from it and which from other health initiatives were responsible for these specific benefits. However, as the community health nurse and staff noted, the Nuxalk Food and Nutrition Program was the only one providing broad-based food or nutrition education or dental health activities in the community at the time. The community's positive response to programme activities, and the improvements documented over the period encouraged community leaders to maintain their commitment to enhancing the use of Nuxalk cultural food resources and traditional health activities while embracing other healthy foods and modern health programmes and services.

Revisiting the Nuxalk Food and Nutrition Program, 2006

The successes documented over the course of the original Nuxalk Food and Nutrition Program were impressive, and led to interest in knowing the longer-term effects that the programme had had on the community. Therefore, in July 2006 – 20 years after the original final assessment – the community's perspective on the programme's impact was investigated as part of the Centre for Indigenous Peoples' Nutrition and Environment (CINE) Global Health Program. Many of the original Nuxalk programme participants were visited, and qualitative open-ended interviews and discussions with community leaders were conducted. Many interviews were included in a film (KP Studios, 2008) examining the longer-term outcomes of the project. In 2006, many of the children, youth and young adults who had participated in the original programme had become parents or grandparents, and many of the elders had passed away.

Table 11.5 Improvements in dental health of Nuxalk children before and after the health promotion intervention programme

Age (years)	1983			1986			Difference
	n	DEF	DMF	n	DEF	DMF	DEF + DMF
2	7	1.4	–	7	3.7	–	+2.3
3	6	7.8	–	11	6.7	–	−1.1
4	10	7.6	–	14	5.9	–	−1.7
5	20	9.2	0.2	21	7.4	–	−2.0
6	10	9.5	0.9	13	6.7	1.2	−2.5
7	15	6.3	2.1	22	5.1	1.6	−1.7
8	21	7.7	4.2	15	4.7	1.9	−5.3
9	14	3.1	4.3	14	4.9	2.4	−0.1
10	20	3.1	5.2	17	2.5	4.2	−1.6
11	9	2.9	6.3	20	2.2	6.3	−0.7
12	14	0.2	10.3	15	0.1	5.6	−4.5
13	10	–	11.0	18	0.3	8.9	−1.8
14	11	–	13.4	8	–	13.1	−0.3
15	7	–	13.6	11	–	14.8	+1.2
16	5	–	18.6	5	–	11.6	−7.0
17	11	–	17.4	7	–	14.4	−3.0
18	5	–	15.0	7	–	17.7	+2.7
19	5	–	14.0	3	–	18.6	+4.6
Total	200			228			

DEF = average number of decayed, extracted or filled deciduous teeth.
DMF = average number of decayed, missing or filled permanent teeth.
Source: Adapted from Kuhnlein, 1987.

Those taking part in the 2006 survey were consulted about their views of the programme and the future for health and nutrition in general. Intervention activities were well remembered, and participants had fond recollections of the feasts that were held, especially for youth, to promote healthy traditional food. People also remembered the spring picnic, when a cottonwood tree was cut down so that elders and youth could harvest the succulent inner bark; many of the elders had not tasted this food for many years. The Moms-and-Tots sessions of instruction and training on traditional food, led by elders, were also remembered with pleasure, and many of the women highlighted the opportunities for learning from elders. Almost everyone still had and used the two handbooks that were published as part of the project (Nuxalk Food and Nutrition Program Staff, 1984;

1985), and plans were under way to republish these in an updated format. One handbook had been repeatedly reprinted over the years for use in local schools.

Most programme activities were remembered and continued to feature in people's lives through their activities and appreciation of traditional foods and the local environment. One indication of this interest was people's participation in a community plant identification and cultural awareness hike along Thorsen Creek, facilitated by traditional Healer Sam Moody and Community Health Representative Thelma Harvey. Participants wanted to know the names of plants, in both English and Nuxalk, and their cultural significance and potential applications as food or medicine. The importance of clean environments and fresh drinking-water was also reinforced, as the group walked alongside the rushing waters of Thorsen Creek and witnessed some of the impacts of industrial logging in the Bella Coola Valley. Frustration and concern were expressed about the declining salmon stocks and disappearance of ooligan from the Bella Coola River (Moody, 2008).

Lessons learned and considerations for the future

The Nuxalk community retains an impressively strong interest in healthy food, including the harvesting and processing of traditional local foods, resulting from its whole-hearted response to the Nuxalk Food and Nutrition Program. Shortages and lack of access to some of these foods make it likely that fewer are now being used regularly, but the youth and young adults of 2006 showed great interest in them. Some of the foods known to elders in the 1980s, such as the inner bark of trees and traditional root vegetables, were virtually unknown to younger people in 2006, but most of the traditional foods documented in the original study were still familiar to many people. Foods that were still being used by the younger generations in Bella Coola included all five species of salmon, steelhead, trout, herring, ooligan, cod species, salmon eggs, crabs, clams, some berries (e.g., blackcaps, wild raspberries, salmonberries, soapberries), thimbleberry shoots, seaweed, Labrador tea, cow parsnip, deer, moose, duck and grouse. People have adapted these foods for modern recipes, such as sushi from Bella Coola fish and special sauces and preparations for marinating and barbecuing fish. In some cases, freezers have replaced smoking and canning for preserving fish and other traditional foods. The 2006 survey made it clear that there was much discussion of food, health and the cultural values of food – reflecting, at least in part, the original Nuxalk Food and Nutrition Program's support and promotion of traditional food.

Local community members and health staff designed and organized a number of activities that – although not directly related to the original project – can be considered as spin-off projects. These include building a community garden, which was planned in conjunction with the day care centre in the older part of the village and set up in the newer Four-Mile housing development. Fresh produce from this garden – peas, carrots, potatoes, beets, spinach and other vegetables – was used by participating organizations and families, and shared with elders and families in need. Another very productive on-reserve garden, set up as part of the current Prenatal Nutrition Program, included traditional medicinal plants. Other activities included food safety classes, the development of a community kitchen, outdoor education tours focusing on traditional foods and medicines, and grocery tours to help people assess the value of different foods in terms of nutrition and cost.

Fitness events and classes were organized, many of them inspired and led by elders. One woman in her late seventies started climbing up and down the 116 steps to the community water reservoir for exercise. Others soon joined her, until as many as 50 or 60 people a day were following her example. Some elders started to use pedometers to measure the distance they walked, and friendly competitions sprang up, with younger people following the elders' lead. A Nuxalk leader remarked that he regularly (almost daily) walked the 4 miles (6.4 km) from his home to the Band Council offices on the reserve: "It saves my life to do this exercise" (A. Pootlass, personal communication, 2008). This is a result of his learning the benefits of fitness in the Nuxalk programme.

Encouragement in the form of contests and prizes made the efforts even more fun. A community fitness centre with weights and exercise equipment became very popular with diabetic individuals and the community at large, and was also used for nutrition education events. In 2006, 23 people from Bella Coola, including two elders in their seventies, took part in the Sun Run, a 10-km run or walk in Vancouver, with more than 59 000 participants.

Nutrition and lifestyle education programmes directed at children and youth were particularly valuable. These community-based programmes stressed not only good nutrition and exercise, but also the mental, emotional and spiritual aspects of health and fitness. A comprehensive wellness plan incorporating all the components of healthy living was being planned in community-led health and nutrition initiatives.

Awareness of the importance of Indigenous Peoples' food for healthy living was recognized and acknowledged by virtually everyone interviewed in 2006 in the Nuxalk community.

Unfortunately, some of the intervention activities that had had initial success in increasing the awareness and use of traditional foods by Nuxalk community members were later eclipsed by negative environmental impacts on some of the traditional food species. Since the end of the original programme in 1986, salmon stocks throughout the west coast have dwindled notably; sockeye and spring salmon in particular are less plentiful than they were 20 years ago. The ooligan, which is of immense cultural importance to the Nuxalk and was a focus of the original project, has declined drastically over the past decade, as its coastal communities and springtime spawning runs on the Bella Coola River have disappeared. The decline of the ooligan has been alarming for the Nuxalk, and for other coastal Indigenous Peoples who relied on grease as an important part of their diets and nutrition. In June 2007, the Nuxalk hosted a mourning feast at which indigenous communities and fishery biologists on the British Columbia coast commemorated the lost ooligan runs (Senkowsky, 2007). Feasting is a long-standing cultural tradition for the Nuxalk and other First Nations in the region, and is used to recognize important occurrences, including memorials of those who have passed away. A feast was therefore a fitting way to mark the passing of the ooligan, and to discuss how it might be restored in the future. Abalone is another traditional food that is no longer available; commercial overharvesting of this valuable shellfish in the 1970s and 1980s resulted in a general collapse of populations, which have still not recovered, despite an ongoing moratorium against abalone harvesting by the Department of Fisheries and Oceans (IUCN, 2008).

The continuing decline of local food traditions also reflects a general, global trend among local and Indigenous Peoples, as more and more people around the world are consuming food that is produced, processed and marketed at a global scale (Nabhan, 2006; Turner and Turner, 2008). As do many other remote food stores, those of Bella Coola now provide a wide range of products from other parts of the world – such as mangoes and packaged cashews – and it is not surprising that people seek out and enjoy these.

In the 1980s interviews, diabetes was emerging as a health problem, but its immensity was not recognized at the time. Now, in the twenty-first century, the incidence of diabetes has grown to almost epidemic proportions among First Nations people, including the Nuxalk. An overall healthy lifestyle promotion programme focusing on children and youth will likely be the most successful strategy for addressing and reversing this situation.

Opportunities for promoting traditional food systems for Indigenous Peoples

Chapter 3 in this volume (Turner, Plotkin and Kuhnlein, 2013) draws attention to the complex environmental concerns that affect the traditional food systems and cultures of Indigenous Peoples around the globe. The Nuxalk and other communities of western Canada are certainly facing severe environmental challenges, including declining populations of some of their key traditional food resources: sockeye salmon, ooligan and abalone. Added to these challenges are concerns about pollution from local sewage outfalls at the Bella Coola River estuary; deforestation and loss of old growth from Nuxalk lands; and the introduction of invasive plant species that might out-compete indigenous species. Degradation of camping sites, impacts from widening roads and increasing traffic, and declines in productivity of berries are other environmental problems that people have noted.

An over-riding concern is global climate change, which is facing Indigenous Peoples and environments everywhere. People on the west coast of Canada have noted many changes in species distributions, seasonal rainfall and snowfall, and local weather patterns in general (Turner and Clifton, 2009). There are indications that the declines in salmon and ooligan may be related, at least in part, to climate regimes. In general, warmer water temperatures reduce the fitness, survival and reproductive success of salmon, facilitating potential long-term population declines (Ministry of Environment, 2002).

There are other challenges to the promotion and use of traditional food for the Nuxalk and other coastal First Nations. The rising costs of fuel and of operating boats and vehicles reduce the opportunities to travel for food procurement and other purposes. Many traditional Nuxalk territories have been privatized and are no longer easily accessible for food harvesting. There are also difficulties in linking children and youth to elders and knowledge holders and in providing appropriate opportunities for passing on important traditional knowledge about food and survival (Turner *et al.*, 2009).

Nuxalk and other First Nations recognize that these obstacles have to be faced and overcome, if they are to maintain their cultural integrity and the health and well-being of their communities (Kuhnlein, 1995; 2001a; 2001b; Parrish, Turner and Solberg, 2007; Senos *et al.*, 2006). There are many ideas for strengthening and supporting healthy traditional food systems (Chapter 14 – Kuhnlein, 2013). However, it is important to remember that a multiplicity of cumulative factors has caused the erosion of traditional food systems, so problems cannot be addressed by only one or two measures; instead, an array of different intervention strategies is needed, for use at different scales and with different audiences. It is important to continue the educational and intervention strategies that showed success in the original Nuxalk Food and Nutrition Program, especially those involving children and youth and hands-on experiential learning, such as harvesting expeditions, science camps and school projects.

The changing of lifestyles and healing of environments take patience and time, and must be evaluated carefully to demonstrate effectiveness and provide guidance to continuing efforts. It is crucially important that successes such as those from the Nuxalk Food and Nutrition Program be recognized and celebrated, to encourage further community development and action. It is only through patience, vigilance and positive, directed action that the transformations necessary for the continued well-being of the Nuxalk community can be realized (Thompson, 2004; Turner and Thompson, 2006; Turner *et al.*, 2009) ✻

Acknowledgements

The authors acknowledge with gratitude the strong participation and support of the many elders, advisers and participants from the Nuxalk Nation who took part in various components of the Nuxalk Food and Nutrition Program and the follow-up work undertaken in 2006. In particular, elders Dr Margaret Siwallace, Felicity Walkus, Alice Tallio, Elsie Jacobs, Willie and Hazel Hans, Lucy Mack, Katie Nelson, Eliza Saunders, Lillian Siwallace and Andy Siwallace are recognized for their vision and immense contributions. Key community researchers were Community Health Nurse Sandy (Moody) Burgess, Community Health Representative Rose Hans, and programme assistants Louise Hilland, Emily Schooner and Grace Hans, who gave immeasurable energy and insights to successful programme activities. Sarah Saunders, Aaron Hans, David Hunt, Theresa Barton, Mary Lynne Siwallace, Don Hood and Karen Anderson contributed their talents and skills. Chiefs Edward Moody, Archie Pootlass and Lawrence Pootlass, and the Nuxalk Nation Band Council members are acknowledged for their support and leadership in ensuring excellent community interest and participation in the research. Appreciation is also expressed to Chief Archie Pootlass and the 2006 Council of the Nuxalk Nation for participating in the film documentation of the Nuxalk traditional food system and the results of various interventions and educational programmes. Participating university graduate students from the 1980s include Dana Lepofsky (Lepofsky, 1985) and Anthea Kennelly (Kennelly, 1986). David Bogoch of Bogie's Fitness organized the fitness awareness programme. Professor Dick Ford (University of Michigan, United States of America) provided encouragement and advice, especially relating to the demonstration of traditional food plant gardens. More recent participants in the programme include Rose Hans, Lillian Siwallace, Andy Siwallace, Grace Hans and Louise Hilland, as well as Chief Archie Pootlass, Jason Moody, Megan Moody, Melvina Mack, Greg Hans and Don Hood. Traditional Healer Sam Moody is acknowledged for his work, and for organizing a community plant identification and cultural awareness hike along Thorsen Creek in 2006.

Initial research was funded by the Natural Sciences and Engineering Research Council, through a grant to Harriet Kuhnlein. Health Canada funded several phases of the actual programme, with grants to Harriet Kuhnlein and Margo Palmer as Health Promotion Contribution Program Officer. The programme evaluation was supported through a grant from the National Health Research and Development Program. Inclusion of the Nuxalk programme in CINE's Global Health Initiative was made possible by funding from the Canadian Institute of Health Research, Institute of Aboriginal Peoples' Health, Institute of Nutrition, Metabolism and Diabetes, and Institute of Population and Public Health.
> **Comments to:** harriet.kuhnlein@mcgill.ca

>> Photographic section **p. XXII**

Chapter 12

Let's Go Local!
Pohnpei promotes
local food production and
nutrition for health

ᘍ LOIS ENGLBERGER[1] ᘍ ADELINO LORENS[1,2] ᘍ PODIS PEDRUS[3] ᘍ KIPED ALBERT[2,3]

ᘍ AMY LEVENDUSKY[1,4] ᘍ WELSIHTER HAGILMAI[5] ᘍ YUMIKO PAUL[6] ᘍ PELIHNA MOSES[3]

ᘍ RALLY JIM[6] ᘍ SOHSE JOSE[7] ᘍ DOUGLAS NELBER[8] ᘍ GIBSON SANTOS[9] ᘍ LAURA KAUFER[10]

ᘍ KATHAY LARSEN[1,4] ᘍ MOSES E. PRETRICK[11,1] ᘍ HARRIET V. KUHNLEIN[10]

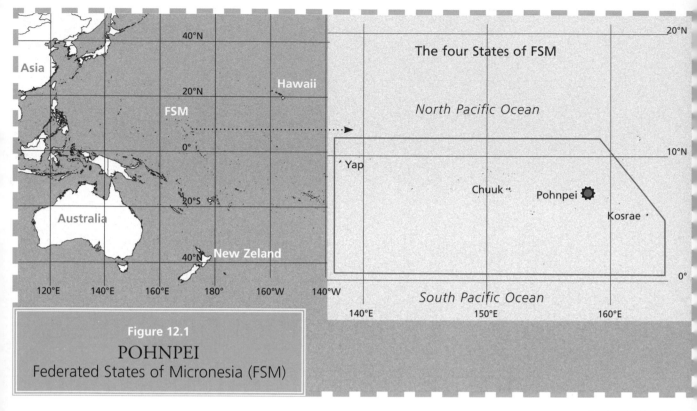

Figure 12.1

POHNPEI
Federated States of Micronesia (FSM)

The four States of FSM

North Pacific Ocean

Yap

Chuuk

Pohnpei

Kosrae

South Pacific Ocean

Data from ESRI Global GIS, 2006.
Walter Hitschfield
Geographic Information Centre,
McGill University Library.

1
Island Food Community of Pohnpei (IFCP), Kolonia, Pohnpei, Federated States of Micronesia

2
Pohnpei Agriculture of the Office of Economic Affairs, Kolonia, Pohnpei, Federated States of Micronesia

3
Community of Mand, Pohnpei, Federated States of Micronesia

4
Peace Corps Micronesia, Kolonia, Pohnpei, Federated States of Micronesia

5
College of Micronesia (COM-FSM) Cooperative Extension Service, Kolonia, Pohnpei, Federated States of Micronesia

6
Pohnpei State Department of Health Services, Kolonia, Pohnpei, Federated States of Micronesia

7
Pohnpei State Department of Education, Kolonia, Pohnpei, Federated States of Micronesia

8
Pohnpei State Department of Land and Natural Resources, Kolonia, Pohnpei, Federated States of Micronesia

9
United States Department of Agriculture, Natural Resources Conservation Service, Kolonia, Pohnpei, Federated States of Micronesia

10
Centre for Indigenous Peoples' Nutrition and Environment (CINE) and School of Dietetics and Human Nutrition, McGill University, Montreal, Quebec, Canada

11
Department of Health and Social Affairs, Palakir, Pohnpei, Federated States of Micronesia

Key words > Pohnpei, Micronesia, traditional food, provitamin A, carotenoids, energy, obesity, diabetes, vitamin A, community, participatory, inter-agency approach, Indigenous Peoples, food security

Photographic section >> XXV

"Let's Go Local! Grow and eat local foods for their 'CHEEF' benefits: culture, health, environment, economics and food security."

Island Food Community of Pohnpei

Abstract

The Pohnpei community intervention programme took place in Mand community, Pohnpei, Federated States of Micronesia (FSM) from September 2005 to June 2007. The programme aimed at increasing the production and consumption of locally grown foods and improving health. A community-based, participatory, inter-agency, multiple-methodology approach was used, with all age groups in the community participating in programme activities. The programme had two phases: phase 1 involved documenting the traditional food system and imported foods, and assessing health status, using the Centre for Indigenous Peoples' Nutrition and Environment methodology; phase 2 involved two sub-phases. In phase 2a, promotion and intervention activities focused on building awareness through workshops, competitions (weight loss, planting and cooking), mass media, posters, billboards, postage stamps, postcards and other materials; the conservation of rare crop varieties; and small-scale food processing. In phase 2b, the impact of promotion and intervention activities carried out in phase 2a was evaluated. Activities for expanding the programme continue.

Phase 1 revealed neglect of the traditional food system, reliance on rice and other imported processed foods, and high incidence and prevalence of overweight, obesity and diabetes among adults, and of stunting, vitamin A deficiency and dental decay among children. Detailed studies initiated in 1998 revealed that local staples, including yellow- and orange-fleshed banana, giant swamp taro, breadfruit and pandanus varieties, are rich in nutrients. They contain substantial levels of provitamin A and other carotenoids, which are important in alleviating vitamin A deficiency and other chronic diseases such as diabetes, heart disease and cancer. Promotional activities based on the campaign slogans – "Let's Go Local" and "Going Yellow" – were adopted widely throughout FSM to promote local foods and yellow- and orange-fleshed staple crop varieties.

The project impact evaluation revealed a significant decrease in rice consumption; an increase in the consumption frequency of local banana varieties, giant swamp taro and vegetables (including green leaves); an increase in the intake of local food diversity and provitamin A carotenoid; and a positive change in attitude towards local food. The Pohnpei Go Local campaign also created interest nationally and regionally.

Background: context of the research site

The Federated States of Micronesia (FSM) is an independent island country located in the western Pacific Ocean. It comprises 607 islands and has a population of 107 434 people (July 2009 estimate) (FSM Department of Economic Affairs, 2002; CIA, 2010). The country is divided into four states: Pohnpei, Chuuk, Yap and Kosrae.

Pohnpei, a mountainous island about 40 km in diameter and 355 km² in area, is the location of the FSM national government. It is situated 6° 55 north latitude and 158° 15 east longitude (Figure 12.1) (CIA, 2010). Pohnpei State (population about 34 500) includes the main island of Pohnpei divided into five municipalities – Nett, U, Madolenihmw, Sokehs and Kitti – and five main outlying low atoll islands with distinct languages and cultures: Sapwuafik, Nukuoro, Kapingamarangi, Mwoakilloa and Pingelap.

Pohnpei Island has rich agricultural resources, while the atolls have hot dry climates and poor sandy soils that make it difficult to grow crops.

Demographic and cultural characteristics of the study site

The intervention study site is Mand community, Madolenihmw Municipality in Pohnpei State. It is a rural community, about a 40-minute drive on a paved road from the town centre of Kolonia. The average annual temperature is about 27 °C, with heavy rainfall and verdant tropical vegetation. Agricultural resources are abundant all year round.

Context of the food system

FSM's economy is based on subsistence farming and fishing. Sources of cash income include formal employment, agriculture, remittances or pensions, and fishing (Drew, 2008). More than 25 percent of the population is considered to be living below the poverty line (Abbott, 2004; CIA, 2010). Because of the availability and convenience of imported processed foods, traditional methods of local food preservation have been greatly neglected, and there has been little uptake of modern methods for the small-scale processing of local foods (Englberger, 2003; Englberger, Marks and Fitzgerald, 2003b).

The main crops of the Pohnpei traditional food system include many varieties of banana, breadfruit, taro and yam, which are consumed with coconut, fish and seafood; foods eaten as snacks include fruits and sugar cane (Merlin *et al.*, 1992; Raynor, 1991). Vegetables (other than the traditional starchy staple foods) have only recently been introduced. Pohnpei society apparently had good nutrition status up to the 1950s, when people consumed mainly traditional staple food crops and had traditional lifestyles with ample physical activity (Murai, Pen and Miller, 1958).

The neglect of Pohnpei's traditional food system and the shift towards processed, less healthy imported foods accelerated in the 1970s. The causal factors for this change include the availability of convenience food; the high status and low cost of imported white rice, flour, sugar, fatty meats and other refined processed foods; changing lifestyles and family structures; the shift from subsistence farming to a market economy and cash employment; inconsistent external and internal government policies and food aid programmes; the large sums of money made available through the Compact of Free Association with the United States of America; and modernization and globalization.

The shift to imported foods in Pohnpei and other parts of Micronesia has been more drastic than in many other parts of the Pacific (Schoeffel, 1992). From 1885 until the end of the Second World War in 1945, Pohnpei was colonized by three colonial powers: Spain, Germany and Japan. In 1945, it and the other islands in what is now FSM became part of the United Nations (UN) Trust Territories of the Pacific Islands under United States of America administration. In 1961, the UN criticized the United States of America for neglecting the islands, and development activities greatly increased soon after.

One set of controversial programmes that greatly influenced FSM food habits were the United States Department of Agriculture (USDA) supplementary feeding programmes. These started in the 1960s, increased in the 1970s and continued into the 1990s. USDA surplus commodities (including rice and tinned foods) provided food for school lunches, needy families, the elderly, and disaster relief in Pohnpei (Schoeffel, 1992; Englberger, Marks and Fitzgerald, 2003b). The 30-year United States School Lunch Program and other food aid programmes introduced rice and processed foods to many children and adults in Pohnpei, establishing new food habits, attitudes and food tastes that persist today.

In 1986, FSM became independent but kept close ties to the United States of America through the Compact of Free Association (CIA, 2010). Large sums of money were provided to FSM communities, giving opportunities for jobs and cash for purchasing store foods that are mainly low in nutrients. The first Compact ended in 2001, but a second Compact renegotiated in 2003 provides large amounts of development funding annually up to 2023.

Overall health and nutrition status

A large health study conducted in Pohnpei in the late 1940s identified no diabetes (Richard, 1957; Hezel, 2004). However, by the late 1980s, the prevalence rates of overweight, obesity and diabetes, along with those of other non-communicable diseases such as heart disease and cancer, were rapidly increasing, and these conditions were becoming problems of epidemic proportion (Coyne, 2000; Elymore *et al.*, 1989; Englberger, Marks and Fitzgerald, 2003b). For example, the STEPS survey[1] of Pohnpei showed that more than 70 percent of the adult population aged 25 to 64 years (both sexes) was either overweight or obese (more than 80 percent of women were classified as overweight) and 32.8 percent of adult participants (both sexes) were diabetic (WHO, 2008).

A global report indicated that FSM had the second highest national prevalence of obesity in the world, ranking below only Nauru, another Micronesian island country (Streib, 2007). Of the ten countries in the world listed as having the highest obesity rates, eight were Pacific Island countries. Although the data considered in this 2007 report vary by sampling procedures, age groupings and year(s) of data collection, making it difficult to interpret country rankings, the situation in Pohnpei and other parts of the Pacific has undeniably become serious. Many families are suffering, and the problem has escalated, as reflected in the STEPS survey (2008). One projection indicates that if behaviour changes in diet and activity are not introduced, by the next quarter century more than half of Pohnpei adults will be diabetic (CDC, 2000).

Pohnpei also has a serious micronutrient deficiency problem. In 1993, more than half of its children under five years of age had vitamin A deficiency (Yamamura *et al.*, 2004). To alleviate this deficiency, a vitamin A supplementation programme was established for all children aged one to 12 years. However, there have been logistical and organizational difficulties with distributing the supplements.

Rationale and objectives

Similar to the situation in other indigenous communities globally, FSM's indigenous foods that are rich in carotenoids and other phytochemicals have been neglected owing to the transformation of food habits. The change in food habits from fresh traditional foods to processed imported foods has been accompanied by high prevalence of overweight, obesity, diabetes, heart disease and cancer among the adult population, while micronutrient deficiencies, such as of vitamin A, are prevalent among children. Responding to growing concern about the emergence of nutrition and health-related epidemics related to change in diets, this project sought to revive the use of neglected traditional foods among the traditional community of Mand in Pohnpei.

The objectives were to:
- improve awareness of the high nutritional values of local foods;
- increase the production and consumption of local Pohnpei foods and varieties, especially those rich in carotenoids and other nutrients;
- evaluate the project using health status measures and awareness indicators, locally, nationally and internationally.

Methodology

A research agreement was established in March 2005, jointly signed by the Mand community leader, the Island Food Community of Pohnpei (IFCP) and CINE (Englberger *et al.*, 2005; 2009b; 2010b).

The Pohnpei case study consisted of two phases, which took place over a period of five years (May 2005 to March 2010):
- phase 1: documentation of baseline data on the contemporary food system (May to August 2005);
- phase 2: intervention activities and evaluation: phase 2a (August 2005 to August 2007) involved intervention and administrative activities, many of which were island-wide, and collection of qualitative data and process indicators (August 2005 to May 2007); and phase 2b was an evaluation of the programme (June to August 2007).

In June 2009 a further assessment of the diet was conducted, following a two-year absence of intervention activities in the target village of Mand.

---

[1] STEPS is a World Health Organization (WHO) research process tool for non-communicable disease risk factor surveillance: www.who.int/chp/steps/manual/en/index.html

Study site

Mand community was chosen because it fulfilled the overall study criteria. These included being a rural indigenous community comprising about 500 people, being accessible for transport, and being willing to participate. A group of settlers from the Pohnpei atoll of Pingelap occupied the village in 1954, and Pingelapese people are still its original inhabitants. An additional selection criterion was the availability of staff from collaborating agencies that had strong linkages with Mand community and could assist the project.

Participatory research

The study adopted a community-based, participatory, inter-agency, multiple-methodology approach, including social marketing. Ethnography (Fitzgerald, 1997)[2] was used for continual assessment of the situation in the community and for considering the intervention approaches that might be most effective. With coordination by IFCP, activities were facilitated by government and non-governmental agencies, including the Pohnpei State Departments of Health, Education, and Land and Natural Resources; the Offices of Economic Affairs, and Social Affairs; the College of Micronesia (COM)-FSM Cooperative Extension Services (CES); USDA's Natural Resources Conservation Service; the Conservation Society of Pohnpei; and Peace Corps Micronesia. Other partners included V6AF Radio, Island Cable Television, Kaselehlie Press, FSM Telecom, Micronesia Seminar and local businesses.[3]

Mand community members in all age groups were well informed about the projects and were encouraged to participate. Some project facilitators were selected because of their close relationship to the community and their commitment to the project. All meetings

and intervention activities were announced on the community hall notice board and at church and other community events. Community leaders and members were trained and fully informed about the project prior to their full engagement in the planning and implementation of appropriate activities. Individual consent for participation in both phases of the study was obtained. This included getting permission for the use of photographs in newspaper articles and film interviews.

As staff from many agencies assisted in this project, an inter-agency approach was also used to prepare this chapter. Many individuals assisted voluntarily, because they were passionate about the project and its importance. Activities were carried out throughout the island of Pohnpei, but the intervention's effects were documented in the community of Mand.

Phase 1: documentation phase

This phase involved documenting the traditional food system through focus group discussions, in-depth interviews of key informants, literature review, photography and observation. These took place during community meetings, home visits and visits to organized meetings in both Mand community and Kolonia, where some community members work and live.

Key informants
These were selected on the basis of their expertise on specific topics, for example individuals (mostly elderly men) with long experience of identifying fish were selected as key informants on fish.

Analysis of neglected local foods
This involved a series of studies, which were published in several papers (Englberger et al., 2003a; 2003b; 2003c; 2006a; 2008; 2009a; Thakorlal, 2009). Sampling and analysis methodologies are described in detail in specific papers.

Cross-section baseline survey
This included the gathering of baseline data from a random sample of households in Mand, using dietary assessment via seven-day food frequency questionnaires

---

2   Ethnography includes such methods as informal focus group discussions, in-depth interviews of key informants, literature review, photography and observation.
3   V6AF radio broadcast many project items; Island Cable Television provided multiple airings of IFCP-supported videos; and Kaselehlie Press published project items in its biweekly newspaper issues. FSM Telecom issued a telephone card promoting the state banana of *Karat*, and published a two-page insert on IFCP's work in its annual telephone directories from 2008 to 2010, including a photograph of this case study. Micronesian Seminar assisted in producing promotional films, and local businesses displayed and distributed IFCP promotional materials in their shops.

(FFQs) and two 24-hour recalls on non-consecutive days; health status assessment by anthropometry, fasting blood glucose (FBG) and blood pressure measurements; and agroforestry and socio-economic questionnaires.

Phase 2: promotion and intervention phase

Activities were promoted and implemented at three major levels: community, state/national, and regional/international. Qualitative assessments, including observation, key informant interviews, focus group discussions, field notes, process indicators, photographs and a Mand log of activities, were used to supplement other data collection methods in both phases and in planning.

Throughout this phase, the focus was on facilitating effective behaviour change, such as by identifying the factors to be considered when designing food-based interventions for Micronesia (Englberger, Marks and Fitzgerald, 2003a); the predisposing, enabling and reinforcing factors of specific issues and problems (Green and Kreuter, 1991); and factors relevant to dietary change and the promotion of traditional food systems (Kuhnlein and Pelto, 1997; Barker, 1996; Pollock, 1992; Shintani et al., 1991).

Activities at the community level

General community meetings. The main purposes of these were to document the contemporary food system and discuss and plan activities appropriate for the intervention phase. Two larger meetings were organized: one in Mand, to allow community participation; and the other in Kolonia, to bring together relevant government and non-governmental agencies and Mand community. Children were encouraged to attend meetings, where they usually viewed films and tasted local food dishes. At these meetings, two major themes were agreed as the focus for all community activities:

- awareness of local foods' value and importance for health;
- training on how to grow and utilize local foods better, including new recipes and appropriate

technology, and involving adults and youth of different ages.

Campaign slogans. The three slogans communicated project concepts in a simple, powerful way, which helped people remember and talk about the project, thereby assisting the process towards behaviour change:

- The project revived the "Let's Go Local" slogan, which was coined in the 1980s by a local leader who encouraged retaining traditional values and customs, including growing and consuming local food. This slogan was used on billboards (see Photographic section, pp. XXV to XXVIII), in songs and videos, on t-shirts, as a title for newspaper articles and a topic for radio programmes, and in face-to-face interactions.
- "Going Yellow" conveyed the "Yellow Varieties Message", emphasizing that yellow-fleshed varieties of fruits and vegetables are rich sources of nutrients that have health benefits. This slogan restored interest in local food varieties that were commonly consumed in the past, but have been neglected in recent years. It was used in local Pohnpei food posters, songs and videos, on t-shirts and in face-to-face discussions.
- "Practise What You Preach" refers to putting into practice the "Let's Go Local" slogan. It encouraged people to prepare and serve local foods and drinks at community functions and meetings.

Mand Community Working Group meetings. General community meetings were followed up by meetings of the Mand Community Working Group, whose members are local leaders interested in participating in the project. The main purposes of these meetings were to plan, decide and implement appropriate approaches and activities for promoting local foods, raising awareness and conducting intervention activities in the Mand community. A total of 500 community members took part in 78 meetings between September 2005 and June 2007, with about 20 to 30 adults attending each. Meetings were usually held in the community hall in the evenings, when people were more likely to be able to attend. Activities included the following:

- *Group interviews:* People were interviewed on the foods they had eaten that day (starchy staples, rice, fruits, vegetables) and there were occasional quizzes on topics related to the theme of the week.
- *Meeting bags* (for new members): The bags contained information materials for the project, including illustrated newspaper articles, colour photographs of members, local food promotion leaflets and a list of project activities. These were given to new members of the Mand Community Working Group, to welcome them and familiarize them with the project.
- *Planning of upcoming activities:* Promotion and intervention activities were planned and endorsed; individual members were consulted and their consensus was sought.
- *Awareness activities:* These included talks, films, weight and waist measurements, field trips and photography.

Invited guests gave brief talks on healthy lifestyles, understanding diabetes, the Yellow Varieties Message, container gardening, weight loss and management, dental care, and breastfeeding. The talks included the use and introduction of teaching materials, such as the Pohnpei local food posters, the Pohnpei bananas booklet, and the Pacific indigenous foods poster.[4] The talks aimed to improve community members' understanding of the relationships among diet, lifestyles and health.

Films shown included *Going Yellow*, to reinforce project messages and provide enjoyment. Local foods and community members were regularly filmed and photographed. Film documentaries were prepared for promoting local food, and one was put online.[5] Others were shown on local television or distributed as videos and DVDs for use at family gatherings. Photographs of Mand community members and their families were used as gifts and souvenirs of project activities, which also reinforced local food promotion messages and generated positive feelings about the project. Photographs were taken for newspaper articles, other publications and presentations at local, national or international meetings; for a recipe collection (Levendusky, 2006); and for local promotion materials, such as three FSM national postage stamps.

Guidelines were given on the use of healthy local refreshments at meetings and in the home. Families were told that all the refreshments they brought to meetings had to use local foods and be prepared hygienically using healthy ingredients (low in salt, fat and sugar). Serving plates and baskets had to be of local biodegradable materials (leaves or woven baskets). Families were told that this saved a lot of money and was also good for the environment. Families providing refreshments for meetings received small payments, as an income-generating activity (this was rotated to give all families an opportunity).

Planting materials, including banana, soursop and citrus seedlings, were distributed to farmers and other interested people. Some varieties of banana (e.g., *Karat* and *Utin Iap*) and coconut (e.g., *Adohl*, which has a sweet husk that can be consumed) are quite rare, and few families had lemon grass, which grows easily and can be made into a tasty hot or cold drink. Provision of these planting materials was important in helping families to start growing these crops.

Demonstrations of ways of minimizing fat, salt and sugar consumption focused on baking, boiling or grilling (versus frying) and the use of natural sweeteners, such as ripe banana, coconut juice or fresh sugar cane. Families were told that excessive sugar and fat contribute to overweight and obesity and lead to specific illnesses, and that sugar contributes to dental decay and salt to high blood pressure, serious health problems and death. Families were encouraged to eat more unprocessed, fresh foods that are low in fat, salt and sugar, and are important for healthy and happy living.

Although recipes were not prepared at the meetings, ways of preparing dishes, with ripe local fruit and nuts for desserts and snacks, were described, so families learned new recipes.

Efforts were made to convey simple health messages, including the following:

- People's health is affected by what they eat.

[4] www.islandfood.org
[5] www.indigenousnutrition.org

- Local food items are rich in essential nutrients, while many processed foods lack nutrients or contain low amounts, and may contain too much fat, salt or sugar.
- Yellow- and orange-fleshed banana, giant swamp taro, breadfruit and pandanus are rich in ß-carotene and other carotenoids, providing health benefits[6] (the Yellow Varieties Message).
- People should avoid eating large amounts and should generally eat less than they desire.
- People need sufficient physical activity to stay healthy.

Cooking, serving and documenting traditional dishes. During a 14-day Expanded Food and Nutrition Education Program course conducted by COM-FSM's CES, community members were trained about the nutritional and health importance of local foods and how to cook foods that are easily available but neglected or underutilized, such as green leafy vegetables, banana blossom and green papaya, as well as introduced vegetables[7] that families can grow easily.

Planting and weight loss competitions. Mand Community Working Group members were selected and trained to monitor planting and plant management activities, and to visit competitors' plantations. The purpose of this activity was to stimulate community members' interest in and commitment to growing and consuming healthy local foods. As part of the weight loss competition, weight, height and waist measurements were recorded. Counselling services were provided to people who required them, especially those found to be overweight, obese and/or suffering from nutrition-related disease. Participants also learned about healthy weights.

Container gardening (vegetables) training and nursery project. This joint project with USDA's Natural Resources Conservation Service (NRCS) was based on 14 demonstration plots and two nurseries, and included seedlings of a rare coconut variety. Demonstrations included the use of composted animal waste as an on-site source for soil improvement, application of mulch for erosion control and moisture retention, and minimal or zero tillage for subsistence agroforestry. The garden produce was served at working group meetings. This activity was linked to the planting and weight loss competitions and was very important in helping participants to grow their own vegetables.

Counselling through home visits. Families were visited in their homes and counselled on the results of their vitamin A and FBG tests. The visits were documented to provide insight for further activities.

Charcoal oven development. Through a week-long workshop, 34 energy-efficient, smokeless charcoal ovens were built and distributed. These were fuelled with charcoal made from coconut shells and readily available fuelwood. The ovens provided an economical, environmentally friendly alternative to kerosene, and a healthier more convenient way of cooking (baking) than traditional earth ovens. Following the workshop, a cooking competition using the charcoal ovens was held. A local carpenter was contacted to build the ovens commercially. He helped improved the design and built an oven for his own use, which he promoted through workshop demonstrations at his own expense. Since January 2009, improved charcoal ovens have been available for purchase from this business.

Youth involvement. The Mand Drama Club involved teenagers and younger children to make them more aware of the values of local food and encourage them to share these messages with others. A COM-FSM drama expert led the children's discussions on local foods and their values, and guided their writing of short pieces to act. Several performances were given, notably one for Easter 2006, which was filmed and raised much interest, demonstrating the value of this activity.

Youth were also involved in activities held with class 4 at Mand elementary school as part of the Youth to Youth project, in collaboration with the Conservation Society

---

6   In-depth talks explained that provitamin A carotenoids protect against vitamin A deficiency disorders and anaemia. Carotenoid-rich foods protect against cancer, heart disease and diabetes.
7   Vegetables included eggplant, bell pepper and Chinese cabbage.

of Pohnpei. Children learned about the importance of rare Pohnpei banana varieties, such as Karat, and how to plant and conserve them and use them in recipes. At the end of the school year, the children performed in the state-wide school fair, where students of other schools performed on other conservation topics.

Pilot farm genebank. This nursery and collection of banana, giant swamp taro and pandanus varieties was established near Mand in 2003 and improved by the project, to provide planting materials for its activities. The genebank was looked after by the Mand youth group, which used it to generate income and educational materials. Working group members made a field trip to the genebank, to obtain and learn about banana and other planting materials.

Mand Breastfeeding Club. About 20 young mothers gathered to talk about breastfeeding, photograph and weigh their babies and themselves, and take part in recreational talks, quizzes and yoga exercises. The focus was on the advantages of breastfeeding and how to produce more milk by stimulating the breast. A strongly held Pohnpei belief is that mothers should wake up to eat during the night, to produce enough milk, and this has often resulted in overeating and overweight. Mothers described "stuffing themselves" even when they were not hungry, as they wanted to help their babies. They expressed relief when they learned that they could stop this practice. A finale was a club picnic held at a small beach park and featuring a talk by the Pohnpei State Breastfeeding Coordinator.

Go Local billboards were installed to share the project message with as many people as possible. The attractive design showed a family planting and preparing foods, and several striking drawings of local foods, including *Karat*. One billboard was placed in Mand and two in Kolonia town, one at the hospital and one near the airport, both strategic and well-frequented sites.

Activities at the state/national level

These aimed to bring the project messages to a broader public. This not only fostered interest among groups

outside the community, but also encouraged the community itself to work hard, as the state and the nation were watching to see the outcome. These activities were therefore of great importance.

Meetings, workshops and gatherings. Table 12.1 presents a summary of selected IFCP activities at state/national events, contributing to the overall Go Local campaign:

- *Annual Farmers' Fair/World Food Day:* This usually takes place in October with about 500 participants. Its main events are food crop competitions, school essay/art competitions, a healthy cooking competition and health screenings. The purpose is to promote crops, including rare yellow-fleshed varieties; healthy cooking of local foods; and art and writing skills and awareness of local food among youth. To help relay the Yellow Varieties Message, banana varieties are categorized, with larger prizes awarded for yellow-fleshed, carotenoid-rich varieties. The healthy cooking competition also has different categories for recipes using Karat and other yellow-fleshed varieties, and criteria include low use of fat, salt and sugar; taste; appearance; and cleanliness.
- *Local food pot luck dinner held by the Ambassador of the United States of America*: The Ambassador supported the Go Local movement and hosted a local food pot luck dinner in 2006 at her residence. Mand community members participated and sang a local food song, and the Let's Go Local High School Club performed a skit using the Pohnpei food posters.
- *Field trips to the outer atoll of Pingelap:* The purposes of these were to document traditional food crops from Pingelap, as Mand community was established by a group of Pingelapese people, and to share the Let's Go Local messages about the benefits of local foods. Two visits were made, focusing on documenting varieties of giant swamp taro and collecting planting material for the genebank.

Table 12.1 Selected Go Local activities and IFCP involvement in state/national events, 2006 to 2008

Event	Date(s)	IFCP involvement	No. participants
Mortlocks taro workshop, Mortlocks, Chuuk	15–18 Mar. 2006	Documenting taro varieties and collecting samples/plantlets	35
Pingelap workshop, Pohnpei	5–8 Jan. 2007	Daily workshops, training on local foods	75
FSM President's Inauguration, Pohnpei	16 July 2007	Display of materials and foods	350
Let's Go Local Club, Ohmine and Kolonia schools, class 6, Pohnpei	21 Sept. 2007	High school students teaching elementary students – Go Local	100
YINEC workshop, Colonia, Yap	26–28 Sept. 2007	Local food/nutrition training	20
FSM consultation on plant genetic resources, Pohnpei	4–6 Feb. 2008	25-minute presentation with Pohnpei students and "Banana Varieties" song	55
IFCP Training Center opening, Kolonia, Pohnpei	21 May 2008	Centre opened by Pohnpei Governor Ehsa	44
COM-FSM Annual Health Fair, Palikir, Pohnpei	30 Apr. 2008	Keynote presentation	50
Upward Bound student workshop, Palikir, Pohnpei	27 June and 11 July 2008	Two 3-hour workshops	80
Camp girls leading our world, Nihco Park, Pohnpei	11 June 2008	60-minute presentation	50
Health values for fish seminar, IFCP Training Center, Kolonia, Pohnpei	2 Sept. 2008	1.5 hour presentation with Japanese team	25
Nukuoro softball teams, IFCP Training Center, Kolonia, Pohnpei	5 and 19 Sept. 2008	40-minute presentation to players. Go Local and poster talk	38
Let's Go Local Club, Nett School class 4, Pohnpei	16 Sept. 2008	1-hour presentation, taught by high school club members	40
Pingelap Green Day, Kolonia, Pohnpei	20 Sept. 2008	Keynote presentation	100

Radio press releases, newspaper articles and e-mail releases were prepared for each activity. The IFCP standard "Go Local" talk was presented at each event, along with promotional materials and, often, fresh samples of rare banana varieties for display and tasting.
YINEC = Yap Interagency Nutrition Education Council.

- *Kolonia fun runs:* From 2007 to 2010, IFCP participated in six fun runs per year, coordinated by the FSM National Olympic Committee (NOC). IFCP gave short and inspiring pre- and post-run health talks, shared local food messages, and provided drinking coconuts and ripe bananas, including rare varieties, as healthy alternatives to imported soft drinks and snacks. In 2010, with help from FSM NOC, IFCP held its first Let's Go Local fun run, which had a record number of participants, with more than 300 youth and adults, and offered prizes relevant to local food production and consumption, such as machetes and shovels for planting, and local food items as raffle prizes.
- *IFCP-coordinated meetings and workshops:* These included strategic planning meetings, charcoal oven and food processing workshops, and meetings where rare banana varieties and other local foods were promoted. In 2009, with support from funding agencies, the Go Local project was expanded to additional communities in Pohnpei and one community in each of the other three FSM states: Kosrae, Chuuk and Yap.

Campaign slogans. The "Let's Go Local", "Going Yellow" and "Practise What You Preach"[8] slogans were used at the state and national levels, through e-mail, newspaper, radio, television and video communications. In 2007, IFCP coined an acronym summarizing the reasons for going local – the CHEEF[9] benefits of local

---

8   Some referred to this as "Walk the Talk".
9   This acronym is of particular relevance in communities where chiefs are prominent in the social organization.

foods are culture, health, environment, economics and food security.

Mass media: The mass media were used to share the promotional messages on local food, health and nutrition with a wider audience than could be reached through face-to-face encounters:

- *Radio* broadcasts reach the entire island and no costs are involved. Press releases were sent to the government radio station, which broadcast them several times in Pohnpeian and English, during news bulletins.
- *Television:* Pohnpei's local television station has a limited broadcasting range to only a few kilometres beyond Kolonia town and does not reach Mand community. Nevertheless, videos provided by the project were frequently broadcast.
- *Video/DVD:* These allow messages to be shared with Pohnpei families and communities that cannot be reached by television, and can be shown at meetings or in homes. Most of the project's eight videos and DVDs are in English, but a Pohnpei version of the main theme video *Going Yellow* was prepared in 2010 and continues to be popular among all ages.
- *Newspapers* are mainly in English with photographs. Although only about 1 100 copies are printed, many people share each copy, and electronic versions are available on national Web sites. More than 160 articles and recipes with photographs were written and published in a column for Kaselehlie Press, Pohnpei's biweekly newspaper, from June 2005 to July 2010. Printed articles were photocopied and distributed to selected locations and people.
- *Go Local e-mail network:* By 2010, the network had more than 700 participants, and sends messages (in English only) to academics, donor and development agencies and family members in all four states of FSM and in many other countries in the Pacific Islands and beyond. The network started in 2005, with updates on nutrition, local foods and activities issued

to a small number of participants in Pohnpei. Comments from more than 200 participants were then gathered and disseminated in a discussion forum. E-mails are also distributed to the Regional Pacific Island Medical Distribution List (with more than 170 members) and PAPGREN News and Biodiversity for Nutrition. Individual and organizational recipients share the messages in their own professional and social networks.[10]
- *Web site and Facebook:* The IFCP Web site[11] was established in late 2005 and provides a wealth of information (mainly in English) on local food, health and the Mand project, providing access to a wider community – locally, nationally and internationally. It shares scientific papers, photos, promotional presentations and IFCP newspaper articles. In 2010, IFCP also established a page on Facebook.
- *"Let's Go Local" song:* This is used at meetings and workshops, and in video, radio and electronic media, including as the background theme to the *Going Yellow* video.[12] In 2009, a second song on the CHEEF benefits of local foods was composed for inclusion in IFCP presentations. The words for both songs were published in an international magazine (Englberger *et al.*, 2010a).

Print and other promotional materials.

- *IFCP local food posters:* These present the Yellow Varieties Message, with photos of varieties, nutrient contents (ß-carotene) and health messages. The posters required off-island printing[13] and became the main teaching tools for the Go Local campaign, through wide distribution (see IFCP Web site). Posters included "Pacific Indigenous Foods", produced by FAO/CINE in 2006,[14] and "No End to the

---

10 One e-mail network participant commented: "I forward these bits of knowledge to my practicum teachers, each adds to the health lessons they are supposed to teach."
11 www.islandfood.org
12 The words are: "Let's go local, let's grow local, let's eat local, let's stay local; Vitamin A, good for eyesight, no heart problems, diabetes; Yellow varieties come from local, *Karat* banana and many others more."
13 The posters involved food analysis at off-island laboratories and photography of rare varieties. The first poster required seven years to complete. Poster printing is now available on the island.

Banana", produced by Bioversity International in 2007[15] and reprinted by IFCP in 2008.

- *FSM postage stamps on local foods:* In collaboration with the FSM Postal Services, three stamp series were developed: i) a commemorative Karat series, released in 2005 at a ceremony with more than 50 invited guests;[16] ii) a series of eight Pohnpei carotenoid-rich banana varieties, released in 2006; and iii) a series on edible coconut items, soursop and banana varieties, released in 2010. Newspaper articles and radio news publicized the development and release of these stamps.

- Local food trends *newsletter:* Articles on the Go Local campaign, other local food activities, recipes and nutrition/health items were published in *Local food trends*, from 2005 to 2010.

- *Local food postcards:* Four postcards were distributed at meetings in Pohnpei and at regional and international meetings. These focused on Karat banana, giant swamp taro, pandanus, and fish and fishing as a healthy food and physical activity.

- *Bumper stickers* on *Karat* and breastfeeding were sold in local shops and offices, and distributed at meetings and workshops. The messages used were "*Karat*: Pohnpei State Banana" and "Breast is Best".

- *Governor's proclamation of* Karat *as the state banana of Pohnpei:* This brought additional prestige to this rare banana, which was very important in the past, including as an infant food. The proclamation, issued in 2005, provided recent scientific findings regarding *Karat's* rich provitamin A carotenoid and other essential nutrient contents, and the related health benefits. Printed copies of the proclamation were distributed for display in offices and homes. *Karat* is now regularly sold at local markets. In 2006, the Banana Market Study documented that in an eight-week period eight of 14 local markets had sold Karat and other rare bananas; in 1998, none of them sold it.

- *Recipe collections and other printed matter:* Recipes collected from Mand families were published in a book (Levendusky, 2006), newspapers and other publications, accompanied by photographs. IFCP continues to receive requests for the recipe book. Other IFCP print materials, including posters, brochures and newsletters were distributed through the IFCP Training Center,[17] meetings and public events, local shops and offices, and the Pohnpei Tourist Office in Kolonia. Materials were provided free of charge for teaching purposes, and sold for personal use.

- *Other materials:* Promotional t-shirts proclaim the project slogans and illustrate valuable but underutilized traditional crops. IFCP staff and partners wear them at project activities, workshops and meetings, where they also serve as competition prizes. One staff member reported in 2010 that "Those t-shirts have become rather popular; people want to wear them." As a result, a local hotel and restaurant purchased 60 t-shirts to sell. Pens with IFCP contact details were also produced, in yellow with positive health messages – such as "Grow and Eat Yellow Varieties" – and in red, with health warnings and information on where to get advice. A heat-sensitive pencil that turns from orange to yellow in the heat (or in the hand) has proved particularly popular. Telephone cards and a two-page colour insert included in the FSM telephone directory also promote the campaign.

Island Food Community of Pohnpei
www.islandfood.org
Be happy - Eat a banana

---

14   This is part of an international series to promote indigenous foods globally, based on Pohnpei photos.
15   This was shown in Ireland, the United Kingdom and the United States of America.
16   Framed copies of the stamps were presented to agencies and are on display in public places. The *Going Yellow* video was premiered at the ceremony.

---

17   The IFCP Training Center, established with extensive support from donor agencies, was opened on 21 May 2008, with an open house display, a charcoal oven demonstration and local food lunch, and an address from the Pohnpei State Governor.

Table 12.2 Selected Go Local activities and IFCP involvement in regional/international events, 2006 to 2008

Event	Date	IFCP involvement	No. participants
30th National Nutrient Databank Conference, Honolulu, Hawaii, USA	19–20 Sept. 2006	Oral presentation on provitamin A carotenoid in bananas and other Micronesian foods	100
1st International Breadfruit Symposium, Suva, Fiji	16–19 Apr. 2007	2 oral presentations	30
Eden Project, Cornwall, UK	16 Aug. 2007	Oral presentation, Go Local initiative, Pohnpei bananas and other local foods	16
Banana and sweet potato study and workshops, Makira, Solomon Islands	2–16 Oct. 2007	Oral presentations at workshops/ACIAR/HarvestPlus, SPC and Solomon Islands activity	700
Pacific Banana Strategy and PAPGREN Meeting, Suva, Fiji	9–16 Nov. 2007	Oral presentation on Pohnpei bananas and Go Local	35
2nd Conference on Health and Biodiversity, Galway, Ireland	25–28 Feb. 2008	Oral presentation on Go Local	200
International Symposium on Underutilized Plants, Arusha, United Republic of Tanzania	3–7 Mar. 2008	Oral presentation on Let's Go Local initiative in Pohnpei	250
CINE Case Studies Meeting, Bellagio, Italy	3–9 May 2008	Oral presentation on Pohnpei case study intervention chapter	30
Banana characterization training, South Johnstone, Australia	28–30 July 2008	Oral presentation on Pohnpei banana promotional materials	35
First Pacific Summit on Diabetes, Saipan, CNMI	8–12 Sept. 2008	Oral presentation on Go Local initiative for diabetes control, and display of promotional materials	150
Sweet potato and banana workshops, North Malaita, Solomon Islands and Lae, Papua New Guinea	2–19 Oct. 2008	Oral presentations at 10 workshops/ACIAR/HarvestPlus and Solomon Islands activity	680

These activities were reported in the *Kaselehlie Press*, on local V6AH radio and in the Island Food Go Local e-mail network. The IFCP standard Go Local talk was presented at each event, along with promotional materials.
CNMI = Commonwealth of the Northern Mariana Islands.

Youth involvement. In 2006, the Upward Bound administration asked IFCP to hold a six-week intensive health and nutrition course (an hour and a half a day, on four days per week) for 25 high school students who were selected based on their school performance and leadership qualities. The course improved the students' understanding of how their health is affected by what they eat, and the importance of local foods. The Upward Bound students went on to form the Let's Go Local High School Club of more than 50 students who are enthusiastic about promoting local foods. In 2007 and 2008 the club gave presentations on the values of local food to the community and in Pohnpei's three elementary schools and a women's technical school.

Assessment of local foods' nutrient contents. The resistant starch content of green banana varieties was analysed in collaboration with the University of Auckland, New Zealand (Thakorlal *et al.*, 2010). Recent studies indicate that resistant starch provides fibre and may help protect against diabetes.

Small-scale processing of local foods. Overseas consultants helped to develop skills and capacity for the drying and blending of local fruits to make fruit nectars. Workshops were held in Kolonia and Mand community, and experiments with solar and charcoal dryers were carried out. Market research was carried out in July 2007 to assess the attitudes and perceptions of market owners, consumers and local food advocates, and to list the product ranges of four food markets and six take-out restaurants. This assisted IFCP and partners in their local food promotion efforts.

Membership drive. In 2009, IFCP established membership rules and annual fees, and recruited more than 150 new members to help promote local foods. Each member receives a membership card and a subscription to the IFCP newsletter. In 2010, they also received IFCP t-shirts. Student, regular, institution and lifetime memberships are available.

Activities at the regional/international level

These led to additional funding support for project activities and encouraged participants by stimulating international recognition of the campaign's importance. Table 12.2 outlines IFCP's involvement in regional/international events, which contributed to the Go Local campaign.

Papers, articles, releases, displays/exhibitions and workshops. These disseminated a wealth of information to diverse audiences. Scientific papers and other materials provided a strong basis for the project's approach and credibility for its activities. Specific activities included:

- the Go Local e-mail network providing short updates on scientific findings and a forum for discussion and experience exchange among members;
- displays on Pohnpei bananas and IFCP's Go Local campaign – held in Cornwall, United Kingdom, and at Bioversity International's No End to the Banana exhibition[18] – which stimulated international interest in valuable FSM local foods, adding to the international prestige;
- scientific papers on findings about the nutrient contents of local foods;
- Pohnpei banana and taro chapters for a Pohnpei ethnobotany book;
- an overview of Pohnpei yam for a regional project;
- articles and releases for development journals and global Web sites;

- Go Local workshops in other Pacific Island countries, reaching more than 1 500 people in remote communities in the Solomon Islands and Papua New Guinea, and led by the Australian Centre for International Agricultural Research (ACIAR), HarvestPlus, and the Secretariat of the Pacific Community Centre for Pacific Crops and Trees;
- presentations at regional and international meetings, including the first Pacific Summit on Diabetes, from 8 to 12 September 2008 at Saipan World Resort, in the Commonwealth of the Northern Mariana Islands, which led to many further requests.

Collaborative research projects. The project collaborated with post-graduate and other university students to investigate the production and consumption of local foods and ways of promoting local foods in Micronesia. Topics included dietary assessment (Corsi, 2004), an assessment of agroforestry relating to diet and health (Shaeffer, 2006), banana marketing (Parvanta, 2006), marketed processed local food (Naik, 2008), Mand project evaluation (Kaufer, 2008; Bittenbender, 2010), challenges to local food availability (Clayton, 2009), youth attitudes and perceptions relating to local food (Greene-Cramer, 2009), resistant starch in Pohnpei banana cultivars (Thakorlal, 2009), diet in times of transition in a remote area of Pohnpei (Emerson, 2009), and food security issues (Del Guercio, 2010; Sears, 2010). Collaborating universities were Emory University, the University of Arizona and the University of Hawaii, all in the United States of America; the University of Auckland in New Zealand; and McGill University in Canada.

Community-level evaluation survey

To evaluate the effect of the promotional and intervention activities discussed in the previous subsections, two cross-sectional surveys were conducted: a baseline survey in June and July 2005; and a major evaluation survey in June and July 2007, after the intervention. A further evaluation focusing on diet was

---

[18] Presented at the Central Library of Leuven in Belgium; the Royal Botanic Garden, Edinburgh, and the Eden Project in the United Kingdom; the National Botanic Gardens of Ireland; the World Bank lobby in Washington, DC; and the Fairchild Tropical Botanic Garden in Florida, United States of America.

conducted in 2009, two years after the interventions had been completed, to determine whether the initial improvements documented in the earlier evaluation had persisted. A standardized protocol and trained interviewers were used.[19] The evaluations assessed changes in the dietary intakes, consumption patterns and health of people in Mand community. Health assessments and dietary interviews took place in Mand Community Hall during Mand Community Working Group meetings. Interviews were primarily in Pohnpeian or Pingelapese, and were transcribed in English. Participants were selected randomly as one adult woman per household. One criterion for inclusion in the 2009 analysis was that households had to have completed the dietary records in both 2005 (baseline) and 2007 (evaluation). The SAS statistical program (SAS Institute Inc., United States of America) was used for statistical analysis. A p value of ≤ 0.05 was considered significant.[20]

Dietary intake was assessed through two non-consecutive 24-hour recalls in 26 out of 44 households. Two individuals from 2005 and five from 2007 were excluded because of underreported data (Goldberg *et al.*, 1991), and 11 lactating women were excluded from nutrient analysis owing to their extreme nutrient requirements. The data were analysed using modified Pacific Island food composition software.[21]

Food frequency was assessed with a seven-day FFQ of 33 food items and 200 sub-items.[22] Data from a total of 40 households were analysed. Each participant was asked to give the number of days in the last seven that a main item had been consumed, and whether sub-items had been consumed at any point during the seven days.

The diversity of foods consumed was assessed systematically. Three scores of dietary diversity were defined and computed: food group diversity (the numbers of total, local and imported food groups consumed); species diversity (the numbers of individual total, local and imported species consumed); and food variety (the numbers of individual total, local and imported varieties consumed).

Anthropometry (weight, height and waist circumference), FGB and blood pressure were measured to assess health status, using standard methods (WHO, 1997).

Additional assessments were carried out to test participants' knowledge, awareness and behaviour patterns regarding project activities. Questions included how and where the participant had heard about the project, who in the family participated, and what lessons had been learned.

The 2009 diet assessment utilized similar methodology as in 2005 and 2007: a seven-day FFQ and two days of 24-hour recalls, collected via door-to-door surveys of the same households as previously studied. Because of migration, changes in household composition and deaths, the number of households surveyed was reduced from 40 to 36.

Results of promotion and intervention activities

Among the many challenges to implementing this project in FSM were the convenience, low cost and high status of imported foods in relation to local foods, and the important role that imported foods and drinks have assumed in people's diets, which makes it difficult to change course. Also important were the lack of awareness that many people have regarding the relationships among diet, lifestyles and health, and the difficulty in storing, transporting and marketing local foods compared with imported processed foods. A dearth of awareness-raising and educational materials relevant to local Pacific Island foods and local varieties led to the use of less relevant and appropriate materials.

FSM faces major challenges owing to its remote location, geographic dispersion, multiple cultures and languages, and the threats of climate change. FSM comprises a small land mass surrounded by a million square miles (about 2.6 million km²) of ocean, so

---

[19] The 2007 research team included 12 officers from eight agencies, nine of whom were in the 2005 team (of four interviewers, two nurses and three research assistants). Three of the four interviewers took part in both 2005 and 2007.
[20] Proc MIXED was used to examine change in dependent variables as continuous quantitative data with a normal distribution. Normality was tested with Proc Univariate, using a Shapiro-Wilk statistic (p value > 0.01 indicated normality). If normality was not met, power transformations were used (in the order of logarithm, square root, cube root, fourth root).
[21] FoodWorks Professional Edition (version 4.0, Xyris Software, Australia).
[22] The FFQ was modified from those previously developed for FSM (Englberger, 2003; Corsi *et al.*, 2008).

national meetings with representatives from all four states are costly, as great distances have to be travelled. English is the government language in all states, but the use of eight official local languages and additional dialects, and cultural differences are challenging for the development of state and national programmes and policies.

Documentation phase

Very little was known about specific varieties of Pohnpei's local foods prior to 1998, when efforts were made to identify local foods that could alleviate the emerging vitamin A deficiency problem. Key informants mentioned *Karat*, an unusual banana variety with deep yellow/orange flesh, which indicates the presence of provitamin A carotenoids. Samples were taken and analysed for provitamin A carotenoids and other nutrients at off-island laboratories, as there are no laboratories in FSM.

These analyses confirmed that *Karat* is rich in ß-carotene, the most important of the provitamin A carotenoids, and other essential nutrients.[23] Other yellow- and orange-fleshed varieties/cultivars of banana, giant swamp taro, breadfruit and pandanus were analysed and identified as containing substantial concentrations of carotenoids, essential minerals and other nutrients (Englberger *et al.*, 2003a; 2003b; 2006a; 2008; 2009a).

There are more than 50 varieties of banana in Pohnpei, with flesh coloration varying from white and cream, to yellow, yellow/orange and orange. In general, the deeper the colour of a variety's flesh the greater the carotenoid content. Similarly, giant swamp taro, breadfruit and pandanus varieties vary by intensity of flesh coloration and associated carotenoid content. Foods rich in ß-carotene and other provitamin A carotenoids protect against vitamin A deficiency disorders (infection and night blindness), anaemia (weak blood) and cancer, heart disease and diabetes (McLaren and Frigg, 2001). The Yellow Varieties Message was developed to relay the concept that consuming these varieties offers important nutrients and health benefits (Englberger *et al.*, 2006b).

As well as documenting carotenoid-rich foods, the project also documented the vast diversity of traditional foods available on Pohnpei (Englberger *et al.*, 2009b).

Results of community-level activities

Meetings in Mand were important for raising awareness about the positive values of local foods and were effective in promoting local foods. Responses indicated that people enjoyed consuming traditional dishes, some of which they had not tasted for a long time.[24]

Attitude towards local foods. The promotion of local foods had a great impact in the community, despite occasional difficulties in mobilizing family members to join activities. There was a clear change in people's attitude towards local foods, as evident from the following comments made by community members:

> People are now talking more about local food at informal gatherings and there is more local food at church feasts, such as Easter.

> There is more local food at cultural events. During the Easter 2008 event, very little rice was brought. In the past it was the major food item. Also people ate more of the local food, leaving the rice. Coconuts were served and there were hardly any soft drinks. Hot dog was not seen. The main protein foods were fish and chicken, whereas previously fatty spare ribs were a main item.

> Since the project I cannot get the coconuts from my Adohl tree. People are always taking them now! I tell you it is a very effective programme, when we have our special gatherings we now have local food dishes, and we say "Go Local". The way they cook the food now is different, and we talk about how local food is good for the body.

> People are starting to say "Where is the local food?" at community events serving food (previously they were happy with rice and other imported foods).

---

23 Analyses also found that *Karat* is rich in riboflavin (vitamin B_2) (Englberger *et al.*, 2006a).

24 "I feel good eating *apior* [edible coconut husk from the *Adohl* variety, tied with pieces of mature coconut]. My grandmother gave me this when I was sick. It is such a long time since I had it." Mand woman.

It is important to teach children the importance and value of their traditional foods.

When we were children, we used to eat kaikes seeds. They tasted good. We need to teach our children today to eat them.

Project leaders indicated other community benefits from the project, such as the carving of three new canoes for fishing, learning the names of more fish, and the provision of training opportunities in areas of interest. Prior to the project, there were no canoes in Mand, and people knew the names of only a few fish. The project documentation phase helped young people to learn the names of many rare fish, while training and awareness-raising opportunities for adults included the container gardening training, the United States Ambassador's local food pot luck dinner, and a half-day planning workshop involving academic and community leaders from CINE's Indigenous Peoples' Food Systems for Health Program, Professor Kuhnlein and Chief Erasmus from Canada.

Dietary intervention

Tables 12.3 to 12.8 summarize the results from the dietary evaluation, which are presented in full by Kaufer (2008) and Kaufer *et al.* (2010). Significant dietary changes were observed in the Mand community. There were a significant reduction in the consumption of rice (Tables 12.3 and 12.5), a significant increase in the intake of provitamin A carotenoid (Table 12.4), increases in the frequencies of consumption of banana,

---

Table 12.3 Top foods consumed by Mand community, Pohnpei, 2005 and 2007

	2005			**2007**	
Source	**Food item**	**Average adult consumption** (g/day)	**Source**	**Food item**	**Average adult consumption** (g/day)
Imported	Rice	846.9	Imported	Rice	544.1*
Local	Banana, all	131.0	Local	Banana, all	170.2
Local	(Fresh) fish	127.8	Imported	Chicken	149.8
Imported	Chicken	111.0	Local	Coconut products	94.2
Local	Breadfruit	80.8	Local	Taro, giant swamp	92.3
Imported	Sugar products	71.8	Local	Breadfruit	88.8
Local	Coconut products	41.5	Local	(Fresh) fish	87.9
Imported	(Canned) fish	39.5	Imported	Sugar products	62.1
Imported	Ramen noodles	35.6	Imported	(Canned) fish	52.9
Local	Taro, giant swamp	30.8	Imported	Ramen noodles	29.6
Local	Local fruit	21.4	Local	Local fruit	22.7
Imported	Canned meat	20.1	Imported	Imported fruit	20.7
Local	Pork	14.4	Imported	Donut	20.1
Imported	Bread	14.2	Imported	Bread	18.1
Imported	Donut	13.4	Local	Pork	16.5
Quantity of local food consumed		471.3	**Quantity of local food consumed**		618.3
Quantity of imported food consumed		1 201.2	**Quantity of imported food consumed**		951.3

Data calculated from two 24-hour recalls on non-consecutive days, from non-lactating women; reported as average daily consumption (26 households per year, one woman per household); collected in same time period both years (June, July), but breadfruit season may vary from year to year.
* Significant decrease from 2005 (*p* = 0.0002).
Canned fish = canned mackerel, tuna and sardines.
Coconut products = cream, flesh, juice and germinating.
Fresh fish = all local fish (tuna, mackerel and reef fish).
Local fruit = excluding banana; including pineapple, pawpaw, pandanus and malay apple.
Sugar products = granulated sugar added to food, and drinks containing sugar.
Source: Kaufer, 2008.

Table 12.4 Average daily energy and nutrient intakes of Mand community, Pohnpei, 2005 and 2007

	2005		2007		
	LS mean*	% energy	LS mean*	% energy	p value
Total					
Energy (kJ)	**9 879.3**		**8 833.4**		**0.04[a]**
Carbohydrate (g)	**354.6**	**59.8**	**303.7**	**56.5**	**0.03**
Protein (g)	100.7	17.0	92.7	17.3	0.39[a]
Fat (g)	61.3	23.3	62.6	26.2	0.82
Vitamin C (mg)	43.2		61.8		0.08[a]
Vitamin A (µg)	176.5		193.2		0.59[a]
Retinol (µg)	176.0		148.1		0.30
ß-carotene equivalents (µg)	**226.6**		**475.7**		**0.02[a]**
		% total[§]		% total[§]	
Local food					
Energy (kJ)	2 286.2	23.2	2 127.6	24.3	0.71[be]
Carbohydrate (g)	51.1	14.8	70.3	24.6	0.24[b]
Protein (g)	31.9	33.9	20.1	23.5	0.06[ce]
Fat (g)	**18.6**	**33.3**	**11.3**	**20.1**	**0.04[b]**
Vitamin C (mg)	42.0	97.6	55.0	97.0	0.21[b]
Vitamin A (µg)	92.3	52.8	80.4	46.1	0.59[b]
Retinol (µg)	**53.4**	**43.1**	**19.0**	**18.1**	**0.02[b]**
ß-carotene equivalents (µg)	**202.1**	**68.6**	**511.6**	**79.7**	**0.02[b]**
Imported food					
Energy (kJ)	7 587.7	76.8	6 624.5	75.7	0.09
Carbohydrate (g)	**294.5**	**85.2**	**216.1**	**75.4**	**0.0007**
Protein (g)	62.2	66.1	65.2	76.5	0.68[a]
Fat (g)	37.1	66.7	45.0	79.9	0.10[a]
Vitamin C (mg)	1.0	2.4	1.7	3.0	0.54[de]
Vitamin A (µg)	82.3	47.2	94.1	53.9	0.55[a]
Retinol (µg)	70.6	56.9	85.8	81.9	0.42[a]
ß-carotene equivalents (µg)	92.3	31.4	130.0	20.3	0.64[e]

Data calculated from two 24-hour recalls on non-consecutive days, from non-lactating women; reported as average daily consumption (26 households per year, one woman per household); collected in same time period both years (June, July), but breadfruit season may vary from year to year.
Bold denotes significant difference.
* Least square mean estimate: standard errors (SEs) of least square mean estimates cannot be obtained for transformed variables, thus SEs are not presented. Variance parameters are provided in the complete evaluation (Kaufer, 2008).
§ Because total, local and imported intakes were analysed separately, the LS means for local and imported do not exactly equal the LS mean for total. For comparison, percentage of total was calculated from the sum of the LS means for local and imported.
[a] Log transformation.
[b] Square root transformation.
[c] Cube root transformation.
[d] Fourth root transformation.
[e] Unable to find power transformation producing normality; used the closest to normality.
Source: Adapted from Kaufer, 2008.

Table 12.5 Frequencies of consumption of selected foods in Mand community, Pohnpei, 2005 and 2007

Food item	Weekly consumption LS means[a] 2005	2007	p value	Food item	Weekly consumption LS means[a] 2005	2007	p value
Local				**Imported**			
Banana, all	**2.6**	**3.9**	**0.0001**	Dairy[d]	0.6	0.8	0.25
Banana, white-fleshed	2.9	2.9	0.86	Drink, imported, with sugar[e]	**2.0**	**3.6**	**≤ 0.0001**
Banana, yellow-fleshed	0.5	0.7	0.13	Egg	**1.1**	**1.6**	**0.03**
Breadfruit	4.0	3.8	0.41	Fish, imported	2.4	2.7	0.46
Coconut fat	1.7	1.3	0.15	Flour products	**4.1**	**5.0**	**0.008**
Drink, local	**2.4**	**3.2**	**0.01**	Fruit, imported	**0.8**	**0.2**	**0.0004**
Fish, local	3.9	4.2	0.42	Meat, imported	**1.7**	**2.6**	**0.003**
Fruit, local[b]	3.5	4.0	0.10	Rice	**6.8**	**6.1**	**≤ 0.001**
Meat, local	1.5	1.1	0.06	Snack, imported	0.3	0.4	0.4
Nuts, local	**0.2**	**0.6**	**0.01**	Sugar, imported products or added to local food[f]	**3.2**	**1.9**	**0.0002**
Pandanus	0.0	0.0	1				
Snack, local	**0.3**	**0.7**	**0.01**	Turkey tail	0.2	0.2	0.65
Starch, other[c]	0.1	0.3	0.07	Vegetable, imported	0.5	0.6	0.32
Taro, giant swamp	**0.2**	**0.9**	**≤ 0.0001**	**Imported and local**			
Vegetable, local	**1.4**	**3.3**	**≤ 0.0001**	Fat, imported/animal	2.1	2.4	0.46
				Fried food	2.1	2.3	0.45
				Fruit, all	**3.0**	**4.5**	**≤ 0.0001**
				Liver	**0.1**	**0.5**	**0.001**
				Vegetable, all	**1.5**	**3.4**	**≤ 0.0001**

Data calculated from a seven-day FFQ, from 40 households per year, from one woman per household.
Bold denotes significant difference.

Source: Kaufer, 2008.

[a] Least square mean estimate: standard error (SEs) of least square mean estimates cannot be obtained for transformed variables, thus SEs are not presented.
[b] Includes ripe banana, excludes pandanus.
[c] Includes dryland taro, yam, cassava, sweet potato.
[d] Includes butter, margarine, cheese, milk.
[e] Includes soft drinks, coffee, tea, Kool-Aid.
[f] Includes donuts and sugar added to local food and/or local drink.

giant swamp taro and vegetables (including green leafy vegetables) (Table 12.5), and an increase in the diversity of local foods (Table 12.6).

Increased dietary diversity was a major achievement, with the total food (local and imported combined) and local food diversity scores increasing in all aspects: by food group, species diversity, and food variety scores (Table 12.6). The mean diversity score for local foods increased between 2005 and 2007, in all three diversity measures.[25]

[25] When the difference in local food diversity scores between 2005 and 2007 was tested with Proc GLIMMIX binomial distribution, all three diversity score groups increased significantly by $p \leq 0.0001$.

Another major achievement was the decrease in rice consumption. The average daily consumption of rice in 2005 was 846 g per person, compared with 544 g in 2007. This reduction was significant ($p < 0.0002$) (Table 12.3). Similarly, the frequency of rice consumption decreased from 6.8 days per week in 2005 to 6.1 days in 2007 ($p < 0.001$) (Table 12.5).

Although the data revealed decreased reliance on imported rice as a food source, there was no significant increase in overall energy intake from local food sources between 2005 and 2007. In 2007, 24.2 percent of energy was from local foods (and 75.7 percent from

Photographic section

Indigenous People who will continue to face the challenge
to protect their traditional knowledge and use of
their local foods for physical, social and environmental health.
We know our work gives you power and strength

Awajún

"Before I didn't eat wel
now I can get fish from my fish pon
until the river is full of fish again an
I can eat the food that I hav
in my farm.

Awajún moth

Chapter 5 >>

Awajún children
(kp-studios.com)

▲
Awajún child
(kp-studios.com)

◀
Palm fruit – *Achu,
Mauritian flexuosa*,
an excellent source
of carotene
(kp-studios.com)

▲
Peach tomato –
Kukush, *Solanum* sp.

Awajún mother
and child

Pumarosa, a popular
Awajún fruit

Pituca – Kiyam
(a tuber)

(kp-studios.com)

⯅⯅
Sra. Agchuin, Awajún Elder
Palm heart – Iju
Irma (Chinita) Tuesta
Fermin Apikai, Awajún leader
Francisco Quiaco, Awajún
community leader
(kp-studios.com)

⯅
Awajún girls
(kp-studios.com)

▶
Sr. Kinin, Awajún Elder
(kp-studios.com)

Dalit

"Today, if I look back, I can sense
 sea-change in my life, and what is
 o exhilarating about it is the feeling
 f control that we are experiencing…"
sheelamma, Raipally village

hapter 6 >> 75

▲
Dalit boys and girls
from Humnapur village,
Nyalkal mandal, Medak district,
Zaheerabad, India
(kp-studios.com)

◄
Drumstick leaf – *Moringa
oleifera* (kp-studios. com)

Sorgum in a mixed farm –
Sorghum vulgare
(kp-studios.com)

Dalit child (kp-studios.com)

Dalit women farmers
weeding and collecting
greens, neer Zaheerabad,
India (H.V. Kuhnlein)

Bachali – Basella sp., a popular l
green vegetable (kp-studios.com

Shakunthalamma, Dalit assistan
in mixed crop field (kp-studios.c

Chinna Narsamma, sorting edible greens collected as "weeds" in fields (H.V. Kuhnlein)

Mayalu, *Basella rubra*, a nutritious green vegetable (kp-studios.com)

Millet, legumes, oil seeds (kp-studios.com)

Dalit women cleaning food grain (kp-studios.com)

Humnapur Laxamamma, Dalit farmer and seed keeper (kp-studios.com)

◀ Young Dalit girls (H.V. Kuhnlein)

Gwich'in

"Through harvesting
traditional foods
you practice your culture
and live your heritage."
Gwich'in community member

Chapter 7 >> 101

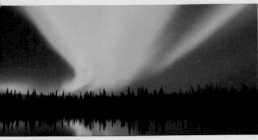

▲ Gwich'in children
(kp-studios.com)

◀ Drying caribou meat and fat
(kp-studios.com)

▲ Northern lights

Winnie Greenland,
Community Health Representative

Elizabeth Vittrekwa,
Gwich'in Project leader

Olive Itsi, Gwich'in assistant

Hazel Nerysoo, Gwich'in leader

(kp-studios.com)

Mary Snowshoe, Gwich'in Elder
(kp-studios.com)

Caribou meat as "stirfry"
with vegetables (H.V. Kuhnlein)

Home freezer with fish and
meat (H.V. Kuhnlein)

Gwich'in girl (kp-studios.com)

Andrew Neyando,
Gwich'in Elder (kp-studios.com)

Alice Andre, Gwich'in Elder,
cutting fish (H.V. Kuhnlein)

Shawn Vittrekwa,
Gwich'in hunter (kp-studios.com)

'I […] promote traditional foods and crops
so that families are able to recover
traditional foods, recipes and drinks,
seeking to make them less dependent on markets
when it comes to health and food."
Libia Diaz, Inga local promoter, San Miguel Indigenous Reserve

Chapter 8 >> 121

Antonia Mutumbajoy, Inga
leader (kp-studios.com)

Eva Yela, Inga leader
(kp-studios.com)

Ana Maria Chaparro, Amazon
Conservation Team, Bogota
(kp-studios.com)

Inga girl (Inga Research Team)

Sonia Caicedo, Amazon
Conservation Team, Bogota
(kp-studios.com)

◄◄
Previous page, Children
squashing chontaduro,
a palm tree fruit (*Bactris
gasipae*s Kunth) for making
chichi (Inga Research Team)

▲
Food plant identification
session (Inga Research Team)

►
Smoked river sardines
(Inga Research Team)

Arazá – *Eugenia stipitata*
McVaugh

Churu – Snails

Cucha – Small fish (Inga

Uchumanga de menudo,
served at culinary festival
(Inga Research Team)

Taita Jose Becerra,
Shaman,
with two young people
(Inga Research Team)

Alvaro Mutumbajoy's
grandson eating fish
soup with vegetables
(Inga Research Team)

Inuit

"Our past is preserved
and explained through the telling of
stories and the passing of information
from one generation to the next…"

Inuit Tapiriit Kanatam

Chapter 9 >> 141

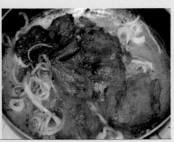

▲
Inuit community of
Pangnirtung, Nunavut,
Canada (kp-studios.com)

▲
Cutting arctic char and caribou
meat with the Inuit *ulu* knife

Caribou meat meal with onions
(kp-studios.com)

◄
Jamesie Mike, Inuit Elder
(kp-studios.com)

▲
Siloah Metuq, Inuit Elder
(kp-studios.com)

◄
Inuit youth and friend
(kp-studios.com)

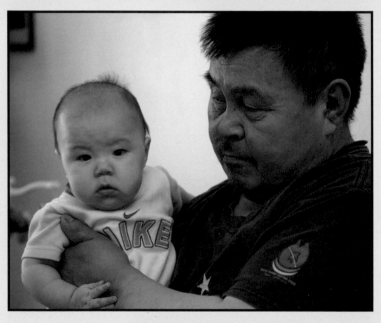

▲
Mahtanah Alivaktuk,
Inuit grandfather and
granddaughter Briana
Alivaktuk (kp-studios.com)

▶
Jaco Ishulutak, Inuit carver
(kp-studios.com)

▲
Joanasie Veevee, Inuit youth
(kp-studios.com)

▲
Johnny Kuluguqtuq, Regional
Community Health
Development Coordinator

Fish soup with market
vegetables

Looee Okalik, Inuit Tapiriit
Kanatami, Ottawa, Ontario

(kp-studios.com)

"We work together.
We understand each other;
much more than when
we started the work."

Huaijeemong Sangkhawimol (Sanephong traditional healer)

▲
Sinee Chotiboriboon and
Mailong-ong Sangkhachalatarn
(kp-studios.com)

Karen family (HV Kuhnlein)

Karen traditional dishes
(Mahidol team)

Harriet V. Kuhnlein and Solot
Sirisai (kp-studios.com)

◄◄
Previous page, Karen vegetables
(Mahidol team)

▲
Sompop Sangkhachalatarn,
Karen community,
Kanchanaburi, Thailand
(H.V. Kuhnlein)

▶
Mainia Sangkhathiti, Karen
community (H.V. Kuhnlein)

◀ Nipaporn Sangkhawimol
and her child (H.V. Kuhnlein)

▲▲ Karen traditional dishes
(Mahidol team)

▲▲ Suttilak Smitasiri, Mahidol
University, Salaya, Thailand
(kp-studios.com)

Prapasri Puwastien,
Mahidol University, Salaya
Thailand (Prangtong
Doungnosaen)

Suaijeemong
Sangkhawimol, Karen
leader (M. Roche)

Anon Setaphan, Karen
leader, Sanephong Village,
(kp-studios.com)

Nuxalk

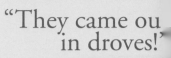

"They came ou
in droves!"

Rose Han

▲
Red elderberries
(kp-studios.com)

◀
Clean, pure and natural
water at the Nuxalk
Nation (kp-studios.com)

◀◀
Richard Pollard on the
Bella Coola River

Red huckleberries,
a summer fruit
(kp-studios.com)

▲
Rose Hans, Nuxalk Elder,
Bella Coola
(kp-studios.com)

Nuxalk Chief Edward
Moody and Jason Moody,
Bella Coola
(kp-studios.com)

Youth barbeque and
feast (H.V. Kuhnlein)

Nuxalk Chief Archie
Pootlass, Bella Coola
(kp-studios.com)

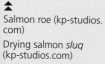
Salmon roe (kp-studios.
com)

Drying salmon *sluq*
(kp-studios.com)

Cow parsnip (*Heracleum
lanatum*), a Nuxalk spring
vegetable (H.V. Kuhnlein)

Filleting spring salmon
(kp-studios.com)

A bountiful catch of
Nuxalk salmon
(kp-studios.com)

Lorraine Tallio cutting
spring salmon
(kp-studios.com)

Pohnpei

'Let's Go Local!
Grow and eat local foods for
their 'CHEEF' benefits:
culture, health, environment,
economics and food security."

Island Food Community of Pohnpei

Chapter 12 >> 191

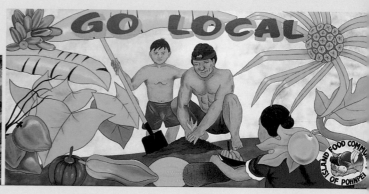

The late Dr Lois Englberger (kp-studios.com)

Kiped Albert (kp-studios.com)

Adelino Lorens (HV Kuhnlain)

Billboard – The "Let's Go Local" campaign slogan was painted or billboards to portray different Pohnpei foods and varieties, including Karat banana (left top) an important fruit to Pohnpei families. (IFCP)

Next page

▶
The late Selihna Johnson (kp-studios.com)

▶▶
Pohnpei child (kp-studios.com)

▲
Pandanus, *Pandanus tectorius*, a rich source of many nutrients (IFCP)

◀◀
Previous page, Pohnpei child (kp-studios.com)

▶
Banana market in Pohnpei (IFCP)

▲
Merlain Poni (kp-studios.com)

Mand school children enjoy learning about bananas in the Youth to Youth project. Here they happily present the carotenoid-rich Karat (middle) and other banana varieties. (IFCP)

Godwin Fritz (kp-studios.com)

Mand Breastfeeding Club (IFCP)

Ainu

"…When the local people
accept Ainu food as a part of local food
there will be no social discrimination
against Ainu people."
Miwako Kaizaw

Chapter 13 >> 22

Miwako Kaizawa, Community Researcher, Ainu project

Koichi Kaizawa, Ainu Community Leader

Millet for tonoto

Masami Iwasaki-Goodman, Ainu Researcher, Hokkai-Gakuen University, Sapporo, Japan (kp-studios.com)

Previous page, Turep, a traditional dried bread prepared from *Lilium cordatum* var. glehni (kp-studios.com)

Pukusakina, *Anemone flaccid* (2 varieties), dried (kp-studios.com)

Ainu Elder (kp-studios.com)

Pukusa, *Allium victorialis*
(2 varieties), dried
(kp-studios.com)

◀ Ainu infant
(kp-studios.com)

▲▲
Taichi Kaizawa

Pukusakina leaves

Yuk (deer) soup

Teaching about the Ainu
traditional food
(kp-studios.com)

Masahiro Nomoto, Ainu
leader (H.V. Kuhnlein)

1. Gail Harrison (Ch. 2, Ch. 7)

2. P.V. Satheesh (Ch. 6) and Salomeyesudas (Dalit, Ch. 6)

3. Martina Schmid (Ch. 6)

4. Jonah Kilabuk (Inuit, Ch.9) and Grace Egeland (Ch. 2, Ch. 6, Ch. 9)

5. Mark Plotkin (Ch. 3)

6. Liliana Madrigal (Ch. 8)

7. Thelma Harvey (Nuxalk, Ch. 11)

8. Marion Roche (Ch. 4)

9. Margaret Mcdonald (Gwich'in, Ch. 7)

10. Andy Siwallace (Nuxalk, Ch. 11)

11. Group photo in Bangkok, 2009 (all chapters)

12. Nancy Turner (Ch. 3, Ch. 11) and Bill Tallio (Nuxalk, Ch. 11)

13. Arjan Sakorn (Ch. 10); Bill Erasmus; Harriet V. Kuhnleinand Suttilak Smitasiri (Ch. 10)

14. Siri Damman (Ch. 15) and Inga Partners (Ch. 8)

15. Hillary Creed-Kanashiro (Ch. 4, Ch. 5) and Irma Tuesta (Awajún, Ch. 5)

16 Peter Slwallace (Nuxalk, Ch. 11)

17, 18. Pohnpei assistants (Pohnpei, Ch. 12)

"The walrus head [in the sculpture]
represents the universe, and we all live
on the earth with the sea connected.
It's telling the story of climate change
occurring in the North. How it's affecting
the animals, the environment and ourselves…
This may only look like a carving, but it depicts
a lot of our lifestyles and environment and animals
which are united. We have to manage the animals
well today because the next generation
will have to have their own meals as well."

Jaco Ishulutak,
Inuit master carver, hunter and cultural specialist,
Pangnirtung, Nunavut, speaking about
his sculpture depicting climate change.

(kp-studios.com, from the DVD film,
The Inuit and their Indigenous Foods, *2008)*

Table 12.6 Dietary diversity in Mand community, Pohnpei, 2005 and 2007

| | 2005 | | 2007 | | | |
	LS means[a]	Range	LS means[a]	Range	p value[b]	p value[c]
Food group score						
Total (n = 14)	10.1	6–13	10.9	6–14	**0.04**	**0.04**
Local (n = 6)	4.8	2–6	5.5	4–6	**0.001**	**≤ 0.0001[d]**
Imported (n = 8)	5.3	2–7	5.4	2–8	0.74	0.74[d]
Species diversity score						
Total (n = 72)	12.4	7–18	18.1	9–29	**≤ 0.0001**	**≤ 0.0001**
Local (n = 51)	12.3	3–11	17.3	5–23	**≤ 0.0001**	**≤ 0.0001**
Imported (n = 21)	5.2	2–8	6.0	3–11	0.14	0.06
Food variety score						
Total (n = 166)	21.3	11–31	32.5	14–66	**≤ 0.0001**	**≤ 0.0001[d]**
Local (n = 100)	11.8	4–19	19.5	8–43	**≤ 0.0001**	**≤ 0.0001[d]**
Imported (n = 66)	9.4	3–16	12.8	5–24	**≤ 0.0001**	**0.0003**

Data calculated from a seven-day FFQ, from 40 households, from one woman per household.
Bold denotes significant difference.
[a] Least square mean estimate: standard errors (SEs) of least square mean estimates cannot be obtained for transformed variables, thus SEs are not presented.
[b] Year effect tested with Proc GLIMMIX, binomial distribution.
[c] Year effect tested with Proc MIXED with arcsine transformation.
[d] Non-normal distribution.
Food group score = number of different food groups consumed by the individual over the reference period,
Local food groups (n = 6) = starchy staples, meat and nuts (including fish), fruit, vegetables, fat, and snacks.
Imported food groups (n = 8) = starchy staples, meat and nuts (including fish), fruit, vegetables, fat, snacks, dairy, and sweets.
Species diversity score = number of unique individual species, excluding cultivars, consumed over the reference period.
Food variety score = number of all food items and sub-items, including cultivars, consumed over the reference period.
Source: Kaufer, 2008.

imported), while in 2005 it was 23.2 percent (Table 12.4). Imported chicken, other imported protein foods, sugar and flour products were major food items in both 2005 and 2007 (Tables 12.3 and 12.5).

Three local foods registered significant increases in their frequencies of consumption: banana increased from 2.6 days/week in 2005 to 3.9 days in 2007; giant swamp taro from 0.2 to 0.9 days/week; and local vegetables, such as chilli leaves, *chaya*, *pele* and Brazilian spinach, from 1.4 to 3.3 days/week ($p \leq 0.0001$) (Table 12.5).

Consumption of ß-carotene equivalents also registered a significant increase, from 202.1 µg in 2005 to 511.6 µg in 2007 ($p \leq 0.02$) (Table 12.4). This included ß-carotene and other provitamin A carotenoids that contribute to vitamin A status.

The evaluation revealed some inconsistencies in the results reported: there was a significant increase in the frequency of consuming some unhealthy food items,

including white flour products and sweet drinks, but a significant decrease in the frequency of items with sugar (Table 12.5), indicating possible underreporting of these items.

The 2007 dietary intake evaluation found that rice was the highest overall contributor of energy (30.4 percent), imported chicken contributed the most protein and fat (39.0 and 34.8 percent, respectively), banana the most vitamin C (29.1 percent), fish the most vitamin A (27.9 percent), and green leafy vegetables the most ß-carotene (55.7 percent) (results not shown in table).

Tables 12.7 and 12.8 show the frequencies of consumption of imported and local foods for 2005 and 2007. Results reveal that the consumption of local foods increased while that of imported foods remained constant. A total of 14 different banana cultivars were consumed and eight green leafy vegetables, while the intake of lemon grass increased between 2005 and

Table 12.7 Dietary diversity in Mand community, Pohnpei, 2005 and 2007

Food group/Imported food	Description	7-day FFQ counts/week		24-hour recalls counts/2 days	
		2005	2007	2005	2007
Starch					
Rice	White	40	38	39	39
Wheat	Ramen, bread, flour	35	40	30	34
Meat					
Chicken	Meat, egg	20	34	27	30
Beef products	Canned meat, hamburger	23	19	13	6
Turkey	Turkey tail	5	7	0	2
Fish					
Mackerel	Canned	22	28	12	13
Sardines	Canned	1	5	0	1
Tuna	Canned	18	21	13	9
Nut					
Peanuts	Whole, butter	0	3	3	4
Dairy					
Dairy products	Milk, ice cream, cheese	10	16	3	3
Vegetable					
Broccoli		n/a	n/a	2	1
Cabbage	European	2	3	3	2
Carrot		1	6	3	2
Chilli		n/a	n/a	1	0
Maize		0	1	4	0
Cucumber		3	0	0	0
Garlic		n/a	n/a	0	4
Lettuce		1	0	0	0
Onion		n/a	n/a	2	7
Potato	Fresh, canned	0	3	3	2
Tomato	Fresh, canned	1	2	1	2
Fruit					
Apple		2	2	0	1
Grapes		1	0	0	0
Guava	Juice	n/a	n/a	0	1
Orange	Whole, juice	1	0	2	1
Pineapple	Canned	8	3	0	2

Data calculated from a seven-day FFQ and two 24-hour recalls, from 40 households per year, from one woman per household; presented as counts per week and counts per two days.
n/a = food did not appear on the FFQ.
Source: Kaufer, 2008.

Table 12.8 Dietary diversity in Mand community, Pohnpei, 2005 and 2007

Common name	Cultivar, description or local names: Pohnpeian (Pingelapese)	Scientific name*	7-day FFQ counts/week		24-hour recalls counts/2 days	
			2005	2007	2005	2007
Starchy staple						
Breadfruit	Mahi (mei)	Artocarpus altilis/mariannensis	38	33	22	22
	Ripe unseeded		16	13	n/c	n/c
	Green unseeded		15	22	n/c	n/c
	Green seeded		0	2	n/c	n/c
	Ripe seeded		0	3	n/c	n/c
Taro, dryland	Sawa (sewa)	Colocasia esculenta	5	4	1	5
Taro, giant swamp	Mwahng (mweiang)	Cyrtosperma merkusii	9	17	6	9
Yellow-fleshed	Pwiliet (Pwilies)		2	5	n/c	n/c
Yellow-fleshed	Simihden		0	2	n/c	n/c
Yellow-fleshed	Sounpwong Weneu		0	1	n/c	n/c
Yellow-fleshed	Tekatek (Sekasek)		0	3	n/c	n/c
Yam	Kehp	Dioscorea spp.	0	6	1	0
Tapioca	Kehp tuhke (dapiohka)	Manihot esculenta	0	2	0	1
Sweet potato	Pidehde	Ipomea batatas	0	3	0	1
Banana	Uht (wis)	Musa spp.	33	36	23	24
White-fleshed	Inahsio (Aroh wis)		2	5	0	2
White-fleshed	Kaimana (Lokoei)		14	15	3	8
White-fleshed	Utin Menihle		7	14	0	1
White-fleshed	Pihsi/Fiji		8	13	1	5
White-fleshed	Utin Ruk (Wis in Ruk)		8	13	19	11
White-fleshed	Utin Wai (Wis in Wai)		n/a	n/a	0	1
Yellow-fleshed	Akadahn (Lakadahn)		2	1	0	1
Yellow-fleshed	Karat (Wis Karat)		0	3	0	0
Yellow-fleshed	Daiwang		6	7	5	2
Yellow-fleshed	Utimwas		0	2	0	0
Yellow-fleshed	Utin lap (Wis in lap)		0	1	0	0
Yellow-fleshed	Utin Kerenis		0	1	0	0
Yellow-fleshed	Utin Rais/Kudud (Sendohki)		2	1	0	1
Nuts						
Chestnut	Mworopw (mwerepw)	Inocarpus fagifer	8	14	1	0
Fish						
Tuna, skipjack, yellowfin	Lesapwil; pweipwei	Katsuwonus pelamis; Thunnus albacares	13	20	17	19
Reef fish	Fresh, dried	More than 100 different fish	28	33	14	20
Mackerel	Double-lined mackerel (pweir)	Grammatorcynus bilineatus	n/a	n/a	6	2

(Continued)

Common name	Cultivar, description or local names: Pohnpeian (Pingelapese)	Scientific name*	7-day FFQ counts/week		24-hour recalls counts/2 days	
			2005	2007	2005	2007
Other seafood						
Crab, mangrove	Elimoang	Scylla sirreda	0	1	1	0
Lobster	Urehna	Panilurus spp.	0	2	0	0
Shrimp	Likedepw	Palaemon serrifer	1	3	0	0
Meat						
Chicken	Malek	Gallus domesticus	5	7	0	1
Dog	Kidi	Canis familiaris	1	0	0	0
Duck	Deki	Aytha fuligula	0	1	0	0
Pork	Pwihk (koaso/pwihk)	Sus scrofa	17	18	14	11
Green leafy vegetables						
Pumpkin	Pwengkin	Cucurbita moschata	0	1	0	2
Chilli	Sele	Capsicum annuum	2	4	0	2
Kangkong, swamp cabbage	Kangkong	Ipomoea aquatica	4	6	0	2
Pele	Bele	Hibiscus manihot	2	8	0	0
Brazilian/Okinawan spinach	Spinis	Alternanthera sissoo; Gynura crepidioides	0	13	3	6
Chinese cabbage	Cabbage	Brassica chinensis	2	14	2	5
Chaya	Chaya	Cnidoscolus chayamansa	3	15	1	3
Drumstick	Moringay (drumstick)	Moringa oleifera	n/a	n/a	0	1
Other vegetables						
Beans	(Pihns)	Vigna sesquipedalis	0	4	0	0
Bell pepper	Bell pepper	Capsicum annuum	1	11	0	1
Cucumber	Kiuhri	Cucumis sativus	7	13	6	3
Eggplant	Nasupi (eggplant)	Solanum melongena	0	7	0	1
Ginger	Sinter (sinser)	Zingiber officinale	n/a	n/a	1	0
Leek	Nira (lihk)	Allium schoenoprasum	0	4	0	0
Onion, green	Nengi	Allium cepa	1	8	0	1
Tomato	Domado	Lycopersicon esculentum	0	4	0	2
Fruit						
Rose apple, bell apple	Apeltik (apolsikisik)	Eugenia jambos	1	6	0	0
Citrus	Karer, karertik (karersik)	Citrus aurantifolia	0	13	9	11
Guava	Kuahpa	Psidium guajava	3	9	2	0
Mango	Kehngid	Mangifera indica	5	2	0	0
Mountain apple	Apel en pohnpei	Syzygium malaccensis	0	20	1	2
Pandanus	Kipar	Pandanus tectorius	2	2	2	0
	Swaipwehpwe		0	1	n/c	n/c
	Aspwihrek		1	0	n/c	n/c

(Continued)

Table 12.8 *(Continued)* Dietary diversity in Mand community, Pohnpei, 2005 and 2007

Common name	Cultivar, description or local names: Pohnpeian (Pingelapese)	Scientific name*	7-day FFQ counts/week		24-hour recalls counts/2 days	
			2005	2007	2005	2007
Fruit *(cont.)*						
Papaya	*Memiap (keiniap)*	*Carica papaya*	4	15	1	2
Pineapple	*Pweinaper (pweiniaper)*	*Ananas comosus*	10	17	8	5
Soursop	*(Sei)*	*Annona muricata*	1	1	0	1
Starfruit	*(Ansu)*	*Averrhoa carambola*	n/a	n/a	1	0
Watermelon	*Sihka (wedamelen)*	*Citrullus vulgaris*	0	6	0	1
Drinks/spices						
Cinnamon	*(Madeu)*	*Cinnamomum carolinense*	3	8	1	0
Lemongrass	Lemon grass	*Cymbopogon citratus*	0	11	1	2
Sugar cane	*Sehu (seu)*	*Saccharum officinarum*	8	14	0	1
Coconut						
Mature coconut, embryo	*Ering, pahr*	*Cocos nucifera*	45	36	30	24

Data calculated from a seven-day FFQ and two 24-hour recalls, from 40 households per year, from one woman per household; presented as counts per week and counts per two days.
n/a = food did not appear on FFQ.
n/c = not captured in 24-hour recall.
Sources: Kaufer, 2008; * species names from Englberger *et al.*, 2009b, verified from The International Plant Names Index and FAO databases.

2007. The change in reliance on imported versus local food was not significant in terms of contribution to daily energy. In 2005, about 23 percent of energy came from local food sources and about 77 percent from imported foods, whereas in 2007, the equivalent figures were about 24 and 76 percent.

Assessment of the diet in 2009 showed that the increase in giant swamp taro consumption, from about zero days a week in 2005 to 1.4 in 2007, had been maintained in the 36 households participating in the survey. Another sign of sustained change in attitudes to local and imported foods and improved understanding of the relationship between food and health, was the ban on soft drinks at community events that the Mand community imposed in 2010.

Health status

Table 12.9 shows results of the health assessments for 2005 and 2007. Overall, there were no significant changes in health indicators. Overweight and abnormal FBG levels are still serious problems in FSM:

- *Body mass index (BMI)*: Only 13 percent of the population had normal BMI in 2005, rising to 19 percent in 2007; there was no underweight, but there was a high prevalence of obesity. Detailed observation revealed that the young adult age group (18 to 29 years) had a mean BMI of 30, indicating high health risk. The mean BMI among women (34) was significantly higher than that among men (29).
- *Waist measurement:* There was no significant difference in average waist circumference between 2005 and 2007. More than 70 percent of the sample population had waist circumferences exceeding 88 cm (the cut-off for women) or 102 cm (the cut-off for men), indicating high risk for obesity-related illnesses (based on sex-specific cut-offs).
- *FBG:* There were no significant differences in FBG concentrations or classification categories (normal, abnormal) between 2005 and 2007, or between genders. However, FBG significantly

Table 12.9 Results from health assessments in Mand community, Pohnpei, 2005 and 2007

Outcome	No. individuals/ year	Descriptive statistics	2005	2007	Statistical analysis — Least square mean[a]	p value[a]
BMI (kg/m²)	68	Median	30.6	31	[c]	
Normal: 18–24.9		Range	18.3–50	20–51.9	Year: 2005 = 31.4; 2007 = 31.4.	Year (p = 0.78)
Overweight: 25–29.9		BMI	n (%)	n (%)	Gender: male = 29.24; female = 33.55	Age (p = 0.46)
Obese: 30–39.9		Normal	9 (13)	13 (19)	Age:[b] A = 30.8; B = 31.6; C = 30.9;	Gender (p = 0.005)
Very obese: ≥ 40		Overweight	21 (31)	16 (24)	D = 30.9; E = 32.9	
		Obese	31 (46)	32 (47)		
		Very obese	7 (10)	7 (10)		
Waist circumference (cm)	42	Median	98	100.5	[c]	
Increased risk:		Range	71.2 – 131.4	71.2 – 129.5	Year: 2005 = 98.12; 2007 = 98.21.	Year (p = 0.94)
Female: > 88		Waist circumference	n (%)	n (%)	Gender: male = 95.43; female = 100.9	Age (p = 0.05)
Male > 102		Increased risk	32 (76)	30 (71)	Age:[b] A = 90.11; B = 99.95; C = 101.35; D = 99.58; E = 99.84	Gender (p = 0.16)
FBG (mg/dl)	108	Median	111	114	[d]	
Normal: < 126		Range	79–436	85–496	Year: 2005 = 128.2; 2007 = 129.1.	Year (p = 0.74)
Abnormal: ≥ 126		FBG	n (%)	n (%)	Gender: male = 124.7; female = 132.8.	Age (p ≤ 0.0001)
		Normal	71 (66)	67 (62)	Age:[b] A = 110.49; B = 110.2;	Gender (p = 0.17)
		Abnormal	37 (34)	41 (38)	C = 142.0; D = 140.8; E = 151.4	
Blood pressure (mmHg)	112	Systolic blood pressure			Systolic[c]	
Optimal/normal: <130/85		Median	110	110	Year: 2005 = 114.8; 2007 = 115.1.	Year (p = 0.85)
High normal: 130–139/85–89		Range	90–195	90–180	Gender: male ≤ 116.6; female ≤ 113.3	Age (p = 0.28)
mild/borderline hypertension: 140–159/90–99					Age:[b] A = 101.4; B = 107.8; C = 114.9; D = 122.2; E = 128.4	Gender (p = 0.0001)
Hypertension: ≥160/100		Diastolic blood pressure			Diastolic[c]	
		Median	70	70	Year: 2005 = 72.6; 2007 = 73.1.	Year (p = 0.64)
		Range	40–120	50–98	Gender: male = 74.1; female = 71.6	Age (p = 0.16)
		Blood pressure	n (%)	n (%)	Age:[b] A = 66.1; B = 68.1; C = 74.3; D = 75.7; E = 80.0	Gender (p ≤ 0.0001)
		Optimal/ normal	93 (83)	89 (80)		
		High normal	8 (7)	9 (8)		
		Mild/ borderline Hypertension	7 (6)	8 (7)		
		Hypertension	4 (4)	6 (5)		

a Proc MIXED (outcomes as continuous variables). Standard errors (SEs) of least square mean estimates cannot be obtained for transformed variables, hence SEs are not presented.
b Age categories: A = 18–29 years; B = 30–39 years; C = 40–49 years; D = 50–59; E 60 years and more.
c No transformation.
d Reciprocal transformation.
Source: Kaufer, 2008.

Table 12.10 Process indicators for intervention activities for Mand community, Pohnpei, 2005 to 2007

Activity	% awareness[a] (n = 42)	% exposure[b] (n = 42)	Duration	Frequency	No. meetings/visits	No. participants	No. regular participants
Community Working Group	93	60	2 years	Weekly/ bimonthly	78	126	11
Youth school education	76	33	8 months	Varied	13	42	n/a
Youth Drama Club	88	29	6 months	Monthly	6 meetings in 2006; 3 performances	20 in first; 10 in second	n/a
Breastfeeding Club	83	33	7 months	Bimonthly	12	43	34
Planting material distribution	86	50	8 months	Throughout	n/a	34	n/a
Home gardening	76	48	12 months	Varied	5	20 at first workshop	n/a
Cooking training	83	50	10 days	Once	10	25	n/a
Charcoal oven	90	50	4 weeks	4 weeks for construction and distribution	8	34	n/a
USA Ambassador's dinner	57	21	Once	Once	1	90 (15 from Mand)	n/a
Planting and weight loss competitions	71	50	2 years	2 planting competitions, 1 weight loss competition	n/a	First planting competition: 23	n/a

Data obtained from sources including IFCP intervention log, newspaper articles and questionnaires.

increased with age, especially for those aged 40 years and more. The prevalence of diabetes (FBG $\geq$ 126mg/dl) was 34 percent in 2005 and 38 percent in 2007.

- *Blood pressure:* More than 80 percent of the sample population had blood pressure measurements in the optimal range, and only 4 to 6 percent were definite cases of hypertension. There was no significant difference in blood pressure measurements between genders, but systolic and diastolic blood pressure (BP) both increased significantly with age ($p < 0.0001$). It is a remarkable "good news paradox" that although the Mand population is seriously obese and prone to diabetes, it has low levels of hypertension.

Awareness of and exposure to project activities and materials

The evaluation showed that there was high awareness of and exposure[26] to project activities and materials (Table 12.10). Of those interviewed, 93 percent knew about the Mand Community Working Group meetings, and 60 percent were directly involved in activities.

Similarly, 90 percent were aware of the charcoal oven project, and 50 percent were directly involved. More than 70 percent of respondents indicated that they were aware of the youth work, the home gardening and cooking training, the distribution of planting materials, the planting/weight loss competitions, and the Breastfeeding Club. More than 50 percent knew about the pot luck dinner at the United States Ambassador's residence, although this single evening event involved only 15 Mand community members.

Of the awareness-raising materials used in the intervention, the Go Local billboard on the main road to Kolonia and the Pohnpei bananas poster displayed in the community meeting hall were the most well-known, by 96 and 95 percent, respectively. The indigenous foods poster was the least known, as it was introduced late in the intervention.

Results of state- and national-level activities

Similar to the project's impact at the community level, there were many indications of changes in attitude towards local foods at the state and national levels.

---

26 Exposure was defined as involvement in a project activity.

Availability of *Karat* and *Daiwang* bananas in markets. Observations indicated that the market availability of *Karat* has been steadily improving since 2006. *Daiwang*, a low-status but tasty banana, was previously described as "the banana that was fed only to pigs" and was not marketed. When analysis showed that it is rich in carotenoids, it started to be promoted as a food for humans, and in 2006 it was sold at four of 14 local markets.

Increase in stalls selling local staple food take-outs. A remarkable increase in local staple food take-outs started in 2005, and has continued. Naik (2008) reports that in July 2007, more than 3 550 kg of cooked local staple foods were sold by a sample of Kolonia markets and shops. These included 1 752 kg of banana, 893 kg of breadfruit and the remainder as taro, yam and cassava. New take-out stands continued to appear. Many people, including market owners, indicated that the Pohnpei local food promotion campaign has contributed much to this increase in the marketing of local foods.

Charcoal oven development. The local carpenter engaged to produce charcoal ovens enthusiastically reported that people were purchasing his ovens: "I sold one oven to a man from Chuuk, where there is a problem with electricity outages. He was so happy and took it with him on the plane. A lady from the Marshall Islands also bought one. I built one for my family. I can bake anything in it, just like a normal electric or gas oven. We save money with it. We don't have to buy fuel."

Interest from other communities. Groups from Madolenihmw, Kitti and FSM's other states asked to replicate the project in their communities. In November 2009, the Go Local Agroforestry and Health Improvements project was initiated in six additional communities in Pohnpei and in communities in other states. Baseline surveys have been conducted and intervention activities started, including encouraging the establishment of local food policies.

Other communities that have initiated noteworthy Go Local activities include the following:

- Saladak and Rohi communities of U Municipality and Sapwohn community of Sokehs Municipality held meetings on local food promotion and hosted several large gatherings that served only local foods.
- Salapwuk community of Kitti Municipality held a workshop on building energy-efficient charcoal ovens and improving local food production skills.
- In Madolenihmw Municipality, through the leadership of Chief Lepen Madau en Metipw, the Metipw community promoted local foods at funerals and traditional gatherings. This included providing food take-outs for up to 1 000 people, using local materials such as banana leaves for wrapping and woven coconut leaves for serving foods, and avoiding plastic containers, which cause environmental problems. This community's activities to increase the use of local resources encouraged neighbouring communities to do the same.

Among the many agencies and events to use the "Go Local" slogan as a theme are Annual Library Week in 2006, COM-FSM's graduation ceremony in May 2006,[27] COM-FSM's 2008 Annual Health Fair[28] and the Pingelap People Organization's Pingelap Green Day in 2008. A youth club in Madolehnihmw Municipality took the name "Go Local", and two local softball teams wore Let's Go Local t-shirts as part of their uniform. The slogan was also used at workshops in Yap, Kosrae and Chuuk States.

Following Mand community's banning of soft drinks from all community functions in 2010, other community organizations, such as Pingelap Peoples' Organization, Inc., have done the same, as publicized in the local newspaper, on the radio and in the Go Local e-mail network.

Many people from all four FSM states have shown great interest in the e-mail network (Englberger *et al.*, 2010c). Comments from members include the following:

I look forward to those e-mails ... I print them out and share them with others too.

---

[27] Instead of the usual hamburgers and soft drinks, the refreshments served were fish sandwiches and drinking coconuts.
[28] A speaker at this event, speaking about the Mand case study, stated "We are all proud of this project."

Because of this campaign, I stopped eating rice and now after about two years my son joined me. I have stopped eating rice and I also discourage my two teenage daughters from eating rice and encourage more local foods. Now my two daughters are complaining of no local foods at home.

When my patients have diabetes, I tell them "go local"… I don't really know nutrition that much so I like this way of talking about taro and local food.

Newspapers have published individuals' comments indicating their appreciation of the Mand project and their interest in the promotion of local foods. The local radio broadcaster reported that following the project's press release, more than ten people had called for more information about local foods.

Results of regional- and international-level activities

The "Go Local" slogan was adopted in the title of an SPC Plant Genetic Resources documentary, and a keynote speaker at one international meeting stated that the Pohnpei Let's Go Local project should be considered a model for promoting local foods (SPC, 2007).

Women's groups in Papua New Guinea have adopted the "Let's Go Local" slogan and song (Anzu, 2008), and the International Centre for Underutilized Crops (ICUC) included a description of the Mand project in its annual symposium report for 2006. In 2008, *Kemelis* – a Pingelapese recipe documented in the Mand project – was selected as the April Recipe of the Month on the ICUC Web site.

In 2009, FAO asked IFCP to assist in developing a booklet on how to carry out a Let's Go Local campaign, and in April 2010, supported by WHO, Ms Englberger was invited to be one of only a few speakers at the Pacific Food Summit in Port Vila, Vanuatu, presenting on Go Local to enhance food security. Other publications include articles on the Pohnpei banana stamp series (Ormerod, 2006) and two chapters on local island food in a Pohnpei ethnobotany book (Balick, 2009).

Lessons learned

An overall lesson learned was that reviving memories and restoring confidence in local foods is important, and touches people's hearts and minds. Participants became more interested in health messages when they learned about their own health problems. The slogans and songs helped greatly in passing on messages, and repetition of the same message in different activities and materials was effective. The project team found that elderly people often had set habits that were hard to change, so it was important to work with youth. A realistic approach was needed, encouraging gradual changes and "practising what you preach" rather than banning imported foods. Project facilitators found that they needed to use humour in their presentations and to make activities fun. Writing and communicating about the project regionally and internationally helped spread the message and resulted in increased local interest.

Two comments from Pohnpei market owners, reported by Naik (2008), provide insight: "Today I think that the education about local food is working … The older generation is going back to eating local and this is influencing the younger people"; and "I have been selling more local food, especially breadfruit, taro and yam, but it is still not common for the Pohnpeians to eat them everyday – they eat rice all the time everyday". Nevertheless, the impressive number of new local food take-outs is promising.

Another factor relevant to the Go Local campaign in Pohnpei is the rise in global food prices. In 2008, this became a major topic throughout the Pacific Islands (Singh, 2008), and the price of rice in Pohnpei has doubled. This could provide a stimulus for growing and consuming more local traditional food. Some have suggested producing rice on Pohnpei, which has been shown to be possible, but past attempts to do this have failed owing to the intensive nature of rice growing, which contrasts with traditional Pohnpei agroforestry. Other barriers have been the need for imported fertilizer and pesticides, and the crop's vulnerability to heavy rains and winds. In addition, processed white rice is known to be nutritionally poor, and high consumption of rice presents health risks.

The inclusion of Pohnpei as a case study in this global project provided a rewarding experience for the project team and participants as they progress towards the goal they share with the other 11 communities in the CINE Food Systems for Health Program.

Conclusions

The health situation in FSM continues to be considered a "state of health emergency", as more and more individuals become ill with non-communicable diseases, including diabetes, heart disease, stroke and cancer (FSM Information Services, 2010). Project activities have been popular and have succeeded in increasing people's awareness of this health crisis, and improving their use of better food, their food diversity and their nutrition. With time, this dietary change seems likely to improve the health of all the people in Pohnpei.

The project found many activities that contribute to healthy diets, and an approach that works. In the past, diets and lifestyles changed towards the use of imported food, but they can also shift back to more local foods, and are already doing so.

The way forward

New goals for the projects are:
- advocacy with Pohnpei leaders (traditional, government, church, and private sector) to promote local food and relevant policies;
- expansion of the project to other communities in Pohnpei and other FSM states;
- expansion to more local, regional and international partners, focusing on local foods and their CHEEF benefits;
- continued development of awareness materials and methods for increasing the production and consumption of local foods and varieties;
- a continuous watch for signs of improved health status that can be attributed to the project.

As one team member pointed out: "Remember that this is a long-term process, we may not see the full change in our lifetime, but it is happening." Another team member pointed out: "It is hard to change but let's hang in there, don't give up!" ✳

Acknowledgements

Thanks are warmly extended to all the members of Mand community, including their leaders, who enthusiastically took part in this challenging project. Their willingness to participate has led to insights that may guide work in improving diets and health in additional communities in Micronesia and possibly other Pacific nations and around the world. The authors acknowledge all partners, funding and support agencies, particularly CINE, kp-studios.com, CINE partners and case studies; Pohnpei State and FSM national government departments, including the Pohnpei Office of Economic Affairs, the Department of Education, the Department of Health Services and the Department of Land and Natural Resources; COM-FSM CES; USDA NRCS; Peace Corps Micronesia; Conservation Society of Pohnpei; FSM National Olympic Committee; Global Environment Facility Small Grant Programme; *Sight and Life*; the embassies of Germany, Australia, New Zealand and Japan; PATS Foundation; Papa Ola Lokahi, Global Greengrants Fund; SPC Land Resources Division, SPC Development of Sustainable Agriculture in the Pacific, SPC Forests and Trees Program, SPC Healthy Pacific Lifestyle Section; WHO; FAO; Emory University, McGill University, University of Hawaii and University of Auckland; ACIAR/HarvestPlus; Forum Secretariat; ICUC; Technical Centre for Agriculture and Rural Cooperation (CTA); Xyris; and Bioversity International.

Warm thanks are extended to FSM President Emanuel Mori and Pohnpei State Governors John Ehsa (present) and Johnny P. David (past) for significant support; Bermin Weilbacher for coining the "Go Local" slogan and sharing its use; Wehns Billen for conceptual design/artwork for the "Go Local" billboard; Rohaizad Suaidi for assistance with the Mand Drama Club; Dr Conrad Perera and Ione deBrum for assistance in small-scale processing of local foods; Capitol Fabricator Inc. for assistance in the charcoal oven project; Alyssa Bittenbender, Dr Doug Taren and Dr Duke Duncan for assistance in the recent dietary assessment in Mand community; and Dirk Schulz, Dr Temo Waqanivalu, Dr Mary Taylor and Dr Graham Lyons for initiating efforts to expand the FSM Go Local experience to other Pacific Island countries. Finally, warm thanks are given to Dr Maureen Fitzgerald, Jim Currie, Michelle Hanson and Anu Mathur for help in reviewing the manuscript, and to all those involved in the study and other help promoting Go Local and contributing to this project.
> **Comments to:** pniagriculture@mail.fm; info@islandfood.org

>> Photographic section **p. XXV**

Tasty *tonoto* and not-so-tasty *tonoto*: fostering traditional food culture among the Ainu people in the Saru River region, Japan

MASAMI IWASAKI-GOODMAN

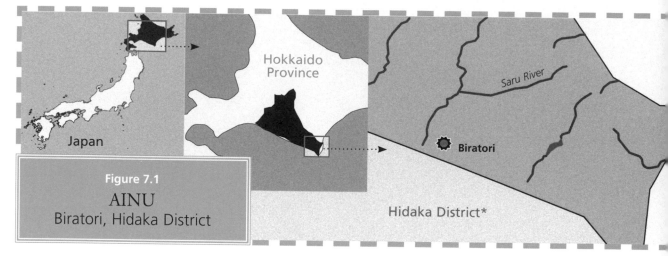

Figure 7.1
AINU
Biratori, Hidaka District

Japan

Hokkaido
Province

Saru River

Biratori

Hidaka District*

Data from ESRI Global
Walter
Geographic Informati
McGill Univers
*Hidaka outline digi
web

1
Faculty of Humanities,
Hokkai-Gakuen University,
Sapporo, Japan

Key words > Ainu, Indigenous Peoples,
food systems, traditional food

Photographic section >> XXVIII

"I would like to see Ainu dishes
served as school lunch
at the local schools.
When the local people accept
Ainu food as a part of local food, there will be
no social discrimination against Ainu people."

Miwako Kaizawa

"Restoring natural environment is the key issue
for preserving Ainu food culture. People should be
able to go out the door and pick wild vegetables
in their backyard, as they did in old days."

Koichi Kaizawa

Abstract

The Ainu community in the Saru River region of Hokkaido has the highest concentration of Ainu population in Japan. In 2004, an Ainu research group led by community leader, Mr Koichi Kaizawa, started the project described in this chapter as part of the international comparative research of the Indigenous Peoples' Food Systems for Health Program coordinated by the Centre for Indigenous Peoples' Nutrition and Environment. The ultimate goal of the project's research and intervention activities was to improve the social and cultural health of Ainu people in the community, by promoting Ainu food culture. The research group conducted four kinds of intervention activities: i) publishing a local community newsletter; ii) holding Ainu cooking lessons; iii) preparing Ainu dishes for ceremonial occasions; and iv) conducting an Ainu cooking project with university students. Over the years since the research began, the research group has observed changes in the community's perception and use of Ainu food. The intervention activities continue, and the researchers anticipate that the Ainu people will integrate their local food culture further into daily life.

[1] The original members of the research group are Koichi Kaizawa (community leader), Miwako Kaizawa (community researcher), Masami Iwasaki-Goodman (anthropologist), Taichi Kaizawa (ethnologist), Satomi Ishii (nutritionist), Hidetomo Iwano (microbiologist) and Hiroki Inoue (microbiologist).

Introduction

At the invitation of Dr Harriet Kuhnlein, Founding Director of the Centre for Indigenous Peoples' Nutrition and Environment (CINE), an Ainu research group[1] led by community leader Mr Koichi Kaizawa joined CINE's Indigenous Peoples' Food Systems for Health Program in the spring of 2004. The CINE-Ainu research group spent the first three years interviewing elders to gather information on traditional food uses, and conducting composition analyses of those food items that the elders identified as being important in their food culture. A summary and the results of this research were published (Iwasaki-Goodman, Ishii and Kaizawa, 2009). During the research period, the group also undertook various intervention activities to promote uses of Ainu food within and outside the community. This chapter summarizes these intervention activities and provides an analysis of their social and cultural implications in the Ainu community of Saru River region.

Improving socio-cultural health among the Ainu in the Saru River region

The research in the Ainu community was expected to generate results on which to base clear project goals. However, during the planning phase in 2004, the research group identified problems with and limitations to conducting this kind of research in the Ainu community. The group identified elements that make this community different from other communities of Indigenous Peoples included in the CINE joint research programme.

First, the Ainu community in Saru River region has consisted of a mixture of Ainu and non-Ainu people since the late nineteenth century, when Japanese people settled in the area and established a township under the Japanese Government (Iwasaki-Goodman, Ishii and Kaizawa, 2009) (Figure 13.1). Following a strong assimilation policy, under which the Government issued numerous laws that affected the Ainu way of life, most Ainu people eventually gave up their traditional hunter-gatherer lifestyle and took up farming. Poverty and racial discrimination seriously affected the Ainu, pushing them to the margins of a society that treated them as second-class members. Facing serious social discrimination, many Ainu people suppressed their ethnic identity and adopted the mainstream Japanese way of life. It is only recently that improved social conditions have given Ainu people the confidence to disclose their ethnic identity, but not all of them have done so. Following this history of distress, the project community in Saru River region is a multi-ethnic community with complex issues of racial discrimination and stigmatized ethnic identity.

The second difference between the Ainu community and the other indigenous communities in CINE's programme is that Ainu people resent academic research, especially when it involves physical and medical examinations. This is mainly because earlier research was conducted unethically (see Lewallen, 2007 for details), and many Ainu people remember their experience of invasive examinations and the accompanying sense of humiliation.

Given that local people had not identified the improvement of physical and medical conditions as a pressing need, and to avoid subjecting them to medical examinations, the research group decided that the CINE global research aim of improving the health of Indigenous Peoples could best be met by focusing on improving Ainu people's social and cultural conditions, rather than their physical and medical ones.

Efforts to improve the Ainu's social and cultural health in Saru River region aimed eventually to resolve social prejudices against Ainu people. The researchers believed that reintroducing traditional Ainu foods and recognizing their nutritional value would promote positive attitudes towards Ainu foods, people and culture, inside and outside the community. This in turn would help create a community in which Ainu people could freely express their ethnic identity with pride. The ultimate aim was for Ainu people to live with dignity as an ethnically distinct group, while maintaining the same social status as non-Ainu members of the community.

A third issue that the research group took into consideration when planning the research is that different age groups of Ainu people hold different levels of cultural knowledge (Iwasaki-Goodman, Ishii and Kaizawa, 2009). Social discrimination following the intensive assimilation policy created negative attitudes towards Ainu ethnicity, including among Ainu people themselves. As a result, many decided not to teach the Ainu language and culture to their children, encouraging them instead to live as their mainstream Japanese neighbours did, speaking Japanese and following Japanese customs. This period of cultural discontinuity created a group of Ainu people with limited understanding of their language and cultural traditions. Starting in the 1980s, however, cultural revitalization among the Ainu led to increasing efforts to reintroduce cultural elements such as language, dance, song and rituals. In the Saru River community, Ainu language classes were held, giving young Ainu an opportunity to learn their language. Rituals that had not been held since before the assimilation period were resumed, as were the prayers, dances, songs and cooking preparations associated with these rituals. Young Ainu people growing up during this period were exposed to various aspects of their own culture. As a result of these developments, the Saru River Ainu community now includes three main age groups of Ainu people: i) a few elders whose first language is Ainu and who have had first-hand experience of Ainu culture; ii) middle-aged people and elders who have had minimum exposure to Ainu culture and who were taught to follow Japanese ways by their parents; and iii) young people who grew up during the cultural revitalization and have learned some aspects of Ainu culture.

A significant difference between the diet changes imposed on Ainu people during the Meiji-Taisho Era (late nineteenth to early twentieth centuries) and those imposed on other Indigenous Peoples, such as in Canada, is that the adoption of Japanese food culture did not result in drastic drops in the nutritional value of the Ainu diet, so there was no major deterioration in health.[2] During the contact period (before the Second World War), the daily diet of Japanese people was almost as simple as the Ainu diet (Ishige, 1979; Iwasaki-Goodman *et al.*, 2005) and contained mainly vegetables with modest amounts of white rice and animal meat. Therefore, the diet shift from Ainu food – salmon, deer and wild vegetables with grains – to Japanese food, such as white rice, pork, chicken and vegetables, did not cause serious negative effects on the health of Ainu people.

However, although the diet change had an insignificant impact on the health and nutrition of Ainu people, it had a serious cultural impact in the minds of both Ainu and non-Ainu people in Saru River region. Mainstream Japanese culture dominated the community, affecting local people's views regarding every aspect of life. Needless to say, the dominant culture was seen as superior, while Ainu culture was regarded as inferior and therefore undesirable. The Government of Japan's powerful assimilation policy added force to social discrimination, pushing Ainu people and their culture to the margins of society.

Many Ainu food items became symbols of the inferiority of Ainu culture. For example, the strong taste and smell of wild onion/garlic (*pukusa* or *kitopiro*)[3] became a target of discrimination; when the rest of the community renamed the onion Ainu *negi* (Ainu onion), Ainu people started to avoid use of both the onion and its new derogatory name, and eventually it became a taboo food. Other traditional food items were classified as undesirable, mainly because of their associations with Ainu people and culture. At the same time, other Ainu food items were gradually integrated into Japanese food because they were not identifiable as distinct Ainu food items. Many Ainu people use Ainu traditional food items without knowing them as such.

This oppression is not unique to the Ainu community of Saru River region and is a feature of the histories of many Indigenous Peoples throughout the world. However, the research group felt that focusing on the historical and current situation of the Ainu community would help identify the social and cultural aspects to consider when revitalizing Ainu food culture. This approach was also adopted by earlier research (Iwasaki-Goodman, Ishii and Kaizawa, 2009).

Interventions

With strong leadership from the community leader, the research group discussed diverse ways of providing people in the community with information and experience of Ainu traditional food, thus reintroducing the food into the local food culture. Intervention activities had four components: i) a community newsletter providing information about traditional food items; ii) a series of cooking lessons; iii) preparation of Ainu dishes for ceremonial occasions; and iv) other activities conducted outside the community.

Ianpero:[4] the local community newsletter

Most Ainu people were unaware of their own use of traditional foods, mainly because these foods had been integrated into mainstream Japanese cuisine. At the start of the research, local people told the research group

---

[2]   However, there are records of health problems among Ainu people during the Meiji Era. For example, a United Kingdom doctor living in Saru region reported that Ainu people were becoming ill because they no longer ate wild game (Biratori Town, 1974).

[3]   *Pukusa* is often referred to as *kitopiro* (*kito* is another Ainu word for *pukusa*).

[4]   "Let us eat" in the Ainu language.

Pukusakina is not well known, but it is an important wild vegetable that Ainu people have been eating for a long time. They dry it and cook it in a soup, called *ohau* or *ruru*, all year round. Because of its frequent use in soup, it is also called *ohau kina* meaning "leaves in soup". When magnolia is in bloom, mothers take their children into the mountains to pick enough *pukusakina* to fill an *icha saranip* (a bag made from the bark of the *shinano-ki*, a Japanese lime), which mothers carry on their backs when harvesting wild vegetables.

Pukusakina has a mild taste and goes well with other ingredients. It is therefore used in many dishes. Its nutritional value is high, and it contains more potassium and phosphorus than *pukusa* (wild garlic). *Pukusakina* is best in *yuku ohau* (deer soup), but is also good in pork soup as deer meat is difficult to get. Fortunately, *pukusakina* is found in many places. It is so abundant that even elders and children can find it easily. Although there are concerns about resource depletion resulting from the booming interest in harvesting wild vegetables, *pukusakina* is abundant and will become popular in the future.

that they no longer ate Ainu food because they had adopted the Japanese way of life. However, it did not take long to discover that they were using traditional Ainu food items in their daily diets, without realizing it. Looking closer at what they eat every day, Ainu people noticed that there are important differences between their food culture and that of non-Ainu people. The most noticeable of these differences is the Ainu's extensive use of various wild vegetables, continuing their ancestral tradition of using wild plants. When planning the intervention activities, the research group decided to focus on creating awareness of traditional food use among Ainu people, and providing information to non-Ainu members of the community. To this end, the group contributed an article on Ainu food use to the community's *Saru unkr newsletter*, for people living in Saru River region.

This monthly newsletter is issued by the Biratori regional branch of the Hokkaido Association of Ainu, and is distributed to all 2 500 households in Biratori. Ms Miwako Kaizawa, a local researcher in the CINE-

Ainu research group, was responsible for writing articles on traditional Ainu foods and their uses, based on information gathered by the research group. Her contributions, each of 700 to 800 words with two or three illustrations, began in April 2005, providing an introduction to the food items and information on harvesting, processing, preservation and cooking methods (Annex 13.1 gives a list of her articles). She took particular care to write in a way that is easy for readers to understand, so that they could harvest the food items and cook the dishes themselves. Box 13.1 gives an example of an article introducing the wild vegetable, *pukusakina*.

The articles also introduced many dishes unique to Ainu food culture. Box 13.2 gives the example of *tonoto*, which is a sacred fermented beverage offered to kamuy (spirits) during rituals.

Although the newsletter reaches every household every month, it is difficult to know how many people read the articles. The research group has to rely on local people's comments as feedback on the intervention. Ms Kaizawa (who writes the articles) has heard that local people enjoy reading them. Some readers have told her that they would like to learn about a greater variety of dishes using Ainu food items, and she has been working on the modification of traditional dishes to meet the tastes of today's local people.

In the first four years of publishing articles in the community newsletter *Ianper*, about 50 articles were written, reaching 2 500 households a month. Ms Kaizawa feels that there have been clear changes in local people's perception of Ainu food. Recently, local non-Ainu women expressed an interest in learning more about Ainu food and culture, and asked her for Ainu cooking lessons. This shows a clearly positive change in attitudes towards Ainu people and their culture.

Cooking lessons

Ainu cooking lessons have been very effective in promoting local people's understanding of Ainu food culture. Since 1996, the local elementary school has regularly held a programme, *Hararaki Time*, to teach children various aspects of Ainu culture (Iwasaki-

Goodman, Ishii and Kaizawa, 2009). Learning how to prepare Ainu dishes is an important and fun part of the programme; the most common recipe learned is the one for *sito* (dumplings). In autumn 2008, the children were given experience of harvesting egg millet, the main ingredient for *sito*, following their Ainu ancestors' traditional method of using *pipa* (freshwater pearly mussel) shells. Including a programme on Ainu culture in the school curriculum is a new approach, demonstrating an interest in sharing Ainu culture and history among both Ainu and non-Ainu community members.

Ainu cooking lessons have also been provided, on request, to other groups. The local CINE-Ainu researcher Ms Kaizawa has taught all of these lessons. In June 2004, about 30 elementary school teachers from Hidaka district gathered at the Biratori community centre to learn how to cook Ainu dishes. Cooking lessons were also held in September 2005, at Rakuno-Gakuen University in Sapporo for about 40 students; in October of the same year for elementary school teachers in Furenai, the neighbouring community; on three occasions in 2006 for Hidaka district school teachers and students; and in 2007, when a new programme of Ainu cooking lessons was combined with a tour of Biratori community for a group of high school students from Hiroshima prefecture in southern Japan, who visited places with Ainu names, learning about the history and diet of Ainu people in the past.

In 2008, a group of local young home-makers – mostly non-Ainu – expressed a keen interest in learning about Ainu food culture and requested a cooking lesson. Ms Kaizawa was the instructor, and she asked other local Ainu women to be her assistants. There were about 16 participants, all in their twenties and thirties, some of whom were born and grew up in Biratori, while others had moved there from outside. They learned how to prepare *kosayo* (bean porridge), *ratashikep* (cooked pumpkin with beans), *yuk ohaw* (deer soup) and *inakibi gohan* (rice cooked with egg millet). They were divided into four groups, with a local Ainu person working with each group to assist with the cooking. The lesson began with explanations of the recipes using printed hand-outs; all the necessary ingredients were provided. The cooking went well, and the young home-makers and local Ainu assistants worked together in a friendly atmosphere. When all the dishes were ready, the participants moved to the dining room and ate them together.

Ainu dishes in ceremonial occasions

The Hokkaido Ainu Association has branch organizations in 48 districts throughout Hokkaido. The branch in Biratori is active in various issues concerning the lives of Ainu people. About ten years ago, as part of efforts to preserve Ainu food culture, a small group of branch members interviewed elders to collect information concerning their memories of traditional food, and prepared a booklet, *Aep* ("What we eat"), which formed the basis for the CINE research when it started in 2005.

As part of the CINE project, the community leader, Mr Koichi Kaizawa, and the local researcher, Ms Kaizawa, as leading members of the Ainu Food Culture

---

Box 13.2
Tonoto

Tonoto is an indispensable offering to *kamuy* (spirits) at Ainu ceremonies, and is always prepared a few days in advance of the ceremony. In the old days, *piyapa* (barnyard millet) was the main ingredient for making *tonoto*, but recently rice has been used instead. Ainu people used to cultivate various kinds of millet, of which barnyard millet was the most common. People grew it to eat in daily meals, and also for making *tonoto*. However, as rice cultivation became more popular, people grew less millet. Nowadays it is not easy to reproduce the taste of traditional *tonoto*. *Piyapa* (barnyard millet), *sipsilep* (egg millet) or *munciro* (foxtail millet) is added to the rice to get the flavour of the *tonoto* of the past. *Ritensipuskep* (sticky egg millet) is usually used in making *tonoto*, because it is the easiest kind of millet to get.

To make *tonoto*, first mix equal amounts of rice and egg millet, and cook them with water to make rice porridge. Cool the porridge until it is about skin temperature. Put the rice porridge into a keg and mix in malted rice until it is smooth. Place a piece of hot charcoal on top of the mixture, asking *Apefuchi Kamuy* (a god of fire) to protect the *tonoto* during fermentation. This takes about two to three days in summer, and a week in winter. Once the *tonoto* is ready, the first drop is offered to the god of fire, with thanks, and then people enjoy it themselves.

Box 13.3
Two *sito* dishes

Traditional *sito*
Ingredients: Rice flour, egg millet flour, water.
- Mix the rice and egg millet flours with water and knead until the dough is as soft as an earlobe.
- Take a portion of this well-mixed, soft dough and roll it into a pancake shape. (The size of the pancake depends on the use of the *sito*. For example, large *sito* [20 cm in diameter] are for offerings, and smaller ones are eaten with other things by the participants of gatherings.)
- Boil the water, and put the dough into the boiling water. Cook until the dough rises to the surface of the boiling water, and wait for another five minutes. Make sure that the *sito* is cooked through in the middle and remove it from the boiling water.
- At the final stage, rinse the *sito* in cold water to get rid of excess flour.

Sito with minced deer meat
Ingredients: Rice flour, egg millet flour, water, deer meat, soy sauce, sugar.
- Chop the deer meat finely, cook it with sugar and soy sauce and set it aside.
- Make the sito dough following the *sito* recipe.
- Mix the cooked deer meat into the *sito* dough and spread the dough on to a flat surface.
- Take a deer-shaped cookie cutter and cut out the *sito* in the shape of deer.
- Put the deer-shaped *sito* into boiling water until they are cooked.
- Take them out of the water to cool.
- Brown both sides of the *sito* in a frying pan, and eat with salt and pepper.

Preservation Group, started the initiative of serving Ainu traditional dishes at ceremonial occasions. In August 2006, *Chipsanke*, the boat launching ceremony, was celebrated with *tonoto* prepared by the group, and gradually the group started to receive requests to prepare Ainu dishes for other local gatherings and ceremonies, and for gatherings in other towns. The group also held cooking lessons for Biratori residents and visitors. As a result, group members have become knowledgeable about Ainu foods and recipes, and are developing an understanding of the cultural aspects of food preparation. For example, preparation of *tonoto* requires an understanding of the role of kamuy (spirits), as guardians of the *tonoto* fermentation process. When

tonoto has fermented, women are responsible for straining it, and praying to the kamuy to thank them for protecting the *tonoto* during the fermentation is an indispensable part of the straining process. Members of the team making *tonoto* take turns to strain it and learn the prayer, which gives them an important introduction to the spiritual side of *tonoto* making.

Through the Ainu Food Culture Preservation Group's efforts, various Ainu rituals held in Saru River region now include the serving of *tonoto* and other Ainu dishes. Ainu foods have therefore become an indispensable part of local rituals.

Interventions outside the community

Ms Masami Iwasaki-Goodman, the author of this chapter, conducted an Ainu cooking project after she and her students worked as volunteers at the International Forum for Indigenous People in 2005, when they tasted Ainu dishes for the first time. The students found *sito* in its original form too filling and too high in calories compared with the snack dishes they are used to eating. They wanted to eat *sito* in smaller portions and with flavours that they like. Since then, the students have developed various *sito* dishes, including with minced deer meat, cheese filling, tomato sauce and various sweet sauces. Box 13.3 gives the original *sito* recipe and the students' recipe for deer *sito*.

Every autumn, during the university festival, the students set up a booth to promote their versions of *sito*, which they cook themselves and sell to friends and festival visitors. Over the three days of the festival, about 300 people buy *sito*. For most of these people, this is their first experience of Ainu food and culture.

The students' *sito* booth provided them with their first experience of introducing an Ainu food. They promote the *sito* enthusiastically, calling it "a tasty dish that you will like at the first bite". The students' positive attitude to making and selling *sito* generates positive feelings about Ainu culture and people; many students have become interested in learning more about Ainu culture, and some have chosen this as the topic for their graduation theses. The students' vociferous promotion of Ainu food at the university festival often provokes unexpected reactions from visitors, who express negative feelings towards Ainu people and their culture. This gives the students direct experience of the social prejudices confronting Ainu people, and an opportunity to think about ethnic issues in their society.

The CINE-Ainu research group is continuing with its interventions in and outside the Ainu community. Group members hope that many more people will learn about the food culture their Ainu ancestors fostered, thus changing attitudes towards Ainu people and their culture.

Cultural implications of intervention activities

K nowledge concerning food is a significant part of culture, similar to language and religion, which people need for their daily lives. People grow up learning the kinds of food and ways of processing them that are culturally acceptable in their community. Intervention activities therefore seek to change the cultural knowledge of people in the community. For the Ainu community in Saru River region the processes of change are complex and difficult to identify because of the multi-cultural nature of the community. However, it is necessary to try to understand these processes, to evaluate the effectiveness of the intervention activities. As cultural changes cannot be examined quantitatively, anthropological analysis is used to examine the change process and current situation in the Ainu community.

Intervention activities as resocialization

Children learn to eat what their family members eat, and gain understanding of various factors surrounding their food, including taste, value and preference, along with the historical and societal relationships of food within the community. Food habits are cultural, in that people eat according to rules that are shared by other members of their group. To understand the implications of the intervention activities that the CINE-Ainu research group conducted, it is necessary to examine the dynamics involved in the acquisition of food habits.

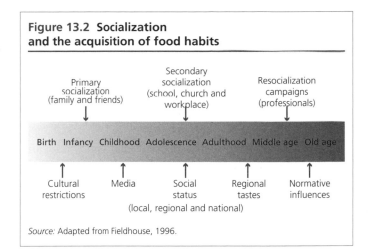

Figure 13.2 Socialization and the acquisition of food habits

Primary socialization (family and friends)

Secondary socialization (school, church and workplace)

Resocialization campaigns (professionals)

Birth Infancy Childhood Adolescence Adulthood Middle age Old age

Cultural restrictions

Media

Social status

Regional tastes

Normative influences

(local, regional and national)

Source: Adapted from Fieldhouse, 1996.

First, people gain knowledge of their food culture through socialization. Figure 13.2 illustrates the process of socialization and acquisition of food habits, based on work by Fieldhouse (1996).

From birth, children learn appropriate food habits through socializing with family and friends. They are fed with locally acceptable foods prepared in ways that are appropriate in the community. Along with food habits, children also learn the attitudes and values associated with food items and their preparation through interactions with family members and friends. Fieldhouse (1996) states that primary socialization is more influential than socialization at later stages in life, because the food habits that people acquire when they are young tend to persist throughout their lives. Individual food habits are established during childhood.

The food habits learned in early childhood undergo changes under the influence of the people who are met in later stages in life. Fieldhouse (1996) calls this secondary socialization, and states that individuals alter their food habits by learning from the people they meet in school and the workplace. Through secondary socialization, people are exposed to different ways of fulfilling their food needs; in some cases, they expand their repertoire of food habits and/or make drastic changes in the food they eat.

Resocialization can occur at any time of life, and is often the result of an organized effort to encourage people to adopt new food habits (Fieldhouse, 1996). Resocialization efforts usually take the form of educational campaigns and interventions, introducing various kinds of information concerning food habits, such as the nutritional values of food, and healthy eating habits.

Other important factors that influence people's food habits are cultural restrictions, the media, social status and regional tastes (Figure 13.2). These factors reflect the normative influences in a specific region; food habits are greatly determined by the social, cultural and historical repertoire of the regional food culture.

The processes of socialization and resocialization described by Fieldhouse refer to normal conditions where no drastic social change is occurring. People live their lives in fairly stable social conditions over generations, and normative influences usually arise at the local, regional and national levels. However, this is often not the case for indigenous communities. For example, for the Ainu, the assimilation policy of the Meiji Era (1868 to 1912) created social conditions in which Ainu food habits were viewed as being less desirable, and in some cases were rejected. To avoid being subject to social prejudice, Ainu adopted Japanese food habits. Although some households appreciated and maintained Ainu food habits, negative social influences generally drove these habits underground and integrated them into Japanese food habits. For example, the many kinds of wild vegetable that continued to be eaten in Ainu households were called by their Japanese names, and were no longer recognized as Ainu food. Some foods were replaced with others; for example deer meat was replaced with pork in many dishes, as the government restricted deer hunting, and pig farming became a common means of livelihood for Ainu people when hunting and fishing were no longer possible. Popular Ainu dishes prepared with deer meat were therefore no longer distinct from pork-based Japanese dishes.

A particularly complicated situation can be observed among middle-aged Ainu people in Saru River region, as described previously. During their primary socialization, these people acquired complex food habits, which were primarily Japanese with some Ainu food incorporated. One of the complexities of these food habits is that they make no reference to the cultural knowledge associated with Ainu food traditions. Ainu people in

this generation pick and eat wild vegetables for their daily meals, but are not familiar with the vegetables' Ainu names. In addition, they have not acquired cultural knowledge regarding the important prayers to and attitudes towards the spiritual beings involved in harvesting and processing certain foods. The normative influences during the adolescence of these people prevented Ainu food habits from being reinforced through secondary socialization. Instead, negative social conditions influenced attitudes to Ainu food habits, making people more willing to shift towards the Japanese way of life.

The younger generation of Ainu also underwent primary socialization without acquiring Ainu food habits. However, in the 1980s, one Ainu elder, Mr Shigeru Kayano, started to teach the Ainu language to small children from both Ainu and non-Ainu families in the community. He took the children into the forest to share his knowledge about trees, plants and aspects of Ainu culture, including the preparation of wild vegetables and other Ainu dishes. Unlike the previous generation, the young Ainu who attended Mr Kayano's classes were exposed to Ainu food habits and resocialized with these (Anderson and Iwasaki-Goodman, 2001). This is consistent with Fieldhouse's explanation that resocialization can occur at any time of life, although he himself focuses on resocialization in middle and old age (Fieldhouse, 1996). Mr Kayano's efforts were especially effective and helped to trigger the revitalization of Ainu culture throughout Hokkaido.

The CINE-Ainu research group results (Iwasaki-Goodman, Ishii and Kaizawa, 2009), together with the intervention activities described in this chapter demonstrate that effective intervention activities in the Ainu community of Saru River region need to be based on resocialization that creates not only changes in individual households, schools and workplaces, but also normative influences at the local, regional and national levels. It is clear that such changes involve people outside the Ainu community even more than people within it. By reintroducing Ainu food culture, the intervention therefore aims to re-establish Ainu people and culture in Japanese society, creating changes at multiple levels, such as households, schools, communities and the nation.

Cultural implications of "good food" and "bad food"

The Ainu people interviewed for the research reported that their attitudes and feelings towards Ainu food and dishes had changed at various points in their lives. Ainu elders recalled the old days when they had felt embarrassed about eating certain Ainu foods, because they were considered tasteless and were less favoured than Japanese foods. For example, the various kinds of millet that Ainu people grew in their backyards as an important part of their daily diet were regarded as being less tasty than the white rice that was part of Japanese food habits, even though millet is nutritionally rich. As Ainu people began to be employed and earn money, they began to buy white rice. Elders recall those days and say "White rice was tasty, and we bought rice when we had money". Eventually, they shifted from growing and eating millet to buying rice for daily meals.

The CINE-Ainu research group was publicizing the nutritional value of millet at the same time as a nationwide healthy food campaign was highlighting millet as an ideal food because of its high fibre content. Scientific knowledge regarding millet was therefore disseminated to both local people in the project area and the general public nationwide. In Saru River region, both Ainu and non-Ainu people have changed their attitudes to millet. One elderly woman informant stated "Strange that my daughters who live in Sapporo now want me to grow millet. I hear that eating millet is healthier than eating rice". This change in attitude extends beyond the food itself, as local people are starting to appreciate the cultural knowledge their Ainu ancestors fostered.

Pukusa (wild onion/garlic) is a symbolic Ainu food that has undergone drastic change from being a "bad" to becoming a "good" food. Most of the Ainu people interviewed by the CINE-Ainu research group reported that they avoided eating *pukusa* because its strong flavour made it a target for social discrimination. However, attitudes changed when recent scientific research revealed the health value of *pukusa*. Initial changes occurred at the national level, with books on *pukusa* and its health value being published and

companies starting to manufacture various health products using *pukusa*. These positive changes eventually reached the local level, and Ainu people in Saru River region began to appreciate *pukusa* again. Every spring, they now go into the forest to harvest it, and cook it for meals. Ainu people do not hesitate to say that they like *pukusa*, and they look forward to the season when they can harvest and enjoy it. They have changed from avoiding *pukusa* to claiming it as one of their foods. The Ainu name (*pukusa*) is now used alongside the Japanese name (*gyoujya ninniku*) and the common nickname (*kitopiro*). A restaurant in Nibutani, Saru River region serves a special noodle dish, *kitopiro ramen* (ramen noodles topped with *pukusa*), which is very popular among local Ainu people and tourists.

The feelings and attitudes associated with food are affected by the experiences of the people who prepare and consume it. When these experiences are positive, the feelings and attitudes are good, while people tend to have negative feelings towards a food if their experience of dealing with it is negative. The increasing popularity of Ainu cooking classes in the community indicates positive changes in people's attitude towards Ainu food culture. In one cooking class, members of the Ainu Food Culture Preservation Group assisted 16 young non-Ainu home-makers while they cooked Ainu dishes. The conversation between the two groups created a friendly atmosphere, and the good experience of the class led to positive changes in the community.

Analysis of changes in food habits shows that they occur through communal experiences. To recognize a food as good or tasty, people need to have shared a similar experience of it, and to have agreed that it has a favourable taste. Lupton (1996) defines taste as "an aesthetic and a moral category", in that "good taste ... is acquired through acculturation into a certain subculture". People recognize a food as good or tasty because it is aesthetically and morally acceptable for their group. Ainu food is accepted as good when the people in the community accept it and incorporate it into their food habits. Such changes are occurring in the Ainu community in Saru River region. One interesting comment was "Now local people can tell the difference between tasty *tonoto* and not-so-tasty *tonoto*. When the *tonoto* is successfully made and tastes good, everyone wants to take home the leftovers from the ceremony. But if it is not so good, no-one wants to take it home". Local people have acquired a taste for *tonoto* and have developed a communally shared way of evaluating its flavour.

Conclusion

The CINE-Ainu research group decided that the project's goal was to improve the socio-cultural health of Ainu people in Saru River region, to create a community where Ainu people are proud of their ethnic background. Four kinds of intervention activity were planned and carried out. The intended cultural changes are a slow and complex process, but positive changes are occurring in the community. The local researcher, Ms Miwako Kaizawa, often says that the ultimate goal is to have Ainu dishes served for lunch in local schools, indicating that Ainu food culture has been integrated into the local food culture. This would demonstrate that the intervention has been effective in creating normative changes. The CINE-Ainu research group is confident that the day is near when this will be the case ✻

Acknowledgements
The author thanks the community leader, Mr Koichi Kaizawa, and the local CINE-Ainu researcher, Ms Miwako Kaizawa, for their contributions to the research and intervention activities, and to the writing of this paper.
> **Comments to:** iwasaki@jin.hokkai-s-u.ac.jp

>> Photographic section **p. XXVIII**

Annex 13.1 Articles in *Ianpero*, the community newsletter (April 2005 to May 2008)

Apr. 2005	*Pukusa, Allium victorialis* var. *platyphyllum* (wild onion/garlic)
May 2005	*Pukusakina, Anemone flaccida* (anemone)
June 2005	*Cimakina, Aralia cordata* (*udo*, spikenard)
July 2005	*Turep, Lilium cordatum* var. *glehnii* (perennial lily)
Aug. 2005	Traditional dishes of Indigenous Peoples around the world (Report of the Nibutani Forum, 2005)
Sept. 2005	*Amam* (grains)
Oct. 2005	Aha, *Amphicarpa bracteata Edgeworthii* var. *japonica* (aha bean)
Nov. 2005	*Kosayo* (porridge with beans)
Dec. 2005	*Citatap* (chopped pork brain with wild onion)
Jan. 2006	*Sito* (dumplings)
Feb. 2006	*Yuk ohaw* (deer soup)
Mar. 2006	*Peneemo* (potato pancakes using frozen potatoes)
Apr. 2006	*Makayo* (a butterbur scape)
May 2006	*Sorma, Matteuccia struthiopteris* (ostrich fern, fiddle head fern)
June 2006	*Korkoni, Petasites japonicus* (Japanese butterbur, coltsfoot)
July 2006	*Nupe* (*veratrum*)
Aug. 2006	Summer fruits: *emauri, turepni* (wild berries)
Sept. 2006	*Chitatapu* (chopped salmon milt with wild onion)
Oct. 2006	Autumn fruits: *matatanpu* (*Actinidia polygama*) and others
Nov. 2006	*Kikuimo*
Dec. 2006	*Ciporusi emo* (potato mixed with salmon roe)
Jan. 2007	*Tonoto* (beverage)
Feb. 2007	Ento, *Elsholtzia ciliata*
Mar. 2007	Haykina, *Urtica platyphylla*
Apr. 2007	*Sito* (adapted *sito* dishes)
May 2007	*Pukusa*, Part 2
June 2007	*Turep*, Part 2
July 2007	*Noya* (mugwort)
Aug. 2007	*Kene* (alder)
Sept. 2007	*Yukkarus* (maitake)
Oct. 2007	*Atane* (turnip)
Nov. 2007	*Sakkabocya* (dried pumpkin), *ratashikepu* (cooked pumpkin)
Dec. 2007	*Huipe* (raw food, liver)
Jan. 2008	*Kankan* (intestines)
Feb. 2008	*Cepohaw* (fish soup)
Mar. 2008	*Nitope* (tree sap)
Apr. 2008	Pickled *pukusa* (pickled wild onion)
May 2008	*Pukusakina* mixed with sesame, vinegared *sorma*

Future directions

What food system intervention strategies and evaluation indicators are successful with Indigenous Peoples?

⚜ HARRIET V. KUHNLEIN[1]

1
Centre for Indigenous
Peoples' Nutrition and
Environment (CINE) and
School of Dietetics and
Human Nutrition, McGill
University, Montreal,
Quebec, Canada

Key words > Indigenous Peoples,
food systems, nutrition intervention, intervention
indicators, intervention evaluation

> "Our elders are our knowledge keepers, who pass it on to the children for their future."
>
> Looee Okalik, Inuk community leader

> "Now our communities are hunger-free and also healthy."
>
> Salome Yesudas, Nutritionist

> "Let's go local, let's stay local."
>
> Song from the Island Food Community of Pohnpei

> "Before, we had to walk very far to gather our traditional vegetables. Now it is no longer necessary; our villagers have grown them. Some people have grown five vegetables; some have grown ten."
>
> Anon Setapan, Karen leader

> "Our forest is our market, our pharmacy. It's everything!"
>
> Irma Tuesta, Awajún leader

Abstract

For decades, food and diet-based intervention strategies have been promoted as a way of solving problems of food insecurity and malnutrition; however, documentation of these strategies' successes has been limited. The success and sustainability of food-based interventions at the community level involves a complex web of issues, including resource allocation, capacity building and behaviour change to improve food availability and use. There is an obvious need to evaluate food-based interventions in ways that capture both these issues and the influence of other contextual factors, but such evaluation is difficult to implement consistently. The Indigenous Peoples' Food Systems for Health Program of the Centre for Indigenous Peoples' Nutrition and Environment has addressed food and diet-based interventions and their evaluation in nine cultures of Indigenous Peoples living in vastly different rural ecosystems in several countries. The issues are surprisingly similar across case studies – financial poverty, disadvantage, discrimination and challenges to well-being. Academic and community partners share a common commitment to improving Indigenous Peoples' health by using local food systems based on culture and nature to the best advantage. Common intervention themes include working with youth, sharing information on the diversity and wholesomeness of communities' local foods, engaging with government offices, empowerment and capacity building, and networking and media activities. The best evaluation indicators are those that are based on improving local food availability and use and that include interdisciplinary partners, both qualitative and quantitative methods in diet and anthropometric evaluations, and reliance on internal community advice about the progress of projects. The programme's ultimate goal was to document successful food-based strategies for protecting the health status of Indigenous Peoples by using their own local food systems.

Background and rationale

It is well recognized that today's globalization of food affects economies and people at all socio-economic levels, in all countries. The forces that drive food globalization are industrialized production and trade of agricultural crops, foreign investment in food processing and retailing, and global advertising (Hawkes, 2006). Globalization leads to the erosion of the food cultures of all people and the simultaneous reduction of food biodiversity, which results from the food industry's demand for standardized and uniform products being passed on to consumers, especially in supermarkets and fast-food chains. Combined with changing lifestyles, loss of livelihoods at all stages in the food production sector, and increasing urbanization and poverty, sedentary lifestyles and new dietary patterns are driving increases in obesity and non-communicable diseases, including micronutrient deficiencies. With unresolved food and nutrition insecurity issues and high incidence of non-communicable diseases, this nutrition transition and the resulting "double burden of malnutrition" are reaching all corners of the globe (Kennedy, Nantel and Shetty, 2004; 2006).

In developed countries, disparities in obesity prevalence are related to the availability and quality of retail food environments, with more overweight and obesity occurring in socio-economically disadvantaged areas, often where there are particular ethnicities. These areas often have poor access to healthy foods and low-quality retail food environments (Ford and Dzewaltowski, 2008). In developing countries and in many areas where Indigenous Peoples live, the retail foods available and purchased have low nutrient densities and lead to poor-quality diets (Kuhnlein *et al.*, 2006a; Chapter 1 in this volume – Kuhnlein and Burlingame, 2013). Many nutrient-rich local indigenous foods still produced at the subsistence level may not be marketed or may be marketed at higher prices than processed imported foods. Direct access to these foods from local ecosystems is often hampered by the environmental damage caused by industrial interests and urbanization (Chapter 3 – Turner, Plotkin and Kuhnlein, 2013).

Broadly speaking, to improve the quality of food available in poor communities, including those of Indigenous Peoples and in both developed and developing countries, it makes sense to: i) improve access to traditional local foods; ii) reduce the prices of healthy foods, with government enforcement; iii) increase education in and demand for "good" food, and understanding of the impact of poor diet on health; and iv) increase the prices of "junk food" (nutrient-poor, high-energy, sweet or fatty foods and drinks), perhaps through increased taxes. Points i), ii) and iv) are usually difficult to implement and require government involvement, whereas point iii) can be supported through community nutrition networks. All four measures will speed up success in food systems predominated by industrialized food markets. Indigenous communities' attention to their food systems may be increased if interventions are based on communities' own knowledge bases, especially if there is limited use of market food and ensured access to traditional food.

Nutrition transition

The contemporary global food security crisis – which includes simultaneous and paradoxical obesity and undernutrition from poor-quality diets – is a public health crisis driven by inequitable access to food through international systems of governance (Traill, 2006; Yngve *et al.*, 2009). This calls for consideration of Indigenous Peoples' food systems, attention to the poor quality of Westernized[1] diets, and action to promote sustainable food systems within the unique local cultures and ecosystems in which Indigenous Peoples live.

Recent moves to disaggregate national health data by ethnicity are uncovering aspects of the nutrition transition that are specific to Indigenous Peoples. For example, research on a large cohort of Inuit children in Nuuk, Greenland, demonstrates that more children are reaching overweight status and at younger ages than in earlier cohorts (1970s versus 1990s) (Schnohr, Petersen and Niclasen, 2008). In a very different part of the world, the socio-economic

---

[1] Derived from or relating to the countries of Europe and North America.

status and obesity levels (measured as percentage of the population obese and as mean body mass index [BMI]) of Surui Indians in Brazil were found to have increased between the 1980s and 2008 – this rapid nutrition transition occurred in the midst of poverty (despite increased income), food insecurity, indicators of poor growth (stunting) and precarious living conditions (Lourenço *et al.*, 2008).

Disparities

Indigenous Peoples face widespread disparities with the larger societies in which they live, in all countries where threats to land, culture and linguistic heritage destabilize identity and self-determination. These circumstances generate health challenges that are specific to Indigenous Peoples (King, Smith and Gracey, 2009; Gracey and King, 2009). In Canada, Indigenous Peoples experience higher rates than the overall national averages for diabetes, heart disease, HIV/AIDS, tuberculosis and many other diseases. Infant mortality rates are higher and life expectancy is lower than in the general population. Suicide rates are two to three times higher overall, and five to six times higher among youth (Reading, 2009; Egeland and Harrison, 2013). Similar and worse circumstances driven by poor access to traditional resources and culture exist in other countries (UNPFII, 2009).

Social practice is recognized as a driving force in dietary strategies (Delormier, Frohlich and Potvin, 2009), so there is need for new, innovative and transformative intervention strategies and policies that use traditional food systems to make long-term improvements in the health of Indigenous Peoples. This is especially so for Indigenous Peoples who recognize that health encompasses physical, mental, social and spiritual dimensions and regard food as affecting all of these dimensions. To address these multiple dimensions, health promotion strategies must therefore be carefully incorporated into the five stages of behaviour change – pre-contemplation, contemplation, preparation, action, and maintenance (Medical College of Georgia, 2005).

Nutrition interventions that stress obesity control and healthy body weight cannot be separated from interventions that support increased physical activity. A study of the physical activity patterns of a large cohort of American Indian and Alaskan Native Peoples concluded that many activities are undertaken within communities, including traditional food harvesting. However half of the respondents (49 percent) reported no vigorous activity at all, and only 23 percent reported up to 30 minutes of moderate to vigorous activity a day. Women had lower activity levels than men; those with vigorous activity had better clinical health characteristics (BMI, blood lipids, etc.) (Redwood *et al.*, 2009). Successful physical activity interventions with American Indians and Alaskan Natives in the United States of America showed significant changes in health knowledge or behaviour, with sustainability being related to capacity building involving locally trained personnel, culturally acceptable programmes, local leadership and stable funding (Teufel-Shone *et al.*, 2009).

There is considerable literature on how to use schools to deliver nutrition education information and improve physical activity. Strategies can be sought from focus groups of teachers, parents and students (Schetzina *et al.*, 2009) examining how to improve school lunch and physical activities. The Pathways Program for American Indian schoolchildren was well constructed for addressing obesity prevention, dietary improvement and the sharing of health messages with families (Caballero *et al.*, 2003; Teufel *et al.*, 1999), and the Native Diabetes Wellness Program of the United States Centers for Disease Control and Prevention (CDC) published the Eagle books for diabetes prevention (CDC, 2009). Obesity prevention in Canadian schoolchildren was researched through an intervention implementation and evaluation programme in Kahnawake, Quebec (Cargo *et al.*, 2003; Jimenez *et al.*, 2003).

Food and nutrition interventions

For decades, national and international health agencies have been promoting food and diet-based approaches to solving nutrition problems, along with nutrient supplementation and food fortification. The food-based approach to alleviating micronutrient deficiencies has received special attention as a relatively easy-to-deliver and cost-effective strategy. However, in the absence of adequate food (energy), supplying micronutrients to improve nutrition status is obviously futile. It has also long been known that nutrition education cannot compensate for poor access to food, and activities for improving access must be integrated into other elements of programmes (Kennedy and Pinstrup-Andersen, 1983). Local food-based strategies are therefore important in contributing food with all the needed nutrients, and they can also provide livelihoods and income (FAO and International Life Sciences Institute, 1997). Community-level production of small animals, poultry and fish can provide necessary animal foods with absorbable micronutrients, as complementary foods for infants and young children. Sustainable food programmes use a combination of approaches directed at increasing the production of animal and plant micronutrient-rich foods, promoting nutrition education and guiding food selection, to increase diet quality through reduced consumption of less nutritious carbohydrates and fats. This helps to prevent both undernutrition, and overweight and obesity (Allen and Gillespie, 2001; Allen, 2008).

Community-based food and nutrition programmes that promote healthy diets have been analysed and summarized in detail to identify the factors that drive success (Gillespie, Mason and Martorell, 1996; FAO, 2003a; 2003b). It is recognized that there is no single way for programmes to succeed, because of cultural diversity and complexity and multiple perceptions of "the good life" and of what constitutes desirable cultural change (Messer, 1999). Review of the contextual factors that are important for programme success suggests that the most important factor is the existence of a nutrition-friendly policy environment at all levels of society in the country (Gillespie, Mason and Martorell, 1996). Infrastructure that supports community nutrition interventions includes charismatic leadership with decision-making skills, particularly among women, and the existence of poverty reduction programmes. The presence of community organizations, non-governmental organizations (NGOs) and infrastructure for basic services is also important, as is a culture that recognizes the need to support child development. Given

the many complexities and local and global contextual factors that affect the food systems of Indigenous Peoples – such as lack of political commitment at the national level, pollution of lands and waters, climate change, and discrimination against women – it is not surprising that there is no single template for improving food security for Indigenous Peoples by ensuring that food resources are available, accessible, acceptable and sustainable. In the context of these pillars of food security, successful programmes for improving health also depend on the application of key components of individual and/or community behaviour change: changing of knowledge, action, attitudes/values and behaviour (Boyle and Holben, 2006).

Climate change, with impacts on many traditional food systems of Indigenous Peoples, is an interfering external contextual factor (Nilsson, 2008). Within the United Nations, considerable attention is paid to global actions for reducing the pollution emissions that drive climate change, with consideration of Indigenous Peoples' issues playing a key role (Damman, 2010). Recently, reports of climate change affecting traditional food have been made for Arctic Peoples (Nuttal, 2008), African pastoralists (Simel, 2008) and small island states in the Pacific (Smallacombe, 2008). The need for resiliency to changes in ecosystems resulting from climate instability or other ecosystem threats focuses Indigenous Peoples' attention squarely on their own management of biodiverse resources to maintain community health.

When nutrition interventions have positive results, their scaling up to other communities and regions is an additional mark of success. Methods for understanding and measuring scale-up can be: i) quantitative, through increasing numbers of participating people/places; ii) functional, through increasing numbers/types of activities within a community; iii) political, with activities moving beyond service delivery to include empowerment and increasing relations with others in the state or region; or iv) organizational, through enhancing an organization's strength to increase the intervention's effectiveness, efficiency and sustainability, such as by improving its finances and income, or introducing appropriate legislation (Uvin, 1999). Scaling up can also be measured with qualitative data that demonstrate improving programme effectiveness in empowering communities to collect data and decide what to do themselves – showing that the programme is "scaling down the top-down management" and using resources generated locally (Messer, 1999).

While the whole world needs improved nutrition with appropriate interventions, Indigenous Peoples are routinely at the bottom of the poverty ladder and especially vulnerable to malnutrition. In many regions there is an overriding assault on "indigeneity" and demands from multinational industries to profit from Indigenous Peoples' land and water resources (oil, gas, minerals, lumber, etc.), ignoring the unique relationships and interlinkages Indigenous Peoples have with their ecosystems, cultures and health, as well as their priorities for the protection and sustainable use of resources. The United Nations actively promotes the human rights to food security and food sovereignty for Indigenous Peoples (FAO, 2009a; 2009b; 2009c). However, successful interventions that provide the necessary resources, promote Indigenous Peoples' priorities, create the social conditions for self-determination and well-being, and develop capacity to build better nutrition and health within communities have still to be identified. The CINE programme is a contribution to this effort.

This chapter revisits the food, nutrition and health interventions described in other chapters. These were conducted by communities of Indigenous Peoples in nine case studies included in the CINE programme. The authors of each chapter share stories, lessons and commonalities that may help identify successful strategies and policies for initiation in other regions.

Basic concepts for creating interventions with communities of Indigenous Peoples

It has long been known that nutrition interventions can be efficacious, but the crucial issue is whether or not they are effective and sustainable in the long run (Gillespie, Mason and Martorell, 1996). Interventions involving education and medical health tend to have the lowest costs, while those that involve feeding or providing food resources to populations can be very expensive (ACC/SCN, 1991). Over the years much has been written about recommended procedures for producing and sustaining food and nutrition programmes (e.g., Caribbean Food and Nutrition Institute and Ministry of Health, 1985; FAO, 1997; 2003a). In the context of Indigenous Peoples, the CINE research team found that it makes a great deal of sense to build community infrastructure so as to make the best use of local food resources, and to protect ecosystems so they can provide sustainably for communities within the local cultural context.

Even in the most favourable conditions, community-based health promotion interventions are not easy to implement and complete successfully. Garnering resources, developing local community members' capacity to conduct intervention activities for behaviour change, evaluating these activities and sustaining them pose immense challenges. Creating food-based community interventions to improve nutrition and health adds another layer of complexity at each of these steps, because getting the desired nutritious food into the community, and creating the formative events that ensure people have access to it and can enjoy and use it effectively and sustainably to improve health with measurable outcomes stretches the research agenda to almost impossible levels. The health promotion effort becomes even more challenging when it depends on creating a successful food-based intervention for Indigenous Peoples living in areas with difficult access, in circumstances of poverty and disadvantage and with multiple nutrition and health issues.

Begin at the beginning – know the resources

The CINE programme research partners recognized that the rural areas where the case study communities are located have the potential for making better use of local food resources to improve food and nutrition security. The partners began by realizing that many people did not know about their local resources, or at least not all of the resources known to the most knowledgeable community residents – the elders. For case study projects that are grounded in knowledge and use of local food it was therefore necessary to ensure community-based knowledge sharing and documentation of local food systems in their full diversity. It was important to understand what people could harvest and purchase locally, what made foods acceptable to different age and gender groups in the community, and what health-giving potential there was for each of these foods. The first step in the overall CINE programme was to document scientific terminology and laboratory-based food compositions, under expert guidance. The methodology used to create this knowledge base is presented by Kuhnlein *et al.* (2006a).

Earlier work with partners demonstrated that a wide variety of intervention approaches are feasible in indigenous communities, and that many indicators can be used to measure the effectiveness of each (Kuhnlein *et al.*, 2006b). However, each community and culture has its own knowledge of benefits for physical, mental, emotional and spiritual health. Each has its own symbolic values for food, its own ways of using these to create identity and culture, and its own holistic approach to well-being (including physical activity, sharing, and community spirit in joint harvesting activities and celebrations). Tapping this knowledge for a health promotion effort requires building a knowledge-sharing consortium in the community. The research team had the benefit of strong community leadership to guide this component in each of the case studies. Applying and disseminating this knowledge in the community, and developing and evaluating health promotion activities depended on building trust and commitment among the community residents, their leaders and the research partners. This required substantial time, and was implemented in unique ways in each of the cultures, with their vastly different ecosystems, local social settings and priorities.

For example, in the Karen case study (Chapter 10 – Sirisai *et al.*, 2013)[2] involving a community of 600 people in western Thai land, the programme explored and documented knowledge of more than 380 different food species/varieties in the rain forest jungle area. The priority was to develop better community experience of how children's health can be fostered using the Karen's own gardening and forestry techniques. Through school and community programmes, children were taught how to grow and cook healthy foods. One outstanding result was that within three years, the community had increased its production of local vegetables and fruits from 81 to 137 different food species, thereby increasing the diversity and variety of family meals, resulting in improved nutrition status (stature) for children.

Common themes for interventions

Programmes must reflect communities' own world views, allowing them to accomplish goals that are within their own priorities and to recognize the values within their own environments. The CINE researchers found that interventions based on local cultural knowledge and focusing on children's health, and food for women and children have universal appeal. Communities welcome scientific knowledge about the physical benefits of good nutrition and health when it is presented sensitively and with meaningful examples, while communities' own understanding and knowledge of food environments offer unique perspectives and indicators for recognizing progress and change in health.

In all of the case studies, it became clear that Indigenous Peoples were keen to document their food systems for the benefit of future generations. However, there was an overarching sense that assessing these food systems was not enough; more had to be done with the information before it could help lead to better lives. The project had to provide not only a research environment and research subjects, but also opportunities for participants and leaders in the community to make decisions and use the information wisely for the community's benefit. Creating research agreements and/or mutual understandings with all those involved helped build the trust that this work required. Capacity building developed the concept of being empowered to express indigenous ways of knowing and indigenous community

---

[2] For the case studies with Indigenous Peoples described in this volume, chapter numbers and authors' names are given only on the first mention. The section on pp. 249 to 255 presents the case studies in alphabetical order of the Indigenous Peoples concerned, and repeats the chapter number and author name(s) for reference.

priorities. As documented with Indigenous Peoples in the United States of America, it was important to create community self-determination that worked from the ground up and was not imposed from the top down (Chino and DeBruyn, 2006).

Partnerships and capacity building are crucial

Canada's ground-breaking public policy of creating guidelines for health research involving aboriginal peoples (Canadian Institutes of Health Research, 2007) and a research partnership methodology involving collective and individual consent were the foundations for community discussions about the projects described in this book. Discussions were held on benefit sharing, protection of cultural knowledge, and joint decisions on the collection, use, storage and secondary use of data and biological samples (when available) (Reading, 2009; Canadian Institutes of Health Research, 2010). Returning results to the community in an understandable format is the key to ensuring communication and trust between outside researchers and the community, while sharing results with the broader research community and the public is important in "translating" the knowledge generated from work carried out in case study communities. CINE's research partners continue to be energized by their work's promotion at the United Nations level, in discussions of the importance and benefits of Indigenous Peoples' food systems, and in publications where results are shared with the world at large (CINE, 2010).

The CINE researchers were constantly aware of the profound disparities that affect indigenous women in particular, and understood that the impact of these disparities must not be underestimated. Indigenous women often face high levels of discrimination, both within their own communities and in external rural and urban areas. Gender disparities may exist in literacy and education levels, access to family planning and other health services, and birth weights, as well as through maternal mortality. Microcredit may be more frequently denied to women than to men. All of these factors can affect capabilities for improving nutrition (Gillespie, Mason and Martorell, 1996). As food and nutrition programmes invariably involve women's participation, the project researchers intended to include and empower women as much as possible, and to provide tools through which women could be the knowledge bearers in social marketing within communities and with community leaders.

Evaluations and sharing of intervention results

There is a huge diversity of ways of measuring and evaluating the success of intervention programmes in indigenous communities (Kuhnlein et al., 2006b). Measurements of knowledge, attitudes and behaviour can be used to identify change in food and diet, and there is compelling logic for measuring food use to evaluate food-based interventions. However, it is also important to assess changes in context, to evaluate processes, and to recognize the potential for measuring change through biological indicators. The many possible confounding, extraneous and interfering effects must be considered. Evaluations must look at the strengths/ diversity of the advocacy efforts managed within the programme. Most important, evaluations must recognize how Indigenous Peoples themselves view improvements through their own lens of understanding and experience, and how this makes self-determination and community development possible. Consideration must be given to the inclusion of at least some of the cultural indicators for food security developed with the United Nations Permanent Forum on Indigenous Issues (UNPFII) and the Convention on Biological (CBD) Diversity Working Group on Article 8(j) (Stankovitch, 2008; FAO, 2008). These include percentages of traditional food or food-related items used in ceremonies, and percentages of households using traditional/subsistence foods regularly.

At the international level, it is widely recognized that evaluations of health promotion programmes can be formative or summative. There can be: i) context evaluation; ii) input evaluation, such as of the adequacy and appropriateness of resources; iii) process evaluation; and/or iv) outcome or impact evaluation. As experienced in the interventions described in this book, it is difficult to distinguish gross outcomes (all the changes taking place in the community during the period) from net changes (only those changes resulting from education or other community intervention activities). In any community, net effects can be influenced or masked by extraneous confounding effects and secular trends, such as increasing income from new employment opportunities in the community. There can be interfering events (climate change, food prices, natural disasters, etc.), design effects and stochastic effects (chance fluctuations). Attention needs to be given to the reliability of measurements, the bias or lack of internal validity that may result from selection bias, information bias, or even the placebo effect of simply having a new activity in the community (Oshaug, 1997).

In evaluations of health programmes in communities of Indigenous Peoples, it is not always possible to survey and analyse populations that are large enough to provide valid findings from statistical methods using quantifiable indicators of health, welfare or environmental improvements. This often limits the researcher to using only descriptive statistics (LaFrance, 2004) and creates the need to use qualitative methodologies that offer alternative ways of determining success, which may be more efficacious. Qualitative methodologies are appropriate for efforts to capture indicators of participatory methodologies, empowerment, community solidarity, and use of culture and traditional foods (Tauli-Corpuz and Tapang, 2006).

Despite the daunting challenges in evaluating interventions, research partners were convinced that their projects in reasonably small community settings were successful, and that evaluations provided meaningful demonstrations of impact measured in many different ways.

Successful strategies and particular challenges

Inter-project communications
Communication among the different case study research partners, community leaders and academic leaders was very important to the successes experienced. It provided rich discussion and cross-fertilization of ideas among interventions, and strategies for implementing and evaluating activities. Annual meetings of research partners created new friendships and helped maintain the interventions' vitality. The CINE research team constantly learned the need for patience and careful listening. By keeping intervention leaders abreast of developments, ensuring opportunities for them to discuss their successes and challenges with colleagues, and providing incentives for revising and bolstering interventions where needed, these meetings were an important part of empowering communities and their leaders to see interventions through. The communities involved were proud that their leaders had been invited to these international meetings.

Understand the basics
It was very important for the food system projects to begin with basic knowledge about the food resources: the species/varieties, their scientific properties and identification, and their use patterns within communities. Interdisciplinary experts such as nutritionists, food scientists, anthropologists, ethnobiologists and public health professionals contributed as needed. This grassroots knowledge made it easy to discuss food resources and ways of maximizing their availability. For example, the Pohnpei State Department of Agriculture of the Office of Economic Affairs, Department of Land and Natural Resources and the College of Micronesia-FSM/Cooperative Extension Service collaborated to provide seeds and seedlings of important traditional species and cultivars, for planting in home areas (Chapter 12 – Englberger et al., 2013; 2010). Karen case study leaders worked with community elders and knowledgeable leaders to teach schoolchildren how to grow many of the species in their food system.

Building pride in the cultural food system and the project
Pride in the local food and culture created enthusiasm and momentum for community leaders and assistants to continue activities. Sharing success stories and showcasing special events and foods within the community and at meetings also provided impetus. Stories were routinely shared with local media, through newsletters, e-mail networks, radio and television, promotional

films, and school class and parent discussions, as available. The Ainu case study issued a weekly newsletter for sharing recipes and stories based on traditional foods (Chapter 13 – Iwasaki-Goodman, 2013); the project succeeded in giving Ainu people pride in and recognition of the particular flavours in their unique traditional preparations. The Nuxalk case study describes the development of user-friendly community food system handbooks presenting photos, names in the local language, identifications, harvest areas and strategies, and recipes (Chapter 11 – Turner *et al.*, 2013). Locally produced books can be expanded and revised periodically, reprinted and shared as a resource for adults and children. The book created for the Awajún case study was deposited in the national library of Peru (Chapter 5 – Creed-Kanashiro *et al.*, 2013). Several case study teams made posters of their foods for display in local community halls and schools.

Focus on children and youth

All the case studies included interventions for children and youth, who represent the future of food system knowledge as it is passed to following generations. The case studies include many examples of food use in child care, and of teaching children to be self-sufficient in food, often using knowledge from elders. For example, elders in the Gwich'in area demonstrated the preparation of dried caribou meat to groups of youth in school. The youth then prepared several boxes of meat for distribution to home-bound elders (Chapter 7 – Kuhnlein *et al.*, 2013). In the Pangnirtung example, elders' stories were recorded on video for sharing with youth in school classes, to stimulate discussions about food, health and the impact of climate change on the availability of local foods (Chapter 9 – Egeland *et al.*, 2013).

Commitment, capacity building and empowerment

Project leaders' commitment and capacity are essential for the success of interventions in communities of Indigenous Peoples. How and by whom activities are organized and carried out are of crucial importance, and there must be genuine community ownership,

for example, involving women's organizations and leaders. People are not just the beneficiaries of programmes; they make programmes happen. Community ownership is developed through the involvement of community partners in programme planning, initiation, organization and implementation. Community leaders recognize what seems reasonable and is operationally realistic for the community. Community nutrition programmes in developing areas are more likely to succeed when intervention activities result from the persistence and persuasiveness of project advocates rather than precisely documented needs (Berg and Muscat, 1971; Gillespie, Mason and Martorell, 1996). Each of the chapters in this book gives evidence of local leadership mobilization and commitment and community empowerment, especially the case studies with the Nuxalk, Awajún and Dalit (Chapter 6 – Salomeyesudas *et al.*, 2013).

Multidisciplinary stakeholders

In all of the case studies, once the basic documentation had been completed and shared with the community, the project team took time to gather and discuss new ideas and strategies from the community. Successful projects have long lists of stakeholders, including intersectoral and multidisciplinary partners, NGOs and government sectors (for capacity building), local research assistants and community volunteer networks. For example, the Inga case study team worked with local schools and health promoters, had immeasurable support from the NGO, Amazon Conservation Team, and built government support for the protection of lands for indigenous food and medicines in Colombia (Chapter 8 – Caicedo and Chaparro, 2013). The Pohnpei project drew on strong support from state, national and other agencies, including those for education, land and natural resources, health and agriculture – particularly the extension service provided through the College of Micronesia-FSM and the United States Department of Agriculture (USDA) Natural Resources Conservation Services Program. All the case studies worked with local professionals to enhance small-scale homestead and community farming of crops and livestock, emphasizing methods that were culturally appropriate and suitable for the local ecosystem.

Rather than basing intervention design on evaluation results, all the case studies adopted multifaceted approaches, which some call the "bottomless pit" of using anything that comes to mind to create awareness and behavioural change to improve food system use. This often brings unexpected rewards, with new ideas and resources coming from unanticipated sources. For example, the extensive communication networks built by the Island Food Community of Pohnpei led to requests for similar initiatives in the rest of the Federated States of Micronesia and on other Pacific islands. The Island Food Community of Pohnpei also worked with the Federated States of Micronesia Philatelic Bureau to develop two series of postage stamps depicting local foods.

Intra-project communications and dialogue to build confidence and trust

Effective project management within communities must meet time-bound goals. In the CINE programme, community leaders played major roles in management, as academic leaders were often distant from the community region (Awajún, Gwich'in, Inuit, Inga). Frequent meetings between academic and community leaders stimulated mutual trust and effective management (Dalit, Nuxalk, Karen, Pohnpei, Ainu). Community steering committees played supportive roles, boosting the confidence of community leaders and assistants, and providing advice and direction for new activities. Community members themselves are the most effective in delivering new information and integrating it to reinforce community knowledge.

Scaling up beyond the initial project communities

Requests for the scaling up of intervention activities can be considered the gold standard of success. The initial Nuxalk programme led to similar programmes in British Columbia and, eventually, the creation of CINE and its Global Health Food Systems Program. The Awajún health promoters are now working in many other communities in the Cenepa River region of Peru. The Ainu education programmes on traditional foods are spreading through universities in Hokkaido. There have been several requests for the Karen programme to be

replicated in other Southeast Asian tribal communities. The Deccan Development Society, an Indian NGO in Andhra Pradesh, has an impressive record of engaging Dalit communities in celebrating their traditional food knowledge through extensive media and community awareness campaigns.

Contextual strengths and weaknesses

Contextual factors have a very strong influence on interventions, as they result in both enabling strengths and disabling weaknesses. For example, the Awajún in Peru faced a serious setback when government policies supported the use of their lands for mining and resource development (*Asociación Interétnica de Desarrollo de la Selva Peruana* Web site).[3] A national socio-economic plan promoting a sufficiency economy and active participation in community development enabled the Karen in Thailand, by procuring support and attention from various development partners at the national and community levels. The Pohnpei case study (Englberger *et al.*, 2010) benefited greatly from a supportive government that ensured attention through the attendance of the President and Governor at programme activities and the creation of national postage stamps, noted earlier. Climate change is a massive contextual factor, with impacts on the availability of traditional food species in all global regions, which are especially noted by Gwich'in and Inuit of Baffin Island. Escalating food prices during project periods, and various local environmental and economic changes were other important contextual features in case studies. All the case studies recognized that evaluations would likely have recorded better intervention results if they had not been carried out so soon after activities were implemented. Despite these caveats, all the case studies were enthusiastic about their projects and the results documented.

Funding constraints

Funding constraints are always a problem when programmes have multifaceted and multisectoral activities. Each of the interventions mentioned in this book was conducted with minimal external budget while

---

3   www.aidesep.org.pe/

stimulating local empowerment and local sustainability. Stimulus funding through CINE was provided by Canada and FAO for all phase 1 activities (developing methods for documenting each of the 12 food systems) and some phase 2 (intervention and evaluation) ones. All of the case studies faced challenges with finding their own resources for phase 2 activities. This sometimes made it difficult to continue an intervention, such as with the Maasai (Oiye *et al.*, 2009), Bhil (Bhattacharjee *et al.*, 2009) and Igbo (Okeke *et al.*, 2009). However, all the other interventions obtained funding through NGOs and a wide variety of local volunteers. Finding this support depends on having committed, energetic and charismatic leaders, networks and a positive attitude to project success. All of the interventions have continued since the evaluations that were carried out in preparation for the chapters in this book.

Evaluation constraints

The constraints to evaluation design considered in these case studies result from activities taking place in small community populations, activities being multifaceted, and appropriate control groups not being available. Another constraint is the use of pre- and post-assessments, which may result in data collection bias and lack of internal validity. As anticipated from the literature, these two- to three-year intervention programmes did not result in anthropometrical changes, reflected as improvements in stunting or obesity control. The exception was growth improvement among Karen children. The CINE programme team used community assistants in the field to collect evaluation data, which increased the burden on communities by requiring them to provide skilled assistants or people who were easily trained. However, communities appreciated this opportunity for capacity building in collecting their own data. Qualitative assessments of communities' knowledge of the changes taking place recorded successes, with notable examples being the Ainu, Pohnpei (Englberger *et al.*, 2010) and Karen case studies.

Resistance to the collection of biological samples (blood, faeces, urine), due to fear of discomfort or superstition, was offset by acceptance of food use and dietary indicators as ways of tracking change resulting from food-based strategies. Dietary indicators were challenged by missing values in the nutrient composition data set for all the species used, and wide data variations within small populations.

Working with government

In some cases, it was possible to work successfully with government agencies as partners, which contributed to intervention success. Although all the case studies had networks within some form of government service or policy setting, governments were not always supportive of efforts to increase Indigenous Peoples' access to local traditional foods. In Pohnpei (Englberger *et al.*, 2010) and for the Karen, both the community and the government felt ownership of the projects and their successes. However, different political priorities led governments or government sectors to avoid direct involvement with the food access strategies of the Awajún, Dalit, Gwich'in or Ainu.

The case studies[4]

Table 14.1 provides a summary of the intervention strategies used by the case studies described in this volume, and Table 14.2 summarizes the evaluation indicators used. The reader is encouraged to read the specific chapters. The following subsections present brief descriptions of the case studies (in alphabetical order), intervention activities, evaluation methods and important contextual features.

Ainu
(Chapter 13 – Iwasaki-Goodman, 2013)

The Ainu are considered Japan's indigenous population, but serious discrimination and assimilation practices have contributed to poor cultural morale among Ainu people. The intervention is being conducted on Hokkaido Island through various projects to promote enjoyment of Ainu traditional food. A monthly newsletter describing harvesting, processing and cooking techniques for Ainu foods is distributed

---

[4] The present tense is used in this section because the interventions described are ongoing.

Table 14.1 Summary of intervention strategies used

Intervention activity	Case studies conducting the activity
Community partner consultation and/or research agreement	Ainu, Awajún, Dalit, Gwich'in, Inga, Inuit, Karen, Nuxalk, Pohnpei
Community steering committee	Gwich'in, Inuit, Nuxalk, Pohnpei
Empowerment in training	Ainu, Awajún, Dalit, Gwich'in, Inga, Inuit, Karen, Nuxalk, Pohnpei
Local resources and traditions	Ainu, Awajún, Dalit, Gwich'in, Inga, Inuit, Karen, Nuxalk, Pohnpei
Employment of community assistants	Ainu, Awajún, Dalit, Gwich'in, Inga, Inuit, Karen, Nuxalk, Pohnpei
Activities in and for schools	Ainu, Awajún, Dalit, Gwich'in, Inga, Inuit, Karen, Nuxalk, Pohnpei
Activities engaging elders	Ainu, Gwich'in, Inga, Inuit, Karen, Nuxalk
Activities for children/youth	Awajún, Gwich'in, Inga, Inuit, Karen, Nuxalk, Pohnpei
Activities for women	Ainu, Awajún, Dalit, Gwich'in, Inga, Karen, Nuxalk, Pohnpei
Presentations of local/traditional food in community settings	Ainu, Awajún, Dalit, Gwich'in, Inga, Karen, Nuxalk, Pohnpei
Reaching diverse segments of communities	Ainu, Awajún, Dalit, Gwich'in, Inga, Inuit, Karen, Nuxalk, Pohnpei
Internal and external communications	Ainu, Awajún, Dalit, Gwich'in, Inga, Inuit, Karen, Nuxalk, Pohnpei
Media activities	Ainu, Dalit, Gwich'in, Inuit, Karen, Nuxalk, Pohnpei
Involvement of the business sector	Karen, Nuxalk, Pohnpei
Engaging partners from government	Dalit, Inga, Inuit, Karen, Pohnpei

to every household in the main community, Biratori, which has a population of 2 500. Two books for local residents have been prepared from interviews with elders, describing different Ainu traditional foods and medicinal plants. Community gatherings now serve Ainu dishes, and traditional prayers and offerings to the gods have been revived. More Ainu rituals have been conducted recently, with elders' participation, and traditional foods are indispensable parts of these occasions. Both Ainu and non-Ainu people living outside Biratori are interested in learning about Ainu foods and dishes, and cooking classes for students are held at Sapporo universities, including Rakuno Gakuen and Hokkai Gakuen. In partnership with elders in Biratori, students have developed a project highlighting new kinds of *shito* (a dumpling made from millet and rice powder). An Ainu graduate student helped to open a *shito* stall for university festivals. Restaurants in the Sapporo Grand Hotel have organized special dinners featuring an Ainu theme, and the Shiraoi Museum has adopted Ainu dishes for its cultural days.

The community and academic leaders view intervention activities as having improved socio-cultural health among Ainu people in Saru River region. The reintroduction of cultural elements associated with food, such as knowledge regarding the harvesting, preservation and cooking of certain foods, and rituals featuring food, has led to cultural revitalization. Ainu people are now proud of their ethnic background and their food culture's integration into local food culture. This is helping to resolve the serious social prejudices faced by Ainu people.

Awajún
(Chapter 5 – Creed-Kanashiro *et al.*, 2013)

The local diet of the Awajún on the River Cenepa in the Amazon region of Peru is almost entirely made up of more than 200 traditional food species, but intakes of micronutrients are low, particularly among children. The intervention aims to increase the accessibility and use of high-quality traditional foods. Through participatory workshops, health promoters have been trained to emphasize traditional food topics and hygiene in food preparation. Use of traditional foods prepared with traditional utensils is promoted through print media, drama, songs and the sharing of recipes for young children's food. Small animal production,

Table 14.2 Summary of evaluation indicators used

Evaluation indicator	Case studies using the indicator
Interdisciplinary partners	Ainu, Awajún, Inga, Karen, Pohnpei
Qualitative methods	Ainu, Awajún, Dalit, Gwich'in, Inga, Inuit, Karen, Nuxalk, Pohnpei
Quantitative methods	Awajún, Dalit, Gwich'in, Inga, Inuit, Karen, Nuxalk, Pohnpei
Dietary indicators	Awajún, Dalit, Gwich'in, Inga, Inuit, Karen, Nuxalk, Pohnpei
Biological indicators	Nuxalk, Pohnpei
Clinical indicators	Awajún, Dalit, Gwich'in, Inga, Karen, Nuxalk
Anthropometry	Awajún, Dalit, Gwich'in, Inga, Inuit, Karen, Nuxalk, Pohnpei
Process indicators	Awajún, Gwich'in, Inga, Karen, Nuxalk, Pohnpei
Employment of community assistants	Awajún, Dalit, Gwich'in, Inga, Karen, Nuxalk, Pohnpei
Training and empowerment for local residents	Ainu, Awajún, Dalit, Gwich'in, Inga, Karen, Nuxalk, Pohnpei
Feedback from community	Ainu, Awajún, Gwich'in, Karen, Pohnpei
Locating local resources for sustainability	Ainu, Awajún, Dalit, Gwich'in, Inga, Inuit, Karen, Nuxalk, Pohnpei
Participation of volunteers	Awajún, Gwich'in, Inga, Karen, Nuxalk, Pohnpei
National awareness of project	Ainu, Awajún, Dalit, Gwich'in, Karen, Nuxalk, Pohnpei
International awareness of project	Ainu, Awajún, Dalit, Gwich'in, Inga, Inuit, Karen, Nuxalk, Pohnpei
Serious contextual constraints	Ainu, Awajún, Gwich'in, Inga, Inuit, Karen, Nuxalk
Policy impact	Ainu, Awajún, Dalit, Pohnpei
Scaling up	Ainu, Awajún, Dalit, Inga, Karen, Nuxalk, Pohnpei

fish farms, fruit tree production, food plant nurseries and medicinal plants have increased. Primary school children are trained in plant nurseries, and school lunches have been improved. Several cultural festivals are held annually. A traditional food reference book with photos, descriptions and nutrient data has been prepared, distributed in the communities and deposited in the national library.

Local health promoters keep records of food and health promotion activities. Surveys have recorded food intakes, mothers' physical activity levels, family food security, infant feeding practices and anthropometry of children. After two years of the intervention, traditional food diversity in the diets of women and children under five years had increased, and young children were consuming more animal-source foods (meat and fish).

External constraints experienced include changes in the local political structure, and a violent land rights conflict with the federal government, which interfered with activities of the local Awajún promoters. In addition, imported food from donation programmes and cash from coca production is beginning to decrease the quality of the local diet.

Dalit
(Chapter 6 – Salomeyesudas *et al.*, 2013)

Although the Dalit are not considered a tribal people,[5] Dalit women are recognized as the most disadvantaged of Indian adults. The local Dalit food system in the Zaheerabad region of Andhra Pradesh contains more than 300 species and has evolved over the thousands of years that these people have lived in the region, working in agriculture. Case study partners have been using participatory methods to promote these foods since the mid-1980s, focusing on locally grown sorghum, millet, legumes, vegetables, fruit and uncultivated green leafy vegetables. A broad spectrum of activities benefit the health of women and children: reclaiming land and using revitalized agricultural methods for food

---

5   The Indigenous Peoples officially noted by the Indian Government.

production by Dalit women farmers; capacity building for women as farmers, advocates, and film and media producers; food preparation for day-care centres; food festivals in communities and educational and industrial settings; film; school curricula; advocacy through radio and television; and work with a policy action group to influence national and state policies. Major impacts noted by the community and partners include enhanced soil fertility and conservation, increased food availability, greater self-reliance, reduced seasonal migration, more animal fodder and livestock, and more children attending school.

The CINE programme has contributed a dietary and nutrition assessment and a socio-cultural evaluation based on interviews. Intervention activities have been conducted through participating women's *sanghams*, and data from participating villages have been compared with data from villages that did not participate in the programme. Women in participating communities have better intakes of sorghum, legumes, vegetables and animal-source foods, and better dietary status; however children's nutrition levels are similar in both intervention and non-intervention villages. The traditional foods protect against chronic energy deficiency and night blindness (Schmid *et al.*, 2006). The evaluation was probably affected by the extensive media exposure of intervention activities in all villages.

Gwich'in
(Chapter 7 – Kuhnlein *et al.*, 2013)

The Gwich'in community of Tetlit Zheh in the Northwest Territories of Arctic Canada have expressed concern that their traditional wildlife foods, particularly caribou, birds and fish, are becoming scarcer and that children are affected by poor-quality diets composed mainly of market food. Local personnel have implemented intervention activities for two years, to improve the use of traditional food and better-quality market food. Activities have included regular radio announcements, school activities with youth, food teaching events involving elders, physical activity events, the creation of a DVD about local food, and production of a Gwich'in food and health book. The

baseline assessment of young women and youth found extensive overweight and obesity and poor diets, with high consumption of sweet, salty and fatty market foods and beverages. Those consuming wildlife foods have better nutrition status, although intakes of some nutrients are still below recommended levels. More than half of households are categorized as food-insecure, and households with less access to traditional foods have lower food security.

Serious external contextual events include climate change impacts on wildlife, and escalating fuel and food prices (market food) driven by external global markets. The local project personnel declined CINE's follow-up assessment with interviews and dietary and health assessment. However, the food and health book has been found to be highly appreciated by the community.

Inga
(Chapter 8 – Caicedo and Chaparro, 2013)

The Inga community members involved in this project are located on several reserves in the Caqueta region of Colombia. Community leaders and leaders of the Tandachiridu Inganokuna Association provide participatory development approaches and planning, in cooperation with the Amazon Conservation Team, an NGO. More than 5 000 Inga from 800 farm areas (covering 94 ha) have participated in the programme to ensure community health, which focuses on traditional food and medicine availability, cultural promotion and primary health care. An Inga traditional food handbook and an agro-ecology calendar have been developed. All the programme materials developed use words and imagery that are common to the Inga and consistent with their official *Plan de Vida*. Traditional foods and medicines have been promoted through family visits, workshops and courses, traditional food seed exchanges, the establishment of community and school gardens for food and medicinal plants, culinary festivals and recipe exchanges, radio programmes and health brigades, which include shamans and women healers.

Quantitative evaluation was based on dietary, anthropometric and clinical nutrition assessments, and qualitative evaluations on individual and family interviews. The proportions of dietary energy, protein, iron and vitamin A in the diet increased over the two-year evaluation period, but there have been no changes in anthropometric indicators. Fewer participating families express insecurity regarding the availability of locally grown food and medicines. Species diversity in farms has increased by 54 percent.

Specific challenges relate to advancing colonization on traditional Inga territory, for logging and seismic exploration for petroleum extraction, and the presence of armed militias and paramilitary groups. These create economic, environmental and social instability, leading to violence against citizens. Coca production and peripheral relations with the narcotics industry have contributed to ecosystem instability.

Inuit
(Chapter 9 – Egeland et al., 2013)

The community of Pangnirtung in the Baffin region of Nunavut in Arctic Canada participate in this project, with community leadership through Inuit Tapiriit Kanatami. Primary traditional food species are caribou, blueberries, seal, Arctic char, clams and local shrimp, and intervention activities focus on elders' traditional food stories, delivered on DVD and through youth radio drama programmes that provide modern-day nutrition and health advice. These build beneficial associations with health and well-being. Inuit traditional knowledge on plants, medicines and foods, and observations on climate change are discussed in these media.

Health surveys of adults and youth for the baseline evaluation found that dietary quality improves with increased intakes of traditional food. More sugared beverages and chips are eaten when traditional food is not, and this contributes greatly to high energy, sugar and saturated fat intakes and resulting overweight. Those consuming even only one serving of traditional food per day have significantly more dietary energy from protein, less carbohydrate and more total fat. Girls, but not boys, who consume traditional food were found to have higher dietary intakes of iron, vitamins A, B and D. Clearly, increasing the proportion of traditional food in food energy would improve nutrition.

As a result of climate change, disturbances in the Arctic ecosystem are happening rapidly, and are likely to continue. Changes in ice formation threaten Inuit traditional food species and food security, and undercut the promotion of greater use of traditional food.

Karen
(Chapter 10 – Sirisai et al., 2013)

The Sanephong Karen community is in a remote village in western Thailand, close to the Myanmar border. The community and academic partners recognize that traditional food availability is deteriorating rapidly, partly because traditional lands have been designated as a national park, thus restricting the Karen's access to wildlife. The intervention has the objective of using traditional food knowledge about more than 300 food species/varieties as a platform for working with the community to improve nutrition and health. Focusing on empowerment and building trust and commitment, partners have worked together to increase awareness of food and water sources; promote production and consumption of these resources in village areas, especially by children; and increase local capacity, knowledge and skills for taking action in children's food and nutrition security. Using a culture-based approach with metaphor and social dialogue, project personnel are growing more food at home, encouraging better food education for schoolchildren, empowering community women, and strengthening community leaders, with input from elders and many stakeholders, including local Buddhist monks and store owners. Interdisciplinary academic participants from Mahidol University include personnel from the Institute of Nutrition, the Faculty of Environment and Resource Studies, the Institute of Language and Culture for Rural Development and the Faculty of Medicine. Staff from the Ministries of Agriculture and Cooperatives and of Health are also involved. A unique aspect of the intervention is its strong empowerment of both the Karen community and local government officials. Local people are encouraged to report on the progress of their community development at all levels, including to Her Royal Highness Crown Princess Mahachakri

Sirindhorn who has visited the community. Many different materials have been created and distributed in the community, including posters and a book highlighting local foods and stressing that "food is part of happiness" and local pride.

Evaluation demonstrated substantial change in the number of household-grown vegetables and fruit species in the community (with more than 50 additional species compared with pre-project numbers); change in opinion, with more household members recognizing the benefits of using traditional vegetables; and schoolchildren's increased capacity for growing, harvesting, cooking and enjoying local traditional plants. The anthropometric status of children under 12 years of age has also improved.

Nuxalk
(Chapter 11 – Turner *et al.*, 2013)

More than 20 years ago, the Nuxalk Nation, an indigenous community on the west coast of British Columbia, Canada, conducted a highly successful intervention stressing traditional food use and well-being. The intervention was based on participatory cooperation with the Nuxalk Nation Council and the local nursing station, with researchers from the University of British Columbia, and stressed the use of traditional food knowledge as a platform for health promotion. Intervention strategies included research and the sharing of results to define the food system, the nutrient composition of key foods and how food use knowledge had changed over three generations of Nuxalk women. Personnel living on the reserve and working in its health centre promoted the use of traditional food and quality market foods. Food teaching events led by elders included the creation of a traditional food and medicine garden, regular information flyers distributed to family mailboxes, fitness events, and school classes on nutrition, dental health and food preparation. A traditional food handbook and a recipe book were distributed to all homes on the reserve and have been reprinted several times (Kuhnlein and Moody, 1989).

Before-and-after evaluation demonstrated increased quantities of traditional food being harvested and used,

and more families participating in these activities after the intervention. Biological evaluations from blood samples showed increased red blood cell folate, serum retinol and carotene in adults and youth, with youth also experiencing better iron status. Process measures documented community participation and tracked successful activities (Kuhnlein and Burgess, 1997).

Key to programme success was regular participation of the Nuxalk Nation Council and community health personnel in traditional food promotion events. A fitness "guru" from Vancouver spent several days in the community, encouraging youth to participate in fitness activities. When the project was revisited 20 years later, the original participants and their families still indicated substantial awareness of the importance of traditional food and an interest in growing garden produce and processing fish and other local foods for their families; they attributed at least part of this interest to the original programme.

Pohnpei
(Chapter 12 – Englberger *et al.*, 2013)

The Mand community of Pohnpei State, one of four states in the Federated States of Micronesia, is the focus of traditional food documentation and food and health promotion activities, which were evaluated two years after they commenced. Activity design and implementation were led by the Island Food Community of Pohnpei and its many collaborators. Documentation of the traditional food system uncovered knowledge and use of 381 species and their varieties/cultivars, and many valuable nutritional properties of these foods. Intervention activities that were community-based are now continuing with interdisciplinary and multisectoral inter-agency support at the state and national government levels and through other NGOs: departments of education, economic affairs, health, lands and natural resources; the College of Micronesia-FSM Cooperative Extension Services; and USDA's Natural Resources Conservation Services. Activities include youth drama clubs, recipe presentations, training in container gardening, cooking classes, newspaper articles, radio, film and television presentations, school and curriculum development, slogans and songs, games, weight loss competitions, and many promotional incentives.

Baseline and two-year evaluations have been successfully conducted with randomly selected households. Significant data differences between the two show that the consumption of refined imported white rice has decreased – this is a major improvement, as white rice has been replacing healthy local foods that are abundant in this rural area. In addition, the consumption of locally grown giant swamp taro and vegetables has increased, as has that of different banana varieties. Dietary diversity and attitudes to local food have improved (Kaufer *et al.*, 2010).

The success of this programme results from several factors, including outstanding and extremely supportive leadership at the local, state, national and NGO levels; the small island situation, creating close proximity among these different levels; extensive communications across sectors and disciplines, including the participation of graduate students from several universities; and the development of imaginative activities that appeal to community leadership. Community and academic leaders are satisfied that the package of project activities has had substantial impact, based on social marketing, education and agricultural provisioning to the community.

What have we learned?
Where do we go from here?

Participants and researchers in all case studies have learned immeasurably about their unique settings, and continue to promote the increasing availability of local food, ensuring that it is accessible and acceptable for all members of the community, and developing programmes with good promise of internal sustainability. These local success stories demonstrate opportunities for scaling up to adjacent communities and wider regions.

Intervention strategies that work intensively at the grassroots level with small communities of Indigenous Peoples and have limited budgets present overwhelming challenges. It is notoriously difficult to use food-based strategies to effect rapid change that can be reflected in quantitative evaluation using statistical methods. There are many justifications for poor returns on public health promotion efforts, such as formidable contextual factors affecting access to land, water and local cultural food resources; the extensive debilities and disparities resulting from neglect and repressive discrimination of indigenous communities, especially women; and the need for time and commitment for capacity building within cultures and communities.

Nevertheless, the research partners described in this book have achieved successes in their work with indigenous communities, and have learned how to create and evaluate effective intervention programmes grounded in the rich diversity and nutrition provisioning available in the local food systems of Indigenous Peoples. Capacity building, participatory decision-making, learning by doing, the use of local cultural and ecosystem knowledge and resources, and networking are key factors for success. These findings contribute to understanding of how governments and advocates can help to create programmes for improving well-being throughout the indigenous world. Building confidence in local food systems and links to cultural and ecosystem integrity develops the commitment to using and protecting these vital resources with community-specific knowledge and methods. Giving credence to these methods in local, national and international communications helps to reduce the disparities in health and well-being experienced by Indigenous Peoples.

The CINE researchers have learned to listen to what people have to say about what is important to them. Definitions of wealth, well-being and happiness are connected to important principles of culture and the food that people enjoy within their local environment. Many indigenous people do not consider themselves poor, even though they are living at the bottom levels of the definitions of poverty accepted by most development agencies. Information exchange has been at the heart of CINE's work, and all partners recognize that development, including food and nutrition security, is more sustainable when they listen and understand each other and work together to build the best strategies for change that is meaningful to the people directly involved.

Despite obvious obstacles, the imperative for United Nations agencies and forward-thinking governments is to recognize the scientific benefits of and the human right to Indigenous Peoples' local food systems, using activities such as those created and described in this book. Programmes that find success with any disadvantaged population do not necessarily work for Indigenous Peoples; additional efforts are needed to address the issues of discrimination, rural inaccessibility, and respect and protection of the culture and ecosystems from which Indigenous Peoples draw their well-being. On the other hand, successful health promotion using local food systems with communities of Indigenous Peoples are very likely to provide important lessons for public health practitioners working with any other community, regardless of whether it is disadvantaged, indigenous or not. ✿

Acknowledgements

This programme is built on partnerships. The research partners, community teams and case study chapter authors are recognized and thanked for their time, careful consideration and suggestions on how best to portray the principles and summaries of intervention strategies and evaluation indicators presented in this chapter.

> **Comments to:** harriet.kuhnlein@mcgill.ca

Human rights implications of Indigenous Peoples' food systems and policy recommendations

SIRI DAMMAN[1] HARRIET V. KUHNLEIN[2] BILL ERASMUS[2, 3]

1
Rainforest Foundation,
Oslo, Norway

2
Centre for Indigenous
Peoples' Nutrition and
Environment (CINE)
and School of Dietetics
and Human Nutrition,
McGill University,
Montreal, Quebec,
Canada

3
Assembly of First Nations,
Ottawa, Ontario, Canada;
Dene Nation, Yellowknife,
Northwest Territories,
Canada

Key words > food security, food system, policy, right to food, human rights-based approach, Indigenous Peoples, traditional food, Pohnpei, Maasai, Awajún, Inga, Inuit

> **"Our elders say we need to have our own food to be healthy and to be who we are."**
>
> Elder Fred Erasmus, Yellowknives Dene First Nation

Abstract

By ratifying the International Covenant on Economic, Social and Cultural Rights, States take on an obligation to ensure the right to adequate food for all. The practical content of this right has recently been concretized through the United Nations and intergovernmental efforts. In some cases, the policy implications of adopting a human rights-based approach to food security may be more substantial than most States realize. The need for such an approach is particularly visible in the case of Indigenous Peoples depending on traditional food systems. This chapter explores the content of the right to food for Indigenous Peoples who rely to a larger or smaller degree on local food systems for their food security.

A right to food-based analysis was applied to five cases described in this book: Pohnpei, Maasai, Awajún, Inga and Inuit. Information was gathered through supplementary questionnaires and interviews. The main findings were that commercial and development activities on indigenous lands and territories pose a threat to Indigenous Peoples' food systems and livelihoods, and thereby their right to food; and encroachments on Indigenous Peoples' lands threaten their food security and nutritional health, and may lead to conflicts and culture loss.

The conclusion was that in many cases, Indigenous Peoples' right to food is inseparable from their right to land, territories and resources, culture and self-determination. An integral human rights-based approach opens constructive dialogue on what policies, regulations and activities are needed to ensure food security for all, regardless of adaptation. Encouraging meaningful participation by all parties may be the key to building trust and resolving ongoing resource conflicts.

A right to food-based analysis

Harvested food is of key importance to the food security of a wide range of Indigenous Peoples worldwide.[1] However, Indigenous Peoples' livelihoods, which include culturally appropriate food harvesting, processing, preserving, preparation and consumption, are under threat. These threats include the expansion of agricultural frontiers, cattle ranching, exploitive industries (mining, gas and oil), excessive hunting, tourism, and other activities where outsiders make use of savannah, tundra, woodland, tropical rain forest and mountain areas that are inhabited and used by Indigenous Peoples and have often been their homes since time immemorial. These activities often threaten Indigenous Peoples' food and nutrition security and the quality of their water sources, their health and their continuous existence as peoples. It is therefore a goal of this book, and a long-term goal of the Indigenous Peoples' Food Systems for Health Program, initiated by McGill University's Centre for Indigenous Peoples' Nutrition and Environment (CINE), to influence national policies in order to improve Indigenous Peoples' access to their territories and food systems, and to improve dietary adequacy, health and well-being. A human rights-based approach to food is a suitable framework for advocacy and policy to that effect.

---

[1] The term "Indigenous Peoples" has not yet been clearly defined internationally. This chapter relates to the description given in International Labour Organization Convention No. 169 (ILO 169) while focusing on selected Indigenous Peoples with a strong link to their territories and local food systems. Indigenous Peoples often refer to themselves as nations with the right to self-determination. In this chapter it is recognized that Indigenous Peoples have specific rights and interests within national and international boundaries that may not yet be generally recognized and implemented.

According to international human rights, indigenous individuals should enjoy the same rights as non-indigenous individuals, while at the same time their right to their own culture is respected and protected. They should enjoy basic human rights such as food and health on equal terms with all citizens. Their right to uphold their distinct cultures often implies having a collective right to self-determination in their territories. Although many countries have accepted – at least on paper, through ratifying human rights treaties – that they have obligations to implement these rights, there tend to be gaps in this implementation. The legal framework is often in place, but lobbying and advocacy work is needed to have the parties to international human right treaties recognize and follow-up their obligations in fact.

Human rights-based advocacy should remind the State of its obligations towards all people, including the Indigenous Peoples under its jurisdiction. Human rights may be threatened by the State itself, or by individuals or entities that the State has an obligation to regulate. The respect and protection of the right to food is key to the future of Indigenous Peoples who rely on their local food system for food security. The right to food should be respected, protected, facilitated and fulfilled by the State. In reality however, the mainstream dominating cultures that States represent are often a threat to the traditional cultures, including the food cultures, of Indigenous Peoples.

This chapter uses a right to food-based analysis to explore some of the obligations that States have towards Indigenous Peoples that rely on land for their food security and livelihoods. It includes data and considerations from five of the CINE case studies recently researched (FAO, 2009b): Pohnpei in the Federated States of Micronesia, Maasai of Kenya, Awajún of Peru, Inga of Colombia and Inuit of Canada.[2] The chapter presents the overall governance issues related to food systems and human rights, followed by a description of each of the five case studies. The conclusions give policy recommendations relevant to each of the case studies, and overall considerations.

Governance issues

Public health nutrition, rights and government responsibilities

Public health nutrition is concerned with promoting good health through improved nutrition, and preventing nutrition-related illnesses in the population (Hughes, 2003). One public health nutrition recommendation resulting from the CINE Indigenous Peoples' Food Systems for Health Program is that Indigenous Peoples' traditional food cultures should be encouraged. This recommendation is based on sound nutrition science. Not only are Indigenous Peoples' traditional diets in the large majority of cases nutritionally superior to market-based diets, but the activities related to providing food through hunting, fishing and various harvesting activities protect against lifestyle-related diseases. As such, they contribute first to the health and well-being of individuals, and second to the health and sustainability of societies (O'Dea, 1992; Uauy, Albala and Kain, 2001; Kuhnlein et al., 2004; Kuhnlein and Receveur, 2007).

There are important similarities between a human rights-based approach and a public health nutrition approach to nutritional health. Both approaches understand nutritional health as being related to larger societal circumstances and skewed access to resources. Both aim to influence policies and provide positive change. However, they are also – as understood by the authors of this chapter – different enough to be complementary and synergistic. Nutrition research provides scientifically based information that is relevant to nutrition and important for policy, while a human rights-based approach provides a suitable legal and normative framework and standards for processes and outcomes. Such an approach focuses explicitly on the role and obligations of governments in addressing nutrition problems and problems related to discrimination and inequalities. By doing so, it provides an objective standard by which civil society may evaluate government performance.

---

2   CINE has developed case studies that strengthen the evidence base of current circumstances surrounding food systems and health for 12 community groups of Indigenous and Tribal Peoples and cultural minorities located in different global regions: Ainu (Japan), Awajún (Peru), Baffin Inuit (Canada), Bhil (India), Dalit (India), Gwich'in (Canada), Igbo (Nigeria), Inga (Colombia), Karen (Thailand), Maasai (Kenya), Nuxalk (Canada), and the people of Pohnpei (Federated States of Micronesia) (FAO, 2009b).

It is the State that is asked to ratify human right treaties, which makes the State the primary duty-bearer to be held accountable for its conduct. In the context of this chapter, it is therefore the role of the State to balance the rights and interests of all individuals and peoples against each other, through appropriate laws and policies, and to regulate the action of non-State actors. However, far too often agricultural, energy and industrial policies, and even national food security and development plans, are poorly adapted to Indigenous Peoples' needs and culture. For example, a food guide in Canada supports indigenous food use, but the issue of access to this food is seldom addressed. Development-related policies may even encroach on and harm Indigenous Peoples' resources, while benefiting the majority population and economic actors. There is little doubt that economic gains are often prioritized over Indigenous Peoples' land rights.

International human rights are created to protect the most vulnerable against violations by the powerful, including the State itself. It may therefore seem a paradox that the State is also the main duty bearer with regard to human rights implementation. Even though States claim their sovereign right to decide in internal matters, many seek to avoid being branded as violators of human rights, particularly if the accusations receive international attention. The international human rights bodies are in regular dialogue with States over their human rights obligations and conduct. Together with national and international civil society organizations, they educate governments on the content of their human rights obligations and stimulate them to take appropriate actions (FAO, 2009a; OHCHR, 2010). At present,

many politicians and civil servants are not aware of the existence of a right to food, nor do they understand their role as duty-bearers within a human right to food-based analysis. This situation is gradually changing, however, as described in the following subsection.

Universal human rights, including the right to food

The "mother document" on human rights, the Universal Declaration of Human Rights, was adopted in 1948. This universal declaration includes civil, political, economic, social and cultural rights, and mentions the right to food and the right to health. Later human rights instruments reconfirm the existence of these rights. The International Covenant on Economic, Social and Cultural Rights (ICESCR) of 1966 recognizes (Article 11.1) "the right of everyone to an adequate standard of living for himself and his family, including adequate food, clothing and housing, and to the continuous improvement of living conditions", and (Article 11.2) "the fundamental right of everyone to be free from hunger". Article 11 also establishes the obligation of States and the international community to realize the right to food: "The States Parties will take appropriate steps to ensure the realization of this right, recognizing to this effect the essential importance of international co-operation based on free consent".

The right to an adequate nutritional situation may be extrapolated from the right to adequate food and the right to health, which are both found in ICESCR Articles 11 and 12 (the right to health). Among the countries in this chapter, only the Federated States of

Table 15.1 **States' ratification and support record: human right instruments relevant to the right to food and Indigenous Peoples' special rights**

Country/ Indigenous People	ICESCR ratified (thereby right to food)	ICCPR ratified	ICERD ratified	ILO 169 ratified	Vote on UNDRIP
Peru/Awajún	Yes	Yes	Yes	Yes	Yes
Canada/Inuit	Yes	Yes	Yes	No	Yes
Colombia/Inga	Yes	Yes	Yes	Yes	Abstained
Kenya/Maasai	Yes	Yes	Yes	No	Abstained
Micronesia/Pohnpei	No	No	No	No	Yes

Micronesia is not a State Party to ICESCR, and thereby bound[3] by it (Table 15.1).

As stated in the Vienna Declaration from the World Conference on Human Rights, which 171 States adopted by consensus: "all human rights are universal, inalienable, indivisible, interdependent and interrelated" (UN, 1993). This principle implies that rights need to be integrated and understood in the light of each other. This is particularly crucial with regard to Indigenous Peoples, who should enjoy their universal human rights without having to relinquish their special rights linked to their collective enjoyment of their culture, which includes the spiritual aspects of the ways that food is collected and used (UNPFII, 2009).

The right to food has recently received extensive international attention. In 1999, an "authoritative interpretation" of the right to food was developed under the Committee on Economic, Social and Cultural Rights (CESCR, 1999). United Nations (UN) declarations from international conferences have recognized and helped clarify the linkages among food, health and human rights (Gruskin and Tarantola, 2002).[4]

The World Food Summit (FAO, 1996) and the World Food Summit: five years later (WFS:fyl) provided momentum for clarifying the content of the right to food, which was called for by the World Food Summit in 1996. This work was taken on by CESCR in Geneva, and resulted in General Comment No. 12 on the right to food (GC 12), presented to WFS:fyl by the High Commissioner for Human Rights, which then requested the development of a more practical tool for national implementation. Under the auspices of FAO, an intergovernmental working group developed voluntary guidelines to support the progressive realization of the right to adequate food in the context of national food security (FAO, 2005). These "right to food guidelines" were developed by States for their own use, representing a breakthrough in terms of international acknowledgement among States that food is a human right. According to the guidelines, the right to food should inform national laws, policies and decision-making related to food security.

According to GC 12, the right to adequate food is realized when every man, woman and child, alone or in community with others, has physical and economic access at all times to adequate food or means for its procurement (CESCR, 1999). A framework that organizes States' human right obligations into levels – the obligation to *respect, protect* and *fulfil* (subdivided into *facilitate* and *provide*) (Eide, 1984; 1989; 2000; ESCCHR, 1999; Oshaug, Eide and Eide, 1994) – is useful and is gradually being applied in human right analysis.

Regarding the right to food, the obligation to *respect* requires States Parties to avoid any measure that results in preventing the access to food that individuals or groups already enjoy. The obligation to *protect* requires States to take measures, in law and in fact, to ensure that enterprises or individuals do not deprive individuals or groups of their access to food. The obligation to *fulfil* in the meaning of to *facilitate* implies that the State must proactively strengthen people's access to and utilization of resources and their means to ensure their own food security. Whenever an individual or group is unable, for reasons beyond his/her/its control, to enjoy the right to adequate food by the means at her/his/its disposal, the State has the obligation to *fulfil* in the meaning of *to provide* that right directly. Food aid should be accompanied by measures that facilitate future self-reliance and food security (CESCR, 1999). At the *fulfil* level, the right to adequate food should not be realized in ways that undermine or hinder the realization of other rights, such as Indigenous Peoples' special rights. The right to adequate food may, for all practical purposes, be considered as a right to food security.

The concept of food security is used at the individual, household, national, regional and global levels. The Plan of Action of the World Food Summit of 1996 states that "food security exists when all people, at all times, have physical and economic access to sufficient, safe and nutritious food to meet their dietary needs and food preferences for an active and healthy life" (FAO, 1996).

---

3   Ratification is the process through which a country becomes a State Party to a covenant or convention (and thereby accepts to be bound by it).
4   These include the 1974 World Food Conference, the 1978 International Conference on Primary Health Care at Alma-Ata in the Union of Soviet Socialist Republics (WHO, 1978) and many large global conferences in the 1990s.

Indigenous Peoples have the same right as others to enjoy the right to health services and to a nutritious diet, but available health and nutrition data indicate that they tend to be worse off than the non-indigenous. The observed disparities are explained by Indigenous Peoples' disadvantaged position in society at large.

By July 2011, ICESCR had been ratified by 160 countries.[5] These countries have (at least on paper) accepted their responsibility for the food security, health and well-being of those under their jurisdiction. Among the countries in this study, Canada, Peru, Colombia and Kenya have ratified the convention and are thereby States Parties to it. The Federated States of Micronesia has so far not done so.

The specific rights of Indigenous Peoples

It is assumed that about 6 percent of the world's population is indigenous (Tomei, 2005; UNPFII, 2007a). This is only a rough estimate however, as there is no official definition of the term "indigenous" (UNPFII, 2007b). The word "peoples" is significant as it points to and relates to the right to self-determination of Indigenous Peoples, each people representing a distinct cultural group. The equal worth and dignity of indigenous individuals are best assured through the recognition and protection of both their rights as individuals and their rights as members of their group (OHCHR, 2006). Culture tends to be shared and constitutes a collective feature, and a people's right to a culture adds an extra dimension to the individual's right to a culture.[6] Indigenous Peoples' collective rights include their collective right to own and use their land, territories and resources, their right to self-determination on their land and territories, and their right to prior consultation and to free, prior and informed consent in matters that may affect them. These collective rights are crucial for the continuation of their cultures.[7]

Both the equal rights of indigenous individuals and Indigenous Peoples' specific rights[8] as a collective are reflected in human rights instruments.

In the Declaration of Atitlán (IITC, 2002) from the Indigenous Peoples' Global Consultation on the Right to Food, the right to food is seen as collective and contextualized within Indigenous Peoples' relationship to land. As expressed in the preamble of the declaration:

> In agreement that the content of the Right to Food of Indigenous Peoples is a collective right based on our special spiritual relationship with Mother Earth, our lands and territories, environment, and natural resources that provide our traditional nutrition; underscoring that the means of subsistence of Indigenous Peoples nourishes our cultures, languages, social life, worldview, and especially our relationship with Mother Earth; emphasising that the denial of the Right to Food for Indigenous Peoples not only denies us our physical survival, but also denies us our social organisation, our cultures, traditions, languages, spirituality, sovereignty, and total identity; it is a denial of our collective indigenous existence…

Indigenous Peoples' right to food is presented here as an integral part of indigenous identity and existence. If Indigenous Peoples are denied the land and the food on the land, their culture will dissolve. This interpretation contrasts with the view often taken by decision-makers within governments. However, it is interesting to note that in this declaration, Indigenous Peoples frame the right to food, formally an individual right, among the collective rights that are fundamental to their identity, culture and existence as peoples.

Several human rights provisions establish Indigenous Peoples' right to uphold control over their territories. These include the International Labour Organization Convention No. 169 (ILO 169) from 1989 and the United Nations Declaration on the Rights of Indigenous Peoples (UNDRIP) from 2007.

---

5   http://treaties.un.org/pages/viewdetails.aspx?src=treaty&mtdsg_no=iv-3&chapter=4&lang=en
6   See article 27 of the International Covenant on Civil and Political Rights (ICCPR) and its General Comment 23.
7   These rights are found in the United Nations Declaration on the Rights of Indigenous Peoples (UNDRIP), ILO 169 on Indigenous and Tribal Peoples (from 1989), and ICCPR article 27 and its General Comment 23.

8   The Universal Declaration of Human Rights, ICESCR and ICCPR are for general application. Other instruments detail the special rights of certain groups that are prone to experience circumstances that make them particularly vulnerable. These include women, children, ethnic, religious or linguistic minorities, and Indigenous Peoples.

Article 27 of the International Covenant on Civil and Political Rights (ICCPR) from 1966, on the right to culture of minorities, also implies States' obligation to ensure the right to land of individuals belonging to Indigenous Peoples.[9]

The right to prior consultation on all legislative and administrative actions that could affect the rights, assets, lives and culture of an Indigenous People is stated in ILO 169. The jurisprudence of UN human rights committees has reiterated this principle. In General Recommendation No. 23 on the rights of Indigenous Peoples, the Committee on the Elimination of Racial Discrimination (CERD) calls on States to "ensure that members of Indigenous Peoples have equal rights in respect of effective participation in public life and that no decisions directly relating to their rights and interests are taken without their informed consent". On several occasions, CESCR[10] too has stressed the need to obtain the consent of Indigenous Peoples in relation to the exploitation of resources. In numerous cases governments have interpreted the principle of prior consultation to mean merely that Indigenous Peoples should be informed about measures that will be taken and that will affect them. The intention, however, is to achieve agreement, as expressed in ILO 169 Article 6.2: "The consultations carried out in application of this convention shall be undertaken, in good faith and in a form appropriate to the circumstances, with the objective of achieving agreement or consent to the proposed measures".

UNDRIP takes the matter a step further by establishing the right to *free and prior informed consent.* This principle opens real dialogue and, by replacing the word "consultation" with "consent", strengthens the case for Indigenous Peoples' influence. UNDRIP is the most progressive of the human rights instruments mentioned here. However, its status as a declaration makes it a political statement by States rather than a self-imposed obligation under international law, as are ILO 169, the International Convention on the Elimination of all forms of Racial Discrimination (ICERD) and ICCPR.

As seen in Table 15.1, four of the five countries in this chapter have ratified ICESCR, thereby acknowledging that everyone has a right to food and health. With regard to Indigenous Peoples' rights, including land rights and the right to prior consultation, both Colombia and Peru have ratified ILO 169. Peru and Micronesia were among the 141 countries that voted for UNDRIP in the United Nations General Assembly in 2007; Canada and three other countries voted against it; and Colombia and Kenya abstained (United Nations General Assembly, 2007). Since then, Canada, the United States of America and Colombia, among other countries, have reversed their position and now support the Declaration.

Regarding land rights, it is significant that Kenya, Canada, Peru and Colombia have ratified ICCPR. During their periodic country reporting to the UN, the Human Rights Committee will question these countries on their conduct with regard to land rights and minorities' rights to culture. Canada has also ratified the Optional Protocol to ICCPR, which makes it possible to complain to the Human Rights Committee if the country threatens, or accepts threats to, Indigenous Peoples' territories, and thereby their right to live according to their culture.[11]

All countries are in one way or another under an obligation to respect Indigenous Peoples' right to prior consultation, which is reflected in several human rights instruments. Peru and Colombia have acknowledged this by ratifying ILO 169. Peru, Colombia, Canada and the Federated States of Micronesia have accepted the principle of free prior and informed consent by voting for UNDRIP, and Kenya Canada, Peru and Colombia through ratifying ICERD.

---

9   General Comment 23 of Article 27 is interpreted as: "... the rights of individuals under that article ... to enjoy a particular culture – may consist in a way of life which is closely associated with territory and use of its resources. This may particularly be true of members of indigenous communities constituting a minority" (Article 27/GC 23). CESR further observes that "Culture manifests itself in many forms, including a particular way of life associated with the use of land resources, especially in the case of Indigenous Peoples. That right may include such traditional activities as fishing or hunting and the right to live in reserves protected by law. The enjoyment of those rights may require positive legal measures of protection and measures to ensure the effective participation of members of minority communities in decisions which affect them" (Article 27/GC 23).

10   CERD and CESCR receive reports on States' progress regarding ICESCR and ICERD (Table 15.1), and represent authoritative sources on how the content of these conventions should be interpreted.

11   This relates to the interpretation of ICCPR Article 27 on minorities and their right to culture, which is explained further in footnote 9.

Indigenous Peoples' right to adequate food

The universality, inalienability, indivisibility, interdependence and interrelatedness of all human rights (UN, 1993) should inform any human rights analysis. This chapter focuses on the right to food, bearing in mind the close links between the right to food and the right to health and, in the context of Indigenous Peoples, to the specific rights of Indigenous Peoples.

A human rights-based approach demands State accountability and transparency, as well as participation and non-discrimination. It focuses on entitlements in concrete terms and identifies who is responsible for ensuring access to these entitlements. The approach exposes the roots of vulnerability and marginalization, expands the range of responses by duty-bearers, and strengthens the ability of indigenous individuals and communities to improve their conditions (FAO, 2009c).

Indigenous Peoples are generally understood to be vulnerable to poverty, malnutrition and disease (PAHO, 2002b; Damman, 2007). Demographic and health data show disparities in life expectancy, nutrition status and disease between indigenous and non-indigenous populations, in both wealthy (Ring and Brown, 2003) and poorer countries (ECLAC, 2005; UNPFII, 2005; PAHO, 2002a; 2002b; WHO, 2007b; Damman, 2005).

Krieger (2001) notes that the ways in which the causes of health problems are conceived and explained are crucial to the way in which the problems are addressed. A human rights-based analysis throws light on the role and obligations of government duty-bearers. The established obligations also provide a framework for advocacy, so that governments may be held accountable for inequalities and failures to respect, protect and fulfil the various aspects of this right.

As seen in Figure 15.1, malnutrition[12] and nutrition-related diseases can be explained by causal factors on several levels. Analyses by health professionals and epidemiologists tend to focus on the immediate and, to some extent, the underlying causes. Minimal attention

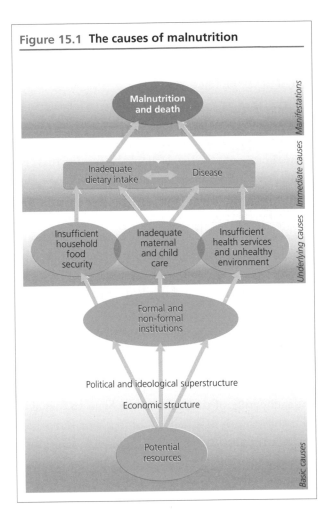

Figure 15.1 The causes of malnutrition

is generally given to the basic causes. Human rights-based analyses focusing on the basic causes of nutrition problems in vulnerable groups, and the way in which resources are managed and allocated, are often considered "political" and at times stir up debate and protest.

It seems that Indigenous Peoples are generally able to maintain a nutritionally adequate diet if they are not denied access to their land and if their traditional food resources are not depleted.

For most practical purposes, ensuring *respect* for and *protection* of the right to food is an obligation on behalf of the State to "do no harm" and to "allow no harm", in that non-government actors should be regulated through the legal system. States are required to do this in ways that involve respecting and protecting Indigenous Peoples' unique cultural identities and special concerns. Both the Right to Food Guidelines and GC 12 stress

---

12 Here understood as undernutrition (low weight-for-age), chronic malnutrition (low height-for-age, or stunting) and wasting (low weight-for-height, or "thinness").

that governments need to give special consideration to Indigenous Peoples' land and traditional food resources when implementing Indigenous People's right to food (CESCR, 1999; FAO, 2005).

Because there may be conflict between the government and other actors, it is particularly important that the respect and protect levels of obligations are meticulously monitored. It should be ensured that laws do not undermine or violate Indigenous Peoples' rights, their own governments or their livelihoods (respect level). Furthermore, everyone should be equal before the law, and the police and the court system should protect indigenous individuals' rights on equal terms (protect level). In addition, Indigenous Peoples' rights should be facilitated and fulfilled on equal terms, but in ways that are in harmony with their culture. This may mean developing – in collaboration with the group itself – unique and culturally sensitive approaches to achieve the end goal of equal rights for all.

Human rights monitoring should be carried out independently of the State, as the State and its allies often have much to gain from disrespecting Indigenous Peoples' right to land and natural resources. Complaint mechanisms should be in place at the local and national levels, as well as internationally. Human rights should be constantly called for so that States are held accountable. Indigenous Peoples and their allies and defenders should demand their human rights (including the right to food) and exert pressure on States and their officials to meet their obligations and commitments in a culturally sensitive way.

Case studies and analytical framework

Sources of information

This chapter is supported by five of the CINE case studies. These have already been presented by Kuhnlein, Erasmus and Spigelski (FAO, 2009b) and in this volume, with a focus on the food systems themselves and on health improvement using the food systems. The studies were not carried out with the analysis presented in this chapter in mind, so additional information related to the human right to

food situation was sought. A questionnaire inquiring about right to food-related issues was sent to the case study focal points on all continents in January 2008. The case study partners responded to this questionnaire, in some cases in consultation with government officials. Responses were returned by e-mail, and follow-up interviews were conducted at gatherings of the partners in 2008.

The questionnaire contained 21 questions with fixed-response categories, and additional space allowing respondents to substantiate their answers. Questions dealt with the local food and nutrition situation and the role taken by the government in respecting, protecting and facilitating the right to food and health. Categories were water and food safety; quality of health services; nutrition status; access to food, water and government assistance; the importance of traditional food and monetary income; signs of climate change; land rights and advocacy; and traditional culture regarding breastfeeding and weaning foods.

The interviews enquired into issues specific to the various case study areas. They were conducted one-on-one and in groups, depending on the participation from the area. Researchers and community partners from nine of the CINE case studies were interviewed: the Maasai of Kenya, the Karen of Thailand, the Awajún of Peru, the Inga of Colombia, the Gwich'in (Tetlit Zheh), Inuit and Nuxalk of Canada, the Dalit of Zaheerabad, India, and the people of Pohnpei in the Federated States of Micronesia. The information was substantiated through available scientific studies and Web-based literature.

The communities

This chapter is based on information from five of the case studies, as it was not possible to present all of the rich information within the limited space available. These five communities are:

- the people of Pohnpei in the community of Mand, Pohnpei, Federated States of Micronesia in the Pacific;
- the Maasai of Enkereyian community in the Kajiado district of Kenya;

- the Awajún of Condorcanqui in the Lower Cenepa region of the Department of the Amazon, Peru;
- the Inga in the State of Caquetá in southern Colombia;
- Inuit of Pangnirtung, Baffin Island, Nunavut, Canada.

The people of Mand, Pohnpei still harvest wild and cultivated food resources from the surrounding area, including food plants, fish and various game. Inuit still hunt caribou and seal. However, new income opportunities and the increased availability of market foods have resulted in lifestyle and diet changes among the people of Pohnpei and Inuit, even though traditional food is still in use. The Inga and the Awajún live in relative isolation in biodiversity-rich rain forest areas. They are offered some government assistance, but their main food sources are still the fish, animals, birds, fruits, tubers, nuts and other plant species harvested in their territories. The Maasai are traditional pastoralists. They now experience serious drought spells, which have made them highly food-insecure and dependent on food assistance. Among the five communities, the Inga and the Awajún have the highest intakes of traditional foods. Inuit also have quite a high intake of fish and game.

The right to food of Indigenous Peoples: five case study examples

Unless otherwise stated, the findings reported for these case studies stem from the e-mailed questionnaires and the follow-up interviews.

Pohnpei

Pohnpei is one of four states in the Federated States of Micronesia (see Figure 12.1, page 192) in the western Pacific. Pohnpei is also the name of the main island in the state of Pohnpei. The population is mainly Micronesian. The Federated States of Micronesia was under United States administration from the Second World War until 1979. It is now a sovereign State in association with the United States of America and

uses the United States dollar as its currency. There are relatively few official data on the Federated States of Micronesia.

The nutrition situation

The CINE case study in Mand on the island of Pohnpei showed that about half of a small sample of children were stunted.[13] The overall stunting[14] rate in the Federated States of Micronesia is not known.[15] Growth stunting may be caused by inadequate weaning foods or poor sanitation and health services, or a combination of several factors. The government is seeking to improve the outreach of health and water services. Water provided by the Public Utility Company is safe, but is not accessible to all; river water tends to be contaminated by pig pens close to rivers, and other waste. According to a Pohnpei state-wide health survey, up to half the adult population is obese (WHO, 2008).

Access to land and resources

In traditional Pohnpei culture, traditional leaders decided how collective resources were to be managed, including where and when fishing was to take place. Local leaders still have the authority to make such decisions, but no longer do so. There is an increased demand for privately owned land, and land owned by smaller family units is outside the control of local leaders.

Recent changes in land-use management have affected the availability of harvested food on the island. With an increasingly cash-based society, many farmers have shifted cultivation to the production of *sakau* (kava). This mild narcotic was traditionally used only for ceremonies, but is now sold daily at markets around the island. This shift has caused many farmers to forgo the planting of traditional crops and has resulted in the clearing of much of the interior

---

[13] The survey, carried out by CINE and the Island Food Community of Pohnpei project, revealed a stunting rate of 46 percent, (< 2 SD, children under five years of age), which is very high. However, the sample size was only 13 children, so results should be interpreted with caution.
[14] Stunting, or low height-for-age, is caused by long-term insufficient nutrient intake and frequent infections. Stunting generally occurs before the age of two years, and its effects are largely irreversible. They include delayed motor development, impaired cognitive function and poor school performance (WHO, 2007a).
[15] The only study referred to in the WHO database, from 2000, shows a lower stunting rate, of 16.7 percent in the states of Kosrae and Yap combined. However, this study covered only 20 percent of the total population of the Federated States of Micronesia, and is therefore not representative (WHO, 2009).

forest. This clearing is causing increases in soil erosion and sedimentation on the reef. The loss of interior forest is also decreasing the island's resilience to such environmental threats as droughts and landslides.

Studies have revealed that development projects such as road construction and increased dredging, coupled with improper waste management have greatly affected the near-shore marine ecosystems. These effects are most clearly seen in decreased health and vigour of coral, destruction of mangrove and sea grass ecosystems, and disruption of the nutrient flow associated with tidal exchange. This destruction of vital marine habitats has greatly reduced Pohnpei's fish and invertebrate numbers. A 2006 study of Pohnpei's fish markets, conducted by Dr Kevin Rhodes, indicated that the island's reefs are being overfished at 149 percent of their healthy capacity.

Food culture and food preferences

Pohnpei is a lush and fertile island where food crops grow readily, and the traditional Pohnpei diet is nutritionally rich. Family gardens are found all over the island, and landowners cultivate bananas, yams, coconut and breadfruit of different varieties, and other species; however, the people of Pohnpei are influenced by United States food culture, and large parts of their caloric needs are provided through imported, processed foods of low nutritional value, such as white rice, white flour products, sugar-rich foods and fatty meat.

These dietary changes are part of a wider set of lifestyle changes and the erosion of traditional culture and heritage. The traditional food resources consumed by adults provide about 25 percent of their total dietary energy (Englberger *et al.*, 2013). Pohnpei inhabitants consider traditional food to be healthy, but it is a public health challenge that traditional food is also seen as being "poor people's food", and most of the population has developed a liking for refined carbohydrates and fatty foods. Unemployment levels are high, but since islanders have enjoyed social security benefits through the Compact of Free Association with the United States of America, their purchasing power for low-cost foods has been ensured, thus contributing to the nutrition transition on the island.

The people of Pohnpei have been hit hard by the nutrition transition. However, measures are now being taken to recuperate and increase pride in the healthier traditional food culture. Government policies encourage local food production and consumption. The elected Pohnpei State Governor has followed a process that included the promotion of local foods. He established a new task force on school snack lunches, which aims to provide meals to primary school students, with a substantial proportion of the meals consisting of local foods. The project based at the Island Food Community of Pohnpei has also been successful in improving attitudes to and increasing the consumption of traditional fruits and vegetables.

Attitudes towards traditional foods are changing, as demonstrated by the increased use of local food during feasts and funerals. The Government of the Federated States of Micronesia has been supportive to the Island Food Community of Pohnpei, through the implementation of policies and media campaigns promoting the harvest and use of local foods. One example is the issuing of a national postage stamp series highlighting the carotenoid-rich *Karat* banana.

The Maasai

The food and nutrition situation

Pastoralist Indigenous Peoples in Kenya depend on land and natural resources for themselves and their herds. The Maa-speaking Enkereyian community is one of many pastoralist Maasai communities in Kenya (Figure 15.2). The areas these communities use today are neglected by the government, and the lack of infrastructure and State services such as health and schooling results in high rates of malnutrition and illiteracy. State policies fail to safeguard the Maasai's interests and protect their rights (Simel, 2008). Stunting rates (-2 SD weight-for-height) among children aged zero to five years are high, at 53 percent in 2003 (World Vision Kenya, 2004), when the national average was 39.4 percent (WHO, 2007b).

Maasai consume traditional food (especially milk and meat) daily if possible. Outside drought periods, they are able to feed themselves from the traditional

Figure 15.2 Enkereyian – Maasai Community Ngong Division, Kajiado District

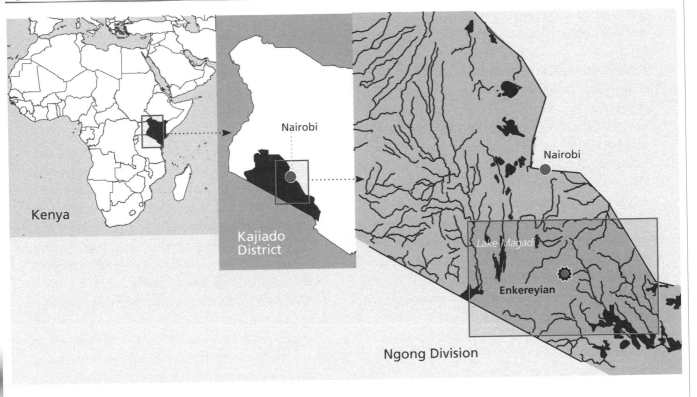

Data from ESRI Global GIS, 2006. Walter Hitschfield Geographic Information Centre, McGill University Library.

food system, but these foods are currently consumed in smaller quantities during most of the year. Amounts of traditional food consumed vary, but generally provide about 10 percent of total food energy (Oiye *et al.*, 2009). The Maasai experience seasonal water and food shortages, and their diets are deficient in several nutrients. Maize and beans are provided through relief programmes. These are important in counteracting famine as a short-term strategy, but are not popular, and are avoided when the situation permits. During drought periods, Maasai try to obtain donated food to avoid having to sell their animals to buy food, and to save their small amounts of money for other needs.

Food donations are problematic. The fact that Maasai and other pastoralists receive food aid allows the government to postpone addressing the underlying land distribution problem. Food aid also creates dependency, and is becoming a permanent condition. Furthermore, the handing out of food aid undermines the Maasai's

cultural and social networks. The Maasai have a sharing ethos, and will take care of the poorest when they are able to. According to the World Food Programme's policy, however, the poorest households are targeted for aid. In accordance with traditional cultural norms, the poorest households then feel obliged to share what they have received with others. In spite of warnings and complaints nothing has been done to address this problem or to find a more culturally sensitive way of providing food assistance.

The Enkereyian community uses the same water source as their animals; this water is a source of diseases, including typhoid fever. Outside the three-month rainy season water is extremely scarce, and women have to trek long distances to get it, which reduces their time for other activities. The government has not been active in improving the water situation, and there are no health services in the community. The nearest health dispensary, 15 to 20 km away, has no drugs and no trained health personnel. The nearest hospital and health centre is 40 km away, and is far too expensive, as it is privately owned (J. Ole Simel, personal communication, 2008).

Access to land and resources

The Maasai have gradually been marginalized and displaced from their land since United Kingdom settlers arrived in Kenya. Under the Maasai-Anglo Treaties (1904 and 1911), the Maasai were removed from their fertile highland areas to arid areas, which led to abject poverty.

Traditional cultural institutions have been dismantled and the Maasai have gradually become assimilated. Traditionally, the Maasai hold their land communally, but the introduction of individual land tenure has contributed to erosion of the collective way of life and pastoralist adaptation. As access to grazing land and necessary social structures are disappearing, the Maasai are losing their identity.

After independence in 1963, the Kenyan Government increased the pressure on Maasai grazing land, and non-Maasai Kenyan farmers have gradually taken over Maasai territories. Fences prevent cattle from reaching grazing land and water sources, sometimes leading to violent clashes (IWGIA, 2007; Kipuri, 2008). The government has facilitated sales of Maasai land to wildlife conservation organizations and the private sector. At present, the Kenyan Government does not encourage or protect Maasai culture and food systems, and collective rights are not acknowledged. As noted by the Maasai leader responding to the interview "We are supposed to all be Kenyans".

Recent droughts have decreased the land's carrying capacity. Seasonal water shortages affect both people and livestock, and cattle inevitably die. The shrinkage of cattle herds makes the Maasai increasingly dependent on food aid. The future of the Enkereyian Maasai looks bleak if the conditions undermining their livelihoods do not change.

Over recent decades, the Maasai have formed organizations and improved their political awareness, lobbying and networking. They now work both nationally and regionally, pushing for recognition of Indigenous Peoples' rights within the African Commission on Human's and Peoples' Rights, and internationally, to strengthen Indigenous Peoples' rights and draw attention to the situation of the Maasai. In spite of strong lobbying, Kenya was one of the few countries that abstained from voting for UNDRIP in 2007 (Kipuri, 2008).

There is little doubt that global warming influences Kenya, especially the Maasai. The rains fail more often, and droughts, which used to strike once a decade – giving herders and herds time to recover – are now far more frequent. The Maasai's pastoralist adaptation is becoming less resilient, and livestock populations are diminishing throughout Maasai areas. Without their livestock, families lack food and money. Poverty makes it difficult for them to pay school fees for their children, or to cover other subsistence needs.

Drought may be accentuated by ongoing deforestation. The root systems of living trees help the land to hold rainwater, and this water feeds rivers. The Mau Complex, Kenya's large mountainous forest, feeds major lakes and provides continuous river flow and favourable microclimate conditions. These are important for medicinal plants, fuelwood and grazing. Massive deforestation has taken place, affecting large-scale agriculture, charcoal production and logging in natural forests. This is already having a tremendous impact on access to water in areas far from the Mau Complex. Lower water levels result in wells and boreholes becoming dry, and rivers carrying less water and drying up earlier (WRM, 2006).

The Kenyan Government's inaction in this grave situation is a serious breach of Indigenous Peoples' specific rights, which are not acknowledged by Kenya, and also of the rights to food, water, health and human life of the Maasai.

The Maasai have made great political progress, and have some hope in the legal system. Kenya's courts deal with many land cases, and positive developments seem to have occurred. A second legal process also inspires hope, as a Constitutional reform may acknowledge collective land rights (Kipuri, 2008). Both processes are crucial opportunities for Kenyan society to rectify previous wrongs committed against traditional herder societies.

The Awajún

The Awajún case study was carried out in six communities: Mamayaque, Tuutin, Cocoaushi (part of Waiwam), Pagki, Nuevo Tutino, and Nuevo Kanam. These hamlets are situated in Condorcanqui, in the Lower Cenepa region of the Amazon tropical rainforest in the northwestern Peruvian Amazon, near the border with Ecuador (see Figure 5.1, pag. 54).

More than 90 percent of the Awajún's food intake (energy) is covered by harvesting local food (Creed-Kanashiro et al., 2009). The percentage is slightly lower among children, who also receive food through government food aid programmes. There is no electricity in the six communities, and Awajún homes are generally built from local trees and plant materials. Traditional medicinal plants and shamanism play an important role.

The food and nutrition situation

Although the Awajún diet is diverse, child malnutrition is a problem. Almost 50 percent of children under six years of age are stunted, and almost 25 percent under two years suffer from wasting.[16] In the CINE study, energy intake seemed adequate in the season evaluated, but these results may be somewhat overestimated (Creed-Kanashiro et al., 2009). The percentage of dietary energy from animal products is relatively low, as are the intakes of fat, protein, iron and zinc, especially among children. River water is likely to be an important cause of malnutrition, because it is used as drinking-water and likely to contain disease vectors causing diarrhoea and parasite infection.

The Awajún do not consider themselves to be poor, owing to the availability of traditional local food and other resources in their natural environment. However, because they generally have little money, they are classified as poor and extremely poor in the national census, and thus by the Peruvian Government. This entitles them to food aid, which they receive through several assistance programmes. At the time of the interviews, government food programmes provided rice, beans, oil and tuna. Recently, the Awajún started to receive a monthly donation of PEN 100 (equivalent to about USD 30.30) from the JUNTOS programme, which encourages education of children and health promotion for mothers and children. A municipal programme provides children with milk and sweetened oats, but not regularly.

Access to land and resources

Traditionally the Awajún lived in widely dispersed houses and hamlets relatively close to game, fishing opportunities and plant food for harvesting. Later, they moved into villages along the river, for transport, schools and missions. This increased the population density, and led to overexploitation of edible birds, game, fish and wild plants in the vicinity of their villages, gradually reducing their access to these resources. This has resulted in reduced consumption of animal products and decreased food variety.

Peru ratified ILO 169 in 1992 and subscribed to UNDRIP in 2007. A large part of indigenous community lands have been demarcated and titled,[17] but the Peruvian State has failed to acknowledge communities' status as Indigenous Peoples with rights to their larger territories. The Awajún territory covers a far larger area than that of the titled communities, and includes a national park and a communal reserve. The Awajún communities are adjacent to each other, separated by untitled free spaces that the population consider very valuable and the property of all Awajún.

Over recent years, the Awajún have been seriously concerned about a gold mining company establishing itself on their land. They fear that their river, water and fish will be contaminated with mercury, as is happening in other Amazon areas.

The six Awajún communities in the study have also found themselves within an oil concession that the Peruvian Government has granted for hydrocarbon (oil and gas) exploitation. Such exploitation in the Bajo Cenepa area could have severe effects on the Awajún's rivers, food security and social situation, as the example

---

[16] Wasting, or low weight-for-height, is a strong predictor of mortality among children under five. It is usually the result of acute significant food shortage and/or disease (WHO, 2007a).

[17] Mamayaque received land title in 1977, Nuevo Tutino in 1998, Tuutin in 1975, Cocoaushi in 1975, and Pagki in 1987. No information was found for Nuevo Kanam, but a map available on the Web shows that the area is registered: www.ibcperu.org/index.php

of the Corrientes River illustrates. Oxydental petroleum and Pluspetrol have been extracting oil and gas in the Corrientes River area for 35 years. Wastewater has contaminated the Corrientes River basin (Agurto, 2008), and the Peruvian Ministry of Health has found very high levels of lead and cadmium in the blood of Achuar people. The surviving birds, game and fish may be contaminated, as may plants used for food. The Achuar are less able to provide themselves with food, and report deaths and illnesses that may be associated with heavy metal poisoning; so far, they have not received any compensation or medical treatment. This has added to the concern of the Awajún and other Indigenous Peoples of Peru, as they fear that extraction activities in their areas could lead to the expropriation of valuable and sacred communal lands, and have severe consequences for communities' health, food security, culture and livelihoods (Achuar inhabitants of the area, personal communications, April 2009).

The case study communities share this destiny with many other indigenous communities in Peru. Peru's current (2010) President, Alan Garcia, is very favourably disposed towards the extractive industries, and the Free Trade Agreement between Peru and the United States of America has substantially increased the number of agreements between international extractive companies and the Government of Peru. During his presidential period, Mr Garcia has increased the proportion of the Peruvian Amazon available for oil and gas prospecting from 20 percent of the total land area to 70 to 80 percent (*Asociación Interétnica de Desarrollo de la Selva Peruana* Web site,[18] 2009; Agurto, 2008). Recently, the government has sought to implement regulations that put at risk both the ongoing land titling processes and the autonomy of Indigenous Peoples to use their land freely. This is counter to the Peruvian Constitution and is being disputed (APRODEH, 1999).

So far, the Awajún's rights to health, food and education have been poorly addressed by the government. Currently, however, the most pressing problem for Awajún is mining and petroleum extraction. Extractive

industries present a real and constant threat to the natural resources that are the basis for the population's subsistence, and thereby its right to food and health. It is worrying that the development policy and legal changes taking place under the current government conflict with Peru's human rights obligations, as has been noted by ILO and the Inter-American Commission on Human Rights (CERD, 2009), among others. By allowing extractive industries on to Awajún land without consultation, the State violates Indigenous Peoples' rights as expressed in ILO 169, UNDRIP and ICERD, among other agreements. This testifies to a failure of the government to take its human rights obligations seriously.

The Inga

The CINE project in Colombia focuses on indigenous territories belonging to the Inga Association (*Asociación de Cabildos Tandachiridu Inganokuna*) (Correal *et al.*, 2009). The project focuses on five Inga territories in Caquetá: Yurayaco, Brisas, San Miguel, Niñeras and Cosumbe (see Figure 8.1, pag. 122). Caquetá is situated in southern Colombia, near the border with Peru and Ecuador, along the northwestern frontier of the Amazon region. Similar to the Awajún, the Inga make use of traditional medicinal plants and shamanism (CINE, 2010).

Interview data established that the communities cultivate food, but land areas have declined and are now too small to produce sufficient food for the people. The further away the communities are from urban areas, the more traditional foods they consume.

Legal framework

Colombia's Constitution and laws have long been considered the most progressive in Latin America. ILO 169 is fully adopted and supported by legislation. Laws specify that communities have autonomous rights to decide over their territories, and that they have to participate in the formulation of policies that may affect them. However, ongoing negotiations linked to the Free Trade Agreement with the United States of America have led to a weakening of certain laws, and new laws have been made. These changes seem to undermine the

---

[18] www.aidesep.org.pe/

rights that Indigenous Peoples in Colombia secured in the 1990s. The Rural Development Statute (Law 1152 of 2007) (Houghton, 2008), which sought to make land available for investments, was declared unconstitutional in March 2009, while the National Development Plan (effective since 2007) seeks to expand the agricultural frontier and invite extractive industries into vast new areas of Colombia (World Bank, 2007).

The fight against illicit drugs

Colombian law appears to provide communities with unprecedented opportunity to control their own territories and food systems. In practice, however, interviewees for this chapter reported that there are several ways in which their control of territories can be overruled. First, military activity violates the land titles and tribal sovereignty previously enjoyed by the Inga. Both the guerrilla organization, Revolutionary Armed Forces of Colombia, and government military forces hinder Inga subsistence production and harvesting. Second, Plan Colombia, the United States-Colombian collaboration to fight illicit drugs, violates the sovereignty of the Inga's territories and food systems in important ways (HREV, 2008). The herbicide Roundup is being sprayed from aeroplanes on areas where plantations of coca (used to make cocaine) and poppy (used to make opium and heroin) are grown, some of which are close to Inga territories. The government has made no effort to protect the Inga against the impact of these herbicides, which cause diarrhoea, fever and other undocumented health effects (Gallardo, 2001). According to interviewees, fields that are sprayed become infertile. A recent scientific study showed that Roundup in residual dosages may cause cell damage, cancer and even death (Otaño, Correa and Palomares, 2010). One of the chemical ingredients was found to cause redness, swelling and blisters, short-term nausea and diarrhoea. Although Roundup is considered harmless for humans, tests have shown that it harms human cells in cell culture, and may cause damage in the concentrations found on herbicide-treated vegetables[19] (Gasnier et al., 2009).

---
[19] Dilutions of 1.100 000.

The effects that regular spraying will have on people and animals living in the vicinity of spraying and on drinking-water from rivers that flow through sprayed areas are unknown. There are no records of health authorities investigating the effects of these chemicals on the health of local inhabitants.

Hydrocarbons

The Inga's land, food security and livelihoods are also threatened by hydrocarbon companies. Although Indigenous Peoples in Colombia have territorial rights, the government continues to own the subsoil resources, including minerals, gas and oil. Up to 70 percent of Colombia, including nearly all indigenous territories, will be granted as concessions to hydrocarbon companies, implying a 50-percent increase in concessions (Houghton, 2008). This extractive policy will most likely violate Indigenous Peoples' rights to land, territories and resources, and is likely to undermine their right to food and water. However, as one interviewee remarked, the armed conflict has so far kept foreign investment/development at bay. If the conflict stops and the area becomes safer, oil drilling may become the major threat to the Inga's food security.

Food security, land and natural resources

The Inga's food security is threatened because food and drinking-water taken from the rivers may not be safe, owing to the Colombian policy of eradicating illicit crops by aerial Roundup spraying. Illicit drug and military activities in the area threaten food and water safety and hamper Indigenous Peoples' access to game animals, fish and other harvested foods. Most likely these activities also reduce the availability of game. In the future, oil and mining companies may start drilling within or near the Inga territories, further threatening their food security and access to safe food and water.

The autonomous indigenous councils try to counteract this by expanding their ancestral territories. For example, in 1999, the Inga – in collaboration with the Amazon Conservation Team (*Instituto de Etnobiológica*) and the National Parks Service – requested the creation of the Alto Fragua Indi Wasi National Natural Park in the southern department of Caquetá (Chapter 8 – Caicedo and Chaparro, 2012), an area rich in biodiversity and adjacent to the Inga's territory. This would protect the area from the migrant farmers who are being displaced from other regions, the pollutants that accompany illegal coca and poppy cultivation, and the herbicides used to eradicate illegal crops.

Although the Inga are in a difficult situation, their autonomous position and the size of their territories offer some protection to their livelihoods. However, there are reasons for claiming that the government has failed to address the armed conflict in the area and to take the necessary action to protect the Inga against the pollutants that are most likely threatening their food and water safety.

Inuit

Pangnirtung is a small Inuit community on Baffin Island, located in the vicinity of Iqaluit, the capital of Nunavut Territory in Canada (see Figure 9.1, pag. 142). Compared with that of most Indigenous Peoples, the living standard is relatively good, but lower than in the rest of Canada. Inuit children have significantly lower education outcomes than average Canadians, housing conditions remain well below national standards, and health indicators continue to lag behind those for the rest of Canada (Simon, 2009).

The quality of health services is also lower in Nunavut, partly because of high turnover of health personnel (Nunavut Tunngavik Incorporated, 2008). A recent report demonstrates that the low standard of housing, and overcrowding are linked to the rate of hospital admissions for infants with respiratory infections, which is the highest in the world (Kovesi *et al.*, 2007). Local water is good, and is regularly tested by government services. Water is provided by truck to homes, and elders continue to use melted ice chunks for their drinking-water and tea.

Traditional food
Inuit in northern Canada, including in Pangnirtung, have experienced reduced availability of harvested meat since giving up their nomadic way of life and being forced to settle in communities in the 1950s. The increased cost of hunting and the increased dependency on snowmobiles and petrol to travel the distance needed to find game have made it difficult for some households to harvest the traditional foods (or "country foods") they need.

Some consider themselves to be deficient in resources, including food and the ability to ensure their own food security. However, the sharing ethos survives, and many Inuit receive country food from relatives and others. Inuit practise and believe in sharing, and people say that when you give, you will get more in return.

Through Health Canada's First Nations and Inuit Health Branch and the Ministry of Indian and Northern Affairs Canada, the Canadian Government encourages the use of traditional food in the North, including in Nunavut and Pangnirtung. While adults and older generations tend to appreciate traditional foods, youth are turning towards market foods (Chapter 9 – Egeland *et al.*, 2013). This may indicate that the food culture is changing. However, it is speculated that the current generation of youth may appreciate country food more when they become older and form their own families (Egeland *et al.*, 2009).

Access to market food
Inuit depend on money to satisfy all their food needs, but incomes are often too low to provide the family with the food it requires. Many people in Pangnirtung receive income support and health care. However, food insecurity is a problem, due to both the costs associated with hunting and fishing and the high prices of airborne perishable market foods in the Arctic. As also occurs in some other northern communities, perishable foods of good nutritional quality have been subsidized through a Government Food Mail Program, but the subsidies are not sufficient to lower food prices to the level enjoyed in southern Canada. Less money is generally available to households, and food prices are higher among Inuit than among Canadians in the south, leading to food insecurity for some. It has been reported that some Inuit skip meals because they lack food or the money to buy it (Johnson-Down and Egeland, 2010).

The nutrition transition

Increased intake of market foods is also associated with overweight, obesity and diabetes (Egeland *et al.*, 2009). Inuit are experiencing a nutrition transition in which a market-based diet is gaining importance, especially among the young (Johnson-Down and Egeland, 2010). Rates of overweight and diabetes are increasing for Inuit internationally (Jørgensen *et al.*, 2003), which – given the obvious lack of food security – makes the paradox of the nutrition transition especially relevant for Inuit children.

Traditional foods are of key importance to Inuit food security, and 70 percent of households consume traditional food. Country food provides 41 percent of dietary energy for adults, but only about 23 percent for youth. Most of the carbohydrates in the diet come from market foods. Unfortunately, carbonated drinks lead to increasing intakes of sugar, while market food is increasing the content of saturated fat in the total diet (Kuhnlein *et al.*, 2004; Egeland *et al.*, 2009). Youth (people under 25 years of age) are a large consumer group, as they represent more than half (56 percent) of the total Inuit population (Statistics Canada, 2006).

Contaminants and climate change

Environmental contaminants in the country food harvested by Inuit have raised concern among inhabitants, the government and researchers. The government monitors levels of contaminants and funds research on adverse substances that may affect human health (Kuhnlein and Chan, 2000). The creation of CINE was largely a consequence of the realization that these matters called for close collaboration with indigenous communities in the Arctic.

It has now been proved beyond reasonable doubt that global warming is affecting the climate in the Arctic, and the ice is melting rapidly. In Pangnirtung, melting of the glacier in the surrounding mountainous terrain has resulted in serious and unprecedented flooding in the community (L. Okalik, personal communication, 2009). Climate change affects the living conditions of the local animal species that Inuit depend on for their food security (Chapter 9 – Egeland *et al.*, 2013). Despite these unsettling developments, however, the

Federal Government does not yet have an overall plan for environmental monitoring (ITK, 2007).

There is certainly local concern about climate change, but this is a global issue and is not restricted to Pangnirtung. Inuit are actively advocating for government and international action against climate change.

Land rights and policy

The Nunavut Land Claim Agreement (1993) gave Inuit of Pangnirtung and the whole Inuit population of Nunavut Territory a form of domestic self-determination. Eighty-five percent of the population of Nunavut is Inuit. Through the Nunavut government and a participatory governance structure, Inuit of Nunavut (including in Pangnirtung) are now able to make important decisions about their common future, but within the wider legal and policy framework provided by the Canadian Federal Government. Closing the gaps in housing standards, education and health services will require substantial public sector investment.

Canada represents Inuit in international climate negotiations, but has so far not played a particularly constructive role in ongoing efforts to reduce carbon emissions from fossil fuels, and thereby end global warming. So far Canada and other Western countries have gained reputations for undermining negotiations to protect the rights of Indigenous Peoples and forest-dependent communities in the face of climate change.

Conclusions and policy recommendations

These five Indigenous Peoples represent diverse traditional adaptations and food systems. Over recent decades, their territories and food systems have, in different ways and with different results, been influenced by State government decisions and economic actors. National governments have all assumed human rights obligations relevant to Indigenous Peoples' right to food. Some governments perform reasonably well in this regard, although they could perform even better. Others fail seriously, and violate the human rights and specific rights of Indigenous Peoples, which are crucial for their food security. Food systems can be undermined in several ways, including through national laws and policies and by unregulated extractive industries. In some cases, government policies and development processes contribute indirectly to nutrition-related disease by not making timely and effective efforts to stimulate the use of nutritionally superior foods, including traditional indigenous foods and diets.

The Federated States of Micronesia has an unimpressive record in ratifying human rights conventions, although it is encouraging that the country voted in favour of UNDRIP in 2007. However, it is puzzling that it should have voted for this, the most progressive human rights instrument on Indigenous Peoples' rights ever made, while failing to ratify ICESCR, ICCPR, ICERD and ILO 169. The human rights situation does not seem to be particularly problematic, and the people of Pohnpei have a government that is not imposed on them and other islanders, unlike most Indigenous Peoples elsewhere. With regard to ethnic descent, the government represents the people of Pohnpei to a large degree.

The transition from collective to individual landownership reduces traditional leaders' authority regarding natural resources. While traditional land management strategies have dwindled, the State seems largely to have failed to fill the void and assume the necessary regulatory responsibility. The degradation of land and the overuse of fish resources currently taking place are clearly unsustainable and challenge future national food security. The strong United States influence on the consumption patterns of Micronesians, and their increased purchasing power due to social security transfers from the United States of America have undoubtedly contributed to the obesity-prone food culture in Pohnpei.

From food security and right to food perspectives, it is promising that recently, when faced with the persistent and increasing obesity problem, the government took action to motivate the population's use of more locally grown and nutritious foods. This initiative should be strengthened, and the sustainability of land-use policies and fishery regulations improved. However, the Federated States of Micronesia should

also consider improving its record for ratifying human rights agreements.

The Inuit in Pangnirtung have an Inuit government at the community level. The government of Nunavut Territory is responsible for an area where most residents are Inuk seeking to meet Inuit interests and needs, but the Canadian Federal Government is the highest authority, and makes decisions regarding the funding of Nunavut social programmes and food subsidy programmes. The Inuit's well-being therefore often depends on Canadian laws and policies, including funding policy. Pangnirtung Inuit are also dependent on the Nunavut Land Claim Agreement. Canada was one of only four countries that voted against the new UNDRIP, although the Canadian Government later reversed this stance and voted for UNDRIP, as did the governments of the United States of America, Australia, New Zealand and Colombia, among others.

It is a concern that Canada does not take a more progressive role in international climate negotiations. Nationally, there is a need for a more decisive stand. Laws and regulations should be enacted to reduce the emissions of climate gases effectively; policies should support these regulations, and should include creating good incentives for the population at large and for industries, making it easy to choose climate-friendly alternatives. Internationally, Canada needs to play a more proactive role in climate negotiations and in negotiations regarding the protection of tropical forests and the rights of traditional peoples who depend on these for their livelihoods.

It is of great concern that Canada, the United States of America, Australia and New Zealand seek to block reference to the protection of Indigenous Peoples' rights in negotiated climate texts, in spite of the obvious and immediate threats that climate change poses to the traditional livelihoods and food security of Inuit and traditional peoples worldwide.

The Maasai are in a dire situation. As a first step, the Kenyan Government needs to acknowledge – in the Constitution and in law – the collective rights of pastoralists and their indigenous specific rights to uphold their herding livelihood (respect level). The Maasai are in critical need of water and grazing land,

and it should be the role of the State to ensure their land rights and to protect them from encroachments on to their land and other violations of their rights. Immediate government action is needed to address the precarious and immediate water, health, food and schooling situation of the Maasai, and to facilitate access to water and grazing land for the animals that they depend on for their livelihoods. Failing to do so is a violation of their rights, including, in some cases, their right to life. The Mau Complex is in urgent need of a sound restoration policy; the protection and restoration of its water-retaining capacity needs to be continued, to benefit the Maasai and the large numbers of other people who depend on the rivers downstream. The Kenyan Government should ensure that the most immediate needs are met. It is also critical that the government enter into dialogue with the Maasai, inviting their opinions and giving serious consideration to their inputs and suggested solutions.

The Awajún and the Inga are in fairly similar situations. Peru and Colombia were formerly progressive countries with regard to indigenous rights, but, (apparently) partly as a consequence of signing free trade agreements with the United States of America, they have started to undermine the progress made, and even their own Constitutions, to provide more possibilities for economic growth through the extraction of natural resources. Both governments are in conflict with their obligations under ILO 169, and both countries also have serious conflicts between their Constitutions and the policies and legal changes being implemented. Given the Peruvian and Colombian governments' unwillingness to respect and protect indigenous rights, there is need for strong international and external pressure. The Inga are now in a "no-win" situation. When or if the violence and military activities in their areas stop, the oil and mining companies are likely to enter.

In all five cases, the causes of food and nutrition-related problems can be found in the interactions between the indigenous community and the larger society, represented by and controlled by national governments. Governments have obligations that most are far from fulfilling, including those regarding the right to food. Inuit and the people of Pohnpei have become increasingly dependent on market food and have entered the nutrition transition. Many Inuit seem to be experiencing food insecurity, due to high food prices, low income and, probably, declining access to country food. The Maasai experience serious full-fledged food insecurity and even starvation, owing to lack of fertile land and water for their cattle herds, and failing income. The Awajún and the Inga have access to land areas with relatively bountiful natural food resources, but their land, water and food resources are threatened by unrest, exploitive industries, pollution and the side-effects of illicit drug cultivation, which also undermine their food security and thereby their right to adequate food.

Climate change is likely to have large repercussions, at least in the short to medium term, for all peoples who depend on nature for their subsistence. Indigenous Peoples are therefore rightfully concerned and are taking an increasingly visible stand internationally. The areas where the Maasai graze their animals have been hit hard by drought, probably partly caused by deforestation but accelerated by the general pattern of climate change. Inuit are experiencing rapid ice melting and their access to wildlife is threatened. In the Federated States of Micronesia, rising sea levels affect agricultural lands. This calls for national adaptation and mitigation strategies in addition to measures to reduce emissions of carbon and other climate gases. There is need for policies and legislation to protect Indigenous Peoples and others whose livelihoods and food systems are likely to suffer as a result of climate change.

This chapter has aimed to stimulate analysis of the wider circumstances surrounding the food and nutrition situation of Indigenous Peoples in various countries and circumstances. As noted by Stavenhagen (2007), there is an implementation gap between the actual situations that Indigenous Peoples live under, and the content of national laws, constitutions and States' international obligations. A human rights-based approach to food will increase awareness of governments' role with regard to food security. It is also a tool for ensuring attention to the need to find policy solutions that ensure social equity. Governments have an obligation to ensure, through laws and policies, that socio-economically and politically marginalized groups do not suffer from a poorer nutrition situation than other population groups.

These problems may be overcome if measures to ensure Indigenous Peoples' food security are planned and carried out with Indigenous Peoples' free, prior and informed consent and in accordance with their rights, including their right to food. In particular, the 2007 UNDRIP provides a good framework for interaction between governments and Indigenous Peoples, especially with regard to food security and nutrition.

UNDRIP is gradually gaining political attention and momentum; the 2007 vote and later updates show that most countries are now willing to accept, at least in theory, not only the existence of Indigenous Peoples' rights, but the need for progressive stands on land rights and self-determination. Reflecting this in national law and policy would be a large step towards equity-based and culturally sensitive food security and health policies. However, for this to happen, governments must first understand the full implications of current laws and policies on Indigenous Peoples' livelihoods and futures, and demonstrate a clearer understanding of their own obligations under human rights law. Not only must the State decide to play an active role, it must also take a stand for equity and cultural diversity, and against the aggressive exploitation and destruction of natural resources and the common global ecosystem heritage that is seen today. The support for UNDRIP suggests that the time is ripe for change. The escalation of food prices triggered by the global financial downturn, and the threat of climate change may increase awareness in both governments and national populations. This may translate into new policy directions, but these are more likely to occur on a large international scale if civil society, the media and other global citizens insist that change at many levels is essential to uphold the human rights of Indigenous Peoples ✿

Acknowledgements

The authors express their sincere gratitude to those who generously shared their time and insights by completing questionnaires, participating in individual and group interviews, and commenting on drafts, both orally and in writing. These people include, but are not limited to, members of the CINE Indigenous Peoples' Food Systems for Health Program, including several authors of the case studies presented in this volume.
Particular thanks to:
- *Inuit:* Looee Okalik, Grace Egeland and Inuit Tapiriit Kanatami;
- *Inga:* Ana María Chaparro, Sonia Caicedo and José Pablo Jaramillo;
- *Awajún:* Pedro (Perico) Garcia, Hilary Creed-Kanashiro, Irma Tuesta and Miluska Carrasco;
- *Pohnpei:* Lois Englberger, Adelino Lorens, Kiped Albert, Gus Kohler, Ricky Cantero, Rufino Mauricio, Dr Jim Rally, Jim Currie and Patterson Shed;
- *Maasai:* Joseph Ole Simel.
> **Comments to:** siri.damman@gmail.com

Chapter 16

Policy and strategies to improve nutrition and health for Indigenous Peoples

HARRIET V. KUHNLEIN[1] BARBARA BURLINGAME[2] BILL ERASMUS[1, 3]

1
Centre for Indigenous
Peoples' Nutrition and
Environment (CINE)
and School of Dietetics
and Human Nutrition,
McGill University,
Montreal, Quebec,
Canada

2
Food and Agriculture
Organization of the
United Nations,
Rome, Italy

3
Dene Nation and
Assembly of First Nations,
Yellowknife and Ottawa,
Canada

Key words > Indigenous Peoples, food systems,
food security, food policy, human rights

> **"Salmon has always been part of our life. It is the essence of our existence."**
>
> Bill Tallio, Nuxalk leader
>
> **"If our culture breaks apart, we will blow away in the wind."**
>
> Mailong-ong Sangkhachalatarn, Karen leader
>
> **"We need our land because it's sacred for our life."**
>
> Antonia Mutumbajoy, Inga leader
>
> **"We need to be able to control our own resources."**
>
> Bill Erasmus, Dene National Chief[1]

Abstract

Understanding how to achieve food and nutrition security for the world's populations is especially important for Indigenous Peoples, who often experience the most severe financial poverty and health disparities, and who often live and depend on ecosystems that are under increasing stress. There is need for targeted strategies and policies that facilitate and foster Indigenous Peoples' use, processing and management of their natural resources for food security and health, through self-determination and autonomy. These policies should be effective at the local, state, national, international and regional levels. This chapter reviews the international policy documents now in place that identify and protect Indigenous Peoples, their food systems and their human rights to adequate food and to the enjoyment of traditional food and food traditions. Basic principles of engagement in research and development activities that have been successful with Indigenous Peoples are explored and described through nine case studies from Indigenous Peoples' rural communities where interventions have taken place to improve local food use and health. Policies are most successful when they stress the importance of using cultural knowledge to advance health promotion activities and improve health and well-being – mental, emotional and spiritual health, as well as physical health – for individuals and communities. Academic and community partners in the case studies met annually for ten years to discuss strategies for documenting local indigenous food systems and ways of promoting them within local cultures and ecosystems. This chapter elaborates on the diversity of strategies and policies operating in unique rural ecosystems with varying degrees of success. Several of the case studies have successfully shared their methods and results for implementation in other communities in their regions. Community and academic partners have communicated widely about their visions and goals, the strategies used and the policies developed for improving food and nutrition security in all of its dimensions for Indigenous Peoples throughout the world.

Background and introduction

The world's attention has been drawn to the plight of Indigenous Peoples as they strive to retain their cultures and protect their ecosystems and food traditions in the face of globalization. The decade of work described in this chapter has documented vast knowledge about food biodiversity in Indigenous Peoples' areas, and the many cultural meanings and spiritual values reflected in these resources. The chapters in this book show how Indigenous Peoples' food systems are critical for health in all its forms. It is therefore logical to complete this research effort with an overview of existing policies surrounding these food systems

---

[1] These quotes can be heard in their original languages in a short film at www.indigenousnutrition.org

and with suggestions for enhancing policies that will promote and protect these resources.

This chapter pulls together experiences and perspectives from 12 case studies that have contributed methodologies and findings to the project on Indigenous Peoples' Food Systems for Health, published by the Centre for Indigenous Peoples' Nutrition and Environment (CINE) and the Food and Agriculture Organization of the United Nations (FAO) in two earlier publications (Kuhnlein *et al.*, 2006; FAO, 2009a) and the current volume. It draws on discussions with indigenous leaders and their academic partners, and with resource people met at The Rockefeller Foundation Bellagio Center in 2004, 2007 and 2008. These discussions were sentinel in formulating this chapter. Collaboration is required not only for success in the research process of documenting how these food systems can improve the health and well-being of Indigenous Peoples, but also for understanding and creating successful policies that involve many actors and many dimensions.

The questions addressed here are:

- Why should Indigenous Peoples' food systems be protected and strengthened?
- What policies will stimulate efforts at the local, national and international levels towards the achievement of food and nutrition security for Indigenous Peoples?

In this book, consideration of the benefits of Indigenous Peoples' food systems is based on empirical knowledge of the biodiversity and nutrient content of the rich resources contained in those food systems. The task has been to merge this information with the many other factors at issue: imperatives for environmental conservation; health challenges for Indigenous Peoples, who often live in financial poverty; Indigenous Peoples' recognition of the many physical, mental, social and spiritual aspects of local food resources; and the human right to enjoy these resources, which are intimately connected to food security, culture, and land and aquatic ecosystems.

The world is struggling under the burden of food insecurity, recently exacerbated by the increasing use of agricultural lands to produce biofuel, food price crises in global markets and lack of adequate policies for

improving the nutrition situation of all citizens (UNS/SCN, 2009). Understanding how the human right to food can be realized for the entire world's population is especially important for Indigenous Peoples, who often experience the most severe poverty and health disparities. In addition, the ecosystems on which many Indigenous Peoples depend are under increasing stress.

According to conservative estimates, Indigenous Peoples number more than 370 million people living in 90 countries around the world (UNPFII, 2009); they use and represent more than 5 000 languages and cultures in diverse ecosystem settings. In seeking an appropriate definition of "indigeneity", important principles are those of self-identification; collective attachment to a distinct geographic territory and the resources therein; separate customary cultural, economic, social or political institutions; and an indigenous language that is often different from the official language of the State (UNPFII, 2009). For example, the 645 scheduled tribes of India, which are considered as Indigenous Peoples by the State of India, comprise 84 million people, or approximately 8.2 percent of the country's total population (Appendix 1). Computations of the indices for human development, human poverty and gender equality, literacy rates and key health indicators all demonstrate that these tribal peoples are far more deprived than the rest of India's population (Sarkar *et al.*, 2008).

Indigenous Peoples face disparities resulting from the serious impacts of colonization, which to varying degrees have influenced not only the ways in which people view their local food resources in contrast to imported foods, but also people's social structures and hierarchies. Indigenous Peoples face disparities in income, access to health care and the provision of services that are often taken for granted in the mainstream societies of the countries in which they live.

The effects of globalization on nutrition and health disparities are far-reaching and reduce exposure to traditional cultural knowledge and the biodiversity of local food resources. Examples are the wide availability of low-quality industrially produced foods, development paths that bring mining and other ecosystem-destroying activities, and forces that compel migrations to cities to seek jobs.

Despite these factors, which drive Indigenous Peoples from their local foods, there is considerable economic rationale for promoting local foods and lifestyles for their health benefits. Not only is local food less expensive from an economic point of view, while harvesting and supplying families with local foods provides many fitness and cultural benefits, but also the costs of poor health and health care can be exceedingly high when nutrition and lifestyle are compromised. Malnutrition in its various forms, including obesity and its consequences (diabetes, heart disease, cancer, etc.), is very costly for tertiary care institutions and in social terms.

There is need for targeted policies that facilitate and foster the conservation, management and sustainable use of Indigenous Peoples' natural resources for food security while also fostering Indigenous Peoples' self-determination and autonomy. Health promotion can be achieved through policies – at the community, local, state, national, international and regional levels – that improve the livelihoods of Indigenous Peoples. These policies may include facilitating the sustainable marketing of foods and medicines derived from the ecosystems where Indigenous Peoples live.

Existing international policy documents that identify and protect Indigenous Peoples and their food systems

In recent years, Indigenous Peoples have become increasingly active in international policy settings to counteract discrimination and other injustices. Although there have been many challenges and conflicting opinions regarding situations and activities, United Nations (UN) agencies and their Member States have developed institutional frameworks to protect Indigenous Peoples' traditional customs, livelihoods and lands. In all regions, development assistance has been offered in recognition of Indigenous Peoples' valuable knowledge and role as custodians of much of the world's food biodiversity. Indigenous Peoples first entered the UN arena in 1982, when the Working Group on the Rights of Minorities of what was then the UN Sub-Commission on the Protection and Promotion of the Rights of Minorities (now the Advisory Group to the United Nations Human Rights Council) invited Indigenous Peoples' representatives to take part in its meetings as observers and to express their views on their own situations. Since then, the United Nations Permanent Forum on Indigenous Issues (UNPFII) has been established and has declared two International Decades of Indigenous Peoples (1995 to 2015), with the current decade having the objectives of: i) promoting non-discrimination and inclusion in national processes affecting Indigenous Peoples; ii) promoting Indigenous Peoples' effective participation in decisions affecting lifestyles, cultural integrity and collective rights, including through free, prior and informed consent; iii) promoting development policies with full equity; iv) adopting targeted policies that focus on indigenous women, children and youth; and v) developing the monitoring of and accountability for national, regional and international policies that affect Indigenous Peoples' lives.[2]

The issues have consistently been discussed in the context of Indigenous Peoples' human rights. In September 2007, the Declaration on the Rights of Indigenous Peoples was adopted by the United Nations General Assembly. This was a landmark for the acceptance of Indigenous Peoples' rights, with the majority of nations as signatories. The declaration enforces Indigenous Peoples' rights to maintain and develop political, economic and social systems, secure in the enjoyment of their own means of subsistence, and to engage freely in traditional and economic activities (Appendix 2). Thus, nation States have the obligation not only to ensure non-discrimination but also to safeguard the distinct cultural identities of Indigenous Peoples.

Indigenous Peoples' right to adequate food

The World Conference on Human Rights (UN, 1993) states that all human rights are universal, inalienable, indivisible, interdependent and interrelated, thereby ensuring that the right to adequate food is understood

---

[2] http://social.un.org/index/indigenouspeoples/aboutusmembers.apsx

in the context of all other human rights, including, when applicable, Indigenous Peoples' rights. The 2002 Declaration of Atitlán from the Indigenous Peoples' Consultation on the Right to Food reflects this understanding (IITC, 2002).

The human right to adequate food can be understood as a right to food security (Damman, Eide and Kuhnlein, 2008). The international legal standards underpinning the right to food for Indigenous Peoples have been described by several authors (Damman, Eide and Kuhnlein, 2008; FAO, 2009b), including in Chapter 15 (Damman, Kuhnlein and Erasmus, 2013) in this volume. The UN Declaration on the Rights of Indigenous Peoples and the International Labour Organization (ILO) Convention on Indigenous and Tribal Peoples' rights (ILO 169 of 1989) underpin Indigenous Peoples' special right to enjoy their specific cultures. This includes their right to enjoy their traditional food, as food traditions are at the core of indigenous identities, cultures and economies (Damman, Eide and Kuhnlein, 2008).

UN agencies have adopted development policies specifically for Indigenous Peoples and issues related to their food systems, nutrition and health. To date, these agencies and organizations include the World Health Organization (WHO, 2010), the Pan-American Health Organization (PAHO),[3] the World Bank (2010), the United Nations Educational, Scientific and Cultural Organization (UNESCO, 2010), the Convention on Biological Diversity (CBD, 2010), Bioversity International (2010) and the Human Rights Council (UNPFII, 2009). The International Fund for Agricultural Development (IFAD) has a funding programme specifically for Indigenous Peoples (IFAD, 2010). FAO has several initiatives relevant to Indigenous Peoples and their food systems (Appendix 4) within the Livelihood Support Programme (FAO, 2010a), the Sustainable Agriculture and Rural Development Programme (FAO, 2010b), the FAO Informal Working Group on Indigenous Issues, and the Right to Food Unit, in collaboration with the Focal Point on Indigenous Issues (FAO, 2010d). FAO's Nutrition and Consumer Protection Division has established publication venues and fora for scholarly and public information on the food diversity and nutrition of Indigenous Peoples, through peer-reviewed processes including the International Network of Food Data Systems (INFOODS) (FAO, 2010c), the International Conference on Dietary and Activity Methods (2009)[4] (Appendix 3) and the Conference on Health and Biodiversity (COHAB Initiative, 2008); and has integrated Indigenous Peoples' nutrition issues into policy instruments and recommendations such as food-based dietary guidelines (Health Canada, 2007) and the AFROFOOD Call for Action (AFROFOOD, 2009). The International Union of Nutritional Sciences (IUNS) and the United Nations System Standing Committee on Nutrition (UNS/SCN) each have task force activities on Indigenous Peoples' food systems (IUNS, 2010; UNS/SCN Web site).[5] FAO has published a previous book from the programme discussed in this volume (FAO, 2009a).

The effectiveness of policies to ensure that nation State duty-bearers take the right to food and food security into account can be explored through the normative framework to "respect, protect and fulfil" these rights (CESCR, 1999). Specific indicators relevant to Indigenous Peoples, the biodiversity in their food systems and their well-being have been reported (Kuhnlein and Damman, 2008; Stankovitch, 2008). To facilitate the interpretation and use of the concept of the right to adequate food, FAO has recently released operational guidelines on Indigenous Peoples' right to food (FAO, 2009c). These are useful in Indigenous Peoples' settings, for advocacy on the right to food; for ensuring that national data are disaggregated for indigeneity, to develop suitable indicators for assessing food security; and for creating human rights-based strategies and policies for the food security of Indigenous Peoples.

Food and nutrition security to improve the health of Indigenous Peoples

Food security exists "when all people at all times, have physical and economic access to sufficient, safe, and nutritious food to meet their dietary needs and

---

[3] www.new.paho.org

[4] www.icdam.org/index.cfm
[5] www.unscn.org

food preferences for an active and healthy life" (FAO, 1996). Food security can be defined for individuals, households/families, communities or larger populations such as nation States. In practical terms, for a household or individual it implies the sustained availability of and access to sufficient, safe and culturally acceptable food from which to prepare nutritious meals that will meet the dietary needs and preferences for maintaining a healthy and active life. Food security is a precondition for nutrition security, which requires simultaneous access to adequate health services, clean water and adequate sanitary conditions, plus adequate care for vulnerable age groups, the sick and the infirm.

For Indigenous Peoples, food security is necessary not only for health, but also for maintaining relationships with the land, resources, values and social organization, and for identification with indigenous culture, including culturally appropriate food (FAO, 2009b). Health is recognized broadly as intertwining with nature and culture for well-being and being articulated through physical, mental, spiritual and social elements, for both individuals and communities. Elders in many cultures recognize that consuming their own indigenous foods is necessary for maintaining health and well-being. Thus, advocacy and promotion of food security and health must include essential aspects of political, economic, social and cultural life, values and world views, to maintain equilibrium and harmony in the community (Cunningham, 2009). This holism requires the integration of local indigenous world views and visions and an interdisciplinary and multisectoral approach from researchers and food security and health promotion agents.

By ensuring that health data are disaggregated by culture and gender to reveal the circumstances faced by Indigenous Peoples in both urban and rural areas, many studies now show that the health circumstances faced by Indigenous Peoples are disturbingly worse than those of their non-indigenous counterparts in the population, in both low- and high-income countries. These disparities are manifest in virtually all health indicators, and predominantly in measures of undernutrition (particularly stunting and wasting) and overweight (obesity and related chronic diseases)

(Damman, 2005; Damman, Eide and Kuhnlein, 2008; UNPFII, 2009; Chapter 2 – Egeland and Harrison, 2013). Data to this effect have been found in many nations and regions, including South, Central and North America (Damman, 2005; PAHO, 2007), India (Sarkar et al., 2008), Venezuela and Guatemala (UNPFII, 2009; International Work Group for Indigenous Affairs, 2004), and Canada (Chapter 2 – Egeland and Harrison, 2013; Gracey and King, 2009; King, Smith and Gracey, 2009).

Such nutrition circumstances and disparities for Indigenous Peoples are rooted in often extreme income poverty. However, perhaps even more important is the poverty that results from poor access to health, social services and education, including a lack of education on indigenous structures and heritage, which seriously hampers access to the ecosystem resources that contribute to food and nutrition security. For example, the Ainu case study in Japan (Chapter 13 – Iwasaki-Goodman, 2013) identifies disparities in both income assistance and education. Among Ainu people in Hokkaido, annual income was lower and fewer students graduated from high school, with many respondents reporting financial hardship as the reason for not attending higher education institutions (Hokkaido University Center for Ainu and Indigenous Studies, 2008). It should be noted that many Indigenous Peoples do not recognize their wealth in terms of financial income.

Indigenous women and children are especially vulnerable to health disparities, with poor health leading to higher morbidity and mortality statistics for indigenous populations. There is need for special attention to indigenous women, who are often targets of discrimination and racism. They are particularly vulnerable during pregnancy and lactation, and are crucial to the healthy growth and development of their young children. Women are the "gatekeepers" of family food provisioning, and have particular need of policies that protect the right to food of both their families and themselves.

A holistic understanding of Indigenous Peoples' food traditions reveals that they are linked to physical, emotional, social and mental health and well-being. The capacity to enjoy their own culture is a human right of all peoples. This leads to consideration of the negative effects that lack of access to traditional food resources will have on cultural morale, identity, and mental as well as physical health. Mental health and suicide statistics demonstrate disparities and health gaps. The intolerably higher rates suffered by Indigenous Peoples are linked to the compounding factors of poor diet and fitness and lack of responsive community health care services (King, Smith and Gracey, 2009).

While Indigenous Peoples in all countries tend to be poorer than their non-indigenous counterparts, and while many suffer from undernutrition, they are also showing increasing obesity and related chronic diseases. Resulting from processes of the nutrition transition, through which poor people consume increasing amounts of poor-quality cheap food, and the reduction of physical activity as living conditions become more sedentary, obesity leads to alarming increases in diabetes and its sequelae among Indigenous Peoples, who may also be more vulnerable owing to genetic circumstances (Damman, Eide and Kuhnlein, 2008; UNPFII, 2009). In the Canadian Arctic, three cultures of Indigenous Peoples consumed from 5 to 40 percent of their dietary energy as traditional food. Even only one serving of traditional food a day led to improved dietary nutrient profiles compared with diets composed of only purchased foods, which were noted as being of low nutrient density (Kuhnlein *et al.*, 2004; Kuhnlein and Receveur, 2007). As noted in the Gwich'in case study in Chapter 7 (Kuhnlein *et al.*, 2013), access to traditional food meant better food security.

Policies to counteract these immense health challenges should be developed with Indigenous Peoples in communities and governments. Properly implemented policies can ensure access to highly nutritious traditional indigenous local foods and reduce incentives for purchasing poor-quality market foods (especially those with high sugar and saturated and trans-fat contents) and other junk foods. The use of healthy foods can be promoted through government subsidies that make them affordable. Policies can also give impetus to the protection and conservation

of traditional food ecosystems by enforcing joint management of these resources between governments and indigenous leaders, and can promote incentives that encourage the harvesting of foods from the land. There is also need for policies that include Indigenous Peoples in the management of their traditional community food resources and the importation of healthy market foods into communities, and that provide training in how to use these appropriately. All such policies will help communities and nations to move forward in enhancing the food security and nutritional health of Indigenous Peoples, who are often the most vulnerable and face the greatest risks to health.

Engagement with Indigenous Peoples for research and development activities: basic principles

Indigenous Peoples themselves are the major participants in the projects reported in this volume. The research was carried out in the expectation that specific findings would lead to positive benefits to improve local circumstances, and that the results of activities would strengthen and reflect the identity of the community concerned. The principles of free, prior and informed consent for research on food systems and activities to enhance their use have therefore been instrumental to success.

Recognizing that cultural sensitivity towards the community's goals, needs, perspective and vision of prime importance, research and development processes should ideally be created and conducted by indigenous researchers and development officers within the communities concerned. Frequently, however, indigenous communities request the assistance of highly trained and respected academic leaders for research into and promotion of food security, nutrition and health, as occurred in the case studies described here. These are built on the principle that Indigenous Peoples must be in equal partnership with academic leaders from the home country of the project. Throughout the activities of the many partners in the overall programme, significant efforts have been made to use participatory research and development practices with the communities involved and their leaders. The programme's academic and indigenous leaders have heralded these principles as guiding the case study process (Kuhnlein, Erasmus and Spigelski, 2009).

In Canada, CINE has been a leader in developing the concept of research agreements with the communities where research is conducted (Sims and Kuhnlein, 2003; Kuhnlein *et al.*, 2006). The Institute of Aboriginal Peoples' Health of the Canadian Institutes of Health Research now has guidelines for health research with Canadian aboriginal communities, which have been developed from this model (Canadian Institutes of Health Research, 2007).

Capacity building is key to successful research and development with Indigenous Peoples. This relates to ensuring the principles of inclusion and self-determination while building essential skills in research, reporting research in peer-reviewed literature and at conferences, and designing and delivering relevant development programmes for the community. Decolonizing methodologies promote culturally sensitive and often unique ways of working with Indigenous Peoples in their communities that support success for better nutrition (Canadian Institutes of Health Research, 2007; Tomaselli, Dyll and Francis, 2008; Tauli-Corpuz and Tapang, 2006).

Identifying partnerships with indigenous communities and including local researchers establishes credibility within the community and contributes to capacity building, inclusion and employment. Throughout the work described in this volume, the leading researchers have been impressed by indigenous women's knowledge of and capacity for research and food promotion activities that emphasize women's knowledge of the foods in their environments and the availability and acceptability of these foods to children and others. Women often oversee the family food supply, and can be encouraged to explore the possibilities for change using their knowledge of their food systems.

A key principle when designing policy is recognizing the need for Indigenous Peoples to have access to their own foods. This recognition can be demonstrated by considering subsistence harvesting permissions, ensuring land access for the agriculture of traditional crops,

promoting unique conservation efforts for Indigenous Peoples' harvest of medicines and food, and other factors. Additional examples are given in Chapter 15 (Damman, Kuhnlein and Erasmus, 2013).

National and international policies and networks discussed in the case studies

Throughout the annual discussions held with case study partners, it became clear that the wealth of knowledge on ecosystem resources that could be used for food security, livelihoods and health in various dimensions is a major part of indigenous identity and that the use of these resources is important for self-determination and cultural morale. There is an obvious need to harness these resources for the betterment of the people directly involved, which underscores the necessity for developing and applying effective policies at all levels to ensure the conservation and sustainability of local food systems.

Over ten years of communications and meetings, the programme created methods for documenting the resources used in food systems (Kuhnlein *et al.*, 2006) and presented documentation of 12 case studies of food system resources (FAO, 2009a). Discussions focused on how these resources could be used to better advantage in health promotion, and on what kinds of local, national and international policies currently existed in the case study environments. Each case study reflected on the existing and hoped-for policies for community, regional (state), national and international collaboration. This section attempts to capture some of these policies and visions on how they would provide greater insurance for food security, nutrition and health in the study areas.

Intersectoral collaboration within governments is a major imperative when dealing with the many influences that affect nutrition within a country, and the many disparities in food access across cultures, economic strata and geopolitical locations. However, the unique issues that Indigenous Peoples experience also need to be discussed and acted on in many other settings – locally and nationally as well as internationally. In the case studies, it was perceived that planning and practical activities should be undertaken by state and federal ministries working in agriculture, health (especially maternal and child health), education, culture (including national history and museums), environment and natural resources, as well as by universities and research institutes, the church(es), local and national media, commerce/trade and economic interests. Thus, a broad spectrum of interests in local and national governments and in the non-governmental organizations (NGOs) and funding agencies working in developed and developing countries must be addressed, to conduct meaningful research and solve problems. The goals established by indigenous leaders and national government authorities must be met, so that effective policies can be established and pursued.

While there is need to recognize and maximize the use of local rural indigenous food resources for Indigenous Peoples, it is also important to understand that an increasing proportion of Indigenous Peoples live in large urban environments. Here, as well as in most rural settings, there is heavy reliance on foods derived from national agricultural production and the globalized food industry. There is therefore need for careful consideration of policies that bring healthy affordable foods into both the rural and the urban areas where Indigenous Peoples live. Efforts to do this have been under way in Australia, with the Remote Indigenous Stores and Takeaways Project,[6] and Canada, with the Healthy Foods North Program[7] and the Food Mail Program,[8] sponsored by the Canadian Government to provide subsidies for transporting healthy foods to indigenous communities in northern Canadian. In Africa, the Rural Outreach Program has worked extensively to encourage small farmers to provide leafy green vegetables to urban areas (Shiundu and Oniang'o, 2007). Thus, there is a broad diversity of possible responses to ensuring food and nutrition security through adequate food supplies in communities.

Various sectors of government need to reflect carefully to understand the origins of malnutrition problems (both undernutrition and the overweight/obesity complex) in indigenous communities, and the best ways of addressing these. For example, the provision of subsidized refined white rice to communities in the Zaheerabad district of Andhra Pradesh in India had the effect of undermining agricultural production of local biodiverse crops, and contributed to lower micronutrient contents in diets. As people gave up the production of local crops and had to find financial means of subsistence, the cost of rice increased, and poverty became worse. Activities for promoting the production and sale of local millets and uncultivated green vegetables required substantial planning and action in several government sectors in the local area (Chapter 6 – Salomeyesudas et al., 2013).

National governments need to reflect on colonization's far-reaching impacts at the local level, and on how to reverse unhealthy food purchasing behaviours and restore access to healthy local foods. This often requires substantial cultural education that gives credence to the traditional knowledge of elders, particularly for the benefit of youth. At the same time, knowledge of health qualities and the preparation of foods available in commercial markets is also needed. One activity that promotes this comes from the case study with Inuit people in Pangnirtung, Baffin Island, Canada (Chapter 9 – Egeland et al., 2013), where recorded stories about traditional food harvests and use were presented on DVDs in classrooms and the media.

In Canada, the Food Security Reference Group of the First Nations and Inuit Health Branch made strides in identifying how to conduct community-level research into the needs for food and nutrition development activities, and on how to stimulate and implement these. Regular meetings were held with indigenous health leaders and government sectors dealing with health, agriculture and Indian affairs, and research activities that reflect local community values and local food resources to improve health were developed (Power, 2008).

School curricula in indigenous areas are successful when policies are in place to incorporate traditional language instruction and cultural knowledge, particularly about traditional food resources. Several of the case studies documented their food systems, including with photographs and text describing different

---

6 www.healthinfonet.ecu.edu.au/health-risks/nutrition/resources/rist
7 www.phac-aspc.gc.ca/publicat/2009/be-eb/nunavut-eng.php
8 www.ainc-inac.gc.ca/nth/fon/fm/index-eng.asp

food species suitable for use in schools so that children and youth can enjoy the benefit of research results. Resource books, posters and videos were prepared in Nuxalk, Gwich'in, Inuit, Awajún, Inga, Pohnpei, Dalit, Ainu and Karen areas. The resource books documenting the food resources of the Awajún (Chapter 5 – Creed-Kanashiro et al., 2013) are deposited in Peru's National Library in Lima.

A national policy for stimulating ecotourism in partnership with indigenous communities can give focus and voice to indigenous cultures. Such policies have the benefit of showcasing indigenous values and can promote appreciation of local foods and their preparation while providing income to local communities. In some areas, demonstrations of food harvesting activities that draw on ecotourism revenues have also been useful. Success in such endeavours has been shown with the Ainu of Hokkaido Island in Japan, the Dene Nation in Canada, the Karen in western Thailand and indigenous people on the island of Pohnpei in the Federated States of Micronesia.

It is useful to consider issues of human rights and the right to food for Indigenous Peoples in the context of food security, with a focus on the right to decide over the use of resources in indigenous territories. Several examples of human rights issues in food systems and their application are given in Chapter 15 (Damman, Kuhnlein and Erasmus, 2013). An example from the Awajún case study shows how media attention to government efforts to sell Awajún land to forest and mine developers stimulated a media backlash that resulted in a reversal of government policy. It is now possible to monitor government policies for the use of land in the Amazon area of Peru, although the threat of negative government policies remains (Asociación Interétnica de Desarrollo de la Selva Peruana Web site).[9]

Academic and community leaders in the Karen case study in Thailand (Chapter 10 – Sirisai et al., 2013) reflected on policies at different levels that have been identified as affecting the Karen and that could be promoted in different parts of government. At the local community level, policy dialogue has considered cultural practices for the health and care of children, income generation and cultural preservation. At the provincial level, successful topics for policy-makers include developing a participatory approach to health care in school curricula, including support to health care workers and measures for preventing and curing undernutrition in Karen communities. At the national level, policy topics directly related to Karen interests include forestry and biodiversity conservation, investments to support children, food safety, equitable sustainability and social movement campaigns. International-level topics that have resonance with Karen priorities are global health, global warming and world biodiversity protection for future generations.

Dam construction has seriously affected the Ainu (Chapter 13 – Iwasaki-Goodman, 2013) through disturbances to local habitat and culture. However, a recent plan to construct a dam near Biratori was assessed for its impact on Ainu culture, as determined by Ainu people themselves. Following their report that there would be a negative effect on Ainu culture, the dam is now on hold.

During programme meetings, many participants reflected on how issues that are important to Indigenous Peoples and that generate favourable government policies also further the goal of general populations in the countries concerned. These issues include unique food resources and their conservation, cultural conservation as part of national heritage, and environmental protection. Indigenous leaders are among the most eloquent voices for mitigating the effects of climate change (Watt-Cloutier, 2009). Programmes that work for Indigenous Peoples can be shared, not only within the indigenous world, but also in other settings, for example, walking/running programmes to prevent diabetes, such as the Zuni Out-Run Diabetes Program, initiated in response to concerns about high-carbohydrate and -sugar foods (United States Department of Health and Human Services, 2010).

It was suggested that governments should take the initiative on policies to prevent the sale of unhealthy foods. This is especially important in communities where there are limited facilities for adult education, where financial poverty prevails, and where there is

[9] www.aidesep.org.pe/

minimum or no competition in local food stores. Such policies follow on from successful prohibition programmes to reduce alcohol, tobacco and drug abuse. Nevertheless, it is difficult to create and implement food and nutrition policies that cut across many government sectors, and to identify where strategic responsibility lies for investments to fund these policies. The imperative to include representatives of the people most directly involved in policy decisions is obvious – especially when answers and actions can be generated in the community.

Policies will be more successful if important national characters share their positive visions in the media. This happened in the Pohnpei case study, when the President of the Federated States of Micronesia promoted local foods as the foods of choice for all official events (Island Food Community of Pohnpei, 2010). Another example is the positive impact for the Karen researchers from Her Royal Highness Crown Princess Sirindhorn's commitment to improving disadvantaged children's nutrition and overall quality of life.

In Colombia, the Amazon Conservation Team, an NGO, has worked in partnership with the Colombian National Park Service to create the Alto Fragua Indi Wasi National Park, a protected area jointly managed by indigenous communities. The NGO has also worked with the park service to establish the Orito Ingi-Ande Plant Sanctuary, which creates a new category of reserve that protects plants of high cultural value to indigenous communities, including those in the Inga case study (Amazon Conservation Team, 2010).

Unfortunately, however, some other national policies are counterproductive to Indigenous Peoples' efforts to improve nutrition and health. Examples include governments' promotion of cash cropping by outsiders in indigenous land areas, fumigation of agricultural lands against illegal crops, as described in the Inga case study from Caqueta Province in Colombia (Correal *et al.*, 2009), and sales of lands for mining and oil harvesting, as described by the Awajún in Peru. Another example is permitting free enterprise and the marketing of poor-quality foods to children, through shelf-stocking procedures that make pop, candy and snacks easily accessible for small children and youth in local stores

in indigenous communities because these practices are prescribed by vendors' central offices (H.V. Kuhnlein, personal observation, 2008). The rice subsidies that undermine local Dalit food production in Hyderabad district of India, and the overwhelming lack of control of climate change effects that have impacts on local food availability and harvests are other examples. Special attention should be directed to policies for indigenous women as the holders of much knowledge of food system diversity, and also as those often discriminated against, with health consequences for themselves and their children. Within the UN system, policies that protect women's rights to food and gender equality are encouraged (FAO, 2009b).

Scaling up case study findings to broader regions: moving the agenda forward

The food system interventions described in this volume demonstrate that Indigenous Peoples' nutrition and health are more likely to improve when aspects of their food systems are promoted. Because of funding constraints, the programme tended to engage with small communities, often of fewer than 1 000 people; however, enthusiasm has been high in most of these communities, and some have requested the scaling up of intervention activities, which is already under way in some cases. The leading researchers recognize the values and benefits of starting on a small scale with intensive work based on participatory processes and community engagement, and then sharing and building on success stories with more communities, using the bottom-up approach.

There are numerous options for broadening intervention activities: expanding to nearby communities; or networking to engage NGOs or government agencies in similar activities across regions where world-views are similar and culturally linked ecosystem food species are known to be available. Community members themselves often tell their neighbours and friends about appealing, helpful and successful community programmes, thereby stimulating a demand for similar activities in nearby villages. With community support and engagement, programmes based on access to and use of local resources will become sustainable at the local level. Indeed, the scaling up of a locally supported programme to other communities can be the ultimate proof of programme sustainability. Various forms of scaling up from the grassroots (and the need for scaling down from top management) are presented by Uvin (1999), with examples.

There are many ways of stimulating scaling up activities. First, the need for the activities must be expressed at the community level, and avenues for the input of new ideas for addressing food, nutrition and health issues must be presented. As well as word-of-mouth exchanges among family and friends, meetings of community leaders also often lead to calls for action. Electronic networks – established through government agencies or others at the local, national and international levels, with publications, Web pages, film and other media presenting the findings from problem assessments – also help to share successful strategies. School curricula can be effective for sharing local food system information, not only within the classroom but also at home, with pupils communicating it to their extended families. Curricula can be developed for both local schools and more central educational planning.

As already noted, UN agencies have developed networks for advocacy and the funding of successful intervention programmes, and have the capacity to develop databases on Indigenous Peoples' food systems and intervention strategies.

There is, of course, no single model for engaging with indigenous communities, or for building the strategies and structures for scaling up to nearby communities or to the regional or national level. Because of diversity in cultures, ecosystems, world views, languages and ways of knowing and doing, the local leadership and circumstances must be respected, to ensure the best strategies for the local setting and successful planning and communications with the people most directly involved. Such local action supplements and operationalizes higher-level government policies that protect Indigenous Peoples' land and food systems to improve families' diets and health.

Indigenous and non-indigenous partners can work together in communication and planning, to move the agenda forward to the benefit of more indigenous communities in similar cultures and ecosystems. However, indigenous methodologies must not be applied inappropriately in larger (non-indigenous) populations. Often, both local and national languages must be used in communications to raise awareness of the needs and challenges for programmes to promote food systems, nutrition and health. Through this, local communities that develop successful programmes can be inspired further by sharing their work with broader audiences.

Case study success stories

The programme case studies were based on the expectation that findings would be shared at the international level. Among the nine studies reported here, several plans have been implemented for scaling up successful intervention strategies to improve food security, nutrition and health. Each success story has been guided by local vision and leadership, to achieve what is most useful for the people involved.

Awajún
(Chapter 5 – Creed-Kanashiro *et al.*, 2013)

The original research on documenting the Awajún food system and the before-and-after evaluations were conducted with six communities in the Cenepa River region of the Amazonas district of Peru. The intervention developed the capacity of 32 health promoters, who worked extensively in 16 regional communities to deliver nutrition messages focusing on high-quality foods for infants and young children throughout the *Organización de Desarrollo de las Comunidades Fronterizas de Cenepa region*. So far, many community food gardens and more than 400 fish farms have been created. Community requests for workshops and activities in food topics, including food production, nutrition and culture, continue to be made through women's groups in the region.

Dalit
(Chapter 6 – Salomeyesudas *et al.*, 2013)

This intervention was conducted and evaluated in selected communities in the Zaheerabad region of Andhra Pradesh, southern India. The Deccan Development Society (DDS), an Indian NGO, continues to conduct activities with *sanghams*, which are regional organizations of Dalit ("untouchable" in the Hindu religion) women farmers. The overall objective is to enhance the food security of Dalit families, with multiple outreach activities emphasizing organic agriculture with local food species. DDS's most significant activities include negotiating funding for the management and cultivation of fallow land (2 675 acres [about 1 083 ha] to date) by poor and illiterate women, and distributing the traditional grains (sorghum, millets) produced throughout the communities, as well as creating job opportunities. DDS continues to develop a broad array of awareness-building activities on the use of local foods, such as films, local radio, cooking classes, food festivals and the provision of local foods in meals at day care centres. More than 3 600 families in 75 villages in Andhra Pradesh participate in these activities, and media distribution has been extensive and highly celebrated.[10]

Gwich'in
(Chapter 7 – Kuhnlein *et al.*, 2013)

The First Nations community of Tetlit Zheh in the Northwest Territories of Canada participated in research activities over several years. The intervention was created to increase the use of traditional Gwich'in food and higher-quality market food available in the community. The most appreciated intervention product was a locally produced traditional food and health book distributed through the community council and the Dene Nation in Yellowknife. Tetlit Zheh's local radio station, CBQM, promotes traditional activities, recipes from the food and health book and additional activities on the land. Provincial and national nutrition agencies

throughout Canada promote use of the local cultural food of Canadian First Nations, Inuit and Métis, with participation from national aboriginal organizations, the Assembly of First Nations and the Inuit Tapiriit Kanatami.

Ingano
(Chapter 8 – Caicedo and Chaparro, 2013)

The leaders of the Tandachiridu Inganokuna Association provided the project's development approach and continue to support its activities. The Amazon Conservation Team (ACT) is an NGO with activities throughout the Amazon region that promote the use of traditional food and medicine. Notable among its successes are educational activities at Inga Yachaicuri School near Caqueta, and primary health care and food security centres staffed by health brigades. ACT Colombia assists five indigenous communities with more than 1 000 traditional crops on 650 acres (245 ha) of land. ACT, together with the Government of Colombia, established the Orito-Ingi Ande Medicinal Flora Sanctuary and the Indi-Wasi National Park for conservation of the biodiversity known to Indigenous Peoples in the region.

Inuit
(Chapter 9 – Egeland *et al.*, 2013)

The Inuit community of Pangnirtung on Baffin Island, Canada is the locus of research and activities to promote traditional Inuit foods in the region. Using radio and film media, educational material on traditional foods described by elders has been broadcast to youth. With support from the Baffin Region Health Promotion Office in the Government of Nunavut at Iqaluit, and the Inuit Tapiriit Kanatami based in Ottawa, programme activities have been widely communicated, along with concerns about climate change effects on the availability of traditional food species, and the impact on food security. Community and project leaders have spoken about the impact of climate change on traditional diets in the Arctic at many international conferences and UN-sponsored meetings.

---

[10]   www.ddsindia.com/www/default.asp

Karen
(Chapter 10 – Sirisai et al., 2013)

After developing a uniquely trusting and sharing relationship, the Karen community leaders and research partners at Mahidol University work together to promote traditional food culture and world views by increasing the cultivation of traditional food species. Focusing on women, young children and strengthening the capacity of local leaders and youth, change agents have spoken eloquently in the Karen and Thai languages at national and international conferences. Important to their success are the involvement of interdisciplinary and multisectoral stakeholders, with community priority at the local level, and the sharing of results among related networks in local and national media, as well as at national and international conferences. With a supportive national socio-economic agenda and the commitment of a passionate national leader, the programme received the attention and participation necessary for development.

Nuxalk
(Chapter 11 – Turner et al., 2013)

The Nuxalk Food and Nutrition Program was conducted more than 20 years ago, but is still having a positive impact in the community and more broadly. In the first programme of its kind in Canada, Nuxalk community leaders and academic partners worked together to improve several aspects of traditional food use, nutrition and health. Through the Assembly of First Nations and the First Nations and Inuit Health Board (FNIHB) of Health Canada, the results of the programme have been shared with many similar programmes developed to improve First Nations people's access to traditional foods. The Nuxalk programme was the stimulus for the larger, multifaceted international CINE programme reported in this volume. The FNIHB Food Security Reference Group has met regularly to promote traditional food use, and recommends assessments of traditional food quality and safety. The traditional food and recipe books created by the programme have been reprinted many times, and are still in use in community schools, universities and indigenous communities. Fitness activities initiated by the programme are still benefiting community members. Some traditional foods promoted by the original project – especially ooligan (a small fish important for food and oil in traditional culture) – are now environmentally threatened, and the original nutrition and health data have been used to raise awareness about the need to protect them. For example, the Nuxalk community hosted a major conference on ooligan conservation in 2007. Many spin-off activities stressing the programme's success continue to involve First Nations in British Columbia, as well as nationally and abroad.

Pohnpei
(Chapter 12 – Englberger et al., 2013)

The case study in Pohnpei describes a broad-based intervention throughout the island of Pohnpei, with before-and-after evaluations conducted in the community of Mand. A major focus is on increasing the use of locally grown foods, many of which are quite abundant but have been neglected, along with collecting information on lifestyle changes and the trend for using more convenient but less healthy processed imported foods. Throughout activities, community and academic partners have developed extensive interdisciplinary and intersectoral collaboration. With frequent communications from the Island Food Community of Pohnpei (IFCP),[11] case study activities have been broadcast throughout the island and into the Pacific region. The programme includes presentations, workshops, videos and films, field trips, drama clubs, school programmes, an e-mail network, many media events, and other activities at both the local and state levels. The successes of the Let's Go Local campaign in promoting increased production and consumption of local food have stimulated much interest in Pohnpei and led to requests for similar interventions in other states of the Federated States of Micronesia and in other Pacific nations, with many adopting the slogan in their own areas. The President of the Federated States of Micronesia and state governors have given support

---

[11] www.islandfood.org

and encouragement for continuing the programme because of the many cultural, health, environmental, economic and food security (CHEEF) benefits of local foods. Advocacy work continues through IFCP and its partners, both government and NGOs, with much international attention to the potential for scaling up this successful programme.

Ainu
(Chapter 13 – Iwasaki-Goodman, 2013)

Successful outreach activities for the Ainu have been conducted on the island of Hokkaido in Japan. The intervention was incorporated with other activities for an Ainu cultural revitalization, including capacity building in Ainu language, dance and other cultural aspects. The reintroduction of Ainu traditional food through regular print media and Ainu food cooking lessons has stimulated many requests for demonstrations in schools, communities and ceremonial settings throughout Hokkaido. Non-Ainu have participated in these events, raising the profile of Ainu cultural activities and thereby not only contributing to increased pride in Ainu food as a significant part of Ainu culture, but also reversing social prejudice.

Concluding comments

The intervention projects created by Indigenous Peoples' leaders with academic partners have resulted in many successes. Although there have been substantial challenges for bringing positive change to Indigenous Peoples' lives by promoting their food systems, persistence and vision have encouraged the continuation of efforts. Partners in the programme have found many ways of addressing local issues and moving forward the imperative to protect and use local resources. Initiatives have resulted in the development of curricula and school resources for teaching youth.

There is increased recognition that many foods of high quality and important cultural value originate in Indigenous Peoples' local ecosystems and cultures, and that these bring great benefit to the entire world. Among other factors, this recognition is the result of greater knowledge of unique foods; increased understanding of the benefits of food biodiversity and the importance of communities in realizing the cultural benefits of food; the imperative to protect the world's fragile ecosystems, many of which are inhabited by Indigenous Peoples; and the realization that the nutrition transition is having a negative impact on people who previously had healthy diets from local food resources.

Policies result from growing public consciousness that changes are needed in government and public activity settings. To this end, the programme's community and academic partners have communicated broadly about their work. Results from the case studies described here have been widely reported to local, regional, national and international audiences. Through this, the programme aims to stimulate further dissemination of the value of local food systems in improving the health of Indigenous Peoples.

Programme partners have produced scientific publications, posters and local communications, held meetings with policy-makers, and responded to local and international media. They have trained 18 M.Sc. and Ph.D. university students, and contributed to the capacity building of hundreds of other students and trainees in case study settings. Community leaders and academic partners have reported findings related to the programme's objectives in:

- more than 200 published works;
- more than 270 presentations at local, national and international conferences and UN events and side-events;
- more than 120 public media reports and audiovisual documents.

They are all proud of these accomplishments, which have been achieved through multiple collaborations with a common vision and goals. They are confident that their projects will continue to foster awareness and policy development at the national and international levels. This will turn the tide of the nutrition transition and improve the health of Indigenous Peoples throughout the world, while giving recognition to their contribution to the health and well-being of all humankind ☼

Acknowledgements

The overall programme was conducted under the auspices of IUNS, with contributions from many partners in two phases: i) documentation of Indigenous Peoples' food systems; and ii) use of local food systems to stimulate nutrition and health improvements at the community level (IUNS, 2010). This chapter draws on experiences from both phases of the case studies. FAO supported the initial studies on methodology development, the publication of case study findings and intervention impacts, and case study partners' travel to international conferences. The authors thank especially the Canadian Institutes of Health Research, the Institute of Aboriginal Peoples' Health and the Institute of Nutrition, Metabolism and Diabetes for their grant programmes supporting research and annual conferences of community leaders and academic partners. Academic partners worked diligently throughout their experience with the programme, conducting research and creating credible peer-reviewed literature to further the goals. However, without doubt, the most heart-felt thanks go to the community members and community leaders represented here, who contributed their time and vision to this work with the full intention of sharing their knowledge and improving the health of Indigenous Peoples everywhere.

> **Comments to:** harriet.kuhnlein@mcgill.ca

References

Chapter 2 | Self-determination

Allen, L., de Benoist, B., Dary, O. & Hurrell, R., eds. 2006. *Guidelines on food fortification with micronutrients*, pp. 6–10. Geneva, WHO and FAO.

Anderson, I., Crengle, S., Kamaka, M.L., Chen, T.H., Palafox, N. & Jackson-Pulver, L. 2006. Indigenous health in Australia, New Zealand, and the Pacific. *Lancet*, 367: 1775–1785.

Australian Bureau of Statistics. 2002. *Deaths Australia, 2001*. Canberra.

Australian Bureau of Statistics. 2003. *Deaths Australia, 2002*. Canberra.

Australian Bureau of Statistics. 2004. *Deaths Australia, 2003*. Canberra.

Australian Bureau of Statistics. 2005. *Deaths Australia, 2004*. Canberra.

Australian Bureau of Statistics. 2007. *Deaths Australia, 2006*. Canberra.

Australian Bureau of Statistics & Australian Institute of Health and Welfare. 2005. *The health and welfare of Australia's aboriginal and Torres Strait Islander peoples 2005*. Canberra, Commonwealth of Australia. 316 pp.

Barnett, J., Dessai, S. & Jones, R.N. 2007. Vulnerability to climate variability and change in East Timor. *Ambio*, 36(5): 372–378.

Bayliss-Smith, T. 2009. Food security and agricultural sustainability in the New Guinea Highlands: vulnerable people, vulnerable places. *Institute of Development Studies Bulletin*, 22(3): 5–11. www3. interscience.wiley.com/journal/122395995/abstract.

Bhattacharjee, L., Kothari, G., Priya, V. & Nandi, B.K. 2009. The Bhil food system: links to food security, nutrition and health. *In* H.V. Kuhnlein, B. Erasmus and D. Spigelski. *Indigenous Peoples' food systems: the many dimensions of culture, diversity and environment for nutrition and health*, pp. 209–229. Rome, FAO.

Bramley, D., Hebert, P., Jackson, R. & Chassin, M. 2004. Indigenous disparities in disease-specific mortality, a cross-country comparison: New Zealand, Australia, Canada, and the United States. *Journal of the New Zealand Medical Association*, 117(1207). www.nzma.org.nz/journal/117-1207/1215/.

Canadian Council on Social Development. 2003. *Aboriginal children in poverty in urban communities: social exclusion and the growing racialization of poverty in Canada*. Notes for presentation to Subcommittee on Children and Youth at Risk of the Standing Committee on Human Resources Development and the Status of Persons with Disabilitieson 19 March 2003. www.ccsd.ca/pr/2003/aboriginal.htm.

Cardoso, A.M., Santos, R.V. & Coimbra Jr., C.E. 2005. [Infant mortality according to race/colour in Brazil: what do the national databases say?]. *Cadernos de Saúde Publica*, 21(5): 1602–1608. (in Portuguese)

Carino, J. 2009. Poverty and well-being. *In* UNPFII. *The state of the world's indigenous peoples,* pp.13–45. New York, United Nations Department of Economic and Social Affairs, Secretariat of the Permanent Forum on Indigenous Issues. 250 pp. www.un.org/esa/socdev/unpfii/documents/sowip_web.pdf.

Casey, P.H., Simpson, P.M., Gossett, J.M., Bogle, M.L., Champagne, C.M., Connell, C., Harsha, D., McCabe-Sellers, B., Robbins, J.M., Stuff, J.E. & Weber, J. 2006. The association of child and household food insecurity with childhood overweight status. *Pediatrics,* 118(5): e1406–e1413.

CDC. 2003. Self-reported concern about food security associated with obesity – Washington, 1995–1999. *Morbidity and Mortality Weekly Report,* 52(35): 840–842.

Chan, H.M., Fediuk, K., Hamilton, S., Rostas, L., Caughey, A., Kuhnlein, H., Egeland, G. & Loring, E. 2006. Food security in Nunavut, Canada: barriers and recommendations. *International Journal of Circumpolar Health,* 65(5): 416–431.

Chotiboriboon, S., Tamachotipong, S., Sirisai, S., Dhanamitta, S., Smitasiri, S., Sappasuwan, C., Tantivatanasathien, P. & Eg-kantrong, P. 2009. Thailand: food system and nutritional status of indigenous children in a Karen community. *In* H.V. Kuhnlein, B. Erasmus and D. Spigelski. *Indigenous Peoples' food systems: the many dimensions of culture, diversity and environment for nutrition and health,* pp. 159–183. Rome, FAO.

Correal, C., Zuluaga, G., Madrigal, L., Caicedo, S. & Plotkin, M. 2009. Ingano traditional food and health: phase 1, 2004–2005. *In* H.V. Kuhnlein, B. Erasmus and D. Spigelski. *Indigenous Peoples' food systems: the many dimensions of culture, diversity and environment for nutrition and health,* pp. 83–108. Rome, FAO.

Creed-Kanashiro, H., Roche, M., Tuesta Cerrón, I. & Kuhnlein, H.V. 2009. Traditional food system of an Awajun community in Peru. *In* H.V. Kuhnlein, B. Erasmus and D. Spigelski. *Indigenous Peoples' food systems: the many dimensions of culture, diversity and environment for nutrition and health,* pp. 59–81. Rome, FAO.

Cunningham, M. 2009. Health. *In* UNPFII. *State of the world's indigenous peoples,* pp. 155–187. New York, United Nations Department of Economic and Social Affairs, Secretariat of the Permanent Forum on Indigenous Issues. www.un.org/esa/socdev/unpfii/documents/sowip_web.pdf.

Dabelea, D. 2007. The predisposition to obesity and diabetes in offspring of diabetic mothers. *Diabetes Care,* 30(suppl 2): S169–S174.

Dinour, L.M., Bergen, D. & Yeh, M.C. 2007. The food insecurity-obesity paradox: a review of the literature and the role food stamps may play. *Journal of the American Dietetic Association,* 107(11): 1952–1961.

Drewnowski, A. 2009. Obesity, diets, and social inequalities. *Nutrition Reviews,* 67(suppl 1): S36–S39.

Drewnowski, A. & Specter, S.E. 2004. Poverty and obesity: the role of energy density and energy costs. *American Journal of Clinical Nutrition,* 79(1): 6–16.

Duhaime, G., Chabot, M. & Gaudreault, M. 2002. Food consumption patterns and socioeconomic factors among the Inuit of Nunavik. *Ecology of Food Nutrition,* 41(2): 91–118.

Durand, Z.W. 2007. Age of onset of obesity, diabetes and hypertension in Yap State, Federated States of Micronesia. *Pacific Health Dialog,* 14(1): 165–169.

Durie, M., Milroy, H. & Hunter, E. 2009. Mental health and the indigenous peoples of Australia and New Zealand. *In* L.J. Kirmayer and G.G. Valaskakis, eds. *Healing traditions: the mental health of aboriginal peoples in Canada,* pp. 33–55. Vancouver, British Columbia, Canada, University of British Columbia Press.

Ebbesson, S.O., Schraer, C.D., Risica, P.M., Adler, A.I., Ebbesson, L., Mayer, A.M., Shubnikof, E.V., Yeh, J., Go, O.T. & Robbins, D.C. 1998. Diabetes and impaired glucose tolerance in three Alaskan Eskimo populations. The Alaska-Siberia Project. *Diabetes Care,* 21(4): 563–569.

Egeland, G.M. & Meltzer, S.J. 2010. Following in mother's footsteps? Mother-daughter risks for insulin resistance and cardiovascular disease 15 years after gestational diabetes. *Diabetic Medicine*, 27(3): 257–265.

Egeland, G.M., Skjaerven, R. & Irgens, L.M. 2000. Birth characteristics of women who develop gestational diabetes: population based study. *British Medical Journal*, 321(7260): 546–547.

Egeland, G.M., Charbonneau-Roberts, G., Kuluguqtuq, J., Kilabuk, J., Okalik, L., Soueida, R. & Kuhnlein, H.V. 2009. Back to the future: using traditional food and knowledge to promote a healthy future among Inuit. *In* H.V. Kuhnlein, B. Erasmus and D. Spigelski. *Indigenous Peoples' food systems: the many dimensions of culture, diversity and environment for nutrition and health*, pp. 9–22. Rome, FAO.

Egeland, G.M., Pacey, A., Cao, Z. & Sobol, I. 2010. Food insecurity among Inuit preschoolers: Nunavut Inuit Child Health Survey, 2007–2008. *Canadian Medical Association Journal*, 182(3): 243–248

Egeland, G.M., Williamson-Bathory, L., Johnson-Down, L. & Sobol, I. 2011. Traditional food and monetary access to market-food: correlates of food insecurity among Inuit preschoolers. *International Journal of Circumpolar Health*, 70(4): 373–383.

Egeland, G.M., Yohannes, S., Okalik, L., Kilabuk, J., Racicot, C., Wilcke, M., Kuluguqtuq, J. & Kisa, S. 2013. The value of Inuit elders' storytelling in health promotion during times of rapid climate change and uncertain food security. *In* H.V. Kuhnlein, D. Spigelski, B. Erasmus and B. Burlingame. *Indigenous Peoples' food systems and well-being: interventions and policies for healthy communities*, Chapter 9. Rome, FAO.

Englberger, L., Lorens, A., Levendusky, A., Pedrus, P., Albert, K., Hagilmai, W., Paul, Y., Nelber, D., Moses, P., Shaeffer, S. & Gallen, M. 2009. Documentation of the traditional food system of Pohnpei. *In* H.V. Kuhnlein, B. Erasmus and D. Spigelski. *Indigenous Peoples' food systems: the many dimensions of culture, diversity and environment for nutrition and health*, pp. 109–138. Rome, FAO.

ESCAP. 2009. *Population and development indicators for Asia and the Pacific, 2009*. Bangkok, Economic and Social Commission for Asia and the Pacific (ESCAP).

FAO. 1996. *World Food Summit Plan of Action*. 13–17 November 1996. Rome.

FAO. 2008a. *Indigenous peoples threatened by climate change. World day highlights fundamental role of indigenous peoples in food security*. Rome, New York Newsroom, 8 August 2008. www.fao.org/newsroom/en/news/2008/1000906/index.html.

FAO. 2008b. *The State of Food and Agriculture 2008. Biofuels: prospects, risks and opportunities*, pp 72–85. Rome. www.fao.org/docrep/011/i0100e/i0100e00.htm.

Federated States of Micronesia. 2002. *Population and housing census report*. Palikir, Department of Health Education and Social Affairs.

Fee, M. 2006. Racializing narratives: obesity, diabetes and the "aboriginal" thrifty genotype. *Social Science and Medicine*, 62(12): 2988–2997.

Freemantle, C.J., Read, A.W., de Klerk, N.H., McAullay, D., Anderson, I.P. & Stanley, F.J. 2006. Patterns, trends and increasing disparities in mortality for aboriginal and non-aboriginal infants born in Western Australia, 1980–2001: population database study. *Lancet*, 367: 1758–1766.

Galloway, T., Young, T.K. & Egeland, G.M. 2010. Emerging obesity among preschool-aged Canadian Inuit children: The Nunavut Inuit Child Health Survey, 2007–2008. *International Journal of Circumpolar Health*, 69(2): 151–157.

Garcia, O.P., Long, K.Z. & Rosado, J.L. 2009. Impact of micronutrient deficiencies on obesity. *Nutrition Reviews*, 67: 559–572.

Garí, J.A. 2001. Biodiversity and indigenous agroecology in Amazonia: the indigenous peoples of Pastaza. *Etnoecológica*, 5(7): 21–37.

Garnelo, L., Brandão, L.C. & Levino, A. 2005. Dimensions and potentialities of the geographic information system on indigenous health. *Revista Saúde Pública,* 39(4): 634–640. www.scielosp.org/pdf/rsp/v39n4/en_25537.pdf.

Glouberman, S. & Millar, J. 2003. Evolution of the determinants of health, health policy, and health information systems in Canada. *American Journal of Public Health,* 93(3): 388–392.

Gracey, M. 1976. Undernutrition in the midst of plenty: nutritional problems of young Australian aborigines. *Australian Paediatric Journal,* 12(3): 180–182.

Gracey, M. 2007. Nutrition-related disorders in Indigenous Australians: how things have changed. *The Medical Journal of Australia,* 186(1): 15–17.

Gracey, M. & King, M. 2009. Indigenous health part 1: determinants and disease patterns. *Lancet,* 374 (9683): 65–75.

Gracey, M., Bridge, E., Martin, D., Jones, T., Spargo, R.M., Shephard, M. & Davis, E.A. 2006. An aboriginal-driven program to prevent, control and manage nutrition-related "lifestyle" diseases including diabetes. *Asia Pacific Journal of Clinical Nutrition,* 15(2): 178–188.

Guerrero, R.T.L., Paulino, Y.C., Novotny, R. & Murphy, S.P. 2008. Diet and obesity among Chamorro and Filipino adults on Guam. *Asia Pacific Journal of Clinical Nutrition,*17(2): 216–222.

Gundersen, C., Lohman, B.J., Garasky, S., Stewart, S. & Eisenmann, J. 2008. Food security, maternal stressors, and overweight among low-income US children: results from the National Health and Nutrition Examination Survey (1999–2002). *Pediatrics,* 122(3): e529–e540.

Hales, C.N., Barker, D.J., Clark, P.M., Cox, L.J., Fall, C., Osmond, C. & Winter, P.D. 1991. Fetal and infant growth and impaired glucose tolerance at age 64. *British Medical Journal,* 303(6809): 1019–1022.

Hamill, P.V., Drizd, T.A., Johnson, C.L., Reed, R.B. & Roche, A.F. 1977. NCHS growth curves for children birth–18 years. United States. *Vital and Health Statistics Series,* 11(165): 1–74.

Harrison, G.G. 2010. Public health interventions to combat micronutrient deficiencies. *Public Health Reviews,* 32(1): 256–266.

Harrison, G.G., Tirado, M.C. & Galal, O.M. 2010. Backsliding against malnutrition. *Asia Pacific Journal of Public Health,* 24 (suppl 3): 246S–253S.

Health Canada. 2005. *First Nations, Inuit and aboriginal health. Life expectancy at birth for overall population.* www.hc-sc.gc.ca/fniah-spnia/diseases-maladies/2005-01_health-sante_indicat-eng.php#life_expect.

Hegele, R.A. 2001. Genes and environment in type 2 diabetes and atherosclerosis in aboriginal Canadians. *Current Atherosclerosis Reports,* 3(3): 216–221.

Horn, O.K., Bruegl, A., Jacobs-Whyte, H., Paradis, G., Ing, A. & Macaulay, A.C. 2007. Incidence and prevalence of type 2 diabetes in the First Nation community of Kahawá:ke, Quebec, Canada, 1986–2003. *Canadian Journal of Public Health,* 98(6): 438–443.

Hughes, R.G. & Lawrence, M.A. 2005. Globalization, food and health in Pacific Island countries. *Asia Pacific Journal of Clinical Nutrition,* 14(4): 298–306.

Huxley, R., Owen, C.G., Whincup, P.H., Cook, D.G., Rich-Edwards, J., Smith, G.D. & Collins, R. 2007. Is birth weight a risk factor for ischemic heart disease later in life? *American Journal of Clinical Nutrition,* 85: 1244–1250.

Indian and Northern Affairs Canada. 2007. *The revised northern food basket.* Ottawa, Aboriginal Affairs and Northern Development Canada. www.ainc-inac.gc.ca/nth/fon/fc/pubs/nfb/nfb-eng.asp.

Indian Health Service. 2006. *Facts on Indian health disparities.* Washington, DC, United States Department of Health and Human Services.

Indian Health Service. 2009. Natality and infant/maternal mortality statistics. *In* United States Department of Health and Human Services. *Trends in Indian Health 2002–2003 Edition*, pp. 34–50. Washington, DC, United States Department of Health and Human Services.

Innis, K.E., Byers, T.E., Marshal, J.A., Barón, A., Olreans, M. & Hamman, R. 2002. Association of a woman's own birthweight with subsequent risk for gestational diabetes. *Journal of the American Medical Association*, 287: 2534–2541.

Ito, M. 2008. Diet officially declares Ainu indigenous. *The Japan Times Online*, 7 June 2008. http://search.japantimes.co.jp/cgi-bin/nn20080607a1.html.

Jamison, D.T., Mosley, W.H., Measham, A.R. & Bobadilla, J.L., eds. 1993. *Disease control priorities in developing countries*. Oxford, UK, Oxford University Press.

Jiménez-Cruz, A., Bacardí-Gascón, M. & Spindler, A.A. 2003. Obesity and hunger among Mexican-Indian migrant children on the US-Mexico border. *International Journal of Obesity and Related Metabolic Disorders*, 27(6): 740–747.

Johnson-Down, L. & Egeland, G.M. 2010. Diet quality and traditional food consumption among Inuit preschoolers: Nunavut Inuit Child Health Survey, 2007–2008. *Journal of Nutrition*, 140(7): 1311–1316.

Jørgensen, M.E., Bjeregaard, P., Borch-Johnsen, K., Backer, V., Becker, U., Jørgensen, T. & Mulvad, G. 2002. Diabetes and impaired glucose tolerance among the Inuit population of Greenland. *Diabetes Care*, 25(10): 1766–1771.

King, M., Smith, A. & Gracey, M. 2009. Indigenous health part 2: the underlying causes of the health gap. *Lancet*, 374(9683): 76–85.

Kirkpatrick, S.I. & Tarasuk, V. 2008. Food insecurity is associated with nutrient inadequacies among Canadian adults and adolescents. *Journal of Nutrition*, 138: 604–612.

Kuhnlein, H.V. & Receveur, O. 2007. Local cultural animal food contributes high levels of nutrients for Arctic Canadian Indigenous adults and children. *Journal of Nutrition*, 137(4): 1110–1114.

Kuhnlein, H.V., Receveur, O., Soueida, R. & Egeland, G.M. 2004. Arctic Indigenous Peoples experience the nutrition transition with changing dietary patterns and obesity. *Journal of Nutrition*, 134(6): 1447–1453.

Lambden, J., Receveur, O. & Kuhnlein, H.V. 2007. Traditional food attributes must be included in studies of food security in the Canadian Arctic. *International Journal of Circumpolar Health*, 66(4): 308–319.

Lambden, J., Receveur, O., Marshall, J. & Kuhnlein, H.V. 2006. Traditional and market food access in Arctic Canada is affected by economic factors. *International Journal of Circumpolar Health*, 65(4): 331–340.

Ledrou, I. & Gervais, J. 2005. Food insecurity. *Health Reports*, 16(3): 47–51.

Lindsay, R.S. & Bennett, P.H. 2001. Type 2 diabetes, the thrifty phenotype – an overview. *British Medical Bulletin*, 60(1): 21–32.

Lippe, J., Brener, N., Kann, L., Kinchen, S., Harris, W.A., McManus, T. & Speicher, N. 2008. Youth risk behavior surveillance – Pacific Island United States Territories, 2007. *Morbidity and Mortality Weekly Report. Surveillance Summaries*, 57(12): 28–56.

Lopez, A.D., Mathers, C.D., Ezzati, M., Jamison, D.T. & Murray, C.J.L., eds. 2006. *Global burden of disease and risk factors*. Washington, DC, World Bank Publications.

Lourenço, A.E., Santos, R.V., Orellana, J.D. & Coimbra Jr., C.E. 2008. Nutrition transition in Amazonia: obesity and socioeconomic change in the Suruí Indians from Brazil. *American Journal of Human Biology*, 20(5): 564–571.

Lucas, A. 1991. Programming by early nutrition in man. *In* G.R. Bock and J. Whelan, eds. *The childhood environment and adult disease.* CIBA Foundation Symposium 156, pp. 38–55. Chichester, UK, Wiley.

McCredie, J. 2008. Aboriginal children are still twice as likely to die as other young Australians. *British Medical Journal,* 337: a1852.

McIntyre, L. & Shah, C.P. 1986. Prevalence of hypertension, obesity and smoking in three Indian communities in northwestern Ontario. *Canadian Medical Association Journal,* 134(4): 345–349.

McIntyre, L., Glanville, N.T., Raine, K.D., Dayle, J.B., Anderson, B. & Battaglia, N. 2003. Do low-income lone mothers compromise their nutrition to feed their children? *Canadian Medical Association Journal,* 168(6): 686–691.

Montenegro, R.A. & Stephens, C. 2006. Indigenous health in Latin America and the Caribbean. *Lancet,* 367(9525): 1859–1869.

Mowbray, M. & WHO Commission on Social Determinants of Health. 2007. *Social determinants and indigenous health: the international experience and its policy implications.* International Symposium on the Social Determinants of Indigenous Health, Adelaide, April 2007. Geneva, World Health Organization (WHO). www.who.int/social_determinants/resources/indigenous_health_adelaide_report_07.pdf.

Naqshbandi, M., Harris, S.B., Esler, J.G. & Antwi-Nsiah, F. 2008. Global complication rates of type 2 diabetes in Indigenous Peoples: a comprehensive review. *Diabetes Research and Clinical Practice,* 82(1): 1–17.

National Institute of Nutrition. 2000. *Health and nutritional status of tribal populations: report of the first repeat survey.* Hyderabad, India, National Institute of Nutrition, Indian Council of Medical Research.

New Zealand Ministry of Health. 1999. *Our health, our future – Hauora Pakari, Koiora Roa – the health of New Zealanders 1999.* Auckland. www.moh.govt.nz/moh.nsf/0/6910156be95e706e4c2568800002e403?opendocument.

New Zealand Ministry of Health. 2002. *Diabetes in New Zealand. Models and forecasts 1996–2011.* Auckland.

New Zealand Ministry of Health. 2008. *A portrait of health. Key results of the 2006/07 New Zealand Health Survey.* Auckland.

New Zealand Ministry of Social Development. 2004. *The social report. Te purongo oranga tangata 2004,* pp. 22–35. Health. Auckland.

Nord, M., Andrews, M. & Carson, S. 2006. *Household food security in the United States, 2005.* Food Assistance and Nutrition Research Report No. 29. Washington, DC, Economic Research Service, United States Department of Agriculture.

Ohenjo, N., Willis, R., Jackson, D., Nettleton, C., Good, K. & Mugarura, B. 2006. Health of Indigenous People in Africa. *Lancet,* 367(9526): 1937–1946.

Oiye, S., Simel, J.O., Oniang'o, R. & Johns, T. 2009. The Maasai food system and food and nutrition security. *In* H.V. Kuhnlein, B. Erasmus and D. Spigelski. *Indigenous Peoples' food systems: the many dimensions of culture, diversity and environment for nutrition and health,* pp. 231–249. Rome, FAO.

Okeke, E.C., Ene-Ebong, H.N., Uzuegbunam, A.O., Ozioko, A., Umeh, S.I. & Chukwuone, N. 2009. The Igbo traditional food system documented in four states. *In* H.V. Kuhnlein, B. Erasmus and D. Spigelski. *Indigenous Peoples' food systems: the many dimensions of culture, diversity and environment for nutrition and health,* pp. 251–281. Rome, FAO.

Paradies, Y.C., Montoya, M.J. & Fullerton, S.M. 2007. Racialized genetics and the study of complex diseases: the thrifty genotype revisited. *Perspectives in Biology and Medicine,* 50(2): 203–227.

Parnell, W.R., Reid, J., Wilson, N.C., McKenzie, J. & Russell, D.G. 2001. Food security: is New Zealand a land of plenty? *The New Zealand Medical Journal*, 114(1128): 141–145.

Pathak, N., Kothari, A. & Roe, D. 2005. Conservation with social justice? The role of community conserved areas in achieving the Millennium Development Goals. *In* T. Bigg and D. Satterthwaite, eds. *How to make poverty history: the central role of local organisations in meeting the MDGs*, pp. 55–78. London, International Institute for Environment and Development.

Pettitt, D.J. & Jovanovic, L. 2007. Low birth weight as a risk factor for gestational diabetes, diabetes, and impaired glucose tolerance during pregnancy. *Diabetes Care*, 30(suppl 2): S147–149.

Pettitt, D.J. & Knowler, W.C. 1998. Long-term effects of the intrauterine environment, birth weight, and breast-feeding in Pima Indians. *Diabetes Care*, 21(suppl 2): B138–B141.

Pettitt, D.J., Aleck, K.A., Baird, H.R., Carraher, M.J., Bennett, P.H. & Knowler, W.C. 1998. Congenital susceptibility to NIDDM. Role of intrauterine environment. *Diabetes*, 37(5): 622–628.

Piperata, B.A. 2007. Nutritional status of Ribeirinhos in Brazil and the nutrition transition. *American Journal of Physical Anthropology*, 133(2): 868–878.

Pohnpei STEPS. 2002. *Preliminary data of the Pohnpei Department of Health and FSM Department of Health, Education, and Social Affairs: WHO STEPwise Approach to NCD Surveillance.* Presented at the FSM Health Symposium, 26–27 January 2007, Kolonia, Pohnpei, Federal States of Micronesia.

Power, E.M. 2008. Conceptualizing food security for aboriginal people in Canada. *Canadian Journal of Public Health*, 99(2): 95–97.

Raschke, V. & Cheema, B. 2008. Colonisation, the New World Order, and the eradication of traditional food habits in East Africa: historical perspective on the nutrition transition. *Public Health Nutrition*, 11(7): 662–674.

Ravussin, E., Valencia, M.E., Esparza, J., Bennett, P.H. & Schulz, L.O. 1994. Effects of a traditional lifestyle on obesity in Pima Indians. *Diabetes Care*, 17(9): 1067–1074.

Reading, J. 2009. *The crisis of chronic disease among aboriginal peoples: a challenge for public health, population health and social policy.* Victoria, Australia, University of Victoria. 185 pp.

Renzaho, A.M. 2004. Food insecurity, malnutrition and mortality in Maewo and Ambae islands, Vanuatu. *Pacific Health Dialogue*, 11(1): 12–21.

Ring, I. & Brown, N. 2003. The health status of indigenous peoples and others. The gap is narrowing in the United States, Canada, and New Zealand, but a lot more is needed. *British Medical Journal*, 327(7412): 404–405.

Romaguera, D., Samman, N., Farfán, N., Lobo, M., Pons, A. & Tur, J.A. 2008. Nutritional status of the Andean population of Puna and Quebrada of Humahuaca, Jujuy, Argentina. *Public Health Nutrition*, 11(6): 606–615.

Rose, D., & Bodor, J.N. 2006. Household food insecurity and overweight status in young school children: results from the early childhood longitudinal study. *Pediatrics*, 117(2): 464–473.

Rose, D. & Oliveira, V. 1997. Nutrient intakes of individuals from food-insufficient households in the United States. *American Journal of Public Health*, 87(12): 1956–1961.

Salomeyesudas, B. & Satheesh, P.V. 2009. Traditional food system of Dalit in Zaheerabad Region, Medak District, Andhra Pradesh, India. *In* H.V. Kuhnlein, B. Erasmus and D. Spigelski. *Indigenous Peoples' food systems: the many dimensions of culture, diversity and environment for nutrition and health*, pp. 185–208. Rome, FAO.

Schulz, L.O., Bennett, P.H., Ravussin, E., Kidd, J.R., Kidd, K.K., Esparza, J. & Valencia, M.E. 2006. Effects of traditional and western environments on prevalence of type 2 diabetes in Pima Indians in Mexico and the US. *Diabetes Care*, 29(8): 1866–1871.

Silverman, B.L., Rizzo, T.A., Cho, N.H. & Metzger, B.E. 1998. Long-term effects of the intrauterine environment. The Northwestern University Diabetes in Pregnancy Center. *Diabetes Care*, 21(suppl 2): B142–B149.

Skalicky, A., Meyers, A.F., Adams, W.G., Yang, Z., Cook, J.T. & Frank, D.A. 2006. Child food insecurity and iron deficiency anemia in low-income infants and toddlers in the United States. *Maternal and Child Health Journal*, 10(2): 177–185.

Smith, J., Cianflone, K., Biron, S., Hould, F.S., Lebel, S., Marceau, S., Lescelleur, O., Biertho, L., Simard, S., Kral, J.G. & Marceau, P. 2009. Effects of maternal surgical weight loss in mothers on intergenerational transmission of obesity. *Journal of Clinical Endocrinology and Metabolism*, 94(11): 4275–4283.

Steele, C.B., Cardinez, C.J., Richardson, L.C., Tom-Orme, L. & Shaw, K.M. 2008. Surveillance for health behaviors of American Indians and Alaska Natives – findings from the behavioral risk factor surveillance system, 2000–2006. *Cancer*, 113(suppl 5): 1131–1141.

Stephens, C., Nettleton, C., Porter, J., Willis, R. & Clark, S. 2005. Indigenous peoples' health – why are they behind everyone, everywhere? *Lancet*, 366(9479): 10–13.

Stephens, C., Porter, J., Nettleton, C. & Willis, R. 2006. Disappearing, displaced, and undervalued: a call to action for Indigenous health worldwide. *Lancet*, 367(9527): 2019–2028.

Thrupp, L.A. 2000. Linking agricultural biodiversity and food security: the valuable role of agrobiodiversity for sustainable agriculture. *International Affairs*, 76(2): 283–297.

Tjepkema, M. 2002. *Health of the off-reserve aboriginal population*. Supplements to Health Reports, volume 13, No 82-003. Ottawa, Statistics Canada.

Townsend, M.S., Peerson, J., Love, B., Achterberg, C. & Murphy, S.P. 2001. Food insecurity is positively related to overweight in women. *Journal of Nutrition*, 131(6): 1738–1745.

Trewin, D. & Madden, R. 2005. *The health and welfare of Australia's aboriginal and Torres Strait Islander peoples*. ABS Cat. No. 4704.0. Canberra, Australian Bureau of Statistics.

Tulchinsky, T.H. 2010. Micronutrient deficiency conditions: global health issues. *Public Health Reviews*, 32(1): 243–255.

UNPFII. 2009. *The State of the World's Indigenous Peoples*. New York, United Nations Department of Economic and Social Affairs, Secretariat of the Permanent Forum on Indigenous Issues. 250 pp. www.un.org/esa/socdev/unpfii/documents/sowip_web.pdf.

Vozoris, N.T. & Tarasuk, V.S. 2003. Household food insufficiency is associated with poorer health. *Journal of Nutrition*, 133: 120–126.

Wahlqvist, M.L. & Lee, M.-S. 2007. Regional food culture and development. *Asia Pacific Journal of Clinical Nutrition*, 16(suppl 1): 2–7.

Whitaker, R.C. & Orzol, S.M. 2006. Obesity among US urban preschool children: relationships to race, ethnicity, and socioeconomic status. *Archives of Pediatrics and Adolescent Medicine*, 160(6): 578–584.

Wilkins, R., Uppal, S., Finès, P., Senécal, S., Guimond, E. & Dion, R. 2008. Life expectancy in the Inuit-inhabited areas of Canada, 1989–2003. *Health Reports*, 19(1): 1–14.

Williams, M.A., Emanuel, I., Kimpo, C., Leisenring, W.M. & Hale, C.B. 1999. A population-based cohort study of the relation between maternal birthweight and risk of gestational diabetes mellitus in four racial/ethnic groups. *Paediatric Perinatal Epidemiology*, 13: 452–465.

Willows, N.D., Veugelers, P., Raine, K. & Kuhle, S. 2009. Prevalence and sociodemographic risk factors related to household food security in aboriginal peoples in Canada. *Public Health Nutrition,* 12(8): 1150–1156.

World Vision Kenya. 2004. *Evaluation of Loodariak area development program.* Nairobi.

Young, T.K. & Sevenhuysen, G. 1989. Obesity in northern Canadian Indians: patterns, determinants, and consequences. *American Journal of Clinical Nutrition,* 49(5): 786–793.

Young, T.K., Reading, J., Elias, B. & O'Niel, J.D. 2000. Type 2 diabetes mellitus in Canada's First Nations: status of an epidemic in progress. *Canadian Medical Association Journal,* 163(5): 561–566.

Zalilah, M.S. & Tham, B.L. 2002. Food security and child nutritional status among Orang Asli (Temuan) households in Hulu Langat, Selangor. *Medical Journal of Malaysia,* 57(1): 36–50.

Zephier, E., Himes, J.H., Story, M. & Zhou, X. 2006. Increasing prevalences of overweight and obesity in Northern Plains American Indian children. *Archives of Pediatrics and Adolescent Medicine,* 160(1): 34–39.

Zinn, C. 1995. Aboriginal health gap widens. *British Medical Journal,* 310(6988): 1157–1158.

Chapter 3 | Environmental challenges

Anderson, M.K. & Barbour, M.G. 2003. Simulated indigenous management: a new model for ecological restoration in national parks. *Ecological Restoration,* 21(4): 269–277.

Anderson, E.N., Dzib Zihum de Cen, A., Tzuc, F.M. & Chale, P.V. 2005. *Political ecology in a Yucatec Maya community.* Tucson, Arizona, USA, University of Arizona Press. 264 pp.

Ashford, G. & Castleden, J. 2001. *Inuit observations on climate change. Final report.* Winnipeg, Manitoba, Canada, International Institute for Sustainable Development. 27 pp. www.iisd.org/publications/pub.aspx?id=410)

Balée, W.L. 1994. *Footprints of the forest: Ka'apor ethnobotany – the historical ecology of plant utilization by an Amazonian people.* New York, Columbia University Press. 396 pp.

Beaton, J. 2004. *Diabetes then and now.* Victoria, British Columbia, Canada, Songhees Nation and University of Victoria, British Colombia. (video)

Bennett, E.L. & Robinson, J.G., eds. 2000. *Hunting for sustainability in tropical forests.* New York, Columbia University Press. 582 pp.

Berkes, F. 2008. *Sacred ecology,* second edition. New York, Routledge. 336 pp.

Berkes, F., Heubert, R., Fast, H., Manseau, M. & Diduck, A., eds. 2005. *Breaking ice: renewable resource and ocean management in the Canadian north.* Calgary, Alberta, Canada, University of Calgary Press. 396 pp.

Berkes, F., Hughes, T.P., Steneck, R.S., Wilson, J.A., Bellwood, D.R., Crona, B., Folke, C., Gunderson, L.H., Leslie, H.M., Norberg, J., Nyström, M., Olsson, P., Österblom, H., Scheffer, M. & Worm, B. 2006. Globalization, roving bandits, and marine resources. *Science,* 311(5767): 1557–1558.

Berti, P.R., Chan, H.M., Receveur, O., Macdonald, C.R. & Kuhnlein, H.V. 1997. Exposure to radioactivity through the consumption of caribou in Indigenous People of the Canadian Subarctic. Presented at the 16th International Congress of Nutrition, Montreal, Quebec, Canada.

Boyd, R., ed. 1999. *Indians, fire and the land in the Pacific Northwest.* Corvallis, Oregon, USA, Oregon State University Press. 320 pp.

Brookfield, H. & Padoch, C. 1994. Appreciating agrodiversity: a look at the dynamism and diversity of indigenous farming practices. *Environment,* 36(5): 6–11, 37–45.

Caicedo, S. & Chaparro, A.M. 2013. Inga food and medicine systems to promote community health. *In* H.V. Kuhnlein, D. Spigelski, B. Erasmus and B. Burlingame. *Indigenous Peoples' food systems and well-being: interventions and policies for healthy communities*, Chapter 8. Rome, FAO.

Carlson, T.J.S. & Maffi, L., eds. 2004. *Ethnobotany and conservation of biocultural diversity.* Advances in Economic Botany No. 15. New York, New York Botanical Garden Press.

CBD. 1992. *Convention on Biological Diversity.* www.biodiv.org/convention/.

CBD. 2010. *Article 8(j): traditional knowledge, innovations and practices.* www.cbd.int/traditional/.

Chan, H.M., Kuhnlein, H.V. & Receveur, O. 2001. Evaluation of dietary contaminant intakes in 18 Inuit communities in northern Canada. Presented at the 17th International Congress of Nutrition, Vienna.

Chan, H.M., El Khoury, M., Sedgemore, M., Sedgemore, S. & Kuhnlein, H.V. 1996. Organochlorine pesticides and polychlorinated biphenyl congeners in ooligan grease: a traditional food fat of British Columbia First Nations. *Journal of Food Composition Analysis,* 9: 32–42.

Chotiboriboon, S., Tamachotipong, S., Sirisai, S., Dhanamitta, S., Smitasiri, S., Sappasuwan, C., Tantivatanasathien, P. & Eg-kantrong, P. 2009. Thailand: food system and nutritional status of indigenous children in a Karen community. *In* H.V. Kuhnlein, B. Erasmus and D. Spigelski. *Indigenous Peoples' food systems: the many dimensions of culture, diversity and environment for nutrition and health,* pp. 159–183. Rome, FAO.

Claxton, E., Sr. & Elliott, J., Sr. 1994. *Reef net technology of the Saltwater People.* Brentwood Bay, British Columbia, Canada, Saanich Indian School Board.

Colfer, C.J.P., Peluso, N. & Chung, C.S. 1997. *Beyond slash and burn. Building on indigenous management of Borneo's tropical rain forests.* Advances in Economic Botany No. 11. New York, New York Botanical Garden Press.

Conservation International. 2007. *Biofuels: the next threat to forests?* www.conservation.org/fmg/articles/pages/biofuels_threaten_forests.aspx.

Correal, C., Zuluaga, G., Madrigal, L., Caicedo, S. & Plotkin, M. 2009. Ingano traditional food and health: phase 1, 2004–2005. *In* H.V. Kuhnlein, B. Erasmus and D. Spigelski. *Indigenous Peoples' food systems: the many dimensions of culture, diversity and environment for nutrition and health,* pp. 83–108. Rome, FAO.

Cox, P.A. 1997. *Nafanua: saving the Samoan rain forest.* New York, WH Freeman. 238 pp.

Creed-Kanashiro, H., Roche, M., Tuesta Cerrón, I. & Kuhnlein, H.V. 2009. Traditional food system of an Awajún community in Peru. *In* H.V. Kuhnlein, B. Erasmus and D. Spigelski. *Indigenous Peoples' food systems: the many dimensions of culture, diversity and environment for nutrition and health,* pp. 59–81. Rome, FAO.

Creed-Kanashiro, H., Carrasco, M., Abad, M. & Tuesta, I. 2013. Promotion of traditional foods to improve the nutrition and health of the Awajún of the River Cenepa in Peru. *In* H.V. Kuhnlein, D. Spigelski, B. Erasmus and B. Burlingame, *Indigenous Peoples' food systems and well-being: interventions and policies for healthy communities,* Chapter 5. Rome, FAO.

Crosby, A.W. 1986. *Ecological imperialism: the biological expansion of Europe, 900–1900.* Cambridge, UK, Cambridge University Press. 368 pp.

Cunningham, A.B. 1997. An Africa-wide overview of medicinal plant harvesting, conservation and health care. *In* G. Bodeker, K.K.S. Bhat, J. Burley and P. Vantomme, eds. *Medicinal plants for forest conservation and health care,* pp. 116–129. Rome, FAO. 160 pp.

Damman, S. 2010. Indigenous peoples, rainforests and climate change. *SCN News,* 38: 63–67.

Davis, W. 1998. *Shadows in the sun: travels to landscapes of spirit and desire.* Washington, DC, Island Press. 292 pp.

Davis, W. 2001. *Light at the edge of the world: a journey through the realm of vanishing cultures.* Vancouver, British Columbia, Canada, Douglas and McIntyre Press, and Washington, DC, National Geographic Society. 180 pp.

Deur, D. & Turner, N.J., eds. 2005. *"Keeping it living": traditions of plant use and cultivation on the northwest coast of North America.* Seattle, Washington, USA, University of Washington Press, and Vancouver, British Columbia, Canada, University of British Columbia Press.

Devereaux, F. & Kittredge, K. 2008. *Feasting for change: Reconnecting to food, land and culture.* Victoria, British Columbia, Canada, Office of the Community Nutritionist for Aboriginal Health.

Dewey, K. 1979. Agricultural development, diet and nutrition. *Ecology of Food and Nutrition,* 8(4): 265–273.

Dewey, K. 1981. Nutritional consequences of the transformation from subsistence to commercial agriculture in Tabasco, Mexico. *Human Ecology,* 9(2): 161–187.

DFO. 2009. *Eulachon – Pacific Region.* Department of Fisheries and Oceans Canada (DFO) – Pacific Region. www.pac.dfo-mpo.gc.ca/fm-gp/commercial/pelagic-pelagique/eulachon-eulakane/index-eng.htm.

Dinar, A., Hassan, R., Mendelsohn, R. & Benhin, J. 2008. *Climate change and agriculture in Africa: impact assessment and adaptation strategies.* London, Earthscan. 206 pp.

Egeland, G.M., Charbonneau-Roberts, G., Kuluguqtuq, J., Kilabuk, J., Okalik, L., Soueida, R. & Kuhnlein, H.V. 2009. Back to the future: using traditional food and knowledge to promote a healthy future among Inuit. *In* H.V. Kuhnlein, B. Erasmus and D. Spigelski, eds. *Indigenous Peoples' food systems: the many dimensions of culture, diversity and environment for nutrition and health,* pp. 9–22. Rome, FAO.

Egeland, G.M., Yohannes, S., Okalik, L., Kilabuk, J., Racicot, C., Wilcke, M., Kuluguqtuq, J. & Kisa, S. 2013. The value of Inuit elders' storytelling in health promotion during times of rapid climate change and uncertain food security. *In* H.V. Kuhnlein, D. Spigelski, B. Erasmus and B. Burlingame. *Indigenous Peoples' food systems and well-being: interventions and policies for healthy communities,* Chapter 9. Rome, FAO.

Englberger, L., Schierle, J., Aalbersberg, W., Hofmann, P., Humphries, J., Huang, A., Lorens, A., Levendusky, A., Daniells, J., Marks, G.C. & Fitzgerald, M.H. 2006. Carotenoid and vitamin content of Karat and other Micronesian banana cultivars. *International Journal of Food Sciences and Nutrition,* 57(5): 399–418.

Englberger, L., Lorens, A., Levendusky, A., Pedrus, P., Albert, K., Hagilmai, W., Paul, Y., Nelber, D., Moses, P., Shaeffer, S. & Gallen, M. 2009. Documentation of the traditional food system of Pohnpei. *In* H.V. Kuhnlein, B. Erasmus and D. Spigelski. *Indigenous Peoples' food systems: the many dimensions of culture, diversity and environment for nutrition and health,* pp. 109–138. Rome, FAO.

Englberger, L., Lorens, A., Albert, K., Pedrus, P., Levendusky, A., Hagilmai, W., Paul, Y., Moses, P., Jim, R., Jose, S., Nelber, D., Santos, G., Kaufer, L., Larsen, K., Pretrick, M. & Kuhnlein, H.V. 2013. Let's go local! Pohnpei promotes local food production and nutrition for health. *In* H.V. Kuhnlein, D. Spigelski, B. Erasmus and B. Burlingame. *Indigenous Peoples' food systems and well-being: interventions and policies for healthy communities,* Chapter 12. Rome, FAO.

Environmental Change Institute. 2007. Indigenous Peoples and Climate Change. Symposium organized by J. Salick and A. Byg, University of Oxford, 12–13 April 2007. Oxford, UK. www.eci.ox.ac.uk/news/events/070412conference.php.

FAO. 1996. *Declaration on world food security.* World Food Summit. Rome.

FAO. 2009. *Indigenous Peoples' food systems: the many dimensions of culture, diversity and environment for nutrition and health,* edited by H.V. Kuhnlein, B. Erasmus and D. Spigelski. Rome. 339 pp.

FAO & IPGRI. 2002. *The role of women in the conservation of the genetic resources of maize. Guatemala.* Rome. 54 pp. Rome, FAO and International Plant Genetic Resources Institute (now Bioversity).

FIAN International. 2008. *Brazil: Impact of agrofuels on human right to food.* FoodFirst Information and Action Network (FIAN), Press Release, 19 April 2008. www.fian.org/news/press-releases/brazil-impact-of-agrofuels-on-human-right-to-food.

Fowler, C. & Mooney, P.R. 1990. *Shattering: food, politics, and the loss of genetic diversity.* Tucson, Arizona, USA, University of Arizona Press. 278 pp.

Garibaldi, A. & Turner, N.J. 2004. Cultural keystone species: implications for ecological conservation and restoration. *Ecology and Society,* 9(3): 1. www.ecologyandsociety.org/vol9/iss3/art1.

George, M., Innes, J. & Ross, H. 2004. *Managing sea country together: key issues for developing co-operative management for the Great Barrier Reef World Heritage Area.* CRC Reef Research Technical Report No. 50. Townsville, Australia, CRC Reef Research Centre.

Graham, L.R. 2008. Wayuu and Xavante meet in Xavante protest over destructive soy cultivation. *Anthropology News,* April 2008, pp. 32–33.

Griffiths, M., Taylor, A. & Woynillowicz, D. 2006. *Troubled waters, troubling trends: technology and policy options to reduce water use in oil and oil sands development in Alberta. Report.* Drayton Valley, Alberta, Canada, Pembina Institute. 157 pp.

Hoyt, E. 1988. *Conserving the wild relatives of crops.* Rome, International Board for Plant Genetic Resources (IBPGR) and FAO; and Gland, Switzerland, World Conservation Union (IUCN) and World Wide Fund for Nature (WWF).

Hunn, E.S., Johnson, D.R., Russell, P.N. & Thornton, T.F. 2003. Huna Tlingit traditional environmental knowledge, conservation, and the management of a "wilderness" park. *Current Anthropology,* 44: S79–S103.

Imhoff, D. 2003. *Farming with the wild. Enhancing biodiversity on farms and ranches.* San Francisco, California, USA, Sierra Club. 176 pp.

International Indian Treaty Council. 2002. *Declaration of Atitlán, Guatemala.* Indigenous Peoples' Consultation on the Right to Food, Atitlán, Sololá, Guatemala, 17–19 April 2002.

IPCC. 2007. *Climate change 2007: The physical science basis.* Geneva, World Meteorological Organization and United Nations Environment Programme (UNEP). www.ipcc.ch/ipccreports/ar4-wg1.htm.

Iwasaki-Goodman, M. 2013. Tasty *tonoto* and not-so-tasty *tonoto*: fostering traditional food culture among the Ainu people in the Saru River Region, Japan. *In* H.V. Kuhnlein, D. Spigelski, B. Erasmus and B. Burlingame. *Indigenous Peoples' food systems and well-being: interventions and policies for healthy communities,* Chapter 13. Rome, FAO.

Iwasaki-Goodman, M., Ishii, S. & Kaizawa, T. 2009. Traditional food systems of Indigenous Peoples: the Ainu in the Saru River region, Japan. *In* H.V. Kuhnlein, B. Erasmus and D. Spigelski. *Indigenous Peoples' food systems: the many dimensions of culture, diversity and environment for nutrition and health,* pp. 139–157. Rome, FAO.

Jackson, J.B.C., Kirby, M.X., Berger, W.H., Bjorndal, K.A., Botsford, L.W., Bourque, B.J., Bradbury, R.H., Cooke, R., Erlandson, J., Estes, J.A., Hughes, T.P., Kidwell, S., Lange, C.B., Lenihan, H.S., Pandolfi, J.M., Peterson, C.H., Steneck, R.S., Tegner, M.J. & Warner, R.R. 2001. Historical overfishing and the recent collapse of coastal ecosystems. *Science,* 293(5530): 629–637.

Johannes, R.E. 2002. Did indigenous conservation ethics exist? *Traditional Marine Resource Management and Knowledge Information Bulletin,* 14: 3–7.

Johannessen, D.I. & Ross, P.S. 2002. *Late-run sockeye at risk: an overview of environmental contaminants in Fraser River salmon habitat.* Canadian Technical Reports of Fisheries and Aquatic Science No. 2429. Ottawa, Fisheries and Ocean Canada. 108 pp.

Jones, O.A.H., Maguire, M.L. & Griffin, J.L. 2008. Environmental pollution and diabetes: a neglected association. *Lancet,* 371(9609): 287–288.

Keskitalo, E.C.H. 2008. *Climate change and globalization in the Arctic: an integrated approach to vulnerability assessment.* London, Earthscan. 262 pp.

Khaniki, G.R.J., Alli, I., Nowroozi, E. & Nabizadeh, R. 2005. Mercury contamination in fish and public health aspects: a review. *Pakistan Journal of Nutrition,* 4(5): 276–281.

Knotsch, C. & Lamouche, J. 2010. *Arctic biodiversity and Inuit health. Inuit Tuttavingat.* Ottawa, National Aboriginal Health Organization. http://www.naho.ca/documents/it/2010_Arctic_Biodiversity.pdf.

kp-studios.com. 2008. *The Inuit and their indigenous foods.* Anacortes, Washington, USA. www.indigenousnutrition.org/. (video)

Krümmel, E.M., Macdonald, R.W., Kimpe, L.E., Gregory-Eaves, I., Demers, M.J., Smol, J.P., Finney, B. & Blais, J.M. 2003. Aquatic ecology: delivery of pollutants by spawning salmon. *Nature,* 425: 255–256.

Krupnik, I. & Jolly, D., eds. 2002. *The earth is faster now: indigenous observations of Arctic environmental change.* Fairbanks, Alaska, USA, Arctic Research Consortium of the United States. 356 pp.

Kuhnlein, H.V. 1989. Factors influencing use of traditional foods among the Nuxalk People. *Journal of the Canadian Dietetic Association,* 50(2): 102–106.

Kuhnlein, H.V. 1992. Change in the use of traditional foods by the Nuxalk Native People of British Columbia. *Ecology of Food and Nutrition,* 27: 259–282.

Kuhnlein H.V. & Chan, H.M. 2000. Environment and contaminants in traditional food systems of northern Indigenous Peoples. *Annual Review of Nutrition,* 20: 595–626.

Kuhnlein, H.V., Chan, A.C., Thompson, J.N. & Nakai, S. 1982. Ooligan grease: a nutritious fat used by Native People of Coastal British Columbia. *Journal of Ethnobiology,* 2(2): 154–161.

Kuhnlein, H.V., Receveur, O., Soueida, R. & Egeland, G.M. 2004. Arctic Indigenous Peoples experience the nutrition transition with changing dietary patterns and obesity. *Journal of Nutrition,* 134: 1447–1453.

Kuhnlein, H.V., Chan, H.M., Receveur, O. & Egeland, G. 2005. Canadian Arctic Indigenous Peoples, traditional food systems and POPs. *In* T. Fenge and D. Downey, eds. *Northern lights against POPs: combating toxic threats at the top of the world,* pp. 22–40. Montreal, Quebec, Canada, McGill-Queens's University Press.

Kuhnlein, H.V., Erasmus, B., Creed-Kanashiro, H., Englberger, L., Okeke, C., Turner, N., Allen, L. & Bhattacharjee, L. 2006. Indigenous Peoples' food systems for health: finding interventions that work. *Public Health Nutrition,* 9(8): 1013–1019.

Kuhnlein, H.V., McDonald, M., Spigelski, D., Vittrekwa, E. & Erasmus, B. 2009. Gwich'in traditional food for health: Phase 1. *In* H.V. Kuhnlein, B. Erasmus and D. Spigelski. *Indigenous Peoples' food systems: the many dimensions of culture, diversity and environment for nutrition and health,* pp. 45–58. Rome, FAO.

Kuhnlein, H.V., Goodman, L., Receveur, O., Spigelski, D., Duran, N., Harrison, G.G., Erasmus, B. & Tetlit Zheh. 2013. The Gwich'in traditional food and health project in Tetlit Zheh, Northwest Territories. In H.V. Kuhnlein, D. Spigelski, B. Erasmus and B. Burlingame. *Indigenous Peoples' food systems and well-being: interventions and policies for healthy communities,* Chapter 7. Rome, FAO.

Kurunganti, K. 2006. *Mass protests against GM crops in India. Report.* Tarnaka, India, Centre for Sustainable Agriculture.

La Duke, W. & Carlson, B. 2003. *Our manoomin, our life: the Anishabeg struggle to protect wild rice.* Ponsford, Minnesota, USA, White Earth Land Recovery Project.

Laird, S.A., ed. 2002. *Biodiversity and traditional knowledge: equitable partnerships in practice.* London, Earthscan. 504 pp.

Lambden, J., Receveur O. & Kuhnlein, H.V. 2007. Traditional food attributes must be included in studies of food security in the Canadian Arctic. *International Journal of Circumpolar Health,* 66(4): 308–319.

Lambden, J., Receveur, O., Marshall, J. & Kuhnlein, H.V. 2006. Traditional and market food access in Arctic Canada is affected by economic factors. *International Journal of Circumpolar Health,* 65(4): 331–340.

Lebel, J., Roulet, M., Mergler, D., Lucotte, M. & Larribe, F. 1997. Fish diet and mercury exposure in a riparian Amazonian population. *Water, Air, and Soil Pollution,* 97: 31–44.

Lichota, G.B., McAdie, M. & Ross, P.S. 2004. Endangered Vancouver Island marmots (*Marmota vancouverensis*): sentinels of atmospherically delivered contaminants to British Columbia, Canada. *Environmental Toxicology and Chemistry,* 23(2): 402–407.

Mackenzie, I. 1993. *Cry of the forgotten land.* Oakland, California, USA, The Endangered Peoples' Project and Gryphon Productions. (video, 26 minutes)

Millennium Ecosystem Assessment. 2005. Overall synthesis, and reports on biodiversity, desertification, business and industry, wetlands and water, and health. www.millenniumassessment.org/en/index.aspx.

Mos, L., Jack, J., Cullon, D., Montour, L., Alleyne, C. & Ross, P. 2004. The importance of marine foods to a near-urban First Nation community in coastal British Columbia, Canada: toward a risk-benefit assessment. *Journal of Toxicology and Environmental Health Part A,* 67(8–10): 791–808(18)

Myers, R.A. & Worm, B. 2003. Rapid worldwide depletion of predatory fish communities. *Nature,* 423: 280–283.

Myers, H., Fast, H., Berkes, M.K. & Berkes, F. 2005. Feeding the family in times of change. *In* F. Berkes, R. Huebert, H. Fast, M. Manseau and A. Diduck, eds. *Breaking ice: renewable resource and ocean management in the Canadian north,* pp. 23–46. Calgary, Alberta, Canada, University of Calgary Press.

Nabhan, G.P. 1986. *Gathering the desert.* Tucson, Arizona, USA, University of Arizona Press. 182 pp.

Nabhan, G.P., ed. 2006. *Renewing Salmon Nation's food traditions.* Portland, Oregon, USA, Ecotrust. 66 pp.

Nabhan, G.P. & Rood, A., eds. 2004. *Renewing America's food traditions (RAFT): bringing cultural and culinary mainstays from the past into the new millennium.* Flagstaff, Arizona, USA, Center for Sustainable Environments at Northern Arizona University. 83 pp.

Nazarea, V.D., ed. 1999. *Ethnoecology: situated knowledge/located lives.* Tucson, Arizona, USA, University of Arizona Press. 299 pp.

Nuttall, M. 2006. The Mackenzie Gas Project – Aboriginal interests, the environment and northern Canada's energy frontier. *Indigenous Affairs,* 2–3: 20–29.

Oiye, S., Simel, J.O., Oniang'o, R. & Johns, T. 2009. The Maasai food system and food and nutrition security. *In* H.V. Kuhnlein, B. Erasmus and D. Spigelski. *Indigenous Peoples' food systems: the many dimensions of culture, diversity and environment for nutrition and health,* pp. 231–249. Rome, FAO.

Okeke, E.C., Ene-Ebong, H.N., Uzuegbunam, A.O., Ozioko, A., Umeh, S.I. & Chukwuone, N. 2009. The Igbo traditional food system documented in four states. *In* H.V. Kuhnlein, B. Erasmus and D. Spigelski. *Indigenous Peoples' food systems: the many dimensions of culture, diversity and environment for nutrition and health,* pp. 251–281. Rome, FAO.

Ommer, R.E. & Coasts Under Stress Research Project Team. 2007. *Coasts Under Stress: restructuring and social-ecological health.* Montreal, Quebec, Canada, McGill-Queen's University Press. 574 pp.

Parrish, C.C., Turner, N.J. & Solberg, S.M., eds. 2007. *Resetting the kitchen table: food security, culture, health and resilience in coastal communities.* New York, Nova Science. 257 pp.

Parrish, C.C., Copeman, L., Van Biesen, G. & Wroblewski, J. 2007. Aquaculture and nearshore marine food webs: implications for seafood quality and the environment north of 50. *In* C.C. Parrish, N.J. Turner and S.M. Solberg, eds. *Resetting the kitchen table: food security, culture, health and resilience in coastal communities,* pp. 33–49. New York, Nova Science.

Pasternak, S., Mazgul, L. & Turner, N.J. 2009. Born from bears and corn: why indigenous knowledge systems and beliefs matter in the debate on GM foods. *In* C. Brunk and H. Coward, eds. *Acceptable genes? Religious traditions and genetically modified foods,* pp. 211–230. New York, State University of New York Press.

Pauly, D., Christensen, V., Froese, R. & Palomares, M.L.D. 2000. Fishing down aquatic food webs. *American Scientist,* 88: 46–51.

Pollen, M. 2006. *The omnivore's dilemma. A natural history of four meals.* New York, Penguin. 450 pp.

Porcupine Caribou Management Board. 2007. *Porcupine Caribou herd size.* www.taiga.net/pcmb/population.html.

Posey, D.A. 1985. Indigenous management of tropical forest ecosystems: the case of the Kayapó Indians of the Brazilian Amazon. *Agroforestry Systems,* 3(2): 139–158.

Pukonen, J.C. 2008. *The* Tl'aaya'as *Project: revitalizing traditional Nuu-chah-nulth root gardens in Ahousaht, British Columbia.* Victoria, British Columbia, Canada, University of Victoria. (M.Sc. thesis)

Rayne, S., Ikonomou, M.G., Ellis, G.M., Barrett-Lennard, L.G. & Ross, P.S. 2004. PBDEs, PBBS, and PCNs in three communities of free-ranging killer whales (*Orcinus orca*) from the northeastern Pacific Ocean. *Environmental Science and Technology,* 38(16): 4293–4299.

Richardson, B. 1991. *Strangers devour the land.* Post Mills, Vermont, USA, Chelsea Green.

Rignell-Hydbom, A., Rylander, L. & Hagmar, L. 2007. Exposure to persistent organochlorine pollutants and type 2 diabetes mellitus. *Human and Experimental Toxicology,* 26(5): 447–452.

Roach, J. 2006. Seafood may be gone by 2048, study says. *National Geographic News,* 2 November 2006. news.nationalgeographic.com/news/2006/11/061102-seafood-threat.html.

Ross, P.S. 2000. Marine mammals as sentinels in ecological risk assessment. *Human and Ecological Risk Assessment,* 6(1): 29–46.

Ross, P.S. 2002. The role of immunotoxic environmental contaminants in facilitating the emergence of infectious diseases in marine mammals. *Human and Ecological Risk Assessment,* 8(2): 277–292.

Ross, P.S. 2006. Fireproof killer whales (*Orcinus orca*): flame-retardant chemicals and the conservation imperative in the charismatic icon of British Columbia, Canada. *Canadian Journal of Fisheries and Aquatic Sciences,* 63: 224–234.

Ross, P.S. & Birnbaum, L.S. 2003. Integrated human and ecological risk assessment: a case study of persistent organic pollutants (POPs) in humans and wildlife. *Human and Ecological Risk Assessment,* 9(1): 303–324.

Ross, P.S., Vos, J.G. & Osterhaus, A.D.M.E. 2003. The immune system, environmental contaminants and virus-associated mass mortalities among pinnipeds. *In* J.G. Vos, G.D. Bossart and M. Fournier. *Toxicology of marine mammals,* pp. 543–557. Washington, DC, Taylor and Francis.

Ross, P.S., Jeffries, S., Yunker, M.B., Addison, R.F., Ikonomou, M.G. & Calambokidis, J. 2004. Harbour seals (*Phoca vitulina*) in British Columbia, Canada, and Washington State, USA, reveal a combination of local and global PCB, PCDD and PCDF signals. *Environmental Toxicology and Chemistry,* 23(1): 157–165.

Roulet, M., Lucotte, M., Farella, N., Serique, G., Coelho, H., Sousa Passos, C.J., De Jesus da Silva, E., Scavone de Andrade, P., Mergler, D., Guimarães, J.-R.D. and Amorim, M. 1999. Effects of recent human colonization on the presence of mercury in Amazonian ecosystems. *Water, Air, and Soil Pollution,* 112: 297–313.

Salick, J. & Ross, N., eds. 2009. Indigenous Peoples and climate change. *Global Environmental Change,* 19 (special issue).

Salmón, E. 2000. Kincentric ecology: indigenous perceptions of the human-nature relationship. *Ecological Applications,* 10(5): 1327–1332.

Salomeyesudas, B. & Satheesh, P.V. 2009. Traditional food system of Dalit in Zaheerabad Region, Medak District, Andhra Pradesh, India. *In* H.V. Kuhnlein, B. Erasmus and D. Spigelski. *Indigenous Peoples' food systems: the many dimensions of culture, diversity and environment for nutrition and health,* pp. 185–208. Rome, FAO.

Salomeyesudas, B., Kuhnlein, H.V., Schmid, M.A., Satheesh, P.V. & Egeland, G.M. 2013. The Dalit food system and maternal and child health in Andhra Pradesh, South India. *In* H.V. Kuhnlein, D. Spigelski, B. Erasmus and B. Burlingame. *Indigenous Peoples' food systems and well-being: interventions and policies for healthy communities,* Chapter 6. Rome, FAO.

Schindler, D.E., Essington, T.E., Kitchell, J.F., Boggs, C. & Hilborn, R. 2002. Sharks and tunas: fisheries impacts on predators with contrasting life histories. *Ecological Applications,* 12(3): 735–748.

Senos, R., Lake, F., Turner, N.J. & Martinez, D. 2006. Traditional ecological knowledge and restoration practice in the Pacific Northwest. *In* D. Apostol, ed. *Encyclopedia for restoration of Pacific Northwest ecosystems,* pp. 393–426. Washington, DC, Island Press.

Shiva, V. 2000. *Stolen harvest: the hijacking of the global food supply.* Cambridge, Massachusetts, USA, South End Press. 150 pp.

Shkilnyk, A.M. 1985. *A poison stronger than love: the destruction of an Ojibwa community.* New Haven, Connecticut, USA, Yale University Press. 276 pp.

Simms, W., Jeffries, S.J., Ikonomou, M.G. & Ross, P.S. 2000. Contaminant-related disruption of vitamin A dynamics in free-ranging harbor seal (*Phoca vitulina*) pups from British Columbia, Canada and Washington State, USA. *Environmental Toxicology and Chemistry,* 19(11): 2844–2849.

Sirisai, S., Chotiboriboon, S., Tantivatanasathien, P., Sangkhawimol, S & Smitasiri, S. 2013. Culture-based nutrition and health promotion in a Karen community. *In* H.V. Kuhnlein, D. Spigelski, B. Erasmus and B. Burlingame. *Indigenous Peoples' food systems and well-being: interventions and policies for healthy communities,* Chapter 10. Rome, FAO.

Strand, P., Balonov, M., Aarkrog, A., Bewers, M.J., Howard, B., Salo, A. & Tsaturov, Y.S. 1998. Radioactivity. *In* Arctic Monitoring and Assessment Programme. *AMAP Assessment Report: Arctic Pollution Issues,* pp. 525–620. Oslo, Arctic Monitoring and Assessment Programme.

Thomas, C.D., Cameron, A., Green, R.E., Bakkenes, M., Beaumont, L.J., Collingham, Y.C., Erasmus, B.F.N., de Siqueira, M.F., Grainger, A., Hannah, L., Hughes, L., Huntley, B., van Jaarsveld, A.S., Midgley, G.F., Miles, L., Ortega-Huerta, M.A., Peterson, A.T., Phillips, O.L. & Williams, S.E. 2004. Extinction risk from climate change. *Nature,* 427: 145–148.

Thompson, S. 2005. Sustainability and vulnerability: Aboriginal Arctic food security in a toxic world. *In* F. Berkes, R. Huebert, H. Fast, M. Manseau and A. Diduck, eds. *Breaking ice: renewable resource and ocean management in the Canadian north,* pp. 47–70. Calgary, Alberta, Canada, University of Calgary Press.

Torys LLP. 2010. Alberta court finds Syncrude guilty in duck deaths. *Torys on Environmental, Health and Safety,* 17(3). www.torys.com/publications/documents/publication%20pdfs/ehs2010-3.pdf.

Turner, N.J. 2005. *The earth's blanket. Traditional teachings for sustainable living.* Vancouver, British Columbia, Canada, Douglas and McIntyre, and Seattle, Washington, USA, University of Washington Press. 298 pp.

Turner, N.J. & Berkes, F. 2006. Coming to understanding: developing conservation through incremental learning in the Pacific Northwest. *Human Ecology,* 34(4): 495–513.

Turner, N.J. & Clifton, H. 2009. "It's so different today." Climate change and indigenous lifeways in British Columbia, Canada. *Global Environmental Change,* 19: 180–190. (Special issue on Indigenous Peoples and climate change edited by J. Salick and N. Ross)

Turner, N.J. & Turner, K.L. 2006. Traditional food systems, erosion and renewal in Northwestern North America. *Indian Journal of Traditional Knowledge,* 6(1): 57–68.

Turner, N.J. & Turner, K.L. 2008. Where our women used to get the food: cumulative effects and loss of ethnobotanical knowledge and practice; case study from coastal British Columbia. *Botany,* 86(1): 103–115.

Turner, N.J. & Wilson, B. (Kii'iljuus). 2006. To provide living plants for study: the value of ethnobotanical gardens and planning the Qay'llnagaay garden of Haida Gwaii. *Davidsonia,* 16(4): 111–125.

Turner, N.J., Marshall, A., Thompson, J.C. (Edosdi), Hood, R.J., Hill, C. & Hill, E.-A. 2008a. "Ebb and flow": transmitting environmental knowledge in a contemporary aboriginal community. *In* J.S. Lutz and B. Neis, eds. *Making and moving knowledge. Interdisciplinary and community-based research in a world on the edge,* pp. 45–63. Montreal and Kingston, Quebec, Canada, McGill-Queen's University Press.

Turner, N.J., Gregory, R., Brooks, C., Failing, L. & Satterfield, T. 2008b. From invisibility to transparency: identifying the implications. *Ecology and Society,* 13(2): 7. www.ecologyandsociety.org/vol13/iss2/art7/.

Turner, N.J., Harvey, T., Burgess, S. & Kuhnlein, H.V. 2009. The Nuxalk Food and Nutrition Program, Coastal British Columbia, Canada: 1981–2006. *In* H.V. Kuhnlein, B. Erasmus and D. Spigelski. *Indigenous Peoples' food systems: the many dimensions of culture, diversity and environment for nutrition and health,* pp. 23–44. Rome, FAO.

Turner, N.J., Tallio, W.R., Burgess, S. & Kuhnlein, H.V. 2013. The Nuxalk Food and Nutrition Program for Health revisited. *In* H.V. Kuhnlein, D. Spigelski, B. Erasmus and B. Burlingame. *Indigenous Peoples' food systems and well-being: interventions and policies for healthy communities,* Chapter 11. Rome, FAO.

UNPFII. 2009. *The State of the World's Indigenous Peoples.* New York, United Nations Department of Economic and Social Affairs, Secretariat of the Permanent Forum on Indigenous Issues. 250 pp. www.un.org/esa/socdev/unpfii/documents/sowip_web.pdf.

Volpe, J. 2007. Salmon sovereignty and the dilemma of intensive Atlantic salmon aquaculture development in British Columbia. *In* C.C. Parrish, N.J. Turner and S.M. Solberg, eds. *Resetting the kitchen table: food security, culture, health and resilience in coastal communities,* pp. 75–86. New York, Nova Science.

Wernham, A. 2007. Inupiat health and proposed Alaskan oil development: results of the First Integrated Health Impact Assessment/Environmental Impact Statement for proposed oil development on Alaska's North Slope. Special Feature: Indigenous Perspectives. *EcoHealth,* 4(4): 500–513.

Wilson, B. (Kii'iljuus) & Turner, N.J. 2004. *K'aaw k'iihl:* a time-honoured tradition for today's world. *In* H. Arntzen, C. Fisher, S. Foster and B. Whittington, eds. *Cycle of life/recycle handbook for educators, national edition,* pp. 257–259. Victoria, British Columbia, Canada, Artists' Response Team (ART).

Wilson, E.O. 1992. *The diversity of life.* Cambridge, Massachusetts, USA, Belknap Press.

Wong, A. 2003. Reversing the high prevalence of obesity, diabetes, hypertension and depression among the Aboriginal People. *In* A. Wong, ed. *First Nations Nutrition and Health Conference Proceedings,* pp. 76–80. Vancouver, British Columbia, Canada, Arbokem Inc.

WWF. 2004. *Living planet report 2004.* Gland, Switzerland, World Wide Fund for Nature (WWF).

WWF. 2008. *Living planet report 2008.* Gland, Switzerland.

Wyllie-Echeverria, S. & Cox, P.A. 2000. Cultural saliency as a tool for seagrass conservation. *Biologia Marina Mediterranea,* 7(2): 421–424.

Chapter 4 | Infant and young child feeding

Canadian Paediatric Society, Dietitians of Canada & Health Canada. 1998. *Nutrition for healthy term infants.* Ottawa, Minister of Public Works and Government Services.

Chotiboriboon, S., Tamachotipong, S., Sirisai, S., Dhanamitta, S., Smitasiri, S., Sappasuwan, C., Tantivatanasathien, P. & Eg-kantrong, P. 2009. Thailand: food system and nutritional status of indigenous children in a Karen community. *In* H.V. Kuhnlein, B. Erasmus and D. Spigelski. *Indigenous Peoples' food systems: the many dimensions of culture, diversity and environment for nutrition and health,* pp. 159–183. Rome, FAO.

Correal, C., Zuluaga, G., Madrigal, L., Caicedo, S. & Plotkin, M. 2009. Ingano traditional food and health: phase 1, 2004–2005. *In* H.V. Kuhnlein, B. Erasmus and D. Spigelski. *Indigenous Peoples' food systems: the many dimensions of culture, diversity and environment for nutrition and health,* pp. 83–108. Rome, FAO.

Creed-Kanashiro, H., Roche, M., Tuesta Cerrón, I. & Kuhnlein, H.V. 2009. Traditional food system of an Awajún community in Peru. *In* H.V. Kuhnlein, B. Erasmus and D. Spigelski. *Indigenous Peoples' food systems: the many dimensions of culture, diversity and environment for nutrition and health,* pp. 59–81. Rome, FAO.

Englberger, L., Marks, G.C. & Fitzgerald, M.H. 2003. Insights on food and nutrition in the Federated States of Micronesia: a review of the literature. *Public Health Nutrition,* 6: 5–17.

Englberger, L., Lorens, A., Levendusky, A., Pedrus, P., Albert, K., Hagilmai, W., Paul, Y., Nelber, D., Moses, P., Shaeffer, S. & Gallen, M. 2009. Documentation of the traditional food system of Pohnpei. *In* H.V. Kuhnlein, B. Erasmus and D. Spigelski. *Indigenous Peoples' food systems: the many dimensions of culture, diversity and environment for nutrition and health,* pp. 109–138. Rome, FAO.

FAO. 2009. *Indigenous Peoples' food systems: the many dimensions of culture, diversity and environment for nutrition and health,* edited by H.V. Kuhnlein, B. Erasmus, and D. Spigelski. Rome. 339 pp.

Gaur, A.H., Dominguez, K.L., Kalish, M.L., Rivera-Hernandez, D., Donohoe, M., Brooks, J.T. & Mitchell, C.D. 2009. Practice of feeding premasticated food to infants: a potential risk factor for HIV transmission. *Pediatrics,* 124(2): 658–666.

Gibson, R.S., Ferguson, E.L. & Lehrfeld, J. 1998. Complementary foods for infant feeding in developing countries: their nutrient adequacy and improvement. *European Journal of Clinical Nutrition,* 52(10): 764–770.

Kazimi, L.J. & Kazimi, H.R. 1979. Infant feeding practices of the Igbo, Nigeria. *Ecology of Food and Nutrition,* 8(2): 111–116.

Kimmons, J.E., Brown, K.H., Lartey, A., Collison, E., Mensah, P.P.A. & Dewey, K.G. 1999. The effects of fermentation and/or vacuum flask storage on the presence of coliforms in complementary foods prepared for Ghanaian children. *International Journal of Food Sciences and Nutrition,* 50(3): 195–201.

Kuhnlein, H.V. & Receveur, O. 1996. Dietary change and traditional food systems of Indigenous Peoples. *Annual Review of Nutrition,* 16: 417–442.

Kuhnlein, H.V., McDonald, M., Spigelski, D., Vittrekwa, E. & Erasmus, B. 2009. Gwich'in traditional food for health: Phase 1. In H.V. Kuhnlein, B. Erasmus and D. Spigelski. *Indigenous Peoples' food systems: the many dimensions of culture, diversity and environment for nutrition and health*, pp. 45–58. Rome, FAO.

Kuperberg, K. & Evers, S. 2006. Feeding patterns and weight among First Nations children. *Canadian Journal of Dietetic Practice and Research*, 67(2): 79–84.

Marshall, L.B. & Marshall, M. 1980. Infant feeding and infant illness in a Micronesian village. *Social Science and Medicine. Part B: Medical Anthropology*, 14(1): 33–38.

Nwankwo, B.O. & Brieger, W.R. 2002. Exclusive breastfeeding is undermined by use of other liquids in rural Southwestern Nigeria. *Journal of Tropical Pediatrics*, 48(2): 109–112.

Okeahialam, T.C. 1986. Breast-feeding practices among Nigerian Igbo mothers. *Journal of Tropical Pediatrics*, 32(4): 154–157.

Okeke, E.C., Ene-Obong, H.N., Uzuegbunam, A.O., Ozioko, A., Umeh, S.I. & Chukwuone, N. 2009. The Igbo traditional food system documented in four states in Southern Nigeria. *In* H.V. Kuhnlein, B. Erasmus and D. Spigelski. *Indigenous Peoples' food systems: the many dimensions of culture, diversity and environment for nutrition and health*, pp. 251–281. Rome, FAO.

PAHO & WHO. 2003. *Guiding principles for complementary feeding of the breastfed child*. Washington, DC, Division of Health Promotion and Protection, Food and Nutrition Program, Pan American Health Organization (PAHO) and World Health Organization (WHO).

Panpanich, R., Vitsupakorn, K. & Chareonporn, S. 2000. Nutritional problems in children aged 1–24 months: comparison of hill-tribe and Thai children. *Journal of the Medical Association of Thailand*, 83(11): 1375–1379.

Pelto, G.H., Zhang, Y. & Habicht, J.-P. 2010. Premastication: the second arm of infant and young child feeding for health and survival? *Maternal and Child Nutrition*, 6: 4–18.

Roche, M.L., Creed-Kanashiro, H.M., Tuesta, I. & Kuhnlein, H.V. 2007. Traditional food system provides dietary quality for the Awajún in the Peruvian Amazon. *Ecology of Food and Nutrition*, 46(5 & 6): 377–399.

Roche M.L., Creed-Kanashiro H.M., Tuesta I. & Kuhnlein H.V. 2010. Infant and young child feeding in the Peruvian Amazon: the need to promote exclusive breastfeeding and nutrient-dense traditional complementary foods. *Maternal and Child Nutrition*, Online 21 January 2010. DOI: 10.1111/j.1740-8709.2009.00234.x

Ruel, M., Brown, K.H. & Caulfield, L.E. 2003. *Moving forward with complementary feeding: indicators and research priorities*. FCND Discussion Paper No. 146. Washington, DC, International Food Policy Research Institute (IFPRI).

Salomeyesudas, B. & Satheesh, P.V. 2009. Traditional food system of Dalit in Zaheerabad Region, Medak District, Andhra Pradesh, India. *In* H.V. Kuhnlein, B. Erasmus and D. Spigelski. *Indigenous Peoples' food systems: the many dimensions of culture, diversity and environment for nutrition and health*, pp. 185–208. Rome, FAO.

Schaefer, O. & Spady, D.W. 1982. Changing trends in infant feeding patterns in the Northwest Territories 1973–1979. *Canadian Journal of Public Health*, 73: 304–309.

Schmid, M.A., Egeland, G.M., Salomeyesudas, B., Satheesh, P.V. & Kuhnlein, H.V. 2006. Traditional food consumption and nutritional status of Dalit mothers in rural Andhra Pradesh, South India. *European Journal of Clinical Nutrition*, 60(11): 1277–1283.

Schmid, M., Salomeyesudas, B., Satheesh, P.V., Hanley, J. & Kuhnlein, H.V. 2007. Intervention with traditional food as a major source of energy, protein, iron, vitamin C and vitamin A for rural Dalit mothers and young children in Andhra Pradesh, South India. *Asia Pacific Journal of Clinical Nutrition*, 16(1): 84–93.

Sellen, D.W. 2001. Comparison of infant feeding patterns reported for nonindustrial populations with current recommendations. *Journal of Nutrition*, 131(10): 2707–2715.

Sharma, R.K. 2007. Newborn health among tribes of Madhya Pradesh – an overview. *Biannual Newsletter of Regional Medical Research Centre for Tribals*, 4(1): 1–5.

Taneja, P.V. & Gupta, N.V. 1998. Feeding practices in infants of Bhil tribe in Jhabua District of Madhya Pradesh. *Indian Pediatrics*, 35(6): 568.

Tienboon, P. & Wangpakapattanawong, P. 2007. Nutritional status, body composition and health conditions of the Karen hill tribe children aged 1–6 years in Northern Thailand. *Asia Pacific Journal of Clinical Nutrition*, 16(2): 279–285.

WHO. 1998. *Complementary feeding of young children in developing countries: a review of current scientific knowledge*. WHO/NUT/98.1. Geneva. 228 pp.

WHO. 2008. *Indicators for assessing infant and young child feeding practices*. Conclusions of a consensus meeting held 6–8 November 2007 in Washington DC. Geneva.

WHO & Lippwe, K. n.d. Micronesia, Federated States of. www.wpro.who.int/nr/rdonlyres/f48f2722-5fe8-496b-abcc-784c5f36d8fb/0/mic.pdf.

Chapter 5 | Awajún

AECI, CIPCA & SAIPE. 2000. *Evaluación participativa de necesidades prioritaria, Distrito El Cenepa, Alto Marañón*. Lima, *Agencia Española de Cooperación Internacional* (AECI), *Centro de Investigación y Promoción del Campesinado* (CIPCA) and *Servicio Agropecuario para la Investigación y Promoción Económica* (SAIPE).

AIDESEP. 2010. Website: www.aidesep.org.pe/. *Asociación Interétnica de Desarrollo de la Selva Perúana* (AIDESEP)

Berlin, E.A. & Markell, E.K. 1977. An assessment of the nutritional and health status of an Awajún Jívaro community, Amazonas, Peru. *Ecology of Food and Nutrition*, 6: 69–81.

Carvalho-Costa, F.A., Gonçalves, A.Q., Lassance, S.L., da Silva Neto, L.M., Salmazo, C.A.A. & Bóia, M.N. 2007. *Giardia lamblia* and other intestinal parasitic infections and their relationships with nutritional status in children in Brazilian Amazon. *Revista do Instituto de Medicina Tropical de São Paolo*, 49(3): 147–153.

Chang, A.S. & Sarasara, A.C. 1987. *Organizaciones sociales y económicas en las comunidades del grupo etnolingüístico Aguaruna*. Lima, Ministry of Agriculture.

Claverías, R. & Quispe, C. 2002. Biodiversidad cultivada: una estrategia campesina para superar la pobreza y relacionarse con el mercado. *In* M. Pulgar-Vidal, E. Zegarra and J. Urrutia, eds. *Perú: el problema agrario en debate*, pp. 180–204. Lima, *Seminario Permanente de Investigación Agraria* (SEPIA IX).

Creed-Kanashiro, H., Roche, M., Tuesta Cerrón, I. & Kuhnlein, H.V. 2009. Traditional food system of an Awajún community in Peru. *In* H.V. Kuhnlein, B. Erasmus and D. Spigelski. *Indigenous Peoples' food systems: the many dimensions of culture, diversity and environment for nutrition and health*, pp. 59–81. Rome, FAO.

Dewey, K.G. & Brown, K.H. 2003. Update on technical issues concerning complementary feeding of young children in developing countries and implications for intervention programs. *Food and Nutrition Bulletin*, 24(1): 5–28.

FAO. 2009. *Indigenous Peoples' food systems: the many dimensions of culture, diversity and environment for nutrition and health,* edited by H.V. Kuhnlein, B. Erasmus and D. Spigelski. Rome, 339 pp. www.fao.org/docrep/012/i0370e/i0370e00.htm.

FAO/WHO. 2002. *Human vitamin and mineral requirements.* Report of a Joint Expert Consultation. Rome and Geneva.

FAO/WHO/UNU. 2004. *Human energy requirements.* Report of a Joint Expert Consultation. Rome and Geneva.

FAO/WHO/UNU. 2007. *Protein and amino acid requirements in human nutrition.* Report of a Joint Expert Consultation. Rome and Geneva.

Huamán-Espino, L. & Valladares, C. 2006. Estado nutricional y características del consumo alimentario de la población Aguaruna. Amazonas, Perú. *Revista Perúana de Medicina Experimental y Salud Pública*, 23(1): 12–21.

Ibáñez, N.H., Jara, C.C., Guerra, A.M. & Diaz, E.L. 2004. Prevalencia del enteroparasitismo en escolares de comunidades nativas del Alto Marañón, Amazonas, Perú. *Revista Perúana de Medicina Experimental y Salud Pública*, 21(3): 126–133.

IIN. 2001. *Food composition tables.* Lima.

IIN, CINE & ODECOFROC. 2005. *Global health case study – Awajún.* www.mcgill.ca/cine/resources/data/awajun/.

INEI. 2001. *Encuesta demográfica y de salud familiar 2000.* Lima, *Instituto Nacional de Estadística e Informática* (INEI).

INEI. 2008. *II Censo de Comunidades Indígenas de la Amazonía Perúana 2007 – Resultados Definitivos.* Lima, *Instituto Nacional de Estadística e Informátic*a (INEI), *Dirección Nacional de Censos y Encuestas.* 1 458 pp.

IPAQ. 2002. *International Physical Activity Questionnaire.* Spanish version. www.ipaq.ki.se/downloads.htm.

Kuhnlein, H.V. & Receveur, O. 1996. Dietary change and traditional food systems of Indigenous Peoples. *Annual Review of Nutrition*, 16: 417–442.

Kuhnlein, H.V., Receveur, O., Soueida, R. & Egeland, G.M. 2004. Arctic indigenous peoples experience the nutrition transition with changing dietary patterns and obesity. *Journal of Nutrition*, 134(6): 1447–1453.

Kuhnlein, H.V., Erasmus, B., Creed-Kanashiro, H., Englberger, L., Okeke, C., Turner, N., Allen, L. & Bhattacharjee, L. 2006. Indigenous peoples' food systems for health: finding interventions that work. *Public Health Nutrition*, 9(8): 1013–1019.

MINSA/OGE. 2002. *Análisis de la situación de salud del pueblo Shipibo-Konibo.* Lima, Ministry of Health.

PAHO/WHO. 2003. *Guiding principles for complementary feeding of the breastfed child.* Washington, DC.

Pelto, G.H., Zhang, Y. & Habicht, J.P. 2010. Premastication: the second arm of infant and young child feeding for health and survival. *Maternal and Child Nutrition*, 6(1): 4–18.

Port Lourenço, A.E., Ventura Santos, R., Orellana, J.D.Y. & Coimbra Jr., C.E.A. 2008. Nutrition transition in Amazonia: obesity and socioeconomic change in the Suruí Indians from Brazil. *American Journal of Human Biology*, 20: 564–571.

Ramos Calderón, R. 1999. *Diagnóstico situacional, sociocultural, económico de los Pueblos Indígenas de la Cuenca del Río Cenepa de la Amazonía Peruana.* Lima, Organización de Comunidades Fronterizas del Cenepa.

Roche, M., Creed-Kanashiro, H., Tuesta, I. & Kuhnlein, H.V. 2007. Traditional food system provides dietary quality for the Awajún in the Peruvian Amazon. *Ecology of Food and Nutrition*, 46: 1–23.

Roche, M., Creed-Kanashiro, H., Tuesta, I. & Kuhnlein, H.V. 2008. Traditional food diversity predicts dietary quality for the Awajún in the Peruvian Amazon. *Public Health Nutrition*, 11(5): 457–465.

Roche, M.L., Creed-Kanashiro, H.M., Tuesta, I. & Kuhnlein, H.V. 2011. Infant and young child feeding in the Peruvian Amazon: the need to promote exclusive breastfeeding and nutrient-dense traditional complementary foods. *Maternal and Child Nutrition*, 7(3): 284–294.

Shell, O. & Wise, M.R. 1971. *Grupos idiomáticos del Perú*, p. 415. Lima, Universidad Nacional Mayor de San Marcos.

Soares Leite, M., Ventura Santos, R., Gugelmin, S.A. & Coimbra Jr., C.E.A. 2006. [Physical growth and nutritional profile of the Xavánte indigenous population in Sangradouro-Volta Grande, Mato Grosso, Brazil]. *Cadernos de Saúde Pública*, 22(2): 265–276. (in Portuguese)

Vargas, S. & Penny, M.E. 2010. Measuring food insecurity and hunger in Peru: a qualitative and quantitative analysis of an adapted version of the USDA's Food Insecurity and Hunger Module. *Public Health Nutrition*, 13(10): 1488–1497.

WHO. 1998. *Complementary feeding of young children in developing countries: a review of current scientific knowledge*. WHO/NUT/98.1. Geneva. 228 pp.

Chapter 6 | Dalit

Belavady, B. 1969. Nutrition in pregnancy and lactation. *Indian Journal of Medical Research*, 57(8): 63–74.

Bhandari, N., Bahl, R., Nayyar, B., Khokhar, P., Rohde, J.E. & Bhan, M.K. 2001. Food supplementation with encouragement to feed it to infants from 4 to 12 months of age has a small impact on weight gain. *Journal of Nutrition*, 131: 1946–1951.

Bhandari, N., Mazumder, S., Bhal, F., Martines, J., Black, R.E. & Bhan, M.K. 2004. An educational intervention to promote appropriate complementary feeding practices and physical growth in infants and young children in rural Haryana. *Indian Journal of Nutrition*, 134: 2342–2348.

Bhaskarachary, Rao, K., Deosthale, D.S.S. & Reddy, Y.G.V. 1995. Carotene content of some common and less familiar foods of plant origin. *Food Chemistry*, 54(2): 189–193.

Castenmiller, J.J. & West, C.E. 1998. Bioavailability and bioconversion of carotenoids. *Annual Review of Nutrition*, 18: 19–39.

Chakravarty, I. & Sinha, R.K. 2002. Prevalence of micronutrient deficiency based on results obtained from the national pilot program on control of micronutrient malnutrition. *Nutrition Reviews*, 60(5): S53–S58.

Christian, P. 2002. Recommendations for indicators: night blindness during pregnancy – a simple tool to assess vitamin A deficiency in a population. *Journal of Nutrition*, 132(9): 2884S–2888S.

FAO. 2002. *Smallholder farmers in India: food security and agricultural policy*, by R.B. Singh, P. Kumar and T. Woodhead. Bangkok, FAO Regional Office for Asia and the Pacific. ftp://ftp.fao.org/docrep/fao/005/ac484e/ac484e00.paf.

Food and Nutrition Board Institute of Medicine. 2001. *Dietary reference intakes: vitamin A, vitamin K, arsenic, boron, chromium, copper, iodine, iron, manganese, molybdenum, nickel, silicon, vanadium, and zinc*, pp. 82–161, 290–393 . Washington, DC, National Academy Press.

Gibson, R.S. 1990. *Principles of nutritional assessment.* New York, Oxford University Press. 691 pp.

Gopalan, C., Rama Sastri, B.V., Balasubramanian, S.C., Narasinga Rao, B.S., Deosthale, Y.G. & Pant, K.C. 1989. *Nutritive value of Indian Foods*. Hyderabad, India, National Institute of Nutrition (NIN), Indian Council of Medical Research.

Gopaldas, T., Patel, P. & Bakshi, M. 1988. Selected socio-economic, environmental, maternal and child factors associated with the nutritional status of infants and toddlers. *Food and Nutrition Bulletin*, 10(4): 29–34.

Haskell, M.J., Jamil, K.M., Hassan, F., Peerson, J.M., Hossain, M.I., Fuchs, G.J. & Brown, K.H. 2004. Daily consumption of Indian spinach (*Basella alba*) or sweet potatoes has a positive effect on total-body vitamin A stores in Bangladeshi men. *American Journal of Clinical Nutrition*, 80: 705–714.

ICMR. 1990. *Nutrient requirements and recommended dietary allowances for Indians: a report of the expert group of the Indian Council of Medical Research*. New Delhi, Indian Council of Medical Research (ICMR). 129 pp.

IHEU. 2010. *The Dalit FAQ*. International Humanist and Ethical Union (IHEU). www.iheu.org/dalitfaq.

IIPS & ORC Macro. 2000. *National Family Health Survey (NFHS–2), 1998–99: Nutrition and the prevalence of anaemia*, pp. 241–277. Mumbai, International Institute for Population Sciences (IIPS).

James, W.P.T., Ferro-Luzzi, A. & Waterlow, J.C. 1988. Definition of chronic energy deficiency in adults: Report of a Working Party of the International Dietary Energy Consultative Group. *European Journal of Clinical Nutrition*, 42: 969–981.

Laxmaiah, A., Rao, K.M., Brahmam, G.N.V., Kumar, S., Ravindranath, M., Kahinath, K., Radhaiah, G., Rao, D.H. & Vijayaraghavan, K. 2002. Diet and nutritional status of rural preschool children in Punjab. *Indian Pediatrics*, 39: 331–338.

Lohman, T.G., Roche, A.F. & Martorell, R. 1988. *Anthropometric standardization reference manual*. Champaign, Illinois, USA, Human Kinetics Books. 184 pp.

McLaren, D.S. & Frigg, M. 2001. *Sight and life manual on vitamin A deficiency disorders (VADD)*, pp. 51–62. Basel, Switzerland, Task Force Sight and Life. 163 pp.

McLean Jr., W.C., de Romana, G.L., Placko, R.P. & Graham, G.G. 1981. Protein quality and digestibility of sorghum in preschool children: balance studies and plasma free amino acids. *Journal of Nutrition*, 111: 1928–1936.

Measham, A.R. & Chatterjee, M. 1999. *Wasting away: the crisis of malnutrition in India*. Washington, DC, World Bank.

Minority Rights Group International. 2010. *World directory of minorities – Dalit*. www.minorityrights. org/?lid=5652&tmpl=printpage.

National Center for Health Statistics. 1977. *1977 NCHS growth charts*. Atlanta, Georgia, USA, United States Department of Health and Human Services, Centers for Disease Control and Prevention.

National Nutrition Monitoring Bureau. 2002. *Diet and nutritional status of rural population*. Technical Report No. 21. Hyderabad, India, NIN, Indian Council of Medical Research.

Puwastien, P., Burlingame, B., Raroengwichit, M. & Sungpuag, P. 2000. *ASEAN food composition tables*. Salaya, India, Institute of Nutrition, Mahidol University.

Rajyalaksmi, P., Venkatalaxmi, K., Venkatalakshmamma, K., Jyothsna, Y., Balachandrammani Devi, K. & Suneetha, V. 2001. Total carotenoid and beta-carotene contents of forest green leafy vegetables consumed by tribals of South India. *Plant Foods for Human Nutrition*, 56: 225–238.

Salomeyesudas, B. & Satheesh, P.V. 2009. Traditional food system of Dalit in Zaheerabad Region, Medak District, Andhra Pradesh, India. *In* H.V. Kuhnlein, B. Erasmus and D. Spigelski. *Indigenous Peoples' food systems: the many dimensions of culture, diversity and environment for nutrition and health*, pp. 185–208. Rome, FAO.

Schmid, M.A. 2005. Traditional food consumption and nutritional status of Dalit mothers and young children in rural Andhra Pradesh, South India. McGill University, Montreal, Canada. (Ph.D. thesis)

Schmid, M.A., Egeland, G.M., Salomeyesudas, B., Satheesh, P.V. & Kuhnlein, H.V. 2006. Traditional food consumption and nutritional status of Dalit mothers in rural Andhra Pradesh, South India. *European Journal of Clinical Nutrition*, 60(11): 1277–1283.

Schmid, M., Salomeyesudas, B., Satheesh, P.V., Hanley, J. & Kuhnlein, H.V. 2007. Intervention with traditional food as a major source of energy, protein, iron, vitamin C and vitamin A for rural Dalit mothers and young children in Andhra Pradesh, South India. *Asia Pacific Journal of Clinical Nutrition*, 16(1): 84–93.

Souci, S.W., Fachmann, W. & Kraut, H. 1994. *Food composition and nutrition tables.* 5th edition. Stuttgart, Germany, Medpharm Scientific Publishers.

Tontisirin, K., Nantel, G. & Bhattacharjee, L. 2002. Food-based strategies to meet the challenges of micronutrient malnutrition in the developing world. *Proceedings from the Nutrition Society*, 61: 243–250.

WHO. 2000. *Nutrition profile of the WHO South-East Asia Region.* New Delhi, World Health Organization, Regional Office for South-East Asia.

WHO Working Group. 1986. Use and interpretation of anthropometric indicators of nutritional status. *Bulletin of the World Health Organization*, 64(6): 929–941.

Women Sanghams of the Deccan Development Society, Satheesh, P.V. & Pimbert, M. 1999. Reclaiming diversity, restoring livelihoods. *Seedling – The Quarterly Newsletter of Genetic Resources Action International.* www.grain.org/seedling/?id=87.

Chapter 7 | Gwich'in

Adams, A., Receveur, O., Mundt, M., Paradis, G. & Macaulay, A.C. 2005. Healthy lifestyle indicators in children (grades 4 to 6) from the Kahnawake Schools Diabetes Prevention Project. *Canadian Journal of Diabetes,* 29(4): 403–409.

American Academy of Pediatrics. 2001. Committee on Public Education. Children, adolescents, and television. *Pediatrics,* 107: 423–426.

Andre, A. & Fehr, A. 2001. *Gwich'in ethnobotany: Plants used by the Gwich'in for food, medicine, shelter and tools.* Tsiigehtchic, NWT, Canada, Gwich'in Social and Cultural Institute and Aurora Research Institute.

Appavoo, D., Kubow, S. & Kuhnlein, H.V. 1991. Lipid composition of indigenous foods by the Sahtú (Hareskin) Dene/Métis of the Northwest Territories. *Journal of Food Composition and Analysis,* 4: 107–119.

Beaton, G.H., Milner, J., Corey, P., McGuire, V., Cousins, M., Stewart, E., de Ramos, M., Hewitt, D., Grambsch, P.V., Kassim, N. & Little, J.A. 1979. Sources of variance in 24-hour dietary recall data: implications for nutrition study design and interpretation. *American Journal of Clinical Nutrition*, 32: 2546–2549.

Bernard, L., Lavallée, C., Gray-Donald, K. & Delisle, H. 1995. Overweight in Cree schoolchildren and adolescents associated with diet, low physical activity, and high television viewing. *Journal of the American Dietetic Association,* 95(7): 800–802.

Bickel, G., Nord, M., Price, C., Hamilton, C.W. & Cook, J. 2000. *Measuring food security in the United States: Guide to measuring household food security.* Washington, DC, USDA.

Canadian Fitness and Lifestyle Research Institute. 2005. Physical Activity Monitor. Ottawa. www.cflri.ca.

CDC. 2000. *CDC growth charts.* Atlanta, Georgia, USA, Centers for Disease Control and Prevention. www.cdc.gov/growthcharts.

CDC. 2004. *A report of the Surgeon General: physical activity and health – adolescents and young adults.* www.cdc.gov/nccdphp/sgr/adoles.htm.

Chan, H.M., Fediuk, K., Hamilton, S., Rostas, L., Caughey, A., Kuhnlein, H.V., Egeland, G. & Loring, E. 2006. Food security in Nunavut, Canada: barriers and recommendations. *International Journal of Circumpolar Health*, 65(5): 416–431.

Chaudhary, N. & Kreiger, N. 2007. Nutrition and physical activity interventions for low-income populations. *Canadian Journal of Dietetic Practice and Research*, 68: 201–206.

Craig, C.L., Marshall, A.L., Sjostrom, M., Bauman, A.E., Booth, M.L., Ainsworth, B.E., Pratt, M., Ekelund, U., Yngve, A., Sallis, J.F. & Oja, P. 2003. International physical activity questionnaire: 12-country reliability and validity. *Medicine and Science in Sports and Exercise*, 35(8): 1381–1395.

FAO. 2008. *The right to food and the impact of liquid biofuels (agrofuels)*, by A. Eide. Rome. 54 pp. www.fao.org/righttofood/publi08/right_to_food_and_biofuels.pdf.

FAO. 2009. *Indigenous Peoples' food systems: the many dimensions of culture, diversity and environment for nutrition and health*, edited by H.V. Kuhnlein, B. Erasmus and D. Spigelski. Rome. 340 pp.

Fernandez, J.R., Redden, D.T., Pietrobelu, A. & Allison, D.B. 2004. Waist circumference percentiles in nationally representative sample of African-American, European-American, and Mexican-American children and adolescents. *Journal of Pediatrics*, 145: 439–444.

First Nations Regional Longitudinal Health Survey, First Nations Centre. 2005. *First Nations Regional Longitudinal Health Survey (RHS) 2002/2003.* Results for adults, youth and children living in First Nations communities. Ottawa, First Nations Centre. www.rhs-ers.ca.

Ford, P.B. & Dzewaltowski, D.A. 2008. Disparities in obesity prevalence due to variation in the retail food environment: three testable hypotheses. *Nutrition Reviews*, 66(4): 216–228.

Furgal, C. & Seguin, J. 2006. Climate change, health, and vulnerability in Canadian Northern Aboriginal communities. *Environmental Health Perspectives*, 114(12): 1964–1970.

Gallagher, D., Heymsfield, S.B., Heo, M., Jebb, S.A., Murgatroyd, P.R. & Sakamoto, Y. 2000. Healthy percentage body fat ranges: an approach for developing guidelines based on body mass index. *American Journal of Clinical Nutrition*, 72(3): 694–701.

Gibson, R.S. 2005. *Principles of nutritional assessment.* Oxford, UK, Oxford University Press.

Godin London Inc. 2007. CANDAT, Nutrient Calculation System. London, Ontario, Canada. (computer software)

Goldberg, G.R., Black, A.E., Jebb, S.A., Cole, T.J., Murgatroyd, P.R., Coward, W.A. & Prentice, A.M. 1991. Critical evaluation of energy intake data using fundamental principles of energy physiology. *European Journal of Clinical Nutrition*, 45: 569–581.

Goodman, L. 2008. *Factors associated with food insecurity among women in a small indigenous Canadian arctic community.* McGill University, Montreal, Quebec, Canada. (M.Sc. thesis)

Guyot, M., Dickson, C., Macguire, K., Paci, C., Furgal, C. & Chan, H.M. 2006. Local observations of climate change and impacts on traditional food security in two northern Aboriginal communities. *International Journal of Circumpolar Health*, 65(5): 403–415.

Health Canada. 2001. *Diabetes among Aboriginal (First Nations, Inuit and Métis) People in Canada: the evidence.* Ottawa.

Health Canada. 2003. *A statistical profile on the health of First Nations in Canada.* Ottawa.

Health Canada. 2007. *Canada's food guide: First Nations, Inuit and Métis.* www.hc-sc.gc.ca/fn-an/pubs/fnim-pnim/index-eng.php.

Heart and Stroke Foundation of Canada. 2008. *High blood pressure.* www.heartandstroke.com/site/?c=ikiqlc mwjte&b=3484023&src=home.

Institute of Medicine of the National Academies. 2006. *Dietary reference intakes.* Washington, DC, National Academies Press. 543 pp.

IPAQ. 2001. International Physical Activity Questionnaire. www.ipaq.ki.se/ipaq.htm.

Kersting, M., Sichert-Hellert, W., Lausen, B., Alexy, U., Manz, F. & Schöch, G. 1998. Energy intake of 1 to 18 year old German children and adolescents. *Z. Ernährungswiss,* 37: 47–55.

Kuhnlein, H.V. 2001. Nutrient benefits of Arctic traditional/country foods. *In* S. Kalhok, ed. *Synopsis of research conducted under the 2000–2001 Northern Contaminants Program,* pp. 56–64. Ottawa, Ministry of Indian Affairs and Northern Development.

Kuhnlein, H.V. & Receveur, O. 1996. Dietary change and traditional food systems of Indigenous Peoples. *Annual Review of Nutrition,* 16: 417–442.

Kuhnlein, H.V. & Receveur, O. 2007. Local cultural animal food contributes high levels of nutrients for Arctic Canadian indigenous adults and children. *Journal of Nutrition,* 137: 1110–1114.

Kuhnlein, H.V., Appavoo, D., Morrison, N., Soueida, R. & Pierrot, P. 1994. Use of nutrient composition of traditional Sahtú (Hareskin) Dene/Métis foods. *Journal of Food Composition and Analysis,* 7: 144–157.

Kuhnlein, H.V., Chan, H.M., Leggee, D. & Barthet, V. 2002. Macronutrient, mineral and fatty acid composition of Canadian Arctic traditional food. *Journal of Food Composition and Analysis,* 15: 545–566.

Kuhnlein, H.V., Receveur, O., Soueida, R. & Egeland, G.M. 2004. Arctic Indigenous Peoples experience the nutrition transition with changing dietary patterns and obesity. *Journal of Nutrition,* 134(6): 1447–1453.

Kuhnlein, H.V., Barthet, V., Farren, A., Falahi, E., Leggee, D., Receveur, O. & Berti, P. 2006. Vitamins A, D, and E in Canadian Arctic traditional food and adult diets. *Journal of Food Composition and Analysis,* 19: 495–506.

Kuhnlein, H.V., McDonald, M., Spigelski, D., Vittrekwa, E. & Erasmus, B. 2009. Gwich'in traditional food for health: phase 1. *In* H.V. Kuhnlein, B. Erasmus and D. Spigelski. *Indigenous Peoples' food systems: the many dimensions of culture, diversity and environment for nutrition and health,* pp. 45–58. Rome, FAO. 340 pp.

Ladouceur, L.L. & Hill, F. 2001. *Results of the survey of food quality in 6 isolated communities in Labrador, March 2001.* Ottawa, Indian and Northern Affairs Canada.

Lambden, J., Receveur, O. & Kuhnlein, H.V., 2007. Traditional food attributes must be included in studies of food security in the Canadian Arctic. *International Journal of Circumpolar Health,* 66(4): 308–319.

Lambden, J., Receveur, O., Marshall, J. & Kuhnlein, H.V. 2006. Traditional and market food access in Arctic Canada is affected by economic factors. *International Journal of Circumpolar Health,* 65(4): 331–340.

Lawn, J. & Harvey, D. 2001. *Change in nutrition and food security in two Inuit communities, 1992 to 1997.* Ottawa, Indian and Northern Affairs Canada.

Lawn, J. & Harvey, D. 2003. *Nutrition and food security in Kugaaruk, Nunavut: Baseline survey for the Food Mail Pilot Project*. Ottawa, Indian and Northern Affairs Canada.

Lawn, J. & Harvey, D. 2004a. *Nutrition and food security in Fort Severn, Ontario: Baseline survey for the Food Mail Pilot Project*. Ottawa, Indian and Northern Affairs Canada.

Lawn, J. & Harvey, D. 2004b. *Nutrition and food security in Kangiqsujuaq, Nunavik*. Ottawa, Indian and Northern Affairs Canada.

McCarthy, H.D., Cole, T.J., Fry, T. Jebb, S.A. & Prentice, A.M. 2006. Body fat reference curves for children. *International Journal of Obesity*, 30: 598–602.

Morrison, N. & Kuhnlein, H.V. 1993. Retinol content of wild foods consumed by the Sahtú (Hareskin) Dene/Métis. *Journal of Food Composition and Analysis*, 6: 10–23.

Nakano, T., Fediuk, K., Kassi, N., Egeland, G.M. & Kuhnlein, H.V. 2005a. Dietary nutrients and anthropometry of Dene/Métis and Yukon children. *International Journal of Circumpolar Health*, 64(2): 147–156.

Nakano, T., Fediuk, K., Kassi, N. & Kuhnlein, H.V. 2005b. Food use of Dene/Métis and Yukon children. *International Journal of Circumpolar Health*, 64(2): 137–146.

National High Blood Pressure Education Program Working Group on High Blood Pressure in Children and Adolescents. 2004. The fourth report on the diagnosis, evaluation, and treatment of high blood pressure in children and adolescents. *Pediatrics*, 114: 555–576.

Natural Resources Canada. 2004. *Climate change impacts and adaptation: A Canadian perspective*. Ottawa, Her Majesty the Queen in Right of Canada. www.environment.msu.edu/climatechange/canadaadaptation.pdf.

NWT Bureau of Statistics. 2007. *Summary of NWT community statistics 2007*. Government of the Northwest Territories. www.stats.gov.nt.ca/profile/profile%20pdf/nwt.pdf.

Ogden, C.L., Flegal, K.M., Carroll, M.D. & Johnson, C.L. 2002. Prevalence and trends in overweight among US children and adolescents, 1999–2000. *Journal of the American Medical Association*, 288: 1728–1732.

Parlee, B., Berkes, F. & Teetl'it Gwich'in. 2005. *Health of the land, health of the people: a case study on Gwich'in berry harvesting in Northern Canada*. EcoHealth. http://umanitoba.ca/institutes/natural_resources/canadaresearchchair/gwichin%20berry%20harvesting%20from%20northern%20canada.pdf.

Physical Activity Guidelines Advisory Committee. 2008. *Physical activity guidelines advisory committee report*. Washington, DC, United States Department of Health and Human Services.

Receveur, O., Boulay, M. & Kuhnlein, H.V. 1997. Decreasing traditional food use affects diet quality for adult Dene/Métis in 16 communities of the Canadian Northwest Territories. *Journal of Nutrition*, 127(11): 2179–2186.

Receveur, O., Boulay, M., Mills, C., Carpenter, W. & Kuhnlein, H.V. 1996. *Variance in food use in Dene/Métis communities*. Ste-Anne-de-Bellevue, Quebec, Canada, CINE. 198 pp.

Sallis, J.F., Buono, M.J., Roby, J.J., Micale, F.G. & Nelson, J.A. 1993. Seven-day recall and other physical activity self-reports in children and adolescents. *Medicine and Science in Sports and Exercise*, 25(1): 99–108.

SAS Institute Inc. 2003. Statistical analysis system, version 9.0. Cary, North Carolina, USA.

Shields, M. 2005. Measured obesity. Overweight Canadian children and adolescents. *Analytical Study Reports*, 1: 1–34. Ottawa, Statistics Canada. Catalogue no. 82-620-MWE2005001.

Shields, M. 2006. Overweight and obesity among children and youth. *Health Reports*, 17(3): 27–42. Ottawa, Statistics Canada. Catalogue no. 82–003.

Shrimpton, R., Prudhon, C. & Engesveen, K. 2009. The impact of high food prices on maternal and child nutrition. *SCN News,* 37: 60–68.

Skinner, K., Hanning, R.M. & Tsuji, L.J.S. 2006. Barriers and supports for healthy eating and physical activity for First Nation youths in northern Canada. *International Journal of Circumpolar Health,* 65(2): 148–161.

Statistics Canada. 2003. *Aboriginal Peoples Survey 2001 – Initial findings: Well-being of the non-reserve Aboriginal Population. 2001 Aboriginal Peoples Survey.* Ottawa.

Statistics Canada. 2006. Measured adult body mass index (BMI), by age group and sex, household population aged 18 and over excluding pregnant females, Canadian Community Health Survey (CCHS 3.1), Canada, every 2 years (CANSIM Table 105-0407). Ottawa.

Statistics Canada. 2008. Children and youth. In *Canada Year Book 2007.* Ottawa. Catalogue no. 11-402-XIE.

WHO. 2000. *Obesity: preventing and managing the global epidemic. Report on a WHO Consultation.* Technical Report Series No. 894. Geneva. 265 pp.

Chapter 8 | Inga

Correal, C., Zuluaga, G., Madrigal, L., Caicedo, S. & Plotkin, M. 2009. Ingano traditional food and health: phase 1, 2004–2005. *In* H.V. Kuhnlein, B. Erasmus and D. Spigelski. *Indigenous Peoples' food systems: the many dimensions of culture, diversity and environment for nutrition and health,* pp. 83–108. Rome, FAO.

Departmental Health Institute of Caquetá. 2006. *Boletín Epidemiológico.* Caquetá, Colombia

ICBF. 2005. *ENSIN Caquetá.* Caquetá, Colombia, *Instituto Colombiano de Bienestar Familiar* (ICBF).

ILO. 1989. *Convention concerning Indigenous and Tribal Peoples in independent countries.* www.ilo.org/wcmsp5/groups/public/---asia/---ro-bangkok/---ilo-jakarta/documents/publication/wcms_124013.pdf

Kuhnlein, H.V., Smitasiri, S., Yesudas, S., Bhattacharjee, L., Dan, L. & Ahmed, S. 2006. *Documenting traditional food systems of Indigenous Peoples: international case studies. Guidelines for procedures.* In collaboration with S. Sirisai, P. Puwastien, L. Daoratanahong, S. Dhanamitta, F. Zhai, P.V. Satheesh, G. Kothari and F. Akhter. www.mcgill.ca/files/cine/manual.pdf.

Ministry of Social Protection. 2008. *Public Health National Plan (2008–2011).* Santa Fe de Bogotá.

PAHO. 2007. *Health in the Americas, 2007. Volume II – Colombia.* www.paho.org/hia/archivosvol2/paisesing/colombia%20english.pdf.

Parra, P.A. 2004. *Plan de desarrollo municipal. Nuestro único compromiso: San José del Fragua 2004–2007.* Santa Fe de Bogotá, Government of Columbia.

Ramírez, G.Z. 2005. Conservation of the biological and cultural diversity of the Colombian Amazon Piedmont: Dr. Schultes' legacy. *Ethnobotany Research and Applications,* 3: 179–188.

Chapter 9 | Inuit

Aars, J., Lunn, N.J. & Derocher, A.E. 2006. *Proceedings of the 14th Working Meeting of the IUCN/SSC Polar Bear Specialist Group.* 20–24 June 2005, Seattle, Washington, USA. Gland, Switzerland, and Cambridge, UK, International Union for Conservation of Nature. www.iucn.org/publications.

Agriculture and Agri-Food Canada. 1998. *Canada's Action Plan for Food Security. In response to the World Food Summit Plan of Action.* Ottawa. www.agr.gc.ca/index_e.php?s1=misb&s2=fsec-seca&page=action.

Alaska Native Science Commission. 2005. Snow change workshop of indigenous observations and ecological and climate change. *Alaska Native Science Commission Newsletter,* 5(2): 10. www.nativescience.org/assets/documents/pdf%20documents/anscvol%205_2.pdf.

AMAP. 2009. *AMAP assessment 2009: human health in the Arctic.* Oslo, Arctic Management and Assessment Program (AMAP). www.amap.no/documents/index.cfm.

Booth, S. & Zeller, D. 2005. Mercury, food webs and marine mammals: implications of diet and climate change for human health. *Environmental Health Perspectives,* 113(5): 521–526.

Carino, J. 2009. Poverty and well-being. *In* United Nations Permanent Forum on Indigenous Issues. *The State of the World's Indigenous Peoples,* pp.13–45. New York, United Nations Department of Economic and Social Affairs, Secretariat of the Permanent Forum on Indigenous Issues. 250 pp. www.un.org/esa/socdev/unpfii/documents/sowip_web.pdf.

Chan, H.M., Fediuk, K., Hamilton, S., Rostas, L., Caughey, A., Kuhnlein, H.V., Egeland, G. & Loring, E. 2006. Food security in Nunavut, Canada: barriers and recommendations. *International Journal of Circumpolar Health,* 65(5): 416–431.

Chen, L., Appel, L.J., Loria, C., Lin, P.H., Champagne, C.M., Elmer, P.J., Ard, J.D., Mitchell, D., Batch, B.C., Svetkey, L.P. & Caballero, B. 2009. Reduction in consumption of sugar-sweetened beverages is associated with weight loss: the PREMIER trial. *American Journal of Clinical Nutrition,* 89(5): 1299–1306.

Dewailly, E., Lévesque, B., Duchesne, J.F., Dumas, P., Scheuhammer, A., Gariépy, C., Rhainds, M. & Proulx, J.F. 2000. Lead shot as a source of lead poisoning in the Canadian Arctic. *Epidemiology,* 11(4): S146.

Dewailly, E., Ayotte, P., Brunneau, S., Lebel, G., Levallois, P. & Weber, J.P. 2001. Exposure of the Inuit population of Nunavik (Arctic Quebec) to lead and mercury. *Archives of Environmental Health,* 56(4): 350–357.

Dewailly, E., Blanchet, C., Gingras, S., Lemieux, S. & Holub, B.J. 2002. Cardiovascular disease risk factors and n-3 fatty acid status in the adult population of James Bay Cree. *American Journal of Clinical Nutrition,* 76(1): 85–92.

Dewailly, E., Ayotte, P., Pereg, D., Dery, S., Dallaire, R., Fontaine, J. & Côté, S. 2007a. *Exposure to environmental contaminants in Nunavik: metals.* Nunavik Inuit Health Survey 2004. Institut national de santé publique du Québec. www.inspq.qc.ca/pdf/publications/661_esi_contaminants.pdf.

Dewailly, E., Ayotte, P., Lucas, M. & Blanchet, C. 2007b. Risk and benefits from consuming salmon and trout: a Canadian perspective. *Food and Chemical Toxicology,* 45(8): 1343–1348.

Dowsley, M. & Wenzel, G. 2008. The time of the most polar bears: a co-management conflict in Nunavut. *Arctic,* 61(2): 177–179.

Egeland, G.M., Faraj, N. & Osborne, G. 2010. Cultural, socioeconomic, and health indicators among Inuit preschoolers: Nunavut Inuit Child Health Survey, 2007–2008. *Rural and Remote Health,* 10: 1365 (online). www.rrh.org.au/publishedarticles/article_print_1365.pdf.

Egeland, G.M. & Middaugh, J.P. 1997. Balancing fish consumption benefits with mercury exposure. *Science,* 278(5345): 1904–1905.

Egeland, G.M., Berti, P., Soueida, R., Arbour, L.T., Receveur, O. & Kuhnlein, H.V. 2004. Age differences in vitamin A intake among Canadian Inuit. *Canadian Journal of Public Health,* 95(6): 465–469.

Egeland, G.M., Charbonneau-Roberts, G., Kuluguqtuq, J., Kilabuk, J., Okalik, L., Soueida, R. & Kuhnlein, H.V. 2009. Back to the future: using traditional food and knowledge to promote a healthy future among Inuit. *In* H.V. Kuhnlein, B. Erasmus and D. Spigelski. *Indigenous Peoples' food systems: the many dimensions of culture, diversity and environment for nutrition and health,* pp. 9–22. Rome, FAO.

Egeland, G.M., Pacey, A., Cao, Z. & Sobol, I. 2010. Food insecurity among Inuit preschoolers: Nunavut Inuit Child Health Survey, 2007–2008. *Canadian Medical Association Journal,* 182(3): 243–248.

Egeland, G.M., Williamson-Bathory, L., Johnson-Down, L. & Sobol, I. 2011. Traditional food and monetary access to market-food: correlates of food insecurity among Inuit preschoolers. *International Journal of Circumpolar Health,* 70(4): 373–383.

FAO. 1996. *Report of the World Food Summit.* Rome. www.fao.org/docrep/003/w3548e/w3548e00. htm#adopt05.

Fediuk, K., Hidiroglou, N., Madère, R. & Kuhnlein, H.V. 2002. Vitamin C in Inuit traditional food and women's diets. *Journal of Food Composition and Analysis,* 15(3): 221–235.

Ferguson, S.H., Stirling, I. & McLoughlin, P. 2005. Climate change and ringed seal (*Phoca hispida*) recruitment in western Hudson Bay. *Marine Mammal Science,* 21(1): 121–135.

Ford, J.D. 2009. Vulnerability of Inuit food systems to food insecurity as a consequence of climate change: a case study from Igloolik, Nunavut. *Regional Environmental Change,* 9: 83–100.

Ford, J. & Berrang-Ford, L. 2009. Food insecurity in Igloolik, Nunavut: a baseline study. *Polar Record,* 45(234): 225–236.

Ford, J.D. & Pearce, T. 2010. What we know, do not know, and need to know about climate change vulnerability in the western Canadian Arctic: a systematic literature review. *Environmental Research Letters,* 5: 014008. http://iopscience.iop.org/1748-9326/5/1/014008/fulltext.

Ford, J.D., Smit, B. & Wandel, J. 2006. Vulnerability to climate change in the Arctic: a case study from Arctic Bay, Canada. *Global Environmental Change,* 16(2): 145–160.

Ford, J.D., Smit, B., Wandel, J. & Macdonald, J. 2006. Vulnerability to climate change in Igloolik, Nunavut: what we can learn from the past and present. *Polar Record,* 42(221): 127–138.

Furgal, C.M., Martin, D. & Gosselin, P. 2002. Climate change and health in Nunavik and Labrador: lessons from Inuit knowledge. *In* I. Krupnik and D. Jolly, eds. *The earth is faster now: indigenous observations of Arctic environmental change,* pp. 266–300. Washington, DC, Arctic Research Consortium of the United States, Arctic Studies Centre, Smithsonian Institution Press.

Giammattei, J., Blix, G., Marshak, H.H., Wollitzer, A.O. & Pettitt, D.J. 2003. Television watching and soft drink consumption: associations with obesity in 11- to 13-year-old schoolchildren. *Archives of Pediatrics and Adolescent Medicine,* 157(9): 882–886.

Gibson, S. 2008. Sugar-sweetened soft drinks and obesity: a systematic review of the evidence from observational studies and interventions. *Nutrition Research Reviews,* 21(2):134–147.

Guyot, M., Dickson, C., Paci, C., Furgal, C. & Chan, H.M. 2006. Local observations of climate change and impacts on traditional food security in two northern Aboriginal communities. *International Journal of Circumpolar Health,* 65(5): 403–415.

Haines, A. & McMichael, A.J. 1997. Climate change and health: implications for research, monitoring and policy. *British Medical Journal,* 315(7112): 870–874.

Health Canada. 2007. *Eating well with Canada's Food Guide. First Nations, Inuit and Métis.* Ottawa.. www.hc-sc.gc.ca/fn-an/alt_formats/fnihb-dgspni/pdf/pubs/fnim-pnim/2007_fnim-pnim_food-guide-aliment-eng.pdf.

Hotez, P.J. 2010. Neglected infections of poverty among Indigenous Peoples of the Arctic. *PLoS Neglected Tropical Diseases,* 4(1): e606.

Humphries, M.M., Umbanhowar, J. & McCann K.S. 2004. Bioenergetic prediction of climate change impacts on northern mammals. *Integrative and Comparative Biology*, 44(2): 152–162.

INAC. 2003. *Nutrition and food security in Kugaaruk, Nunavut – baseline survey for the Food Mail Pilot Project*. Ottawa, Indian and Northern Affairs Canada (INAC).

INAC. 2004. *Nutrition and food security in Kangiqsujuaq, Nunavik – baseline survey for the Food Mail Pilot Project*. Ottawa, Indian and Northern Affairs Canada (INAC).

INAC. 2007. *The Revised Northern Food Basket*. Ottawa, Indian and Northern Affairs Canada. www.ainc-inac.gc.ca/nth/fon/fc/pubs/nfb/nfb-eng.asp.

Institute of Medicine. 2000. *Dietary reference intakes: applications in dietary assessment*. Washington, DC, National Academies Press.

Jacobson, J.L., Jacobson, S.W., Muckle, G., Kaplan-Estrin, M., Ayotte, P. & Dewailly, E. 2008. Beneficial effects of a polyunsaturated fatty acid on infant development: evidence from the Inuit of Arctic Quebec. *Journal of Pediatrics*, 152(3): 356–364.

Johnson-Down, L. & Egeland, G.M. 2010. Diet quality and traditional food consumption among Inuit preschoolers: Nunavut Inuit Child Health Survey, 2007–2008. *Journal of Nutrition*, 140(7): 1311–1316.

Jørgensen, M.E., Bjeregaard, P., Borch-Johnson, K., Backer, V., Becker, U., Jørgensen, T. & Mulvad, G. 2002. Diabetes and impaired glucose tolerance among the Inuit population of Greenland. *Diabetes Care*, 25(10): 1766–1771.

Kaufman, D.S., Schneider, D.P., McKay, N.P., Ammann, C.M., Bradley, R.S., Briffa, K.R., Miller, G.H., Otto-Bliesner, B.L., Overpeck, J.T., Vinther, B.M. & Arctic Lakes 2k Project Members. 2009. Recent warming reverses long-term arctic cooling. *Science*, 325(5945): 1236–1239.

King, M., Smith, A. & Gracey, M. 2009. Indigenous health part 2: the underlying causes of the health gap. *Lancet*, 374(9683): 76–85.

kp-studios.com. 2009. *Inuit and their indigenous foods*. www.indigenousnutrition.org/inuit.html.

Kraemer, L.D., Berner, J.E. & Furgal, C.M. 2005. The potential impact of climate on human exposure to contaminants in the Arctic. *International Journal of Circumpolar Health*, 64: 498–508.

Krupnik, I. & Jolly, D., eds. 2002. *The earth is faster now: Indigenous observations of Arctic environmental change*. Fairbanks, Alaska, USA, Arctic Research Consortium of the United States. 384 pp.

Kuhnlein, H.V. & Receveur O. 1996. Dietary change and traditional food systems of Indigenous Peoples. *Annual Review of Nutrition*, 16: 417–442.

Kuhnlein, H.V. & Receveur, O. 2007. Local cultural animal food contributes high levels of nutrients for Arctic Canadian Indigenous adults and children. *Journal of Nutrition*, 137(4): 1110–1114.

Kuhnlein, H.V. & Soueida, R. 1992. Use and nutrient composition of traditional Baffin Inuit foods. *Journal of Food Composition and Analyses*, 5: 112–126.

Kuhnlein, H.V., Chan, H.M., Leggee, D. & Barthert, V. 2002. Macronutrient, mineral and fatty acid composition of Canadian Arctic traditional food. *Journal of Food Composition and Analyses*, 15: 545–566.

Kuhnlein, H.V., Receveur, O., Soueida, R. & Egeland, G.M. 2004. Arctic Indigenous Peoples experience the nutrition transition with changing dietary patterns and obesity. *Journal of Nutrition*, 124: 1447–1453.

Kuhnlein, H.V., Barthet, V., Farren, E., Falahi, E., Leggee, D., Receveur, O. & Berti, P. 2006. Vitamins A, D, and E in Canadian Arctic traditional food and adult diets. *Journal of Food Composition and Analysis*, 19: 495–506.

Lambden, J., Receveur, O., Marshall, J. & Kuhnlein, H.V. 2006. Traditional and market food access in Arctic Canada is affected by economic factors. *International Journal of Circumpolar Health*, 65(4): 331–340.

Ledrou, I. & Gervais, J. 2005. Food insecurity. *Health Reports*. 16(3): 47–51.

MacKenzie, D. 2007. Lack of sea ice devastates seal populations. *New Scientist*, 18: 32. www.newscientist.com/article/dn11489-lack-of-sea-ice-devastates-seal-populations.html.

Marks, G.C., Hughes, M.C. & van der Pols, J.C. 2006. Relative validity of food intake estimates using a food frequency questionnaire is associated with sex, age, and other personal characteristics. *Journal of Nutrition*, 136(2): 459–465.

McLaughlin, J.B., Sobel, J., Lynn, T., Funk, E. & Middaugh, J.P. 2004. Botulism type E outbreak associated with eating beached whale, Alaska. *Emerging Infectious Disease*, 10(9): 1685–1686. www.cdc.gov/ncidod/eid/vol10no9/pdfs/04-0131.pdf.

McLaughlin, J.B., Depoala, A., Bopp, C.A., Martinek, K.A., Napolilli, N., Allison, C.G., Murray, S.L., Thompson, E.C., Bird, M.M. & Middaugh, J.P. 2005. Emergence of *Vibro parahaemolyticus* gastroenteritis associated with consumption of Alaskan oysters and its global implications. *New England Journal of Medicine*, 353(14): 1463–1470.

Meier, H.E.M., Döscher, R. & Halkka, A. 2004. Simulated distributions of Baltic Sea-ice in warming climate and consequences for the winter habitat of the Baltic ringed seal. *AMBIO: A Journal of the Human Environment*, 33(4): 249–256.

Mozaffarian, D. & Rimm, E.B. 2006. Fish intake, contaminants, and human health: evaluating the risks and the benefits. *Journal of the American Medical Association*, 296(15): 1885–1899.

Nuttall, M., Berkes, F., Forbes, B., Kofinas G., Vlassova, T. & Wenzel, G. 2010. Climate change impacts on Canadian Inuit in Nunavut. *In* C.J. Cleveland, ed. *Encyclopedia of Earth*. Washington, DC, Environmental Information Coalition, National Council for Science and the Environment. www.eoearth.org/article/climate_change_impacts_on_canadian_inuit_in_nunavut.

Parkinson, A.J. & Evengård, B. 2009. Climate change, its impact on human health in the Arctic and the public health response to threats of emerging infectious diseases. *Global Health Action*, 2: 10.3402/gha.v2i0.2075.

Proulx, J.F., MacLean, J.D., Gyorkos, T.W., Leclair, D., Richter, A-K., Serhir, B., Forbes, L. & Gajadhar, A.A. 2002. Novel prevention programs for trichinellosis in Inuit communities. *Clinical Infectious Diseases*, 34(11): 1508–1514.

Richmond, C.A. & Ross, N.A. 2009. The determinants of First Nation and Inuit health: a critical population health approach. *Health Place*, 15(2):403–411.

Richmond, C., Ross, N. & Egeland, G.M. 2007. Societal resources and thriving health: A new approach for understanding the health of Indigenous Canadians. *American Journal of Public Health*, 97(10): 1827–1833.

Sanigorski, A.M., Bell, A.C. & Swinburn, B.A. 2007. Association of key foods and beverages with obesity in Australian schoolchildren. *Public Health Nutrition*, 10: 152–157.

Simmonds, M.P. & Isaac, S.J. 2007. The impacts of climate change on marine mammals: early signs of significant problems. *Oryx*, 41(1): 19–26.

Simon, M. 2009. *Sovereignty begins at home: Inuit and the Canadian Arctic*. Inuit Tapiriit Kanatami President's Speech. www.itk.ca/media-centre/speeches/sovereignty-begins-home-inuit-and-canadian-arctic.

Standing Committee on Aboriginal Affairs and Northern Development. 2007. *No higher priority: Aboriginal post-secondary education in Canada. Report of the Standing Committee on Aboriginal Affairs and Northern Development*. Ottawa, House of Commons.

Stoll, H.M. 2006. Climate change: the Arctic tells its story. *Nature*, 441: 579–581.

Tome, D. 2004. Protein, amino acids and the control of food intake. *British Journal of Nutrition,* 92 (suppl 1): S27–S30.

Van Dolah, F.M. 2000. Marine algal toxins: origins, health effects, and their increased occurrence. *Environmental Health Perspectives,* 108(suppl 1): 133–141.

Veugelers, P.J., Yip, A.M. & Mq, D. 2001. The north-south gradient in health: analytic applications for public health. *Canadian Journal of Public Health,* 92(2): 95–98.

Vors, L.S. & Boyce, M.S. 2009. Global declines of caribou and reindeer. *Global Change Biology,* 15(11): 2626–2633.

Wilkins, R., Uppal, S., Finès, P., Senècal, S., Guimond, E. & Dion, R. 2008. Life expectancy in the Inuit-inhabited areas of Canada, 1989 to 2003. *Health Reports,* 19(1): 7–19.

Yalowitz, K.S., Collins, J.F. & Virginia, R.A. 2008. *The Arctic Climate Change and Security Policy Conference. Final report and findings.* 1–3 December 2008. Hanover, New Hampshire, USA, Dartmouth College.

Yohannes, S. 2009. Traditional food consumption, anthropometry, nutrient intake and the emerging relationship between Inuit youth and traditional knowledge in a Baffin Island community. Montreal, Quebec, Canada, McGill University. (M.Sc. thesis)

Young, T.K. & Bjerregaard, P., eds. 2008. *Health transitions in Arctic populations.* Toronto, Ontario, Canada, University of Toronto Press. 496 pp.

Chapter 10 | Karen

Arimond, M. & Ruel, M.T. 2004. Dietary diversity is associated with child nutritional status: evidence from 11 demographic and health surveys. *Journal of Nutrition,* 134: 2579–2585.

Banjong, O., Menefee, A., Sranacharoenpong, K., Chittchang, U., Eg-kantrong, P., Boonpraderm, A. & Tamachotipong, S. 2003. Dietary assessment of refugees living in camps: a case study of Mae La Camp, Thailand. *Food and Nutrition Bulletin,* 24(4): 360–367.

Bialeschki, M.D., Henderson, K.A. & James, P.A. 2007. Camp experiences and developmental outcomes for youth. *Child and Adolescent Psychiatric Clinics of North America,* 16: 769–788.

Bohm, D. 1996. *On dialogue.* London, Routledge. 128 pp.

Changbumrung, S., ed. 2003. [*Dietary reference intake for Thais 2003*]. Bangkok, Express Transportation Organization Printing Press. 347 pp. (in Thai)

Chotiboriboon, S., Tamachotipong, S., Sirisai, S., Dhanamitta, S., Smitasiri, S., Sappasuwan, C., Tantivatanasathien, P. & Eg-kantrong, P. 2009. Thailand: food system and nutritional status of indigenous children in a Karen community. *In* H.V. Kuhnlein, B. Erasmus and D. Spigelski. *Indigenous Peoples' food systems: the many dimensions of culture, diversity and environment for nutrition and health,* pp. 159–183. Rome, FAO.

Damman, S., Eide, W.B. & Kuhnlein, H.V. 2008. Indigenous Peoples' nutrition transition in a right to food perspective. *Food Policy,* 33(2): 135–155.

Ganjanaphan, A., Laungaramsri, P., Jatuworapruek, T., Hengsuwan, P., Rukyutitum, A., Unprasert, V., Nutpoolwat, S., Jamroenprucksa, M., Soontornhou, P. & Onprom, S. 2004. *Swidden farming systems: status and changes.* Chiangmai, Thailand, Faculty of Social Science, Chiangmai University.

Gleason, G. & Scrimshaw, N.S. 2007. An overview of the functional significance of iron deficiency. *In* K. Kraemer and M.B. Zimmermann, eds. *Nutritional anemia,* pp. 45–58. Basel, Switzerland, Sight and Life Press.

Goldstein, S., Japhet, G., Usdin, S. & Scheepers, E. 2004. Soul City: a sustainable edutainment vehicle facilitating social change. *Health Promotion Journal of Australia*, 15: 114–120.

Grenier, L. 1998. *Working with the indigenous knowledge: a guide for researchers.* Ottawa, International Development Research Centre.

Henderson, K.A., Bialeschki, M.D. & James, P.A. 2007. Overview of camp research. *Child and Adolescent Psychiatric Clinics of North America*, 16: 755–767.

Isaacs, W. 1999. *Dialogue and the art of thinking together: a pioneering approach to communicating in business and in life.* New York, Doubleday. 448 pp.

Israel, B.A., Schulz A.J., Parker, E.A. & Becker, A.B. 1998. Review of community-based research: assessing partnership approaches to improve public health. *Annual Review of Public Health*, 19: 173–202.

Khor, G.L. 2008. Food-based approaches to combat the double burden among the poor: challenges in the Asian context. *Asia Pacific Journal of Clinical Nutrition*, 17(S1): 111–115.

Kuhnlein, H.V. & Receveur, O. 1996. Dietary change and traditional food systems of Indigenous Peoples. *Annual Review of Nutrition*, 16: 417–442.

Kuhnlein, H.V., Smitasiri, S., Yesudas, S., Bhattacharjee, L., Dan, L. & Ahmed, S. 2006a. *Documenting traditional food systems of Indigenous Peoples: international case studies. Guidelines for procedures.* In collaboration with S. Sirisai, P. Puwastien, L. Daoratanahong, S. Dhanamitta, F. Zhai, P.V. Satheesh, G. Kothari and F. Akhter. Montreal, Quebec, Canada, McGill University. www.mcgill.ca/files/cine/manual.pdf.

Kuhnlein, H.V., Erasmus, B., Creed-Kanashiro, H., Englberger, L., Okeke, C., Turner, N., Allen, L. & Bhattacharjee, L. 2006b. Indigenous peoples' food systems for health: finding interventions that work. *Public Health Nutrition*. 9(8): 1013–1019.

Lakoff, G. & Johnson, M. 1980. *Metaphors we live by.* Chicago, Illinois, USA, University of Chicago Press. 256 pp.

Langness, L.L. & Frank, G. 1981. *Lives: an anthropological approach to biography.* Novato, California, USA, Chandler and Sharp Publishers. 221 pp.

Nonaka, I. 1998. The knowledge-creating company. In *Harvard Business Review on Knowledge Management*, pp. 21–45. Boston, Massachusetts, Harvard Business School Publishing.

Pellegrini, G.B. 2006. Examples of metaphors from fauna and flora. *In* G. Sanga and G. Ortalli, eds. *Nature knowledge: ethnoscience, cognition, and utility*, pp. 185–190. Oxford, UK, Berghahn Books.

Quisumbling, A.R., Brown, L.R., Feldstein, H.S., Haddad L. & Pena, C. 1995. *Women: the key to food security.* Washington, DC, International Food Policy Research Institute.

Wasi, P. 2009. *Khwam suk khong khon thung mual. [Happiness for all].* www.prawase.com. (in Thai).

Wheatley, M.J. 2002. *Turning to one another: simple conversations to restore hope to the future.* San Francisco, California, USA, Berrett-Koehler Publishers.

Yankelovich, D. 1999. *The magic of dialogue: transforming conflict into cooperation.* New York, Simon and Schuster. 240 pp.

Chapter 11 | Nuxalk

BC Stats. 2006. *British Columbia statistical profile of aboriginal peoples 2006. Aboriginal peoples compared to the non-aboriginal population.* Victoria, British Columbia, Canada.

Boyd, R.T. 1990. Demographic history, 1774–1874. *In* W. Suttles, ed. *Northwest coast. Volume 7. Handbook of North American Indians*, pp. 135–148. Washington, DC, Smithsonian Institution Press.

Census of Canada. 1981. *Enumeration area statistics*. Vancouver, British Columbia, Statistics Canada.

IUCN. 2008. *IUCN Red List of threatened species*. Haliotis kamtschatkana – *Endangered*. International Union for the Conservation of Nature (IUCN), Species Survival Commission. www.iucnredlist.org/search/details.php/61743/all.

Kendall, P.R.W. 2002. *The health and well-being of aboriginal people in British Columbia*. Provincial Health Officer's annual report, 2001. Victoria, British Columbia Ministry of Health Planning.

Kennedy, D.I.D. & Bouchard, R.T. 1990. Bella Coola Indians. *In* W. Suttles, ed. *Northwest coast. Volume 7. Handbook of North American Indians*, pp. 323–339. Washington, DC, Smithsonian Institution Press.

Kennelly, A.C. 1986. *A nutrient evaluation of selected Nuxalk salmon preparations*. Vancouver, British Columbia, Canada, University of British Columbia. (M.Sc. thesis)

KP Studios. 2008. *The Nuxalk and their traditional food*. DVD production. Bella Coola, British Columbia, Canada, Nuxalk Nation Council, and Montreal, Quebec, Canada, CINE, McGill University.

Kuhnlein, H.V. 1984. Traditional and contemporary Nuxalk foods. *Nutrition Research*, 4: 789–809.

Kuhnlein, H.V. 1986. The Nuxalk Food and Nutrition Program – overview and objectives. *Nutrition Newsletter*, 7: 26–34.

Kuhnlein, H.V. 1987. *Final report. Nuxalk Food and Nutrition Program*. Submitted to Health Canada, National Health Research Development Program and Health Promotion Contribution Program.

Kuhnlein, H.V. 1989. Factors influencing use of traditional foods among the Nuxalk people. *Journal of the Canadian Dietetic Association*, 50(2): 102–108.

Kuhnlein, H.V. 1992. Change in the use of traditional food by the Nuxalk Native people of British Columbia. *In* G.H. Pelto and L.A.Vargas, eds. Perspectives on dietary change. Studies in nutrition and society. Special Issue, *Ecology of Food and Nutrition*, 27(3–4): 259–282.

Kuhnlein, H.V. 1995. *Changing patterns of food use by Canadian Indigenous Peoples*. 3rd International Conference on Diabetes and Indigenous Peoples, Winnipeg, Manitoba, Canada.

Kuhnlein, H.V. 2001a. Improving nutritional status with traditional Nuxalk food. FAO/CINE/Mahidol Workshop on Traditional Food Systems of Indigenous Peoples in Asia, Salaya, Thailand.

Kuhnlein, H.V. 2001b. Promoting the nutritional and cultural benefits of traditional food systems of Indigenous Peoples. Keynote lecture. 17th International Congress of Nutrition, Vienna.

Kuhnlein, H.V. 2013. What food system intervention strategies and evaluation indicators are successful with Indigenous People? *In* H.V. Kuhnlein, D. Spigelski, B. Erasmus and Barbara Burlingame. *Indigenous Peoples' food systems and well-being: interventions and policies for healthy communities*. Chapter 14. Rome. FAO.

Kuhnlein, H.V. & Burgess, S. 1997. Improved retinol, carotene, ferritin and folate status in Nuxalk teens and adults following a health promotion program. *Food and Nutrition Bulletin*, 18(2): 202–210.

Kuhnlein, H.V. & Moody, S.A. 1989. Evaluation of the Nuxalk Food and Nutrition Program: traditional food use by a Native Indian group in Canada. *Journal of Nutrition Education*, 21(3): 127–132.

Kuhnlein, H.V., Turner, N.J. & Kluckner, P.D. 1982. Nutritional significance of two Important "root" foods (Springbank clover and Pacific silverweed). *Ecology of Food and Nutrition*, 12: 89–95.

Kuhnlein, H.V., Chan, A.C., Thompson, J.N. & Nakai, S. 1982. Ooligan grease: A nutritious fat used by Native people of Coastal British Columbia. *Journal of Ethnobiology*, 2(2): 154–161.

Kuhnlein, H.V., Yeboah, F., Sedgemore, M., Sedgemore, S. & Chan, H.M. 1996. Nutritional qualities of ooligan grease: a traditional food fat of British Columbia First Nations. *Journal of Food Composition and Analysis*, 9: 18–31.

Lepofsky, D. 1985. *An integrated approach to studying settlement systems on the Northwest Coast: The Nuxalk of Bella Coola, B.C.* Vancouver, British Columbia, Canada. University of British Columbia. (M.A. thesis)

Lepofsky, D., Turner, N.J. & Kuhnlein, H.V. 1985. Determining the availability of traditional wild plant foods: An example of Nuxalk foods, Bella Coola, British Columbia. *Ecology of Food and Nutrition,* 16: 223–241.

McIlwraith, T.F. 1948. *The Bella Coola Indians.* (2 volumes). Toronto, Ontario, Canada, University of Toronto Press.

Ministry of Environment. 2002. Climate change and freshwater ecosystems. In *Indicators of climate change for British Columbia 2002.* Victoria, British Columbia, Ministry of Environment, Environmental Protection Division.

Moody, M.F. 2008. *Eulachon past and present.* Vancouver, British Columbia, Canada, University of British Columbia. (M.Sc. thesis)

Nabhan, G.P. 2006. *Renewing Salmon Nation's food traditions.* Corvallis, Oregon, USA, Oregon State University Press.

Nuxalk Food and Nutrition Program Staff. 1984. *Nuxalk Food and Nutrition Handbook: a practical guide to family foods and nutrition using native foods.* Richmond, British Columbia, Canada, Malibu Printing. 116 pp. (reprinted 1985)

Nuxalk Food and Nutrition Program Staff. 1985. *Kanusyam a snknic.* [*Real good food*] Vancouver, British Columbia, Canada, University of British Columbia. 124 pp.

Parrish, C.C., Turner, N.J. & Solberg, S., eds. 2007. *Resetting the kitchen table: food security, culture, health and resilience in coastal communities.* New York, Nova Science Publishers.

Senkowsky, S. 2007. A feast to commemorate – and mourn – the eulachon. *BioScience,* 57(8): 720. www.bioone.org/doi/pdf/10.1641/b570815.

Senos, R., Lake, F., Turner, N. & Martinez, D. 2006. Traditional ecological knowledge and restoration practice in the Pacific Northwest. *In* D. Apostol, ed. *Encyclopedia for restoration of Pacific Northwest ecosystems,* pp. 393–426. Washington, DC, Island Press.

Thommasen, H.V. & Zhang, W. 2006. Impact of chronic disease on quality of life in the Bella Coola Valley. *Rural and Remote Health – The International Electronic Journal of Rural and Remote Health Research, Education, Practice and Policy.* www.rrh.org.au/articles/subviewnthamer.asp?articleid=528.

Thompson, J.C. 2004. *Gitga'at plant project: the intergenerational transmission of traditional ecological knowledge using school science curricula.* Victoria, British Columbia, Canada, University of Victoria. (M.Sc. thesis)

Turner, N.J. & Clifton, H. 2009. "It's so different today": climate change and indigenous lifeways in British Columbia, Canada. *Global Environmental Change,* 19(2): 180–190.

Turner, N.J., Plotkin, M. & Kuhnlein, H.V. 2013. Global environmental challenges to the integrity of Indigenous Peoples' food systems. *In* H.V. Kuhnlein, D. Spigelski, B. Erasmus and Barbara Burlingame. *Indigenous Peoples' food systems and well-being: interventions and policies for healthy communities.* Chapter 3. Rome. FAO.

Turner, N.J. & Thompson, J.C., eds. 2006. *Plants of the Gitga'at People. 'Nwana'a lax Yuup.* Hartley Bay, British Columbia, Gitga'at Nation and Coasts Under Stress Research Project, and Victoria, British Columbia, Canada, Cortex Consulting.

Turner, N.J. & Turner, K.L. 2008. "Where our women used to get the food": cumulative effects and loss of ethnobotanical knowledge and practice; case studies from coastal British Columbia. *Botany,* 86(1): 103–115.

Turner, N.J., Harvey, T., Burgess, S. & Kuhnlein, H.V. 2009. The Nuxalk food and nutrition program, Coastal British Columbia, Canada: 1981–2006. *In* H.V. Kuhnlein, B. Erasmus and D. Spigelski. *Indigenous Peoples' food systems: the many dimensions of culture, diversity and environment for nutrition and health,* pp. 23–44. Rome, FAO.

Chapter 12 | Pohnpei

Abbott, D. 2004. *Asian Development Bank Pacific Department: The Federated States of Micronesia hardship and poverty status discussion paper.* Presented at the FSM Participatory Assessment on Poverty and Hardship Workshop, 19 January 2004. Kolonia, Pohnpei, FSM.

Anzu, S. 2008. Let's go local. *National,* Papua New Guinea newspaper, 15 October 15 2008. Laie, Papua New Guinea, National Agricultural Research Institute.

Balick, M.J., ed. 2009. *Ethnobotany of Pohnpei: plants, people, and island culture.* Honolulu, Hawaii and New York, USA, University of Hawaii Press and New York Botanical Garden. 608 pp.

Barker, C. 1996. *Review of the Family Food Production and Nutrition Project.* Prepared for the UNICEF Pacific Programme Office, Suva, Fiji.

Bittenbender, A. 2010. *Evaluation of the Mand Nutrition and Local Food Promotion Project: Pohnpei, Federated States of Micronesia.* Tucson, Arizona, USA, University of Arizona. (Master of Public Health thesis)

CDC. 2000. *Diabetes: a serious public health problem, at-a-glance 2000.* Atlanta, Georgia, USA.

CIA. 2010. *The World Factbook: Federated States of Micronesia.* Washington DC. Central Intelligence Agency (CIA).

Clayton, S. 2009. *Factors influencing the production and availability of local foods on Pohnpei, Federated States of Micronesia.* Atlanta, Georgia, USA, Emory University. (Master of Public Health thesis)

Corsi, A. 2004. *An exploratory study of food and nutritional beliefs and practices in Pohnpei, Federated States of Micronesia.* Atlanta, Georgia, USA, Emory University. (Master of Public Health thesis)

Corsi, A., Englberger, L., Flores, R., Lorens, A. & Fitzgerald, M.H. 2008. A participatory assessment of dietary patterns and food behavior in Pohnpei, Federated States of Micronesia. *Asia Pacific Journal of Clinical Nutrition,* 17(2): 309–316.

Coyne, T. 2000. *Lifestyle diseases in Pacific communities.* Noumea, New Caledonia, SPC.

Del Guercio, K. 2010. *Assessment of household food security in a Polynesian community living in Pohnpei, Micronesia.* Atlanta, Georgia, USA, Emory University. (Master of Public Health thesis)

Drew, W.M. 2008. *Socioeconomic analysis of agroforestry and livelihoods on a small island developing state: a case study of Pohnpei, Federated States of Micronesia.* Gainesville, Florida, USA, University of Florida. (Ph.D. thesis)

Elymore, J., Elymore, A., Badcock, J., Bach, F. & Terrell-Perica, S. 1989. *The 1987/88 national nutrition survey of the Federated States of Micronesia.* Technical report prepared for the Government and Department of Human Resources of the FSM. Noumea, New Caledonia, SPC.

Emerson, K. 2009. *Salapwuk study: promote local foods during times of change.* Peilapalap, Pohnpei, FSM, Kaselehlie Press. 15 pp.

Englberger, L. 2003. *A community and laboratory-based assessment of the natural food sources of vitamin A in the Federated States of Micronesia.* Brisbane, Australia, University of Queensland. (Ph.D. thesis)

Englberger, L. in press. *"Let's Go Local" guidelines: promoting Pacific Island foods.* Apia, Samoa, FAO Sub-regional Office for the Pacific Islands.

Englberger, L., Marks G.C. & Fitzgerald, M.H. 2003a. Factors to consider in Micronesian food-based interventions: a case study of preventing vitamin A deficiency. *Public Health Nutrition,* 7(3): 423–431.

Englberger, L., Marks, G.C. & Fitzgerald, M.H. 2003b. Insights on food and nutrition in the Federated States of Micronesia: a review of the literature. *Public Health Nutrition,* 6(1): 5–17.

Englberger, L., Aalbersberg, W., Ravi, P., Bonnin, E., Marks, G.C., Fitzgerald, M.H. & Elymore, J. 2003a. Further analyses on Micronesian banana, taro, breadfruit and other foods for provitamin A carotenoids and minerals. *Journal of Food Composition and Analysis,* 16(2): 219–236.

Englberger, L., Schierle, J., Marks, G.C. & Fitzgerald, M.H. 2003b. Micronesian banana, taro, and other foods: newly recognized sources of provitamin A and other carotenoids. *Journal of Food Composition and Analysis,* 16(2): 219–236.

Englberger, L., Aalbersberg, W., Fitzgerald, M.H., Marks, G.C. & Chand, K. 2003c. Provitamin A carotenoid content and cultivar differences in edible pandanus fruit. *Journal of Food Composition and Analysis,* 16(2): 237–247.

Englberger, L., Albert, K., Levendusky, A., Paul, Y., Hagilmai, W., Gallen, M., Nelber, D., Alik, A., Shaeffer, S. & Yanagisaki, M. 2005. *Documentation of the traditional food system of Pohnpei: a project of the Island Food Community of Pohnpei, Community of Mand, and Centre for Indigenous Peoples' Nutrition and Environment.* Kolonia, Pohnpei, FSM, IFCP.

Englberger, L., Schierle, J., Aalbersberg, W., Hofmann, P., Humphries, J., Huang, A., Lorens, A., Levendusky, A., Daniells, J., Marks, G.C. & Fitzgerald, M.H. 2006a. Carotenoid and vitamin content of *Karat* and other Micronesian banana cultivars. *International Journal of Food Science and Nutrition,* 57: 399–418.

Englberger, L., Lorens, A., Albert, K., Levendusky, A., Alfred, J. & Iuta, T. 2006b. *Micronesian staple foods and the "Yellow Varieties Message".* Rome, Bioversity International.

Englberger, L., Schierle, J., Kraemer, K., Aalbersberg, W., Dolodolotawake, U., Humphries, J., Graham, R., Reid, A.P., Lorens, A., Albert, K., Levendusky, A., Johnson, E., Paul, Y. & Sengebau, F. 2008. Carotenoid and mineral content of Micronesian giant swamp taro (*Cyrtosperma*) cultivars. *Journal of Food Composition and Analysis,* 21: 93–106.

Englberger, L., Schierle, J., Hoffman, P., Lorens, A., Albert, K., Levendusky, A., Paul, Y., Lickaneth, E., Elymore, A., Maddison, M., deBrum, I., Nemra, J., Alfred, J., Vander Velde, N. & Kraemer, K. 2009a. Carotenoid and vitamin content of Micronesian atoll foods: pandanus (*Pandanus tectorius*) and garlic pear (*Crataeva speciosa*) fruit. *Journal of Food Composition and Analysis,* 22(1):1–8.

Englberger, L., Lorens, A., Levendusky, A., Pedrus, P., Albert, K., Hagilmai, W., Paul, Y., Nelber, D., Moses, P., Shaeffer, S. & Gallen, M. 2009b. Documentation of the traditional food system of Pohnpei. *In* H.V. Kuhnlein, B. Erasmus and D. Spigelski. *Indigenous Peoples' food systems: the many dimensions of culture, diversity and environment for nutrition and health,* pp. 109–138. Rome, FAO.

Englberger, L., Joakim, A., Larsen, K., Lorens, A. & Yamada, L. 2010a. "Go Local" in Micronesia: Promoting the "CHEEF" benefits of local foods. *Sight and Life,* 1/2010: 40–44.

Englberger, L., Kuhnlein, H.V., Lorens, A., Pedrus, P., Albert, K., Currie, J., Pretrick, M., Jim, R. & Kaufer, L. 2010b. Pohnpei, FSM case study in a global health project documents and successfully promotes local food for health. *Pacific Health Dialog,* 16(1): 121–128.

Englberger, L., Lorens, A., Pretrick, M., Spegal, R. & Falcam, I. 2010c. Using email networking to promote island foods for their health, biodiversity, and other "CHEEF" benefits. *Pacific Health Dialog,* 16(1): 41–47.

Fitzgerald, M.H. 1997. Ethnography. *In* J. Higgs, ed. *Qualitative research: discourse on methodologies,* pp. 48–60. Sydney, Australia, Hampden Press.

FSM Department of Economic Affairs. 2002. *2000 Population and housing census report: national census report.* Palikir, Pohnpei, FSM National Government.

FSM Information Services. 2010. *State of health emergency declared by the Pacific Island Health Officers Association.* Press Release No. 0510-15, 20 May 2010, Palikir, Pohnpei, FSM.

Goldberg, G.R., Black, A.E., Jebb, S.A., Cole, T.J., Murgatroyd, P.R., Coward, W.A. & Prentice, A.M. 1991. Critical evaluation of energy intake data using fundamental principles of energy physiology. *European Journal of Clinical Nutrition,* 45: 569–581.

Green, L.W. & Kreuter, M.W. 1991. *Health promotion planning: an educational and environmental approach.* Mountain View, California, USA, Mayfield Publishing Company.

Greene-Cramer, B. 2009. *Youth health behaviors in Pohnpei, Federated States of Micronesia.* Atlanta, Georgia, USA, Emory University. (Master of Public Health thesis)

Hezel, F.X. 2004. Health in Micronesia over the years. *Micronesian Counselor,* 53: 2–15.

Kaufer, L.A. 2008. *Evaluation of a traditional food for health intervention in Pohnpei, Federated States of Micronesia.* Montreal, Quebec, Canada, McGill University. (M.Sc. thesis)

Kaufer, L., Englberger, L., Cue, R., Lorens, A., Albert, K., Pedrus, P. & Kuhnlein, H.V. 2010. Evaluation of a traditional food for health intervention in Pohnpei, Federated States of Micronesia. *Pacific Health Dialog,* 16(1): 61–73.

Kuhnlein, H.V. & Pelto, G.H. 1997. *Culture, environment and food to prevent vitamin A deficiency.* Boston, Maryland, USA, International Nutrition Foundation for Developing Countries.

Levendusky, A. 2006. *Women of Mand share local recipes.* Kolonia, Pohnpei, FSM, IFCP.

McLaren, D.S. & Frigg, M. 2001. *Sight and Life manual on vitamin A deficiency disorders (VADD),* 2nd edition. Basel, Switzerland, Task Force Sight and Life. 163 pp.

Merlin, M., Jano, D., Raynor, W., Keene, T., Juvik, J. & Sebastian, B. 1992. *Tuhke en Pohnpei: plants of Pohnpei.* Honolulu, Hawaii, USA, East-West Center.

Murai, M., Pen, F. & Miller, C.D. 1958. *Some tropical South Pacific Island foods: description, history, use, composition, and nutritive value.* Honolulu, Hawaii, USA, University of Hawaii Press.

Naik, R. I. 2008. *An assessment of local food production in Pohnpei, Federated States of Micronesia.* Atlanta, Georgia, USA, Emory University. (Master of Public Health thesis)

Ormerod, A. 2006. The case of the yellow bananas. *Eden Project Friends,* 23: 6–7.

Parvanta, A. 2006. *Report on a banana volume market study and health education/awareness campaign.* Kolonia, Pohnpei, FSM, IFCP.

Pollock, N.J. 1992. *These roots remain: food habits in islands of the central and eastern Pacific since Western contact.* Laie, Hawaii, USA, Institute for Polynesian Studies.

Raynor, B. 1991. *Agroforestry systems in Pohnpei – practices and strategies for development.* Pohnpei, FSM: RAS/86/036 Field Document No. 4. FAO/UNDP South Pacific Forestry Development Programme.

Richard, D.E. 1957. *United States Naval Administration of the Trust Territory of the Pacific Islands. Volume III. The trusteeship period 1947–1951.* Washington, DC, Office of the Chief of Naval Operations.

Schoeffel, P. 1992. Food, health and development in the Pacific Islands: policy implications for Micronesia. *ISLA: A Journal of Micronesian Studies,* 1(2): 223–250.

Sears, C.D. 2010. *An assessment of household food security in Kosrae, Federated States of Micronesia.* Atlanta, Georgia, USA, Emory University (Master of Public Health thesis)

Shaeffer, S. 2006. *Assessment of the agroforestry system under changing diet and economy on Pohnpei, Federated States of Micronesia.* Atlanta, Georgia, USA, Emory University (Master of Public Health thesis)

Shintani, T.T., Hughes, C.K., Beckham, S. & O'Connor, H.K. 1991. Obesity and cardiovascular risk intervention through the adlibitum feeding of traditional Hawaiian diet. *American Journal of Clinical Nutrition*, 53: 1647S-1651S.

Singh, S. 2008. *South Pacific: food crisis, an opportunity for change?* Suva, Fiji, Inter-Press Service News Agency.

SPC. 2007. *PAPGREN (Pacific Agriculture Plant Genetic Resources Network) 2007 Meeting Report.* 12–16 November 2007. Suva, Fiji, SPC Nabua.

Streib, L. 2007. World's fattest countries. *Forbes Magazine Online.* www.forbes.com/forbeslife/2007/02/07/worlds-fattest-countries-forbeslife-cx_ls_0208worldfat.html.

Thakorlal, J. 2009. *Resistant starch, proximal composition and ultra structure of banana cultivars from Micronesia.* Auckland, New Zealand, University of Auckland. (B.Sc. honours thesis)

Thakorlal, J., Perera, C.O., Smith, B., Englberger, L. & Lorens, A. 2010. Resistant starch in Micronesian banana cultivars offers health benefits. *Pacific Health Dialog*, 16(1): 49–59.

WHO. 1997. *Obesity: preventing and managing the global epidemic.* Report of a WHO Consultation on Obesity, 3–5 June 1997, Geneva.

WHO. 2008. *Federated States of Micronesia (Pohnpei) NCD risk factors STEPS report.* Suva, Fiji, WHO Western Pacific Region.

Yamamura, C., Sullivan, K.M., van der Haar, F., Auerbach, S.B. & Iohp, K.K. 2004. Risk factors for vitamin A deficiency among preschool aged children in Pohnpei, Federated States of Micronesia. *Journal of Tropical Pediatrics*, 50: 16–19.

Chapter 13 | Ainu

Anderson, F. & Iwasaki-Goodman, M. 2001. Language and culture revitalization in a Hokkaido Ainu community. *In* M.G. Noguchi and S. Fotos, eds. *Studies in Japanese bilingualism,* pp. 45–67. New York, Multilingual Matters Ltd.

Biratori Town. 1974. *Biratori Choushi [Biratori town history].* Biratori, Japan.

Fieldhouse, P. 1996. *Food and nutrition: customs and culture.* London, Chapman and Hall. 253 pp.

Ishige, N. 1979. *Kuichinbo no Minzokugaku [Ethnology of gourmand].* Tokyo, Heibonsha.

Iwasaki-Goodman, M., Ishii, S. & Kaizawa, T. 2009. Traditional food systems of Indigenous Peoples: the Ainu in the Saru River region, Japan. *In* H.V. Kuhnlein, B. Erasmus and D. Spigelski. *Indigenous Peoples' food systems: the many dimensions of culture, diversity and environment for nutrition and health,* pp. 139–157. Rome, FAO.

Iwasaki-Goodman, M., Ishii, S., Iwano, H., Kaizawa, M. & Inoue, H. 2005. Applied research on the Ainu traditional food system in the Saru River region. *Annual Bulletin of the New Humanities*, 2: 118–179.

Lewallen, A. 2007. Bones of contention: negotiating anthropological ethics within fields of Ainu refusal. *Critical Asian Studies*, 39(4): 509–540.

Lupton, D. 1996. *Food, the body and the self.* London, SAGE Publications Ltd. 192 pp.

ACC/SCN. 1991. *Managing successful nutrition programmes*. ACC/SCN State-of-the-Art Series, Nutrition Policy Discussion Paper No. 8. Geneva, Administrative Committee on Coordination/Standing Committee on Nutrition (ACC/SCN) of the United Nations.

Allen, L.H. 2008. To what extent can food-based approaches improve micronutrient status? *Asia Pacific Journal of Clinical Nutrition*, 17(S1): 103–105.

Allen, L.H. & Gillespie, S.R. 2001. *What works? A review of the efficacy and effectiveness of nutrition interventions*. ACC/SCN Nutrition Policy Papers No. 19. Manila, United Nations Standing Committee on Nutrition and Asian Development Bank. 123 pp.

Berg, A. & Muscat, R. 1971. Nutrition program planning: an approach. *In* A. Berg, N.S. Scrimshaw and D.L. Call, eds. *Nutrition, national development and planning*, pp. 247–274. Cambridge, Massachusetts, USA, MIT Press. 401 pp.

Bhattacharjee, L., Kothari, G., Priya, V. & Nandi, B.K. 2009. The Bhil food system: links to food security, nutrition and health. *In* H.V. Kuhnlein, B. Erasmus and D. Spigelski. *Indigenous Peoples' food systems: the many dimensions of culture, diversity and environment for nutrition and health*, pp. 209–229. Rome, FAO.

Boyle, M.A. & Holben, D.H. 2006. *Community nutrition in action: an entrepreneurial approach*, p. 565. Belmont, California, USA, Thomson Wadsworth.

Caballero, B., Clay, T., Davis, S.M., Ethelbah, B., Rock, B.H., Lohman, T., Norman, J., Story, M., Stone, E.J., Stephenson, L. & Stevens, J. 2003. Pathways: a school-based, randomized controlled trial for the prevention of obesity in American Indian schoolchildren. *American Journal of Clinical Nutrition*, 78(5): 1030–1038.

Caicedo, S. & Chaparro, A.M. 2013. Inga food and medicine systems to promote community health. *In* H.V. Kuhnlein, D. Spigelski, B. Erasmus and B. Burlingame. *Indigenous Peoples' food systems and well-being: interventions and policies for healthy communities*, Chapter 8. Rome, FAO.

Canadian Institutes of Health Research. 2007. *Canadian Institutes of Health research guidelines for health research involving aboriginal people*. Ottawa. www.cihr-irsc.gc.ca/e/29134.html.

Canadian Institutes of Health Research. 2010. *Ethics of health research involving First Nations, Inuit and Métis people*. www.cihr-irsc.gc.ca/e/29339.html.

Cargo, M., Levesque, L., Macaulay, A.C., McComber, A., Desrosiers, S., Delormier, T., Potvin, L. & Kahnawake Schools Diabetes Prevention Project (KSDPP) Community Advisory Board. 2003. Community governance of the Kahnawake Schools Diabetes Prevention Project, Kahnawake Territory, Mohawk Nation, Canada. *Health Promotion International*, 18: 177–187.

Caribbean Food and Nutrition Institute & Ministry of Health. 1985. *Nutrition handbook for community workers*. Kingston. 190 pp.

CDC. 2009. *Diabetes projects: Native Diabetes Wellness Program*. www.cdc.gov/diabetes/projects/diabetes-wellness.htm.

Chino, M. & DeBruyn, L. 2006. Building true capacity: indigenous models for indigenous communities. *American Journal of Public Health*, 96(4): 596–599.

CINE. 2010. *CINE Global Health Meeting 2007, 2008*. www.mcgill.ca/cine/events/bellagio2007 and www.mcgill.ca/cine/events/bellagio2008.

Creed-Kanashiro, H., Carrasco, M., Abad, M. & Tuesta, I. 2013. Promotion of traditional foods to improve the nutrition and health of the Awajún of the Cenepa River in Peru. *In* H.V. Kuhnlein, D.

Spigelski, B. Erasmus and B. Burlingame. *Indigenous Peoples' food systems and well-being: interventions and policies for healthy communities,* Chapter 5. Rome, FAO.

Damman, S. 2010. Indigenous peoples, rainforests and climate change. *SCN News,* 38: 63–67.

Delormier, T., Frohlich, K.L. & Potvin, L. 2009. Food and eating as social practice – understanding eating patterns as social phenomena and implications for public health. *Sociology of Health and Illness,* 31(2): 215–228.

Egeland, G.M. & Harrison, G.G. 2013. Health disparities: promoting Indigenous Peoples' health through traditional food systems and self-determination. *In* H.V. Kuhnlein, D. Spigelski, B. Erasmus and B. Burlingame. *Indigenous Peoples' food systems and well-being: interventions and policies for healthy communities,* Chapter 2. Rome, FAO.

Egeland, G.M., Yohannes, S., Okalik, L., Kilabuk, J., Racicot, C., Wilcke, M., Kuluguqtuq, J. & Kisa, S. 2013. The value of Inuit elders' storytelling in health promotion during times of rapid climate change and uncertain food security. *In* H.V. Kuhnlein, D. Spigelski, B. Erasmus and B. Burlingame. *Indigenous Peoples' food systems and well-being: interventions and policies for healthy communities,* Chapter 9. Rome, FAO.

Englberger, L., Kuhnlein, H.V., Lorens, A., Pedrus, P., Albert, K., Currie, J., Pretrick, M., Jim, R. & Kaufer, L. 2010. Pohnpei, FSM case study in a global health project documents and successfully promotes local food for health. *Pacific Health Dialog,* 16(1): 121–128.

Englberger, L., Lorens, A., Albert, K., Pedrus, P., Levendusky, A., Hagilmai, W., Paul, Y., Moses, P., Jim, R., Jose, S., Nelber, D., Santos, G., Kaufer, L., Larsen, K., Pretrick, M. & Kuhnlein, H.V. 2013. Let's go local! Pohnpei promotes local food production and nutrition for health. *In* H.V. Kuhnlein, D. Spigelski, B. Erasmus and B. Burlingame. *Indigenous Peoples' food systems and well-being: interventions and policies for healthy communities,* Chapter 12. Rome, FAO.

FAO. 1997. *Improving household food security and nutrition in Kano State: a training manual for agriculture, health, education and community development extension workers,* edited by H. Abubakar. Rome.

FAO. 2003a. *Community-based food and nutrition programmes: what makes them successful; a review and analysis of experience,* by S. Ismail, M. Immink, I. Mazar and G. Nantel. Rome. 74 pp. ftp://ftp.fao.org/docrep/fao/006/y5030e/y5030e00.pdf.

FAO. 2003b. *Promoting healthy diets through schools,* pp. 3–5. Food Nutrition and Agriculture No. 33. pp. 3–5. Rome. http://www.fao.org/docrep/006/j0243m/j0243m00.htm.

FAO. 2008. *Cultural indicators of Indigenous Peoples' food and agro-ecological systems,* by E. Woodley, E. Crowley, J.D. dePryck and A Carmen. Rome. 104 pp. www.fao.org/sard/common/ecg/3045/en/cultural_indicators_paperapril2008.pdf.

FAO. 2009a. *Indigenous and tribal peoples: building on biological and cultural diversity for food and livelihood security,* by S. Battistelli. Rome. 63 pp. www.fao.org/docrep/011/i0838e/i0838e00.htm.

FAO. 2009b. *The right to adequate food and indigenous peoples. How can the right to food benefit indigenous peoples?* by L. Knuth. Rome. 56 pp.

FAO. 2009c. *The right to food guidelines and indigenous peoples. An operational guide.* by L. Knuth. Rome. 40 pp. www.fao.org/righttofood/publi09/rtf_guidelines.pdf.

FAO & International Life Sciences Institute. 1997. *Preventing micronutrient malnutrition: a guide to food-based approaches. Why policy makers should give priority to food-based strategies.* Washington, DC, International Life Sciences Institute Press. www.fao.org/docrep/x0245e/x0245e00.htm#topofpage.

Ford, P.B. & Dzewaltowski, D.A. 2008. Disparities in obesity prevalence due to variation in the retail food environment: three testable hypotheses. *Nutrition Reviews*, 66(4): 216–228.

Gillespie, S.R., Mason, J. & Martorell, R. 1996. *How nutrition improves.* ACC/SCN Nutrition Policy Discussion Paper No. 15. Geneva, United Nations Administrative Committee on Coordination/Standing Committee on Nutrition.

Gracey, M. & King, M. 2009. Indigenous health part 1: determinants and disease patterns. *Lancet*, 374(9683): 65–75.

Hawkes, C. 2006. Uneven dietary development: linking the policies and processes of globalization with the nutrition transition, obesity and diet-related chronic diseases. *Globalization and Health*, 2: 4. www.globalizationandhealth.com/content/2/1/4.

Iwasaki-Goodman, M. 2013. Tasty *tonoto* and not-so-tasty *tonoto*: fostering traditional food culture among the Ainu people in Saru River region, Japan. *In* H.V. Kuhnlein, D. Spigelski, B. Erasmus and B. Burlingame. *Indigenous Peoples' food systems and well-being: interventions and policies for healthy communities,* Chapter 13. Rome, FAO.

Jimenez, M.M., Receveur, O., Trifonopoulos, M., Kuhnlein, H.V., Paradis, G. & Macaulay, A.C. 2003. Comparison of the dietary intakes of two different groups of children (grades 4 to 6) before and after the Kahnawake School Diabetes Prevention Project. *Journal of the American Dietetic Association*, 103: 1191–1194.

Kaufer, L., Englberger, L., Cue, R., Lorens, A., Albert, K., Pedrus, P. & Kuhnlein, H.V. 2010. Evaluation of a traditional food for health intervention in Pohnpei, Federated States of Micronesia. *Pacific Health Dialog*, 16(1): 61–73.

Kennedy, G., Nantel, G. & Shetty, P. 2004. Globalization of food systems in developing countries: a synthesis of country case studies. *In* FAO. *Globalization of food systems in developing countries: impact on food security and nutrition,* pp. 1–25. FAO Food and Nutrition Paper No. 83. Rome, FAO.

Kennedy, G., Nantel, G. & Shetty, P. 2006. Assessment of the double burden of malnutrition in six case study countries. *In* FAO. *The double burden of malnutrition. Case studies from six developing countries,* pp. 1–20. FAO Food and Nutrition Paper No. 84. Rome, FAO.

Kennedy, E.T. & Pinstrup-Andersen, P. 1983. *Nutrition-related policies and programs: past performances and research needs.* Washington, DC, IFPRI.

King, M., Smith, A. & Gracey, M. 2009. Indigenous health part 2: the underlying causes of the health gap. *Lancet*, 374(9683): 76–85.

Kuhnlein, H.V. & Burgess, S. 1997. Improved retinol, carotene, ferritin and folate status in Nuxalk teenagers and adults after a health promotion programme. *Food and Nutrition Bulletin*, 18(2): 202–210.

Kuhnlein, H.V. & Burlingame, B. 2013. Why do Indigenous Peoples' food and nutrition interventions for health promotion and policy need special consideration? *In* H.V. Kuhnlein, D. Spigelski, B. Erasmus and B. Burlingame. *Indigenous Peoples' food systems and well-being: interventions and policies for healthy communities,* Chapter 1. Rome, FAO.

Kuhnlein, H.V. & Moody, S.A. 1989. Evaluation of the Nuxalk food and nutrition program: traditional food use by a Native Indian group in Canada. *Journal of Nutrition Education*, 21(3): 127–132.

Kuhnlein, H.V., Erasmus, B., Creed-Kanashiro, H., Englberger, L., Okeke, C., Turner, N., Allen, L. & Bhattacharjee, L. 2006a. Indigenous peoples' food systems for health: finding interventions that work. *Public Health Nutrition.* 9(8): 1013–1019.

Kuhnlein, H.V., Smitasiri, S., Yesudas, S., Bhattacharjee, L., Dan, L. & Ahmed, S. 2006b. *Documenting traditional food systems of Indigenous Peoples: international case studies. Guidelines for procedures.* In collaboration with S. Sirisai, P. Puwastien, L. Daoratanahong, S. Dhanamitta, F. Zhai, P.V. Satheesh, G. Kothari and F. Akhter. Montreal, Quebec, Canada, CINE. www.mcgill.ca/files/cine/manual.pdf.

Kuhnlein, H.V., Goodman, L., Receveur, O., Spigelski, D., Duran, N., Harrison, G.G., Erasmus, B. & Tetlit Zheh. 2013. Gwich'in traditional food and health in Tetlit Zheh, Northwest Territories. *In* H.V. Kuhnlein, D. Spigelski, B. Erasmus and B. Burlingame. *Indigenous Peoples' food systems and well-being: interventions and policies for healthy communities,* Chapter 7. Rome, FAO.

LaFrance, J. 2004. Culturally competent evaluation in Indian Country. *New Directions for Evaluation,* 2004(102): 39–50.

Lourenço, A.E., Santos, R.V., Orellana, J.D. & Coimbra Jr., C.E. 2008. Nutrition transition in Amazonia: obesity and socioeconomic change in the Suruí Indians from Brazil. *American Journal of Human Biology,* 20(5): 564–571.

Medical College of Georgia. 2005. *Stage process intervention.* www.georgiahealth.edu/medecine/fmfacdev/hp_stagestable.html.

Messer, E. 1999. Community-based assessment of nutritional problems: scaling up local actions; scaling down top-down management. *In* T.J. Marchione, ed. *Scaling up, scaling down. Overcoming malnutrition in developing countries,* pp. 179–201. Amsterdam, Netherlands, Gordon and Breach Publishers. 292 pp.

Nilsson, C. 2008. Climate change from an indigenous perspective: key issues and challenges. *Indigenous Affairs,* 1–2: 8–15.

Nuttall, M. 2008. Climate change and the warming politics of autonomy in Greenland. *Indigenous Affairs,* 1–2: 44–52.

Oiye, S., Simel, J.O., Oniang'o, R. & Johns, T. 2009. The Maasai food system and food and nutrition security. *In* H.V. Kuhnlein, B. Erasmus and D. Spigelski. *Indigenous Peoples' food systems: the many dimensions of culture, diversity and environment for nutrition and health,* pp. 231–249. Rome, FAO.

Okeke, E.C., Ene-Obong, H.N., Uzuegbunam, A.O., Ozioko, A., Umeh, S.I. & Chukwuone, N. 2009. The Igbo traditional food system documented in four states in Southern Nigeria. *In* H.V. Kuhnlein, B. Erasmus and D. Spigelski. *Indigenous Peoples' food systems: the many dimensions of culture, diversity and environment for nutrition and health,* pp. 251–281. Rome, FAO.

Oshaug, A. 1997. Evaluation of nutrition education programmes: implications for programme planners and evaluators. *In* FAO. *Nutrition education for the public. Discussion papers of the FAO Expert Consultation (Rome, Italy 18–22 September 1995),* pp. 151–178. FAO Food and Nutrition Paper No. 62. Rome, FAO.

Reading, J. 2009. *The crisis of chronic disease among Aboriginal Peoples: a challenge for public health, population health and social policy,* pp. 1–4. Victoria, British Columbia, Canada, University of Victoria.

Redwood, D., Schumacher, M.C., Lanier, A.P., Ferucci, E.D., Asay, E., Helzer, L.J., Tom-Orme, L., Edwards, S.L., Murtaugh, M.A. & Slattery, M.L. 2009. Physical activity patterns of American Indian and Alaskan Native people living in Alaska and the southwestern United States. *American Journal of Health Promotion,* 23(6): 388–395.

Salomeyesudas, B., Kuhnlein, H.V., Schmid, M.A., Satheesh, P.V. & Egeland, G.M. 2013. The Dalit food system and maternal and child nutrition in Andhra Pradesh, South India. *In* H.V. Kuhnlein, D. Spigelski, B. Erasmus and B. Burlingame. *Indigenous Peoples' food systems and well-being: interventions and policies for healthy communities,* Chapter 6. Rome, FAO.

Schetzina, K.E., Dalton III, W.T., Lowe, E.F., Azzazy, N., vonWerssowetz, K.M., Givens, C. & Stern, H.P. 2009. Developing a coordinated school health approach to child obesity prevention in rural Appalachia: results of focus groups with teachers, parents, and students. *Rural and Remote Health – The International Electronic Journal of Rural and Remote Health Research, Education, Practice and Policy,* 9: 1157. www.rrh.org.au/publishedarticles/article_print_1157.pdf.

Schmid, M.A., Egeland, G.M., Salomeyesudas, B., Satheesh, P.V. & Kuhnlein, H.V. 2006. Traditional food consumption and nutritional status of Dalit mothers in rural Andhra Pradesh, South India. *European Journal of Clinical Nutrition,* 60(11): 1277–1283.

Schnohr, C.W., Petersen, J.H. & Niclasen, B.V.L. 2008. Onset of overweight in Nuuk, Greenland: a retrospective cohort study of children from 1973 to 1992. *Obesity,* 16(12): 2734–2738.

Simel, J.O. 2008. The threat posed by climate change to pastoralists in Africa. *Indigenous Affairs,* 1–2: 34–43.

Sirisai, S., Chotiboriboon, S., Tantivatanasathien, P., Sangkhawimol, S. & Smitasiri, S. 2013. Culture-based nutrition and health promotion in a Karen community. *In* H.V. Kuhnlein, D. Spigelski, B. Erasmus and B. Burlingame. *Indigenous Peoples' food systems and well-being: interventions and policies for healthy communities,* Chapter 10. Rome, FAO.

Smallacombe, S. 2008. Climate change in the Pacific: a matter of survival. *Indigenous Affairs,* 1–2: 72–78.

Stankovitch, M., ed. 2008. *Indicators relevant for Indigenous Peoples: A resource book.* Baguio City, Philippines, Tebtebba Foundation. 463 pp.

Tauli-Corpuz, F. & Tapang Jr., B. 2006. In the foot prints of our ancestors. *In* Tebtebba Foundation. *Good practices on indigenous peoples' development,* pp. 3–33. New York, Tebtebba Foundation and United Nations Permanent Forum on Indigenous Issues.

Teufel, N.I., Perry, C.L., Story, M., Flint-Wagner, H.G., Levin, S., Clay, T.E., Davis, S.M., Gittelsohn, J., Altaha, L. & Pablo, J.L. 1999. Pathways family intervention for third-grade American Indian children. *American Journal of Clinical Nutrition,* 69(4): 803S–809S.

Teufel-Shone, N.I., Fitzgerald, C., Teufel-Shone, L. & Gamber, M. 2009. Systematic review of physical activity interventions implemented with American Indian and Alaska Native populations in the United States and Canada. *American Journal of Health Promotion,* 23(6): S8–S32.

Traill, W.B. 2006. Trends towards overweight in lower- and middle-income countries: some causes and economic policy options. *In* FAO. *The double burden of malnutrition. Case studies from six developing countries,* pp. 305–325. FAO Food and Nutrition Paper No. 84. Rome, FAO.

Turner, N.J., Plotkin, M. & Kuhnlein, H.V. 2013. Global environmental challenges to the integrity of Indigenous Peoples' food systems. *In* H.V. Kuhnlein, D. Spigelski, B. Erasmus and B. Burlingame. *Indigenous Peoples' food systems and well-being: interventions and policies for healthy communities,* Chapter 3. Rome, FAO.

Turner, N.J., Tallio, W.R., Burgess, S. & Kuhnlein, H.V. 2013. The Nuxalk Food and Nutrition Program for Health revisited. *In* H.V. Kuhnlein, D. Spigelski, B. Erasmus and B. Burlingame. *Indigenous Peoples' food systems and well-being: interventions and policies for healthy communities,* Chapter 11. Rome, FAO.

UNPFII. 2009. *The State of the World's Indigenous Peoples.* New York, United Nations Department of Economic and Social Affairs, Secretariat of the Permanent Forum on Indigenous Issues. 250 pp. www.un.org/esa/socdev/unpfii/documents/sowip_web.pdf.

Uvin, P. 1999. Scaling up, scaling down: NGO paths to overcoming hunger. *In* T.J. Marchione, ed. *Scaling up, scaling down. Overcoming malnutrition in developing countries,* pp.71–95. Amsterdam, Netherlands, Gordon and Breach Publishers. 292 pp.

Yngve, A., Margetts, B., Hughes, R. & Tseng, M. 2009. Editorial on the occasion of the International Congress of Nutrition. World hunger: a good fight or a losing cause? *Public Health Nutrition*, 12(10): 1685–1686.

Chapter 15 | Human rights

Agurto, J. 2008. Peru. *In* K. Wessendorf, ed. *The indigenous world 2008*, pp. 157–169. Copenhagen, International Work Group for Indigenous Affairs.

APRODEH. 1999. *Todos los pueblos pueden disponer libremente de sus riquezas y recursos naturales.* Lima, Informe anual derechos económicos, sociales y culturales 1999: trabajo, salud y educación: deudas del Tercer Milenio/Asociación Pro Derechos Humanos (APRODEH), CEDAL, FIDH. www.aprodeh.org.

Caicedo, S. & Chaparro, A.M. 2013. Inga food and medicine systems to promote community health. *In* H.V. Kuhnlein, D. Spigelski, B. Erasmus and B. Burlingame. *Indigenous Peoples' food systems and well-being: interventions and policies for healthy communities,* Chapter 8. Rome, FAO.

CERD. 2009. *Follow up regarding the urgent situation of the Achuar People of the Rio Corrientes region of Peru (Annex G and H) (75th session).* Geneva, United Nations Committee on the Elimination of Racial Discrimination, Treaties and Commission Branch, OHCHR.

CESCR. 1999. *The General Comment No 12 on the right to adequate food.* E/C.12/1999/5. Geneva.

CINE. 2010. *Global health case study – Ingano.* www.mcgill.ca/cine/resources/data/ingano.

Correal, C., Zuluaga, G., Madrigal, L., Caicedo, S. & Plotkin, M. 2009. Ingano traditional food and health: phase 1, 2004–2005. *In* H.V. Kuhnlein, B. Erasmus and D. Spigelski. *Indigenous Peoples' food systems: the many dimensions of culture, diversity and environment for nutrition and health*, pp. 83–108. Rome, FAO.

Creed-Kanashiro, H., Roche, M., Tuesta Cerrón, I. & Kuhnlein, H.V. 2009. Traditional food system of an Awajún community in Peru. *In* H.V. Kuhnlein, B. Erasmus and D. Spigelski. *Indigenous Peoples' food systems: the many dimensions of culture, diversity and environment for nutrition and health*, pp. 59–81. Rome, FAO.

Damman S. 2005. Nutritional vulnerability in indigenous children of the Americas – a human rights issue. *In* R. Eversole, J.-A. McNeish and A. Cimadamore, eds. *Indigenous Peoples and poverty. An international perspective*, Chapter 5. CROP International Studies in Poverty Research Series. London, Zed Books,

Damman S. 2007. Indigenous vulnerability and the process towards the Millennium Development Goals. Will a human rights-based approach help? *International Journal on Minority and Group Rights*, 14(4): 489–539.

ECLAC. 2005. *The Millennium Development Goals: a Latin American and Caribbean perspective.* Prepared by A. León in collaboration with E. Espíndola. New York, United Nations, Economic and Social Council Commission on Human Rights (ECLAC). 35 pp.

Egeland, G.M., Charbonneau-Roberts, G., Kuluguqtuq, J., Kilabuk, J., Okalik, L., Soueida, R. & Kuhnlein, H.V. 2009. Back to the future: using traditional food and knowledge to promote a healthy future among Inuit. *In* H.V. Kuhnlein, B. Erasmus and D. Spigelski. *Indigenous Peoples' food systems: the many dimensions of culture, diversity and environment for nutrition and health*, pp. 9–22. Rome, FAO.

Egeland, G.M., Yohannes, S., Okalik, L., Kilabuk, J., Racicot, C., Wilcke, M., Kuluguqtuq, J. & Kisa, S. 2013. The value of Inuit elders' storytelling in health promotion during times of rapid climate change and uncertain food security. *In* H.V. Kuhnlein, D. Spigelski, B. Erasmus and B. Burlingame. *Indigenous Peoples' food systems and well-being: interventions and policies for healthy communities,* Chapter 9. Rome, FAO.

Eide, A. 1984. The international human rights system. *In* A. Eide, W.B. Eide, S. Goonatilake, S. Gussow and J. Omawale, eds. *Food as a human right*, pp. 152–161. Tokyo, United Nations University.

Eide, A. 1989. *Right to adequate food as a human right.* Study Series No. 1. Geneva and New York, United Nations Centre for Human Rights.

Eide, A. 2000. Universalization of human rights versus globalization of economic power. *In* F. Coomans, F. Grünfeld, I. Westendorp and J. Willems, eds. *Rendering justice to the vulnerable: liber amicorum in honour of Theo van Boven*, pp. 99–119. The Hague, Kluwer Law International.

Englberger, L., Lorens, A., Albert, K., Pedrus, P., Levendusky, A., Hagilmai, W., Paul, Y., Moses, P., Jim, R., Jose, S., Nelber, D., Santos, G., Kaufer, L., Larsen, K., Pretrick, M. & Kuhnlein, H.V. 2013. Let's go local! Pohnpei promotes local food production and nutrition for health. *In* H.V. Kuhnlein, D. Spigelski, B. Erasmus and B. Burlingame. *Indigenous Peoples' food systems and well-being: interventions and policies for healthy communities,* Chapter 12. Rome, FAO.

ESCCHR. 1999. *The right to adequate food and to be free from hunger.* E/CN.4/Sub.2/1999/12. Updated study on the right to food submitted by A. Eide in accordance with Sub-Commission Decision 1998/106. Economic and Social Council Commission on Human Rights (ESCCHR) background document. Rome, FAO.

FAO. 1996. *Rome Declaration on World Food Security and World Food Summit Plan of Action.* Rome. www.fao.org/docrep/003/w3613e/w3613e00.htm.

FAO. 2005. Voluntary guidelines to support the progressive realization of the right to adequate food in the context of national food security. Adopted by the 127th session of the FAO Council, November 2004. Rome. www.fao.org/docrep/meeting/009/y9825e/y9825e00.htm.

FAO. 2009a. *Guide to conducting a right to food assessment.* Rome. www.fao.org/docrep/011/i0550e/i0550e00.htm.

FAO. 2009b. *Indigenous Peoples' food systems: the many dimensions of culture, diversity and environment for nutrition and health,* edited by H.V. Kuhnlein, B. Erasmus and D. Spigelski. Rome. 339 pp.

FAO. 2009c. *The right to adequate food and indigenous peoples. How can the right to food benefit indigenous peoples?* by L. Knuth. Rome. 56 pp.

Gallardo L. 2001. Aerial herbicide impact on farmers in Ecuador. *Pesticide News*, 54: 8.

Gasnier, C., Dumont, C., Benachour, N., Clair, E., Chagnon, M. & Séralini, G.E. 2009. Glyphosate-based herbicides are toxic and endocrine disruptors in human cell lines. *Toxicology*, 262(3): 184–191.

Gruskin, S. & Tarantola, D. 2002. Health and human rights. *In* R. Detels, J. McEwen, R. Beaglehole and H. Tanaka, eds. *Oxford textbook of public health*, 4th edition, pp 311–336. Oxford, UK, Oxford University Press. 1956 pp.

Houghton, J. 2008. Colombia. *In* K. Wessendorf, ed. *The indigenous world 2008*, pp. 125–146. Copenhagen, International Work Group for Indigenous Affairs.

HREV. 2008. *Cultivos ilícitos. Megaproyecto. Tierra profanada: impacto de los megaproyectos en territorios indígenas de Colombia.* Human Rights Everywhere (HREV). www.hrev.org.

Hughes, R. 2003. Definitions for public health nutrition: a developing consensus. *Public Health Nutrition*, 6(6): 615–620.

IITC. 2002. *Declaration of Atitlán, Guatemala.* Indigenous Peoples' Consultation on the Right to Food: A Global Consultation. Atitlán, Sololá, Guatemala, 17–19 April. International Indian Treaty Council (IITC). www.treatycouncil.org/new_page_5241224.htm.

ITK. 2007. Inuit: the bedrock of Arctic sovereignty. *Globe and Mail*, 26 July 2007. Inuit Tapiriit Kanatami (ITK).

IWGIA. 2007. Kenya. *In* S. Stidsen, ed. *The indigenous world 2007*, pp. 468–476. Copenhagen, International Work Group for Indigenous Affairs (IWGIA).

Johnson-Down, L. & Egeland, G.M. 2010. Adequate nutrient intakes are associated with traditional food consumption in Nunavut Inuit chjildren aged 3–5 years. *Journal of Nutrition*, 140: 1311–1316.

Jørgensen, M.E., Glümer, C., Bjerregaard, P., Gyntelberg, F., Jørgensen, T. & Borch-Johnsen, K. 2003. Obesity and central fat pattern among Greenland Inuit and a general population of Denmark (Inter99): Relationship to metabolic risk factors. *International Journal of Obesity*, 27: 1507–1515.

Kipuri, N. 2008. Kenya. *In* K. Wessendorf, ed. *The indigenous world 2008*, pp. 415–425. Copenhagen, International Work Group for Indigenous Affairs.

Kovesi, T., Gilbert, N.L., Stocco, C., Fugler, D., Dales, R.E., Guay, M. & Miller, J.D. 2007. Indoor air quality and the risk of lower respiratory tract infections in young Canadian Inuit children. *Canadian Medical Association Journal*, 177(2): 155–160.

Krieger, N. 2001. Theories for social epidemiology in the 21st century: an ecosocial perspective. *International Journal of Epidemiology*, 30(4): 668–677.

Kuhnlein, H.V. & Chan, L. 2000. Environment and contaminants in traditional food systems of northern Indigenous Peoples. *Annual Review of Nutrition*, 20: 595–626.

Kuhnlein, H.V. & Receveur, O. 2007. Local cultural animal food contributes high levels of nutrients for Arctic Canadian indigenous adults and children. *Journal of Nutrition*, 137(4): 1110-1114.

Kuhnlein, H.V., Receveur, O., Soueida, R. & Egeland, G.M. 2004. Arctic Indigenous Peoples experience the nutrition transition with changing dietary patterns and obesity. *Journal of Nutrition*, 134(6): 1447–1453.

Nunavut Tunngavik Incorporated. 2008. *Nunavut's health system.* A report delivered as part of Inuit obligations under article 32 of the Nunavut Land Claims Agreement, 1993. Annual report on the state of Inuit culture and society 07-08. Iqaluit, Nunavut Tunngavik Incorporated. www.tunngavik.com.

O'Dea, K. 1992. Diabetes in Australian aborigines: impact of western diet and life style. *Journal of Internal Medicine*, 232(2): 103–117.

OHCHR. 2006. *Frequently asked questions on a human rights-based approach to development cooperation.* New York and Geneva, United Nations. 40 pp.

OHCHR. 2010. What we do. www.ohchr.org/en/aboutus/pages/whatwedo.aspx.

Oiye, S., Simel, J.O., Oniang'o, R. & Johns, T. 2009. The Maasai food system and food and nutrition security. *In* H.V. Kuhnlein, B. Erasmus and D. Spigelski. *Indigenous Peoples' food systems: the many dimensions of culture, diversity and environment for nutrition and health,* pp. 231–249. Rome, FAO.

Oshaug, A., Eide, W.B. & Eide A. 1994. Human rights: a normative basis for food and nutrition-relevant policies. *Food Policy*, 19(6): 491–516.

Otaño, A., Correa, B. & Palomares, S. 2010. *Water Pollutants Investigation Committee – First Report.* www.gmwatch.eu.

PAHO. 2002a. *Health in the Americas.* Scientific and Technical Publication No. 587, volume 1. Washington, DC, Pan-American Health Organization (PAHO).

PAHO. 2002b. *Health in the Americas.* Scientific and Technical Publication No. 587, volume 2. Washington, DC, Pan-American Health Organization (PAHO).

Ring, I. & Brown, N. 2003. The health status of indigenous peoples and others. The gap is narrowing in the United States, Canada, and New Zealand, but a lot more is needed. *British Medical Journal*, 327(7412): 404–405.

Simel, J.O. 2008. The indigenous peoples' movement in Kenya. *Indigenous Affairs,* 3–4: 10–19.

Simon, M. 2009. Sovereignty begins at home: Inuit and the Canadian Arctic. Inuit Tapiriit Kanatami President's Speech. www.itk.ca.

Statistics Canada. 2006. *2006 census: Aboriginal peoples in Canada in 2006: Inuit, Métis and First Nations, 2006 census: Inuit.* www12.statcan.gc.ca/census-recensement/2006/as-sa/97-558/p6-eng.cfm.

Stavenhagen, R. 2007. *Report of the Special Rapporteur on the situation of human rights and fundamental freedoms of indigenous people.* Human Rights Council, Fourth Session. Item 2 of the provisional agenda. A/HRC/4/32, 27 February 2007. New York, United Nations.

Tomei, M. 2005. *Indigenous and tribal peoples: an ethnic audit of selected poverty reduction strategy papers.* Geneva, ILO.

Uauy, R., Albala, C. & Kain, J. 2001. Obesity trends in Latin America: transiting from under- to overweight. *Journal of Nutrition*, 131: 893S–899S.

UN. 1993. *Vienna Declaration and Programme of Action.* Adopted by the World Conference on Human Rights in Vienna on 25 June 1993. UN Doc. A/CONF.157/23. New York. 71 pp. www2.ohchr.org/english/law/vienna.htm.

UNICEF. 1990. *Strategy for improved nutrition of children and women in developing countries.* A UNICEF policy review. New York, United Nations Children's Fund (UNICEF).

United Nations General Assembly. 2007. *General Assembly adopts declaration on rights of indigenous peoples.* Sixty-first General Assembly GA /10612, Plenary 107th and 108the Meetings (AM and PM), 13 September, 2007. New York, United Nations.

UNPFII. 2005. *Report on the Fourth Session.* 15–26 May 2005, Economic and Social Council Official Records, Supplement No. 23, E/2005/43, UN Doc. E/C.19/2005/9. New York, United Nations Permanent Forum on Indigenous Issues (UNPFII).

UNPFII. 2007a. *About UNPFII and a brief history of indigenous peoples and the international system.* United Nations Permanent Forum on Indigenous Issues (UNPFII). www.un.org.

UNPFII. 2007b. *Who are indigenous peoples?* Factsheet. www.un.org/esa/socdev/unpfii/documents/5session_factsheet1.pdf.

UNPFII. 2009. *The state of the world's indigenous peoples.* New York, United Nations Department of Economic and Social Affairs, Secretariat of the Permanent Forum on Indigenous Issues, United Nations Permanent Forum on Indigenous Issues (UNPFII). 250 pp. www.un.org/esa/socdev/unpfii/documents/sowip_web.pdf.

WHO. 1978. *Declaration of Alma-Ata.* www.who.int/publications/almaata_declaration_en.pdf.

WHO. 2007a. *Global Database on Child Growth and Malnutrition.* Geneva.

WHO. 2007b. *Global Database on Child Growth and Malnutrition.* Kenya. www.who.int/nutgrowthdb/database/countries/who_standards/ken.pdf.

WHO. 2008. *Federated States of Micronesia (Pohnpei) NCD risk factors STEPS report.* Suva, WHO Western Pacific Region.

WHO. 2009. *Global Database on Child Growth and Malnutrition: Federated States of Micronesia.* www.who.int/nutgrowthdb/database/countries/nchs_reference/fsm.pdf.

World Bank. 2007. *Colombia 2006–2010: a window of opportunity.* Washington, DC.

World Vision Kenya. 2004. *Evaluation of Loodariak area development program.* Nairobi.

WRM. 2006. Kenya: the Mau Forest Complex threatened. *World Rainforest Movement Bulletin*, Issue No. 113, December 2006. World Rainforest Movement (WRM)

AFROFOOD. 2009. *Call For Action from the Door of Return for Food Renaissance in Africa.* www.fao.org/infoods/afrofood%20call%20and%20appel.pdf.

Amazon Conservation Team. 2010. *Colombia Program.* www.amazonteam.org/index.php/245/colombia_program.

Bioversity International. 2010. *Communities and livelihoods.* www.bioversityinternational.org.

Caicedo, S. & Chaparro, A.M. 2013. Inga food and medicine systems to promote community health. *In* H.V. Kuhnlein, D. Spigelski, B. Erasmus and B. Burlingame. *Indigenous Peoples' food systems and well-being: Interventions and policies for healthy communities,* Chapter 8. Rome, FAO.

Canadian Institutes of Health Research. 2007. *Canadian Institutes of Health Research guidelines for health research involving Aboriginal People.* Ottawa. www.cihr-irsc.gc.ca/e/29134.html.

CBD. 2010. *Traditional knowledge information portal.* www.cbd.int/tk/.

CESCR. 1999. *The General Comment No 12 on the Right to Adequate Food.* E/C.12/1999/5. Geneva, Committee on Economic, Social and Cultural Rights.

COHAB Initiative. 2008. *Second International Conference on Health and Biodiversity, 2008.* www.cohabnet.org/cohab2008/index.htm.

Correal, C., Zuluaga, G., Madrigal, L., Caicedo, S. & Plotkin, M. 2009. Ingano traditional food and health: phase 1, 2004–2005. *In* H.V. Kuhnlein, B. Erasmus and D. Spigelski. *Indigenous Peoples' food systems: the many dimensions of culture, diversity and environment for nutrition and health,* pp. 83–108. Rome, FAO.

Creed-Kanashiro, H., Carrasco, M., Abad, M. & Tuesta, I. 2013. Promotion of traditional foods to improve the nutrition and health of the Awajún of the Cenepa River in Peru. *In* H.V. Kuhnlein, D. Spigelski, B. Erasmus and B. Burlingame, *Indigenous Peoples' food systems and well-being: interventions and policies for healthy communities,* Chapter 5. Rome, FAO.

Cunningham, M. 2009. Health. *In* UNPFII. *The State of the World's Indigenous Peoples.* pp. 156–187. New York, United Nations Department of Economic and Social Affairs, Secretariat of the Permanent Forum on Indigenous Issues.

Damman, S. 2005. Nutritional vulnerability in indigenous children of the Americas – a human rights issue. *In* R. Eversole, J.A. McNeish and A. Cimadamore, eds. *Indigenous Peoples and poverty: An international perspective,* pp. 69–93. CROP International Studies in Poverty Research Series. London, Zed Books.

Damman, S., Eide, W.B. & Kuhnlein, H.V. 2008. Indigenous Peoples' nutrition transition in a right to food perspective. *Food Policy,* 33(2): 135–155.

Damman, S., Kuhnlein, H.V. & Erasmus, B. 2013. Human rights implications of Indigenous Peoples' food systems and policy recommendations. *In* H.V. Kuhnlein, D. Spigelski, B. Erasmus and B. Burlingame. *Indigenous Peoples' food systems and well-being: interventions and policies for healthy communities,* Chapter 15. Rome, FAO.

Egeland, G.M. & Harrison, G.G. 2013. Health disparities: promoting Indigenous Peoples' health through traditional food systems and self-determination. *In* H.V. Kuhnlein, D. Spigelski, B. Erasmus and B. Burlingame. *Indigenous People's food systems and well-being: interventions and policies for healthy communities,* Chapter 2. Rome, FAO.

Egeland, G.M., Yohannes, S., Okalik, L., Kilabuk, J., Racicot, C., Wilcke, M., Kuluguqtuq, J. & Kisa, S. 2013. The value of Inuit elders' storytelling in health promotion during times of rapid climate change and uncertain food security. *In* H.V. Kuhnlein, D. Spigelski, B. Erasmus and B. Burlingame. *Indigenous*

Peoples' food systems and well-being: interventions and policies for healthy communities, Chapter 9. Rome, FAO.

Englberger, L., Lorens, A., Albert, K., Pedrus, P., Levendusky, A., Hagilmai, W., Paul, Y., Moses, P., Jim, R., Jose, S., Nelber, D., Santos, G., Kaufer, L., Larsen, K., Pretrick, M. & Kuhnlein, H.V. 2013. Let's go local! Pohnpei promotes local food production and nutrition for health. *In* H.V. Kuhnlein, D. Spigelski, B. Erasmus and B. Burlingame. *Indigenous Peoples' food systems and well-being: interventions and policies for healthy communities,* Chapter 12. Rome, FAO.

FAO. 1996. *Rome Declaration on World Food Security and World Food Summit Plan of Action.* Rome. www.fao.org/docrep/003/w3613e/w3613e00.htm.

FAO. 2009a. *Indigenous Peoples' food systems: the many dimensions of culture, diversity and environment for nutrition and health,* edited by H.V. Kuhnlein, B. Erasmus and D. Spigelski. Rome. 339 pp. www.fao.org/docrep/012/i0370e/i0370e00.htm.

FAO. 2009b. *The right to adequate food and indigenous peoples. How can the right to food benefit indigenous peoples?* by L. Knuth. Rome. 56 pp. www.fao.org/righttofood/publi09/ind_people.pdf.

FAO. 2009c. *The right to food guidelines and Indigenous Peoples. An operational guide,* by L. Knuth. Rome. 40 pp. www.fao.org/righttofood/publi09/rtf_guidelines.pdf.

FAO. 2010a. *FAO Livelihood Support Programme.* www.fao.org/sd/dim_pe4/pe4_040501_en.htm.

FAO. 2010b. *Sustainable Agriculture and Rural Development Initiative.* www.fao.org/sard/en/init/2224/index.html.

FAO. 2010c. *The International Network of Food Data Systems.* www.fao.org/infoods/index_en.stm.

FAO. 2010d. *The right to food.* www.fao.org/righttofood/.

Gracey, M. & King, M. 2009. Indigenous health part 1: determinants and disease patterns. *Lancet,* 374(9683): 65–75.

Health Canada. 2007. *Eating well with Canada's food guide – First Nations, Inuit and Metis.* www.hc-sc.gc.ca/fn-an/pubs/fnim-pnim/index-eng.php.

Hokkaido University Center for Ainu and Indigenous Studies. 2008. *Center for Ainu and indigenous studies.* www.sustain.hokudai.ac.jp/2007/06/center_for_ainu_and_indigenous.php.

IFAD. 2010. *Indigenous People.* www.ifad.org/english/indigenous/index.htm.

IITC. 2002. *Declaration of Atitlán, Guatemala.* Indigenous Peoples' Consultation on the Right to Food: A Global Consultation. Atitlán, Sololá, Guatemala, 17–19 April 2002. International Indian Treaty Council (IITC). www.treatycouncil.org/new_page_5241224.htm.

International Work Group for Indigenous Affairs. 2004. South America. *In* D. Vinding, ed. *The indigenous world 2004,* pp. 119–194. Copenhagen.

IUNS. 2010. *Task Forces.* www.iuns.org/taskforces.htm.

Iwasaki-Goodman, M. 2013. Tasty *tonoto* and not-so-tasty *tonoto*: fostering traditional food culture among the Ainu people in the Saru River region, Japan. *In* H.V. Kuhnlein, D. Spigelski, B. Erasmus and B. Burlingame. *Indigenous Peoples' food systems and well-being: interventions and policies for healthy communities,* Chapter 13. Rome, FAO.

King, M., Smith, A. & Gracey, M. 2009. Indigenous health part 2: the underlying causes of the health gap. *Lancet,* 374(9683): 76–85.

Kuhnlein, H.V. & Damman, S. 2008. Considering indicators of biodiversity for food security of Indigenous Peoples. *In* M. Stankovitch, ed. *Indicators relevant for Indigenous Peoples: a resource book,* pp. 337–350. Baguio City, Philippines, Tebbtebba Foundation.

Kuhnlein, H.V., Erasmus, B. & Spigelski, D. 2009. Acknowledgements. *In* H.V. Kuhnlein, B. Erasmus and D. Spigelski. *Indigenous Peoples' food systems: the many dimensions of culture, diversity and environment for nutrition and health,* pp. vii–viii. Rome, FAO.

Kuhnlein, H.V. & Receveur, O. 2007. Local cultural animal food contributes high levels of nutrients for Arctic Canadian indigenous adults and children. *Journal of Nutrition,* 137(4): 1110–1114.

Kuhnlein, H.V., Receveur, O., Soueida, R. & Egeland, G.M. 2004. Arctic Indigenous Peoples experience the nutrition transition with changing dietary patterns and obesity. *Journal of Nutrition,* 134: 1447–1453.

Kuhnlein, H.V., Smitasiri, S., Yesudas, S., Bhattacharjee, L., Dan, L. & Ahmed, S. 2006. *Documenting traditional food systems of Indigenous Peoples: international case studies. Guidelines for procedures.* In collaboration with S. Sirisai, P. Puwastien, L. Daoratanahong, S. Dhanamitta, F. Zhai, P.V. Satheesh, G. Kothari and F. Akhter. www.mcgill.ca/files/cine/manual.pdf.

Kuhnlein, H.V., Goodman, L., Receveur, O., Spigelski, D., Duran, N., Harrison, G.G., Erasmus, B. & Tetlit Zheh. 2013. The Gwich'in traditional food and health project in Tetlit Zheh, Northwest Territories. In H.V. Kuhnlein, D. Spigelski, B. Erasmus and B. Burlingame. *Indigenous Peoples' food systems and well-being: interventions and policies for healthy communities,* Chapter 7. Rome, FAO.

PAHO. 2007. *Health in the Americas, 2007. Volume I – regional health.* Washington, DC. www.paho.org.

Power, E.M. 2008. Conceptualizing food security for aboriginal people in Canada. *Canadian Journal of Public Health,* 99(2): 95–97.

Salomeyesudas, B., Kuhnlein, H.V., Schmid, M.A., Satheesh, P.V. & Egeland, G.M. 2013. The Dalit food system and maternal and child health in Andhra Pradesh, south India. *In* H.V. Kuhnlein, D. Spigelski, B. Erasmus and B. Burlingame. *Indigenous Peoples' food systems and well-being: interventions and policies for healthy communities,* Chapter 6. Rome, FAO.

Sarkar, S., Mishra, S., Dayal, H. & Nathan, D. 2008. Development and deprivation of Indigenous Peoples/scheduled tribes in India: what the figures tell. *In* M. Stankovitch, ed. *Indicators relevant for indigenous peoples: a resource book,* pp. 295–315. Baguio City, Philippines, Tebtebba Foundation.

Shiundu, K.M. & Oniang'o, R.K. 2007. Marketing African leafy vegetables: challenges and opportunities in the Kenyan context. *African Journal of Food Agriculture Nutrition and Development,* 7(4) online.

Sims, J. & Kuhnlein, H.V. 2003. *Indigenous Peoples' participatory health research, planning and management. Preparing research agreements.* Geneva, WHO and Centre for Indigenous Peoples' Nutrition and Environment. 35 pp. www.who.int/ethics/indigenous_peoples/en/index1.html.

Sirisai, S., Chotiboriboon, S., Tantivatanasathien, P., Sangkhawimol, S & Smitasiri, S. 2013. Culture-based nutrition and health promotion in a Karen community. *In* H.V. Kuhnlein, D. Spigelski, B. Erasmus and B. Burlingame. *Indigenous Peoples' food systems and well-being: interventions and policies for healthy communities,* Chapter 10. Rome, FAO.

Stankovitch, M., ed. 2008. *Indicators relevant for indigenous peoples: a resource book.* Baguio City, Philippines, Tebtebba Foundation. 463 pp.

Tauli-Corpuz, F. & Tapang Jr., B. 2006. In the foot prints of our ancestors. *In* Tebtebba Foundation. *Good practices on indigenous peoples' development,* pp. 3–33. New York, Tebtebba Foundation and UNPFII.

Tomaselli, K.G., Dyll, L. & Francis, M. 2008. "Self" and "other": auto-reflexive and indigenous ethnography. *In* N.K. Denzin, Y.S. Lincoln and L.T. Smith, eds. *Handbook of critical and indigenous methodologies,* pp. 347–372. Thousand Oaks, California, USA, Sage Publications.

Turner, N.J., Tallio, W.R., Burgess, S. & Kuhnlein, H.V. 2013. The Nuxalk Food and Nutrition Program for Health revisited. *In* H.V. Kuhnlein, D. Spigelski, B. Erasmus and B. Burlingame. *Indigenous*

Peoples' food systems and well-being: interventions and policies for healthy communities, Chapter 11. Rome, FAO.

UN. 1993. *Vienna Declaration and Programme of Action.* Adopted by the World Conference on Human Rights in Vienna on 25 June 1993. UN Doc. A/CONF.157/23. New York. 71 pp.

UNESCO. 2010. *UNESCO and Indigenous Peoples: partnership for cultural diversity.* Geneva.

United States Department of Health and Human Services. 2010. *Indian Health Service.* www.ihs.gov.

UNPFII. 2009. *The State of the World's Indigenous Peoples.* New York, United Nations Department of Economic and Social Affairs, Secretariat of the Permanent Forum on Indigenous Issues. 250 pp. www.un.org/esa/socdev/unpfii/documents/sowip_web.pdf.

UNS/SCN. 2009. *The sixth report on the world nutrition situation.* Geneva, WHO. www.unscn.org/en/publications/rwns.

Uvin, P. 1999. Scaling up, scaling down: NGO paths to overcoming hunger. *In* T.J. Marchione, ed. *Scaling up, scaling down. Overcoming malnutrition in developing countries,* pp. 179–201. Amsterdam, Netherlands, Gordon and Breach Publishers. 292 pp.

Watt-Cloutier, S. 2009. Our indigenous voice to change Copenhagen. Keynote Presentation at Indigenous Peoples' Global Summit on Climate Change, Plenary Session. Anchorage, Alaska, United States, 20 April 2009.

WHO. 2010. *The health and human rights of Indigenous Peoples.* www.who.int/hhr/activities/indigenous/en/.

World Bank. 2010. *Indigenous Peoples.* http://web.worldbank.org/wbsite/external/topics/extsocialdevelopment/extindpeople/0,,menupk:407808~pagepk:149018~pipk:149093~thesitepk:407802,00.html.

Appendices

Appendix 1
Identifying Indigenous Peoples[1]

The concept of Indigenous Peoples

In the forty-year history of indigenous issues at the United Nations, and its even longer history at the ILO, considerable thinking and debate have been devoted to the question of the definition or understanding of "indigenous peoples". But no such definition has ever been adopted by any United Nations-system body.

One of the most cited descriptions of the concept of "indigenous" was outlined in the José R. Martínez Cobo's Study on the Problem of Discrimination against Indigenous Populations. After long consideration of the issues involved, Martínez Cobo offered a working definition of "indigenous communities, peoples and nations". In doing so, he expressed a number of basic ideas forming the intellectual framework for this effort, including the right of indigenous peoples themselves to define what and who indigenous peoples are. The working definition reads as follows:

Indigenous communities, peoples and nations are those which, having a historical continuity with pre-invasion and pre-colonial societies that developed on their territories, consider themselves distinct from other sectors of the societies now prevailing on those territories, or parts of them. They form at present non-dominant sectors of society and are determined to preserve, develop and transmit to future generations their ancestral territories, and their ethnic identity, as the basis of their continued existence as peoples, in accordance with their own cultural patterns, social institutions and legal system.

This historical continuity may consist of the continuation, for an extended period reaching into the present of one or more of the following factors:

a   Occupation of ancestral lands, or at least of part of them

b. Common ancestry with the original occupants of these lands

c. Culture in general, or in specific manifestations (such as religion, living under a tribal system, membership of an indigenous community, dress, means of livelihood, lifestyle, etc.)

d. Language (whether used as the only language, as mother-tongue, as the habitual means of communication at home or in the family, or as the main, preferred, habitual, general or normal language)

e. Residence in certain parts of the country, or in certain regions of the world

f. Other relevant factors.

On an individual basis, an indigenous person is one who belongs to these indigenous populations through self-identification as indigenous (group consciousness) and is recognized and accepted by these populations as one of its members (acceptance by the group).

This preserves for these communities the sovereign right and power to decide who belongs to them, without external interference.[2]

During the many years of debate at the meetings of the Working Group on Indigenous Populations, observers from indigenous organizations developed

---

1   United Nations Secretariat of the Permanent Forum on Indigenous Issues. 2009. *The State of the World's Indigenous Peoples*. New York. www.un.org/esa/socdev/unpfii/documents/SOWIP_web.pdf

2   Martinez Cobo, J. 1986/7. *Study of the Problems of Discrimination against Indigenous Populations*. UN Doc. E/CN.4/Sub.2/1986/7 and Add. 1–4. paras 379–383.

a common position that rejected the idea of a formal definition of indigenous peoples at the international level to be adopted by states. Similarly, government delegations expressed the view that it was neither desirable nor necessary to elaborate a universal definition of indigenous peoples. Finally, at its fifteenth session, in 1997, the Working Group concluded that a definition of indigenous peoples at the global level was not possible at that time, and this did not prove necessary for the adoption of the Declaration on the Rights of Indigenous Peoples.[3] Instead of offering a definition, Article 33 of the United Nations Declaration on the Rights of Indigenous Peoples underlines the importance of self-identification, that indigenous peoples themselves define their own identity as indigenous.

Article 33

1. Indigenous peoples have the right to determine their own identity or membership in accordance with their customs and traditions. This does not impair the right of indigenous individuals to obtain citizenship of the States in which they live.

2. Indigenous peoples have the right to determine the structures and to select the membership of their institutions in accordance with their own procedures.

ILO Convention No. 169 also enshrines the importance of self-identification. Article 1 indicates that self-identification as indigenous or tribal shall be regarded as a fundamental criterion for determining the groups to which the provisions of this Convention apply.

Furthermore, this same Article 1 contains a statement of coverage rather than a definition, indicating that the Convention applies to:

a) tribal peoples in independent countries whose social, cultural and economic conditions distinguish them from other sections of the national community and whose status is regulated wholly or partially by their own customs or traditions or by special laws or regulations;

b) peoples in independent countries who are regarded as indigenous on account of their descent from the populations which inhabited the country, or a geographical region to which the country belongs, at the time of conquest or colonization or the establishment of present state boundaries and who irrespective of their legal status, retain some or all of their own social, economic, cultural and political institutions.

The concept of indigenous peoples emerged from the colonial experience, whereby the aboriginal peoples of a given land were marginalized after being invaded by colonial powers, whose peoples are now dominant over the earlier occupants. These earlier definitions of indigenousness make sense when looking at the Americas, Russia, the Arctic and many parts of the Pacific. However, this definition makes less sense in most parts of Asia and Africa, where the colonial powers did not displace whole populations of peoples and replace them with settlers of European descent. Domination and displacement of peoples have, of course, not been exclusively practised by white settlers and colonialists; in many parts of Africa and Asia, dominant groups have suppressed marginalized groups and it is in response to this experience that the indigenous movement in these regions has reacted.

It is sometimes argued that all Africans are indigenous to Africa and that by separating Africans into indigenous and non-indigenous groups, separate classes of citizens are being created with different rights. The same argument is made in many parts of Asia or, alternatively, that there can be no indigenous peoples within a given country since there has been no large-scale Western settler colonialism and therefore there can be no distinction between the original inhabitants and newcomers. It is certainly true that Africans are indigenous to Africa and Asians are indigenous to Asia, in the context of European colonization. Nevertheless, indigenous identity is not exclusively determined by European colonization.

[3]   Working Group on Indigenous Populations (WGIP). 1996a. Working Paper by the Chairperson-Rapporteur, Mrs. Erica-Irene A. Daes, on the concept of "indigenous people". UN Doc. E/CN.4/Sub.2/AC.4/1996/2, 10 June 1996; Working Group on Indigenous Populations (WGIP). 1996b. Report of the WGIP on its fourteenth session. UN Doc. E/CN.4/Sub.2/1996/21, 16 August 1996.

The Report of the Working Group of Experts on Indigenous Populations/Communities of the African Commission on Human and Peoples' Rights therefore emphasizes that the concept of indigenous must be understood in a wider context than only the colonial experience.

The focus should be on more recent approaches focusing on self-definition as indigenous and distinctly different from other groups within a state; on a special attachment to and use of their traditional land whereby ancestral land and territory has a fundamental importance for their collective physical and cultural survival as peoples; on an experience of subjugation, marginalization, dispossession, exclusion or discrimination because these peoples have different cultures, ways of life or modes of production than the national hegemonic and dominant model.[4]

In the sixty-year historical development of international law within the United Nations system, it is not uncommon that various terms have not been formally defined, the most vivid examples being the notions of "peoples" and "minorities". Yet the United Nations has recognized the right of peoples to self-determination and has adopted the Declaration on the Rights of Persons Belonging to National or Ethnic, Religious and Linguistic Minorities. The lack of formal definition of "peoples" or "minorities" has not been crucial to the Organization's successes or failures in those domains nor to the promotion, protection or monitoring of the rights accorded to these groups. Nor have other terms, such as "the family" or "terrorism" been defined, and yet the United Nations and Member States devote considerable action and efforts to these areas.

In conclusion, in the case of the concept of "indigenous peoples", the prevailing view today is that no formal universal definition of the term is necessary, given that a single definition will inevitably be either over- or under-inclusive, making sense in some societies but not in others. For practical purposes, the commonly accepted understanding of the term is that provided in the Martínez Cobo study mentioned above.

---

4   African Commission on Human and Peoples' Rights (ACHPR). 2005. *Report of the African Commission's Working Group of Experts on Indigenous Populations/communities*. Banjul and Copenhagen: ACHPR and IWGIA.

Appendix 2
United Nations Declaration on the Rights of Indigenous Peoples

Annex to Human Rights Council Resolution 2006/2

Affirming that indigenous peoples are equal to all other peoples, while recognizing the right of all peoples to be different, to consider themselves different, and to be respected as such,

affirming also that all peoples contribute to the diversity and richness of civilizations and cultures, which constitute the common heritage of humankind,

affirming further that all doctrines, policies and practices based on or advocating superiority of peoples or individuals on the basis of national origin, racial, religious, ethnic or cultural differences are racist, scientifically false, legally invalid, morally condemnable and socially unjust,

eeaffirming also that indigenous peoples, in the exercise of their rights, should be free from discrimination of any kind,

concerned that indigenous peoples have suffered from historic injustices as a result of, inter alia, their colonization and dispossession of their lands, territories and resources, thus preventing them from exercising, in particular, their right to development in accordance with their own needs and interests,

recognizing the urgent need to respect and promote the inherent rights of indigenous peoples which derive from their political, economic and social structures and from their cultures, spiritual traditions, histories and philosophies, especially their rights to their lands, territories and resources,

Further recognizing the urgent need to respect and promote the rights of indigenous peoples affirmed in treaties, agreements and other constructive arrangements with States,

Welcoming the fact that indigenous peoples are organizing themselves for political, economic, social and cultural enhancement and in order to bring an end to all forms of discrimination and oppression wherever they occur,

Convinced that control by indigenous peoples over developments affecting them and their lands, territories and resources will enable them to maintain and strengthen their institutions, cultures and traditions, and to promote their development in accordance with their aspirations and needs,

Recognizing also that respect for indigenous knowledge, cultures and traditional practices contributes to sustainable and equitable development and proper management of the environment,

Emphasizing the contribution of the demilitarization of the lands and territories of indigenous peoples to peace, economic and social progress and development, understanding and friendly relations among nations and peoples of the world,

Recognizing in particular the right of indigenous families and communities to retain shared responsibility for the upbringing, training, education and well-being of their children, consistent with the rights of the child,

Recognizing also that indigenous peoples have the right freely to determine their relationships with States in a spirit of coexistence, mutual benefit and full respect,

Considering that the rights affirmed in treaties, agreements and constructive arrangements between States and indigenous peoples are, in some situations, matters of international concern, interest, responsibility and character,

Also considering that treaties, agreements and other constructive arrangements, and the relationship they represent, are the basis for a strengthened partnership between indigenous peoples and States,

Acknowledging that the Charter of the United Nations, the International Covenant on Economic, Social and Cultural Rights and the International Covenant on Civil and Political Rights affirm the fundamental importance of the right of self-determination of all peoples, by virtue of which they freely determine their political status and freely pursue their economic, social and cultural development,

Bearing in mind that nothing in this Declaration may be used to deny any peoples their right of self-determination, exercised in conformity with international law,

Convinced that the recognition of the rights of indigenous peoples in this Declaration will enhance harmonious and cooperative relations between the State and indigenous peoples, based on principles of justice, democracy, respect for human rights, non-discrimination and good faith,

Encouraging States to comply with and effectively implement all their obligations as they apply to indigenous peoples under international instruments, in particular those related to human rights, in consultation and cooperation with the peoples concerned,

Emphasizing that the United Nations has an important and continuing role to play in promoting and protecting the rights of indigenous peoples,

Believing that this Declaration is a further important step forward for the recognition, promotion and protection of the rights and freedoms of indigenous peoples and in the development of relevant activities of the United Nations system in this field,

Recognizing and reaffirming that indigenous individuals are entitled without discrimination to all human rights recognized in international law, and that indigenous peoples possess collective rights which are indispensable for their existence, wellbeing and integral development as peoples,

Solemnly proclaims the following United Nations Declaration on the Rights of Indigenous Peoples as a standard of achievement to be pursued in a spirit of partnership and mutual respect,

Article 1

Indigenous peoples have the right to the full enjoyment, as a collective or as individuals, of all human rights and fundamental freedoms as recognized in the Charter of the United Nations, the Universal Declaration of Human Rights and international human rights law.

Article 2

Indigenous peoples and individuals are free and equal to all other peoples and individuals and have the right to be free from any kind of discrimination, in the exercise of their rights, in particular that based on their indigenous origin or identity.

Article 3

Indigenous peoples have the right of self-determination. By virtue of that right they freely determine their political status and freely pursue their economic, social and cultural development.

Article 4

Indigenous peoples, in exercising their right to self-determination, have the right to autonomy or self-government in matters relating to their internal and local affairs, as well as ways and means for financing their autonomous functions.

Article 5

Indigenous peoples have the right to maintain and strengthen their distinct political, legal, economic, social and cultural institutions, while retaining their rights to participate fully, if they so choose, in the political, economic, social and cultural life of the State.

Article 6
Every indigenous individual has the right to a nationality.

Article 7
1. Indigenous individuals have the rights to life, physical and mental integrity, liberty and security of person.
2. Indigenous peoples have the collective right to live in freedom, peace and security as distinct peoples and shall not be subjected to any act of genocide or any other act of violence, including forcibly removing children of the group to another group.

Article 8
1. Indigenous peoples and individuals have the right not to be subjected to forced assimilation or destruction of their culture.
2. States shall provide effective mechanisms for prevention of, and redress for:
   (a) Any action which has the aim or effect of depriving them of their integrity as distinct peoples, or of their cultural values or ethnic identities;
   (b) Any action which has the aim or effect of dispossessing them of their lands, territories or resources;
   (c) Any form of forced population transfer which has the aim or effect of violating or undermining any of their rights;
   (d) Any form of forced assimilation or integration by other cultures or ways of life imposed on them by legislative, administrative or other measures;
   (e) Any form of propaganda designed to promote or incite racial or ethnic discrimination directed against them.

Article 9
Indigenous peoples and individuals have the right to belong to an indigenous community or nation, in accordance with the traditions and customs of the community or nation concerned. No discrimination of any kind may arise from the exercise of such a right.

Article 10
Indigenous peoples shall not be forcibly removed from their lands or territories. No relocation shall take place without the free, prior and informed consent of the indigenous peoples concerned and after agreement on just and fair compensation and, where possible, with the option of return.

Article 11
1. Indigenous peoples have the right to practice and revitalize their cultural traditions and customs. This includes the right to maintain, protect and develop the past, present and future manifestations of their cultures, such as archaeological and historical sites, artefacts, designs, ceremonies, technologies and visual and performing arts and literature.
2. States shall provide redress through effective mechanisms, which may include restitution, developed in conjunction with indigenous peoples, with respect to their cultural, intellectual, religious and spiritual property taken without their free, prior and informed consent or in violation of their laws, traditions and customs.

Article 12
1. Indigenous peoples have the right to manifest, practice, develop and teach their spiritual and religious traditions, customs and ceremonies; the right to maintain, protect, and have access in privacy to their religious and cultural sites; the right to the use and control of their ceremonial objects; and the right to the repatriation of their human remains.
2. States shall seek to enable the access and/or repatriation of ceremonial objects and human remains in their possession through fair, transparent

and effective mechanisms developed in conjunction with indigenous peoples concerned.

Article 13

1. Indigenous peoples have the right to revitalize, use, develop and transmit to future generations their histories, languages, oral traditions, philosophies, writing systems and literatures, and to designate and retain their own names for communities, places and persons.

2. States shall take effective measures to ensure this right is protected and also to ensure that indigenous peoples can understand and be understood in political, legal and administrative proceedings, where necessary through the provision of interpretation or by other appropriate means.

Article 14

1. Indigenous peoples have the right to establish and control their educational systems and institutions providing education in their own languages, in a manner appropriate to their cultural methods of teaching and learning.

2. Indigenous individuals, particularly children, have the right to all levels and forms of education of the State without discrimination.

3. States shall, in conjunction with indigenous peoples, take effective measures, in order for indigenous individuals, particularly children, including those living outside their communities, to have access, when possible, to an education in their own culture and provided in their own language.

Article 15

1. Indigenous peoples have the right to the dignity and diversity of their cultures, traditions, histories and aspirations which shall be appropriately reflected in education and public information.

2. States shall take effective measures, in consultation and cooperation with the indigenous peoples concerned, to combat prejudice and eliminate discrimination and to promote tolerance,

understanding and good relations among indigenous peoples and all other segments of society.

Article 16

1. Indigenous peoples have the right to establish their own media in their own languages and to have access to all forms of non-indigenous media without discrimination.

2. States shall take effective measures to ensure that State-owned media duly reflect indigenous cultural diversity. States, without prejudice to ensuring full freedom of expression, should encourage privately-owned media to adequately reflect indigenous cultural diversity.

Article 17

1. Indigenous individuals and peoples have the right to enjoy fully all rights established under applicable international and domestic labour law.

2. States shall in consultation and cooperation with indigenous peoples take specific measures to protect indigenous children from economic exploitation and from performing any work that is likely to be hazardous or to interfere with the child's education, or to be harmful to the child's health or physical, mental, spiritual, moral or social development, taking into account their special vulnerability and the importance of education for their empowerment.

3. Indigenous individuals have the right not to be subjected to any discriminatory conditions of labour and, inter alia, employment or salary.

Article 18

Indigenous peoples have the right to participate in decision-making in matters which would affect their rights, through representatives chosen by themselves in accordance with their own procedures, as well as to maintain and develop their own indigenous decision-making institutions.

Article 19

States shall consult and cooperate in good faith with

the indigenous peoples concerned through their own representative institutions in order to obtain their free, prior and informed consent before adopting and implementing legislative or administrative measures that may affect them.

Article 20

1. Indigenous peoples have the right to maintain and develop their political, economic and social systems or institutions, to be secure in the enjoyment of their own means of subsistence and development, and to engage freely in all their traditional and other economic activities.
2. Indigenous peoples deprived of their means of subsistence and development are entitled to just and fair redress.

Article 21

1. Indigenous peoples have the right, without discrimination, to the improvement of their economic and social conditions, including, inter alia, in the areas of education, employment, vocational training and retraining, housing, sanitation, health and social security.
2. States shall take effective measures and, where appropriate, special measures to ensure continuing improvement of their economic and social conditions. Particular attention shall be paid to the rights and special needs of indigenous elders, women, youth, children and persons with disabilities.

Article 22

1. Particular attention shall be paid to the rights and special needs of indigenous elders, women, youth, children and persons with disabilities in the implementation of this Declaration.
2. States shall take measures, in conjunction with indigenous peoples, to ensure that indigenous women and children enjoy the full protection and guarantees against all forms of violence and discrimination.

Article 23

Indigenous peoples have the right to determine and develop priorities and strategies for exercising their right to development. In particular, indigenous peoples have the right to be actively involved in developing and determining health, housing and other economic and social programmes affecting them and, as far as possible, to administer such programmes through their own institutions.

Article 24

1. Indigenous peoples have the right to their traditional medicines and to maintain their health practices, including the conservation of their vital medicinal plants, animals and minerals. Indigenous individuals also have the right to access, without any discrimination, to all social and health services.
2. Indigenous individuals have an equal right to the enjoyment of the highest attainable standard of physical and mental health. States shall take the necessary steps with a view to achieving progressively the full realization of this right.

Article 25

Indigenous peoples have the right to maintain and strengthen their distinctive spiritual relationship with their traditionally owned or otherwise occupied and used lands, territories, waters and coastal seas and other resources and to uphold their responsibilities to future generations in this regard.

Article 26

1. Indigenous peoples have the right to the lands, territories and resources which they have traditionally owned, occupied or otherwise used or acquired.
2. Indigenous peoples have the right to own, use, develop and control the lands, territories and resources that they possess by reason of traditional ownership or other traditional occupation or use, as well as those which they have otherwise acquired.
3. States shall give legal recognition and protection to these lands, territories and resources. Such

recognition shall be conducted with due respect to the customs, traditions and land tenure systems of the indigenous peoples concerned.

Article 27

States shall establish and implement, in conjunction with indigenous peoples concerned, a fair, independent, impartial, open and transparent process, giving due recognition to indigenous peoples' laws, traditions, customs and land tenure systems, to recognize and adjudicate the rights of indigenous peoples pertaining to their lands, territories and resources, including those which were traditionally owned or otherwise occupied or used. Indigenous peoples shall have the right to participate in this process.

Article 28

1. Indigenous peoples have the right to redress, by means that can include restitution or, when this is not possible, of a just, fair and equitable compensation, for the lands, territories and resources which they have traditionally owned or otherwise occupied or used, and which have been confiscated, taken, occupied, used or damaged without their free, prior and informed consent.

2. Unless otherwise freely agreed upon by the peoples concerned, compensation shall take the form of lands, territories and resources equal in quality, size and legal status or of monetary compensation or other appropriate redress.

Article 29

1. Indigenous peoples have the right to the conservation and protection of the environment and the productive capacity of their lands or territories and resources. States shall establish and implement assistance programmes for indigenous peoples for such conservation and protection, without discrimination.

2. States shall take effective measures to ensure that no storage or disposal of hazardous materials shall take place in the lands or territories of indigenous peoples without their free, prior and informed consent.

3. States shall also take effective measures to ensure, as needed, that programmes for monitoring, maintaining and restoring the health of indigenous peoples, as developed and implemented by the peoples affected by such materials, are duly implemented.

Article 30

1. Military activities shall not take place in the lands or territories of indigenous peoples, unless justified by a significant threat to relevant public interest or otherwise freely agreed with or requested by the indigenous peoples concerned.

2. States shall undertake effective consultations with the indigenous peoples concerned, through appropriate procedures and in particular through their representative institutions, prior to using their lands or territories for military activities.

Article 31

1. Indigenous peoples have the right to maintain, control, protect and develop their cultural heritage, traditional knowledge and traditional cultural expressions, as well as the manifestations of their sciences, technologies and cultures, including human and genetic resources, seeds, medicines, knowledge of the properties of fauna and flora, oral traditions, literatures, designs, sports and traditional games and visual and performing arts. They also have the right to maintain, control, protect and develop their intellectual property over such cultural heritage, traditional knowledge, and traditional cultural expressions.

2. In conjunction with indigenous peoples, States shall take effective measures to recognize and protect the exercise of these rights.

Article 32

1. Indigenous peoples have the right to determine and develop priorities and strategies for the

development or use of their lands or territories and other resources.

2. States shall consult and cooperate in good faith with the indigenous peoples concerned through their own representative institutions in order to obtain their free and informed consent prior to the approval of any project affecting their lands or territories and other resources, particularly in connection with the development, utilization or exploitation of their mineral, water or other resources.

3. States shall provide effective mechanisms for just and fair redress for any such activities, and appropriate measures shall be taken to mitigate adverse environmental, economic, social, cultural or spiritual impact.

Article 33

1. Indigenous peoples have the right to determine their own identity or membership in accordance with their customs and traditions. This does not impair the right of indigenous individuals to obtain citizenship of the States in which they live.

2. Indigenous peoples have the right to determine the structures and to select the membership of their institutions in accordance with their own procedures.

Article 34

Indigenous peoples have the right to promote, develop and maintain their institutional structures and their distinctive customs, spirituality, traditions, procedures, practices and, in the cases where they exist, juridical systems or customs, in accordance with international human rights standards.

Article 35

Indigenous peoples have the right to determine the responsibilities of individuals to their communities.

Article 36

1. Indigenous peoples, in particular those divided by international borders, have the right to maintain and develop contacts, relations and cooperation, including activities for spiritual, cultural, political, economic and social purposes, with their own members as well as other peoples across borders.

2. States, in consultation and cooperation with indigenous peoples, shall take effective measures to facilitate the exercise and ensure the implementation of this right.

Article 37

1. Indigenous peoples have the right to the recognition, observance and enforcement of Treaties, Agreements and Other Constructive Arrangements concluded with States or their successors and to have States honour and respect such Treaties, Agreements and other Constructive Arrangements.

2. Nothing in this Declaration may be interpreted as to diminish or eliminate the rights of Indigenous Peoples contained in Treaties, Agreements and Constructive Arrangements.

Article 38

States in consultation and cooperation with indigenous peoples, shall take the appropriate measures, including legislative measures, to achieve the ends of this Declaration.

Article 39

Indigenous peoples have the right to have access to financial and technical assistance from States and through international cooperation, for the enjoyment of the rights contained in this Declaration.

Article 40

Indigenous peoples have the right to have access to and prompt decision through just and fair procedures for the resolution of conflicts and disputes with States or other parties, as well as to effective remedies for all infringements of their individual and collective rights. Such a decision shall give due consideration to the customs, traditions, rules and legal systems of the indigenous peoples concerned and international human rights.

Article 41

The organs and specialized agencies of the United Nations system and other intergovernmental organizations shall contribute to the full realization of the provisions of this Declaration through the mobilization, inter alia, of financial cooperation and technical assistance. Ways and means of ensuring participation of indigenous peoples on issues affecting them shall be established.

Article 42

The United Nations, its bodies, including the Permanent Forum on Indigenous Issues, and specialized agencies, including at the country level, and States, shall promote respect for and full application of the provisions of this Declaration and follow up the effectiveness of this Declaration.

Article 43

The rights recognized herein constitute the minimum standards for the survival, dignity and well-being of the indigenous peoples of the world.

Article 44

All the rights and freedoms recognized herein are equally guaranteed to male and female indigenous individuals.

Article 45

Nothing in this Declaration may be construed as diminishing or extinguishing the rights indigenous peoples have now or may acquire in the future.

Article 46

1. Nothing in this Declaration may be interpreted as implying for any State, people, group or person any right to engage in any activity or to perform any act contrary to the Charter of the United Nations.

2. In the exercise of the rights enunciated in the present Declaration, human rights and fundamental freedoms of all shall be respected. The exercise of the rights set forth in this Declaration shall be subject only to such limitations as are determined by law, in accordance with international human rights obligations. Any such limitations shall be non-discriminatory and strictly necessary solely for the purpose of securing due recognition and respect for the rights and freedoms of others and for meeting the just and most compelling requirements of a democratic society.

3. The provisions set forth in this Declaration shall be interpreted in accordance with the principles of justice, democracy, respect for human rights, equality, non-discrimination, good governance and good faith.

JOINT BRIEF

THE RIGHT TO FOOD AND
INDIGENOUS PEOPLES

Indigenous Peoples' Right to Food: Legal foundation

Indigenous peoples, like everyone else, have a right to adequate food and a fundamental right to be free from hunger. This is stipulated in Article 11 of the *International Covenant on Economic, Social and Cultural Rights* (ICESCR) of 1966 and constitutes binding international law. This means States Parties to the ICESCR are obliged to implement the right to food domestically, ensuring that it becomes part of their national legal system.

The right to food entitles every person to an economic, political, and social environment that will allow them to achieve food security in dignity through their own means. Individuals or groups who do not have the capacity to meet their food needs for reasons beyond their control, such as illness, discrimination, age, unemployment, economic downturn, or natural disaster, are entitled to be provided with food directly. The obligation to ensure a minimum level necessary to be free from hunger is one of immediate effect.

Various other binding and non-binding international legal instruments protect indigenous peoples' right to food, directly or indirectly.

The *UN Declaration on the Rights of Indigenous Peoples*[2], adopted in September 2007, is a comprehensive statement addressing the human rights of indigenous peoples. It emphasizes the rights of indigenous peoples to live in dignity, to maintain and strengthen their own institutions, cultures and traditions and to pursue their self-determined development, in keeping with their own needs and aspirations.

Photo by S. Wren

The Declaration contains provisions on land, natural resources and subsistence activities, which are highly relevant for the realization of the right to food, recognizes indigenous peoples' collective rights and stresses cultural rights.

The *Voluntary Guidelines to Support the Progressive Realization of the Right to Adequate Food in the Context of National Food Security (Right to Food Guidelines)* provide policy orientation in a number of areas and can be used by indigenous peoples as a tool for advocacy. They refer to indigenous communities in the context of access to resources and assets. In addition, provisions relating to vulnerable groups and disaggregation of data, among others (3.3 on strategies; 7.2 on legal framework; 8.2 and 8.3 on access to resources and assets; 12.3 on national financial resources; 13 on support for vulnerable groups; 14.4 on safety nets; 15.1 on international food aid; 17.2 and 17.5 on monitoring, indicators and benchmarks) are of particular relevance to indigenous peoples.

The Right to Food Guidelines, *adopted by the FAO Council in November 2004, are a practical tool reflecting the consensus among FAO members on what needs to be done in all of the most relevant policy areas to promote food security using a human rights based approach.*

The Right to Food Guidelines are available on the FAO right to food website: www.fao.org/righttofood.

The right to food as a collective right

The UN Declaration on the Rights of Indigenous Peoples states that indigenous peoples have the right to fully enjoy as a collective or as individuals, all human rights and fundamental freedoms. With particular regard to the right to food, indigenous representatives on the occasion of the 2002 Global Consultation signed the Declaration of Atitlán[3], stating that they were: "... in agreement that the

THE RIGHT TO FOOD AND
INDIGENOUS PEOPLES

content of the right to food of indigenous peoples is a collective right".

The adoption of the *UN Declaration on the Rights of Indigenous Peoples* brought the universal recognition of the right to food as a collective right one step forward. This is already reflected in the preamble of the Declaration which states that "… Indigenous Peoples possess collective rights which are indispensable for their existence, well-being and integral development as peoples." The right to food is one such indispensable right.

Cultural dimension of the right to food

Indigenous peoples' right to food has a particular cultural dimension which is relevant in terms of food choices, food preparation and acquisition. Culturally appropriate foods and the activities to obtain them, such as agriculture, hunting and fishing, form an important part of cultural identity.[4] Furthermore, cultural acceptability of food is an element of the normative content of the right to food, and is of particular relevance to indigenous peoples. Right to Food Guideline 10.10 on nutrition highlights the cultural aspects of nutrition and pertains to indigenous peoples in particular. It translates this principle into a practical policy recommendation by reminding States "...of the cultural values of dietary and eating habits in different cultures..."

Photo by P K Mahanand

How can the right to food benefit indigenous peoples?

The right to food may be violated in case of denial of access to land, fishing or hunting grounds, deprivation of access to adequate and culturally acceptable food and contamination of food sources. Some court cases, in which indigenous peoples have been involved, have already illustrated and proven that the right to food provides indigenous peoples with an additional legal argument when claiming their rights or challenging decisions or omissions before administrative authorities or courts.

Indeed, States have particular obligations concerning the right to food of indigenous peoples. These include respecting indigenous peoples' traditional ways of living, strengthening traditional food systems and protecting subsistence activities such as hunting, fishing and gathering.

The respect, protection and fulfilment of the right to food as a collective right has an additional value in comparison to individual rights. This additional value is related to the fact that some property rights to lands, territories and resources are held collectively, and subsistence based activities carried out collectively are not only part of indigenous peoples' cultural identity, but are often essential for their very existence. The right to food, in its collective dimension, is clearly supplementary to the individual one. A collective right to food may imply, for example:

• An obligation by the State to respect collective property rights over lands, territories and resources, the right to culture and the right to self determination (including the right to pursue own economic, cultural and social development)

• An obligation by the State to protect certain activities that are essential to obtaining food (e.g. agriculture, hunting, fishing); and

• An obligation by the State to provide or ensure a minimum level of essential food that is culturally appropriate.

Under the right to food, States are also responsible for ensuring the application of general human rights principles to indigenous peoples, both in their food and nutrition security policies and other policies that may affect their access to food. The right to food does not only address the final outcome of eliminating hunger and ensuring food security, but provides a holistic tool and approach for indigenous peoples to improve their food security situation. The rights based approach, normatively based on international human rights standards, determines in the food security context the relationship between indigenous groups and individuals as rights holders and the State with correlative obligations as a duty bearer. These human rights principles include participation, accountability, non-discrimination, transparency, human dignity, empowerment and the rule of law. A rights based approach requires particular attention to indigenous peoples' specific circumstances and concerns. Applied to the right to food in practice, this means that indigenous peoples must be engaged and particularly supported in processes that determine food security and related policies, legislation and decisions. States should provide space for participation in the setting of verifiable targets and benchmarks for subsequent monitoring and accountability of food security.

With regards to non-discrimination, governments should also ensure that data is disaggregated by age, sex and ethnicity. This information should then be used for the development, design, implementation and monitoring of more appropriate food and nutrition policies, which address the needs of all groups, including indigenous peoples.

What is the relation between the right to food and food sovereignty?

The concepts of the right to food and food sovereignty are related. The clarification of their content is particularly necessary in the context of indigenous peoples because these terms are often mentioned together and interchangeably in different statements and documents.[6]

The right to food is a legal concept a human right and in the case of its violation, remedies can be claimed where available. Food sovereignty is a political concept; there is no existing international human right corresponding to the right to food sovereignty. However, the two concepts have some common elements, and food sovereignty includes calls for the realization of the right to food. The claim of food sovereignty campaigners is for emphasis on local and national production and the right of peoples to freely define their own food and agricultural policies.

"Ogoni Case": SERAC (The Social and Economic Rights Action Centre) and CESR (The Center for Economic and Social Rights) v. Nigeria

One illustrative case for the significance of the right to food for empowerment, advocacy and litigation is a decision by the African Commission on Human and Peoples' Rights regarding a complaint brought by *SERAC and CESR against Nigeria*.[5]

The complaint alleged the military Government of Nigeria of violating human rights of the Ogoni people. The Nigerian National Petroleum Company (NNPC), the State oil company, formed a joint venture with Shell Petroleum Development Corporation (SPDC) whose activities in the Ogoni region allegedly caused environmental degradation, health problems among the Ogoni people and a destruction of food sources, resulting from the contamination of soil, water and air. In its decision, the African Commission found several violations of the African Charter, including a violation of the right to food, which is implicit in the right to life, the right to health and the right to economic, social and cultural development. The African Commission argued that the minimum core of the right to food requires the Nigerian Government to not destroy or contaminate food sources. Furthermore it found that the Government has a duty to protect its citizens, not only through appropriate legislation and effective enforcement but also by protecting them from damaging acts that may be perpetrated by private parties and by preventing peoples' efforts to feed themselves.

Food sovereignty and the right to food are often complementary. For example, the right to food as a human right implies the application of the participatory approach to food security on the basis of the human rights principle of participation. This signifies that participating population groups and individuals, including indigenous peoples, can shape strategies, policies and programmes promoting the realisation of right to food. Thus right to food mechanisms can be used for the promotion of food sovereignty claims when this contributes to regular, permanent and unrestricted access to quantitatively and qualitatively adequate and sufficient food.

The right to food has both a cultural and a sustainability dimension. While it can be argued that local production is more sustainable, the right to food does not otherwise prescribe methods of production or of trade.

What are some of the challenges that need to be addressed?

The process of clarifying the implications of the right of indigenous peoples to adequate food has just started. Laws and policies need to be put into practice to ensure that indigenous peoples fully enjoy the right to food. Indigenous peoples must assert their human rights, including the right to food and exert pressure on States and their officials to meet human rights obligations and commitments. States on the other hand are required to respect and protect indigenous peoples' unique cultural identities and special concerns when realising their right to food. Increased awareness and capacity of both rights holders and duty bearers is necessary for rights to be realized.

The right to food is an important tool for indigenous peoples to bring about real change in their lives and to negotiate power structures.

Endnotes

[1] See also definition of right to adequate food in General Comment 12. Committee on Economic, Social and Cultural Rights (twentieth session 1999). 12 May 1999. UN doc. E/C.12/1999/5, para. 8. The definition of the right to food builds on the now commonly used definition of food security of the World Food Summit 1996 and its four main pillars (availability, access, utilization and stability).

[2] General Assembly resolution A/RES/61/295, 13 September 2007.

[3] Indigenous Peoples' Consultation on the Right to Food. A Global Consultation. Atitlán, Sololá, Guatemala April 17-19 2002.

[4] In 2004, the International Indian Treaty Council conducted a survey amongst indigenous peoples to ascertain cultural indicators for sustainable agricultural development. The survey showed the importance of traditional foods to indigenous peoples cultures and identities.

[5] Social and Economic Rights Action Center & the Center for Economic and Social Rights v. Nigeria. Cited as: Communication No. 155/96.

[6] See Declaration of the Forum for Food Sovereignty, Nyéléni 2007, available at http://www.nyeleni2007.org/spip.php?article290. It states that food sovereignty means defending and recovering the territories of Indigenous Peoples and ensuring fishing communities' access to and control over their fishing area and ecosystems. The declaration defines food sovereignty as the peoples', countries' or state unions' "right" to define their agricultural and food policy, without any dumping vis-à-vis third countries.

Layout by Tomaso Lezzi

Food and Agriculture Organization
of the United Nations, Right to Food Unit
Economic and Social Development Department
viale delle Terme di Caracalla, 00153 Rome, Italy
Tel: +39 06 57055149 Fax: +39 06 57053712
www.fao.org/righttofood Email: righttofood@fao.org

Secretariat of the Permanent Forum on Indigenous Issues
United Nations, 2 UN Plaza
Room DC2-1454, New York, NY, 10017
Tel: 1 917 367 5100 Fax: 1 917 367 5102
http://un.org/esa/socdev/unpfii/ Email:
indigenouspermanentforum@un.org

Prepared by Food and Agriculture Organization
of the United Nations, Right to Food Unit
in collaboration with
Division for Social Policy and Development,
Secretariat of the Permanent Forum of Indigenous People
FAO and UN
Rome and New York 2008

© FAO and UNPFII

Appendix 4

 AFROFOODS

CALL FOR ACTION FROM THE DOOR OF RETURN FOR FOOD RENAISSANCE IN AFRICA

Dakar - 10th December 2009.
Human Rights Day

We, the participants at the 5th AFROFOODS Sub-regional Data Center Coordinators Meeting held in Dakar, Senegal, on 9 –11 December 2009,

- **Note** that the degradation of ecosystems and the loss of food biodiversity is contributing greatly to the increases in poverty and malnutrition in Africa;
- **Recognize** that returning to local crops and traditional food systems is a prerequisite for conservation and sustainable use of biodiversity for food and nutrition;
- **Acknowledge** that local foods are the basis for African sustainable diets;
- **Urge** that food composition data be emphasized as the fundamental information underpinning almost all activities in the field of nutrition;
- **Call upon** the sectors of public health, agriculture, and environment and food trade to help reinforce and assist with the improvement of food composition data, particularly on local foods;
- **Request** that the contribution of food composition be credited as one of the most important components for action in nutrition and food quality, food safety, and food and nutrition security;

We **invite** all sectors to place AFROFOODS on the national, regional and international agenda for all food and nutrition activities in Africa through interdisciplinary strategic plans for achieving the relevant MDGs; and therefore, from the **Door of Return** of the House of the Slaves of Gorée-Dakar, we accept the challenge ourselves and send this **call for action** to our colleagues, as well as to governments, the private sector and financial entities, to strengthen AFROFOODS activities in a renewed commitment to an African food renaissance.

Index

nutritional status 97

protein intake 91

protein-energy malnutrition 97–8

stunting **84**, 85, 97, **99**, 100

underweight **84**, 85, **99**

vitamin A deficiency 97, **98**

vitamin A intake 91, 92–3

vitamin C intake 92

wasting **84**, 85

community interventions 79–80

complementary feeding 43–4, 91, 98

conjunctival xerosis

children 98–9

mothers **86**, 93

Deccan Development Society initiatives 78–9, 292

sanghams 79, 80, 81

discrimination against 78

energy

intake by mothers 85, 86–7, 88–9

requirements of mothers 88

sources 85, 91

energy deficiency 13

mothers 93, **95**, 96, 97, 100

fallow land reclamation 79

food frequency questionnaire 81

food insecurity 78

food resource plants 79

interventions 251–2

iron

intake by children 91–2

intake by mothers 95

requirements of mothers 89–90

sources 86, *88*

iron deficiency

children **84**, 85, 97, **98**, **99**

mothers 96

pregnancy 96

land reclamation/redistribution 79

mothers

activity levels 87–8

Bitot's spots **86**, 93

conjunctival xerosis **86**, 93

diet correlates with health 96–7

dietary intake 85–91

energy deficiency 93, **95**, 96, 97, 100

energy intake 85, 86–7, 88–9

energy requirements 88

health 93–7

iron deficiency 96

iron intake 95

iron requirements 89–90

night blindness 93–4, 95–6, 97, 100

nutrient intakes 81–2

protein intake 86–7, 88–9

protein requirements 88

vitamin A deficiency 93–5, 96, 100

vitamin A requirements 90–1

vitamin A status 96

vitamin C intake 85, **87**, 90

night blindness 83, 93–4, 95–6

children 98–9

mothers 100

pregnancy 96

nutrient intakes

children 91–3

mothers 81–2

participatory research

methods 81–5

statistical analysis 84–5

pregnancy

iron deficiency 96

night blindness 96, 97

protein

maternal requirements 88

sources 85

protein intake

children 91

mothers 86–7, 88–9

rice subsidies 78–9, 288, 290

strategy success 292

territory *76*

traditional cultural food use 100

untouchables 78

vitamin A

intake of children 91, 92–3

maternal requirements 90–1

maternal status 96

social support networks 21
soft drink consumption
    Inuit youth 147, 148, 149, 275
    Pohnpei 211
        ban 218
soil erosion **26**
    climate change 154
Solanum macrocarpum (eggplant) 45
sorghum (*Sorghum vulgare*) 44
    rural India 77
Sphenostylis stenocarpa (yam bean) 46
storytelling, Inuit 149–50, 288, 292
    elders 149–50, 155, 156, 288, 292
    youth radio drama 150
Straits Salish First Nation (Canada)
    reefnet fishing 35
    seafood contamination 30
stunting, children
    Awajún 271
    Dalit **84**, 85, 97, **99**, 100
    Ibo 13
    Maasai 13
    Pohnpei 267
subsistence crops 29
subsistence farming
    Awajún 55
    Pohnpei 194
suicide rate 286
Suruí (Amazon) 14
    obesity 241
SWA 165–6, 173, 174
sweet potato (*Ipomea batatas*) 43

tar sands (Alberta, Canada) 31–2
taro gardens, destruction by sea level rise 32
taro, giant swamp (*Cyrtosperma chamissonis*) 43, 210, 211, 215
Task Force on Indigenous Peoples' Food Systems and Nutrition of the International Union of Nutritional Sciences (IUNS) 5

territory
    access to 259
    control over 263–4
    ties to 37
Thaleichthys pacificus (ooligan) *see* ooligan (*Thaleichthys pacificus*)
thrifty gene hypothesis 14–16
thrifty phenotype 15
tonoto 226, **227**
    preparation and use 228
Torres Strait Islanders 13
trace metals, food web 154
traditional food systems 5–6
    access to 287
    cultural identity 21
    environmental challenges to integrity 25–38
    environmental impacts **26**
    interventions 5–6
        context 6–7
        implications for policy 8
        participatory processes 7
        themes 7–8
    promotion 21
    research 36–7
    resource documentation 287
    retention 20
    scientific credibility building 7
    threats to 18
traditional knowledge retention 20
tuna, persistent organic pollutants 31
Twa (Uganda), infant mortality 16
typhoid, water contamination 30

undernutrition 12–17, 286
    food insecurity association 17–18
    Maasai 13
    origins of problems 288
underweight 12–14
    Dalit children **84**, 85, **99**
unemployment, Nuxalk 183
United Nations
    involvement of Indigenous Peoples 283